AF751431

Industrial and Applied Mathematics

Series Editors

G. D. Veerappa Gowda, TIFR Centre for Applicable Mathematics, Bengaluru, Karnataka, India

Pammy Manchanda, Guru Nanak Dev University, Amritsar, India

S. Kesavan, The Institute of Mathematical Sciences, Chennai, India

Associate Editors

Akhtar A. Khan, Rochester Institute of Technology, Rochester, NY, USA

Firdous A. Shah, Department of Mathematics, University of Kashmir, Kashmir, Jammu and Kashmir, India

K. Balachandran, Bharathiar University, Coimbatore, Tamil Nadu, India

M. Zuhair Nashed, University of Central Florida, Orlando, FL, USA

Editorial Board

Noore Zahra, Princess Nourah Bint Abdulrahman University, Riyadh, Saudi Arabia

René Pierre Lozi, University Côte d'Azur, Nice, France

Industrial and Applied Mathematics, a Scopus-indexed book series, publishes high-quality research-level monographs, lecture notes and contributed volumes focusing on areas where mathematics is used in a fundamental way, such as industrial mathematics, bio-mathematics, financial mathematics, applied statistics, operations research and computer science.

Anita Tomar · Manish Jain
Editors

Applications of Fixed-Point Theorem

Extending Classical Results to Modern Frameworks

Springer

Editors
Anita Tomar
Sri Dev Suman Uttarakhand University
Rishikesh, Uttarakhand, India

Manish Jain
Department of Mathematics
Ahir College
Rewari, Haryana, India

ISSN 2364-6837 ISSN 2364-6845 (electronic)
Industrial and Applied Mathematics
ISBN 978-981-95-5159-0 ISBN 978-981-95-5160-6 (eBook)
https://doi.org/10.1007/978-981-95-5160-6

Mathematics Subject Classification: 47H10, 37F10, 54H25, 28A10, 31E05, 37F46

© The Editor(s) (if applicable) and The Author(s), under exclusive license to Springer Nature Singapore Pte Ltd. 2026

This work is subject to copyright. All rights are solely and exclusively licensed by the Publisher, whether the whole or part of the material is concerned, specifically the rights of translation, reprinting, reuse of illustrations, recitation, broadcasting, reproduction on microfilms or in any other physical way, and transmission or information storage and retrieval, electronic adaptation, computer software, or by similar or dissimilar methodology now known or hereafter developed.
The use of general descriptive names, registered names, trademarks, service marks, etc. in this publication does not imply, even in the absence of a specific statement, that such names are exempt from the relevant protective laws and regulations and therefore free for general use.
The publisher, the authors and the editors are safe to assume that the advice and information in this book are believed to be true and accurate at the date of publication. Neither the publisher nor the authors or the editors give a warranty, expressed or implied, with respect to the material contained herein or for any errors or omissions that may have been made. The publisher remains neutral with regard to jurisdictional claims in published maps and institutional affiliations.

This Springer imprint is published by the registered company Springer Nature Singapore Pte Ltd.
The registered company address is: 152 Beach Road, #21-01/04 Gateway East, Singapore 189721, Singapore

If disposing of this product, please recycle the paper.

Preface

Fixed-point theory has emerged as a foundational pillar in modern mathematics, with far-reaching applications across engineering, economics, computer science and the natural sciences. At its core, the concept of a fixed point provides a powerful lens for analyzing the stability, convergence and structure of mathematical models, particularly those involving nonlinear equations, dynamical systems, and iterative processes.

This volume presents a collection of contemporary research that highlights both the theoretical depth and practical utility of fixed-point techniques. Chapters in the book traverse a wide spectrum of mathematical spaces—including metric, b-metric, partial metric, fuzzy metric, cone b-metric, and S-metric spaces—demonstrating the richness and versatility of fixed-point theory.

The book begins with interdisciplinary applications such as traffic flow optimization and stock market modeling, showing how fixed-point methods can effectively address real-world challenges. Subsequent chapters dive into advanced contraction principles, convergence algorithms, and generalized mappings that reflect the current frontier of mathematical research. Topics such as probabilistic contractions, hybrid-interpolative methods, and weak integral contractive conditions underscore the evolving landscape of fixed-point analysis.

Each contribution has been carefully peer-reviewed and revised to ensure scholarly rigor. In response to editorial feedback, an additional round of independent peer reviews by international experts has been conducted to further strengthen the academic integrity of the volume. We thank all reviewers for their constructive feedback and insights.

The book is intended for graduate students, researchers, and professionals working in pure and applied mathematics, optimization theory, and computational sciences. Whether seeking foundational understanding or advanced developments, readers will find value in both the structure and substance of the contributions.

We extend our sincere gratitude to the contributing authors for their high-quality work and to the publishing team for their dedication throughout the editorial process. It is our hope that this volume inspires new investigations and serves as a lasting reference in the vibrant field of fixed-point theory.

Rishikesh, India Anita Tomar
Rewari, India Manish Jain

Contents

Chapter 1
Handling the Chaos and Improving the Traffic Control Systems

Vipul Kumar, Anita Tomar, U. S. Rana, Mohammad Sajid, and Manish Jain

Abstract This paper aims to develop a technique to handle chaos using Ishikawa iteration and stabilizing the unstable periodic and fixed states that are the cause of chaotic behavior in a one-dimensional discrete system. In particular, we study theoretical and numerical simulation to demonstrate the influence and efficiency of the presented Ishikawa technique by analyzing the Lyapunov exponent property and period-doubling bifurcation. Further, we discuss the discrete traffic flow problem of converting an unstable approach to traffic into a nontraffic and stable region.

Keywords Fixed points · Ishikawa iteration · Lyapunov exponent · Periodic points · Traffic flow problem

AMS Subject Classification 37C25 · 39B12 · 93D05 · 60K30

1.1 Introduction

Chaos is the characteristic of a complex system whose behavior is so random that it appears to be unpredictable. In fact, such a system shows great sensitivity to even small changes in the conditions (see Chaudhary et al. 2022a, b, 2023; Devaney 1992, and so on). Mainly, four straightforward concepts delineate chaos, which can be described as follows:

V. Kumar · U. S. Rana
D.A.V. College Dehradun, Dehradun, Uttarakhand, India

A. Tomar (✉)
Pt.L.M.S. Campus, Sri Dev Suman Uttarakhand University, Rishikesh, India
e-mail: anitatmr@yahoo.co.in

M. Sajid
Department of Mechanical Engineering, College of Engineering, Qassim University, Buraydah, Saudi Arabia

M. Jain
Ahir College, Rewari, India

© The Author(s), under exclusive license to Springer Nature Singapore Pte Ltd. 2026

A. Tomar and M. Jain (eds.), *Applications of Fixed-Point Theorem*, Industrial and Applied Mathematics, https://doi.org/10.1007/978-981-95-5160-6_1

First, chaotic systems are deterministic, which means that they follow certain straightforward rules. This generally indicates that we can anticipate their behavior for brief periods. Second, because chaotic systems are sensitive to initial conditions, we are unable to anticipate their behavior over long periods. Thirdly, attractors are underlying patterns that resides typically in chaotic systems. Finally, chaos is typically regarded as low-dimensional, implying that it can be summed up using a few straightforward variables.

To prevent major changes in the natural dynamics of a system, control of chaos entails the stabilization of unstable periodic orbits via modest system perturbations that are extremely small in comparison to the size of the attractor of the system. The goal is to produce chaotic motion, which is frequently useful, more stable, and predictable. The Ott et al. (1990) technique and Pyragas (1992) continuous controlling, call for the assurance of the unstable periodic orbits to design the algorithm that controls a chaotic system. These are two basic approaches that have served as the foundation for numerous techniques for controlling chaos. An infinite number of unstable periodic orbits make a chaotic attractor. In state space, an attractor is a set toward which trajectories converge over time. In a chaotic attractor, there are innumerable unstable periodic orbits such as fixed points, period-2 points, and so on (see Feketa et al. 2020; Guo & Qi 2021; Jahanzaib 2020; Kumar et al. 2020 and references therein). A generic trajectory will eventually get as close as possible to any one of these orbits if the attractor is ergodic. A chaotic attractor is exponentially sensitive to small perturbations.

The motion of chaotic dynamics follows the pattern where the system state shifts for a while near one of these orbits, it drifts toward another unstable periodic orbit where it stays for a while and goes. Over extended periods, this causes complicated and unpredictable drifting. Many systems, including turbulent fluids, signal transmission, oscillating chemical reactions, coding–decoding procedure magneto-mechanical oscillators, population dynamics, cardiac tissues, and traffic flow problems have been experimentally controlled for chaos by these techniques.

Small changes to a chaotic system can control chaos, such as directing chaotic trajectories to desired locations or stabilizing an unstable periodic orbit. There are two fundamental, interdependent characteristics of chaos: the butterfly effect (or exponential sensitivity to minor changes), and complex orbital structure. The largest Lyapunov exponent typically defines the first characteristic, which can be utilized in the creation of chaos control strategies (Ma et al. 2023; Yi et al. 2020).

As the most important problem in non-linear dynamical systems, the idea of controlling chaos has been the subject of extensive research over the years. The use of feedback techniques is one of the key methods for controlling chaos, which involves reducing the state variables on a fixed and recurring basis. Several control strategies have also been developed in response to the irregular dynamical behavior of the traffic flow to establish stable traffic on the road. Liz (2010), Braverman and Liz (2012) used proportional feedback control, to investigate the stabilization of one-dimensional discrete systems in periodic and fixed conditions. In recent times, Cao and Chugh (2019) utilized the fixed point feedback technique to establish its significance to the discrete traffic flow model and investigate the dynamical characteristics of the logistic system. This paper aims to discuss the control of chaos and its usefulness to traffic flow models using the Ishikawa iteration in a one-dimensional discrete system. For different traffic flow models, refer to Kurganov and Polizzi (2009), Li (2013), and references therein.

1.2 Preliminaries

Consider a self-mapping f defined on some domain $\mathbb{X}$, say. A point $x^* \in \mathbb{X}$ is an equilibrium or a fixed point if $x^* = f(x^*)$. Generally, a small deviation from the fixed point is taken into account when studying the stability of fixed points of mappings. Define $dx_n = x_n - x^*$. Then, by applying Taylor's series to the mapping in the neighborhood of x^*, we obtain

$$x^* + dx_{n+1} = f(x^* + dx_n) = f(x^*) + f'(x^*)dx_n + \cdots$$

where f' equals $\frac{df}{dx}$. This can be reduced to $dx_{n+1} = f'(x^*)dx_n + \cdots$ Since x^* is the fixed point of f, the above Taylor's series expansion truncates for sufficiently small dx_n. Now, we have three important cases discussed below:

1. If $|f'(x^*)| < 1$, then $|dx_{n+1}|$ is smaller than $|dx_n|$, indicating that the distance from the fixed point diminishes as we iterate the map. As a result, the fixed point is stable. Noticeably, the sign of dx_n may change from iterate to iterate when $f'(x^*)$ is negative but does not affect the overall conclusion.
2. If $|f'(x^*)| > 1$, the fixed point is unstable as the distance from the fixed point grows with each iteration.
3. If $|f'(x^*)| = 1$, the stability to linear order cannot be decided.

A point x^* is of period k of the mapping f, if $f^k(x^*) = x^*$, where f^k is the k-fold composition of the mapping f, that is, $f^k(\alpha) = f(f(...(f(\alpha))))$. In this case, x^* is said to be a periodic k-point. Sometimes, it is also termed as a periodic fixed point of order k, where $k \in \mathbb{N}$. If x^* is a periodic k-point, at every k-step, the value gets repeated.

For a dynamical system, a periodic orbit is a unique type of solution that repeats itself over time. An oscillator is a term used to describe a dynamical system with a stable, periodic orbit.

The Banach contraction principle (Banach 1922) guarantees the existence and uniqueness of fixed points for contraction mappings on complete metric spaces, providing a theoretical foundation for the stability analysis of iterative systems. The Picard iteration scheme can be defined as $x_{n+1} = f(x_n)$.

Definition 1.1 (Ishikawa 1974) Let $\{x_n\}$ be the sequence of iterates for any initial point $x_0 \in \mathbb{X}$ and $f : \mathbb{X} \to \mathbb{X}$ be a mapping on $\mathbb{X}$ so that

$$y_n = x_n + \beta(f(x_n) - x_n),$$
$$x_{n+1} = x_n + \alpha(f(y_n) - x_n),$$

where $\alpha, \beta \in (0, 1)$ and $n \in \mathbb{N} \cup \{0\}$. Then, $\{x_n\}$ defined above is said to form an Ishikawa orbit.

It can be seen that for $\beta = 0$, it reduces to Mann (1953) orbit. For $\alpha = 1$ and $\beta = 0$, it reduces to $x_{n+1} = f(x_n)$, that is, Picard (1890) orbit.

Definition 1.2 (Holmgren 1994) A fixed point $x \in \mathbb{X}$ of the mapping $f : \mathbb{X} \to \mathbb{X}$ is a stable fixed point if it satisfies $|f'(x)| < 1$ and an unstable fixed point if $|f'(x)| > 1$ and x is called a neutral fixed point if $|f'(x)| = 1$.

Also, it is quite noteworthy that any fixed point is also a point of period k-orbit for each $k \geq 1$. Hence, it can be easily said that an orbit of period 2 is also an orbit of period $2n$, a period 3 orbit is also an orbit of period $3n$, and so on. Therefore, several lower-period orbits will typically also be found when period k orbits are discovered. With these lower-period orbits, we should have k points that are all part of the same orbit. Period k points belonging to n different orbits would result from the rare discovery of multiple (true) period k orbits. As a matter of fact, the formulae $x_{n+k} = f_k(x_n)$ itself is a mapping. Because period k points on this mapping are fixed points, we can use the same methods to study the stability of periodic orbits as usually used for fixed point stability. Each of an orbit's period k points will have the same stability properties, so it is sufficient to investigate just one of them. We describe a control concept to overcome chaos and to get the kind of dynamic behavior one wants, like a stable fixed point or a stable periodic orbit. Methods for stabilizing unstable periodic orbits and linear feedback control are two examples of the control ideas discussed here. There are numerous potential applications for controlling chaotic states. It is noteworthy to mention here that not every periodic orbit has a period of $2n$. For instance, the logistic map $x_{n+1} = \lambda x_n(1 - x_n)$ has a stable period 3 orbit in a restricted set of values of λ. It has already been demonstrated that the primary characteristic is the sensitive relationship between chaos on initial conditions. Technically, this is the difference $d_n = y_n - x_n$ between the two-time series that grow exponentially if we take two initial points that are very close to each other such as x_0 and $y_0 = x_0 + d_0$ with d_0 very small, and iterate them under the map. The Lyapunov exponent is the name given to the constant μ. This behavior occurs for large n, as indicated by the similarity symbol. Chaotic systems exhibit positive Lyapunov exponents, whereas the negative Lyapunov exponent refers to the stability of the states of the system.

1.3 Main Results

1.3.1 Controlling Chaos in Ishikawa Orbit

Stability Examination of Fixed Point States

Motivated by distinct applications of the Ishikawa iterative scheme, in this section, we design a technique to control the chaos by stabilizing the unstable fixed points and unstable periodic fixed points of a given dynamical framework which is non-linear in nature. Consider

$$x_{n+1} = f(x_n), \tag{1.1}$$

where f represents the original Picard system having high chaotic behavior which is defined on $[a, b]$. Now using Definition 1.1 with the original Picard one-dimensional system,

$$\overline{f}(y_n) = x_n + \beta(f(x_n) - x_n), \tag{1.2}$$

$$\overline{\overline{f}}(x_{n+1}) = x_n + \alpha(\overline{f}(y_n) - x_n), \tag{1.3}$$

where $n \in \mathbb{N}$ and $\alpha, \beta \in (0, 1)$ are the two control parameters, the above system is Ishikawa (1974) fixed point system.

Theorem 1.3 *The fixed points in the Ishikawa system (1.3) are identical to those in the original Picard system (1.1). That is,* $f(x) = x$ *if and only if* $\overline{\overline{f}}(x) = x$ *for each* $x \in [a, b]$.

Proof Suppose x is a fixed point of mapping f defined on $[a, b]$, then using Eq. (1.2), we have

$$\begin{aligned}\overline{f}(y) &= x + \beta(f(x) - x)\\ &= x,\end{aligned}$$

which in turn on using Eq. (1.3) yields

$$\begin{aligned}\overline{\overline{f}}(x) &= x + \alpha(\overline{f}(y) - x)\\ &= x,\end{aligned}$$

implying that x is also a fixed point for an Ishikawa system $\overline{\overline{f}}$.

Conversely, let x be a fixed point for the system $\overline{\overline{f}}$, then using Eq. (1.2)

$$\begin{aligned}x + \alpha(\overline{f}(y) - x) &= x\\ \alpha(\overline{f}(y) - x) &= 0,\end{aligned}$$

but since $\alpha \in (0, 1)$, we get

$$\overline{f}(y) = x,$$

then using equation (1.2)

$$\begin{aligned} x + \beta(f(x) - x) &= x \\ \beta(f(x) - x) &= 0, \end{aligned}$$

since $\beta \in (0, 1)$, we can get $f(x) = x$.

Theorem 1.3 establishes a correlation between the original Picard system and the preferred Ishikawa system (which may sometimes be called a two-step superior system or modified system) for the fixed points which are in fact the points that are pinned in the space considered and hence, Theorem 1.3 can help explain the fundamental properties of the non-linear dynamical systems at fixed points. □

Theorem 1.4 *There always exists an effectual range (or potential range) for controlling parameters α, β in the Ishikawa system (1.3) for a fixed point that is unstable in an original Picard system $f(x_n) = x_{n+1}$ such that $|\overline{\overline{f}}'(x)| < 1$. That is, for $|f'(x)| > 1$, there always exists an effectual range for $\alpha \in (\alpha_{\min}, \alpha_{\max})$ and $\beta \in (\beta_{\min}, \beta_{\max})$ such that $|\overline{\overline{f}}'(x)| < 1$.*

Proof Let x be a fixed point that is unstable in the original Picard system (1.1). Then, by Ishikawa system (1.2) and (1.3), we have

$$\overline{f}'(y) = 1 + \beta(f'(x) - 1), \tag{1.4}$$

$$\overline{\overline{f}}'(x) = 1 + \alpha(\overline{f}'(y) - 1). \tag{1.5}$$

We now discuss the following possible cases:

Case 1 x is an unstable fixed point of the mapping f and $f'(x) > 1$.

Using Eq. (1.4), $\overline{f}'(y) > 1$ and consequently,

$$\begin{aligned} 1 + \beta(f'(x) - 1) &> 1, \\ that\ is, \beta(f'(x) - 1) &> 0, \\ that\ is, \beta &> 0. \end{aligned}$$

Now, when x is a stable fixed point of the preferred Ishikawa system, $|\overline{\overline{f'}}(x)| < 1$, then, either $\overline{\overline{f'}}(x) < 1$, or $\overline{\overline{f'}}(x) > -1$.

On taking $\overline{\overline{f'}}(x) < 1$, and using Eqs. (1.4) and (1.5), we get

$$\begin{aligned} 1 + \alpha(\overline{f}'(y) - 1) &< 1, \\ that\ is, \alpha &< 0. \end{aligned}$$

Now, on considering $\overline{\overline{f}}'(x) > -1$ and using Eqs. (1.4) and (1.5), we get

$$1 + \alpha(\overline{f}'(y) - 1) > -1,$$
$$that\ is, \alpha > \frac{-2}{\overline{f}'(y) - 1}.$$

Hence, the effectual (potential) ranges of α and β for which the preferred Ishikawa system attains the stability of the unstable fixed point x when $f'(x) > 1$ are given by $(\frac{-2}{\overline{f}'(y)-1}, 0)$ and $(0, 1)$, respectively, that is, $\alpha_{\min} = \frac{-2}{\overline{f}'(y)-1}$, $\alpha_{\max} = 0$ and $\beta_{\min} = 0$, $\beta_{\max} = 1$.

Case 2 x is an unstable fixed point of the mapping f and $f'(x) < -1$.

Using the Eq. (1.4), we can get $\overline{f}'(y) < 1$. As a consequence of which, we obtain

$$1 + \beta(f'(x) - 1) < 1,$$
$$that\ is, \beta(f'(x) - 1) < 0,$$
$$that\ is, \beta > 0.$$

Now, for the stability of fixed points of the preferred Ishikawa system, $|\overline{\overline{f}}'(x)| < 1$, so that either $\overline{\overline{f}}'(x) < 1$ or $\overline{\overline{f}}'(x) > -1$.

First, on taking $\overline{\overline{f}}'(x) < 1$ and using Eqs. (1.4) and (1.5), we get

$$1+\alpha(\overline{f}'(y) - 1) < 1,$$
$$that\ is, \alpha > 0.$$

In the second case on taking $\overline{\overline{f}}'(x) > -1$ and using Eqs. (1.4) and (1.5), we attain

$$1 + \alpha(\overline{f}'(y) - 1) > -1,$$
$$that\ is, \alpha(\overline{f}'(y) - 1) > -2,$$
$$that\ is, \alpha < \frac{2}{1 - \overline{f}'(y)}.$$

Hence, the effectual (potential) ranges of α and β for which the preferred Ishikawa system attains the stability of the unstable fixed point x when $f'(x) < -1$ are given by $(0, \frac{2}{1-\overline{f}'(y)})$ and $(0, 1)$, respectively; that is, $\alpha_{\min} = 0$, $\alpha_{\max} = \frac{2}{1-\overline{f}'(y)}$ and $\beta_{\min} = 0$, $\beta_{\max} = 1$.

Controlling chaotic behavior is the most important aspect of a dynamical system which can be done by shifting the unstability of fixed points toward stable states. □

Example 1.5 Let $f(x) = 5x(1 - x)$ be the original system with two unstable fixed points $x_1 = 0$ and $x_2 = \frac{4}{5}$ for which $|f'(x)| > 1$. We find the potential range of control parameters α and β for the fixed points in the preferred Ishikawa system.

Proof Here we provide a proper graphical and analytical solution to the proposed problem. We have generated Fig. 1.1a for the original system $f(x) = 5x(1 - x)$ and the chaotic behavior is shown in Fig. 1.2f for an initial value of $x = 0.1689$. Using the preferred Ishikawa-type superior state system, we shall manage it in the following way. Using Eq. (1.3),

$$\overline{\overline{f}}(x) = x + \alpha(\overline{f}(y) - x). \tag{1.6}$$

Since $f'(0) = 5 > 1$, Theorem 1.4 implies that for any $\beta \in (0, 1)$, there does not exist any $\alpha \in (0, 1)$ for which any effectual regime exists. For example, on choosing $\beta = 0.1$, we get

$$\begin{aligned} \overline{f'}(y) &= 1 + 0.1(f'(0) - 1) \\ &= 1 + 0.1(5 - 1) \\ &= 1.4, \end{aligned}$$

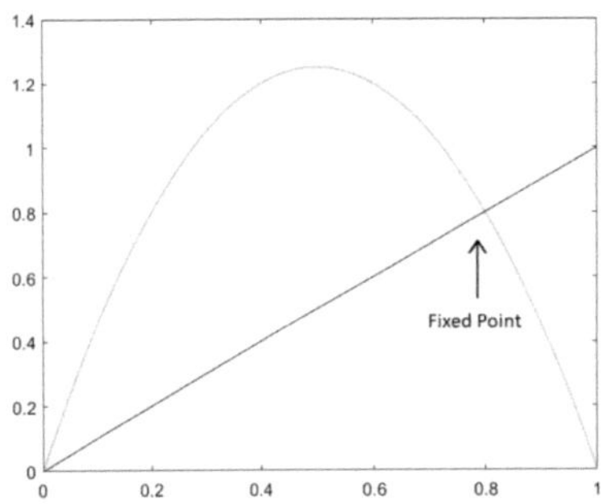

(a) Diagram for the original Picard system $f(x)$ when $x \in [0, 1]$

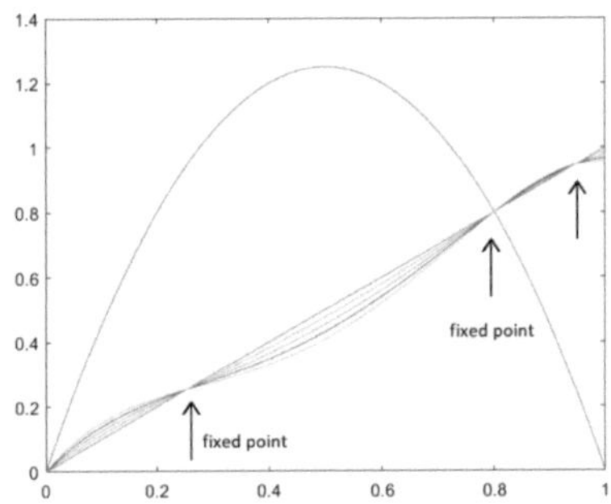

(b) $\beta = 0.11$

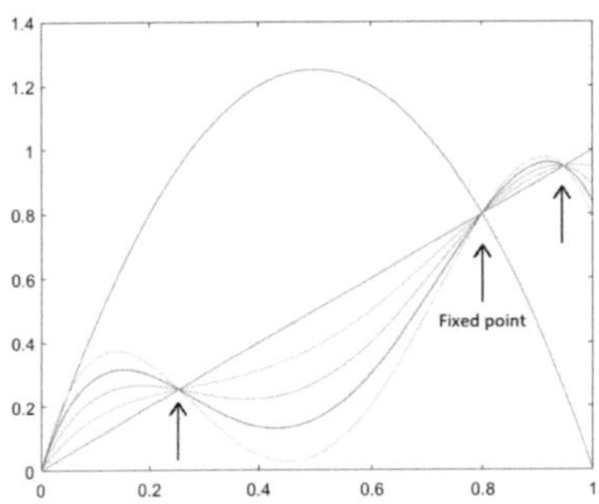

(c) $\beta = 0.55$

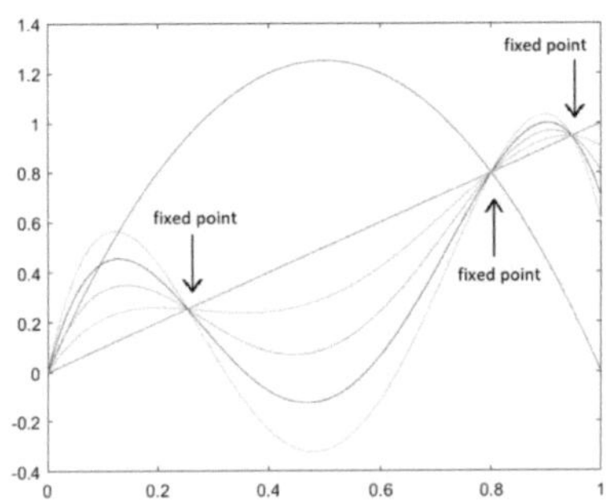

(d) $\beta = 0.99$

Fig. 1.1 Picard system $f(f(x) = 5x(1 - x)$ and Ishikawa control for varying α and β

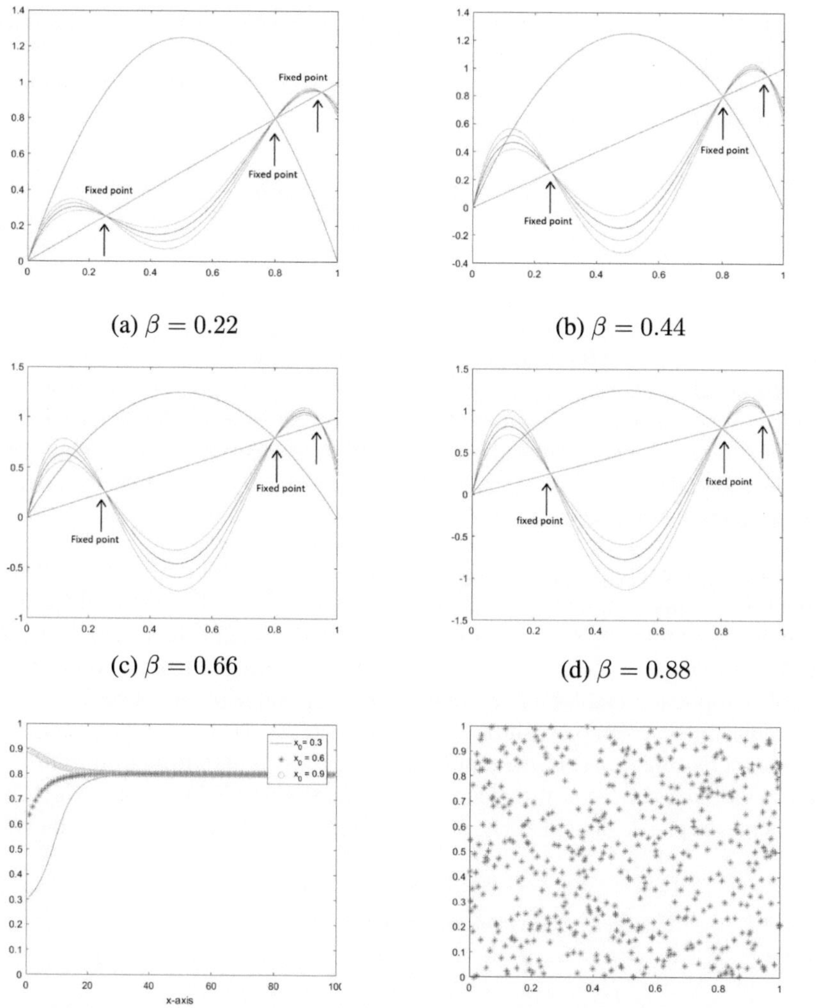

(a) $\beta = 0.22$

(b) $\beta = 0.44$

(c) $\beta = 0.66$

(d) $\beta = 0.88$

(e) Stabilized trajectories for the Ishikawa system for $\alpha = 0.02$ and $\beta = 0.99$

(f) Unstable original Picard system $f(x)$ for $x_0 = 0.1689$ and N = 500

Fig. 1.2 Ishikawa dynamics for varying α and β, stabilized trajectories, and original Picard chaos

so that, $\overline{f'}(y) - 1 = 0.4$ implying $\alpha \in (-5, 0)$, which cannot be considered since α has a restricted range between 0 and 1.

Again since $f'(\frac{4}{5}) = -3 < -1$, $\beta \in (0, 1)$ in the superior Ishikawa system.

Considering $\beta = 0.99$, we get

$$\begin{aligned}\overline{f}'(y) &= 1 + 0.99(f'(\frac{4}{5}) - 1)\\ &= 1 + 0.99(-3 - 1)\\ &= -2.96.\end{aligned}$$

So that $1 - \overline{f}'(y) = 3.96$, implying $\alpha \in (0, 0.505)$. Since $f'(x_1), f'(x_2) \notin (-1, 1)$, both the fixed points x_1 and x_2 are unstable. Figure 1.1a represents the original system in [0, 1]. The impact of the values of α and β for the values in Table 1.1 for S. No. b, c, and d are plotted graphically in Fig. 1.1b, c, d, respectively. The blue, green, purple, and yellow lines correspond to $\alpha = 0.4, 0.3, 0.2$, and 0.1, respectively, for the preferred Ishikawa system. □

Table 1.1 and Fig. 1.1b, c, d suggest that for a fixed value of β and varying α from 0 to the maximum range of α toward the fixed point, the graphical spectrums become asymptotically stable. Further, as the value of parameter α decreases, lines get closer to the line $y = x$. A comparative study of various Figures, viz., Fig. 1.1b, c, d also implicates that for a fixed value of α, as the value of parameter β increases, the graphical spectrum gets shifted toward unstability. Hence, it can be concluded that on increasing α and β, the graphical spectrum gets stabilized toward the fixed point. A similar pattern is also recognized by the study of Table 1.2 and Fig. 1.2b, c, d, whereas Fig. 1.2e signifies that the trajectories with different initial values of x,

Table 1.1 The parameters utilized in visualizing Fig. 1.1

Sr. no.	α_1	α_2	α_3	α_4	β
a	–	–	–	–	–
b	0.1	0.2	0.3	0.4	0.11
c	0.1	0.2	0.3	0.4	0.55
d	0.1	0.2	0.3	0.4	0.99

Table 1.2 The values of parameters utilized in visualizing Fig. 1.6

Sr. no.	α_1	α_2	α_3	α_4	β
a	0.6	0.7	0.8	0.9	0.22
b	0.6	0.7	0.8	0.9	0.44
c	0.6	0.7	0.8	0.9	0.66
d	0.6	0.7	0.8	0.9	0.88

viz., 0.3, 0.6, and 0.9 converge to the fixed point $x_2 = \frac{4}{5}$. The yellow, purple, green, and blue lines correspond to $\alpha = 0.9, 0.8, 0.7$, and 0.6, respectively, for the preferred Ishikawa system.

Stability Examination of the Periodic States

Although periodic fixed points in dynamical systems have significant applications across various fields of engineering, technology, and nature, these dynamical systems unfortunately lack generic properties for periodic fixed points of orders greater than one. Specifically, for the values of k greater than one, the equations $f^k(x) = x$ and $\overline{\overline{f}}^k(x) = x$ do not exhibit stable solutions.

Next, we focus on the examination of the stability of periodic fixed points via the preferred Ishikawa-type control system. Analytically, it is now required to fabricate a preferred superior Ishikawa control system that provides superiority over the given original Picard state system, providing direct headways toward the stability of periodic fixed points. Now, we design the preferred Ishikawa control state system to explore the stability of periodic fixed points, as follows:

$$\overline{f}^k(y) = x + \beta(f^k(x) - x), \tag{1.7}$$

$$\overline{\overline{f}}^k(x) = x + \alpha(\overline{f}^k(y) - x), \tag{1.8}$$

where $x \in [a, b]$ and $f^k(x)$ are the kth iterations of original Picard system.

Theorem 1.6 *The original Picard system f^k and the Ishikawa system $\overline{\overline{f}}^k$ both have the same periodic fixed points of period k, that is, $f^k(x) = x$ if and only if $\overline{\overline{f}}^k(x) = x$.*

Proof Suppose x is the periodic fixed point for period k of the original Picard control system f, that is, $f^k(x) = x$. Using Eq. (1.7) of the Ishikawa control system, we get

$$\begin{aligned}\overline{f}^k(y) &= x + \beta[f^k(x) - x),\\ &= x.\end{aligned}$$

Now, utilizing this in Eq. (1.8), we have

$$\begin{aligned}\overline{\overline{f}}^k(x) &= x + \alpha(\overline{f}^k(y) - x),\\ &= x + \alpha(x - x),\\ &= x.\end{aligned}$$

Conversely, consider $\overline{\overline{f}}^k(x) = x$, then

$$\begin{aligned} \overline{\overline{f}}^k(x) &= x, \\ \text{that is, } x + \alpha(\overline{f}^k(y) - x) &= x, \\ \text{that is, } \alpha(\overline{f}^k(y) - x) &= 0, \\ \text{that is, } \overline{f}^k(y) &= x, \end{aligned}$$

then, by using this calculation in Eq. (1.8), we have

$$\begin{aligned} x + \beta(f^k(x) - x) &= x, \\ \text{that is, } f^k(x) &= x. \end{aligned}$$

□

Theorem 1.7 *There always exists an effectual range (potential range) for controlling parameters* α, β *in the Ishikawa control system (1.8) for a periodic fixed point which is unstable in an original Picard control system* $f^k(x)$ *for period k such that* $|\overline{\overline{f^k}}'(x)| < 1$. *That is, for* $|f^{k'}(x)| > 1$, *there always exists an* $\alpha \in (\alpha_{\min}, \alpha_{\max})$ *and* $\beta \in (\beta_{\min}, \beta_{\max})$ *such that* $|\overline{\overline{f^k}}'(x)| < 1$.

Proof Suppose x is a periodic fixed point for the mapping f which is unstable in the original Picards system (1.1). Now differentiate equations (1.7) and (1.8), then

$$\overline{f'}^k(y) = 1 + \beta(f'^k(x) - 1), \tag{1.9}$$

$$\textit{and} \quad \overline{\overline{f'}}^k(x) = 1 + \alpha(\overline{f'}^k(y) - 1). \tag{1.10}$$

We now discuss the following possible cases:

Case 1 When x is an unstable periodic fixed point of the mapping f and $f'^k(x) > 1$. Then, using Eq. (1.9), $\overline{f'}^k(y) > 1$ and consequently,

$$\begin{aligned} 1 + \beta(f'^k(x) - 1) &> 1, \\ \text{that is, } \beta(f'^k(x) - 1) &> 0, \\ \text{that is, } \beta &> 0. \end{aligned}$$

Now, when x is a stable periodic fixed point of the preferred Ishikawa system (superior system), $|\overline{\overline{f'}}^k(x)| < 1$,
then, we have either $\overline{\overline{f'}}^k(x) < 1$, or $\overline{\overline{f'}}^k(x) > -1$.

On taking $\overline{\overline{f'}}^k(x) < 1$, and using Eqs. (1.9) and (1.10), we get

$$1+\alpha(\overline{f'}^k(y) - 1) < 1,$$
$$that\ is, \alpha < 0.$$

Now, on considering $\overline{\overline{f'}}^k(x) > -1$ and using Eqs. (1.9) and (1.10), we get

$$1 + \alpha(\overline{f'}^k(y) - 1) > -1,$$
$$\text{that is,}\quad \alpha > \frac{-2}{\overline{f'}^k(y) - 1}.$$

Hence, the effectual (potential) ranges of α and β for which the preferred Ishikawa system attains the stability of the unstable periodic fixed point x when $f'^k(x) > 1$ are given by $(\frac{-2}{\overline{f'}^k(y)-1}, 0)$ and $(0, 1)$, respectively; that is, $\alpha_{\min} = \frac{-2}{\overline{f'}^k(y)-1}$, $\alpha_{\max} = 0$ and $\beta_{\min} = 0$, $\beta_{\max} = 1$.

Case 2 When x is an unstable periodic fixed point for the mapping f and $f'^k(x) < -1$. Then, using Eq. (1.9), we can get $\overline{f'}^k(y) < 1$. As a consequence of which,

$$1 + \beta(f'^k(x) - 1) < 1,$$
$$\text{that is,}\quad \beta(f'^k(x) - 1) < 0,$$
$$\text{that is,}\quad \beta > 0.$$

Now, for the stability of fixed points of the preferred Ishikawa system, $|\overline{\overline{f'}}^k(x)| < 1$, so that either $\overline{\overline{f'}}^k(x) < 1$ or $\overline{\overline{f'}}^k(x) > -1$.

First, on taking $\overline{\overline{f'}}^k(x) < 1$ and using Eqs. (1.9) and (1.10), we get

$$1+\alpha(\overline{f'}^k(y) - 1) < 1,$$
$$\text{that is,}\quad \alpha > 0.$$

In the second case of taking $\overline{\overline{f'}}^k(x) > -1$ and using Eqs. (1.9) and (1.10), we attain

$$1 + \alpha(\overline{f'}^k(y) - 1) > -1,$$
$$\text{that is,}\quad \alpha(\overline{f'}^k(y) - 1) > -2,$$
$$\text{that is,}\quad \alpha < \frac{2}{1 - \overline{f'}^k(y)}.$$

Hence, the effectual (potential) ranges of α and β for which the preferred Ishikawa system attains the stability of the unstable fixed point x when $f'^k(x) < -1$ are given by $(0, \frac{2}{1-f'^k(y)})$ and $(0, 1)$, respectively; that is, $\alpha_{\min} = 0$, $\alpha_{\max} = \frac{2}{1-f'(y)}$ and $\beta_{\min} = 0$, $\beta_{\max} = 1$. □

Example 1.8 Let $f(x) = 6x(1-x)$ be the original system with two unstable fixed points $x_1 = 0$ and $x_2 = \frac{5}{6}$ for which $|f'(x)| > 1$. Next, we check the stability of period-2 fixed points of f^2 for an effectual range (potential range) of the values of parameters in the Ishikawa control system (1.8).

Proof Now $f^2(x) = 36x(1-x)(1-6x(1-x))$ is an original system, which has four fixed points $x_1 = 0$, $x_2 = 0.201452$, $x_3 = 0.833333$, and $x_4 = 0.965215$ for $x \in [0, 1]$, where x_1 and x_2 are trivial and the other two are non-trivial for period 2. Figure 1.3a is a graph for the function of period 2.

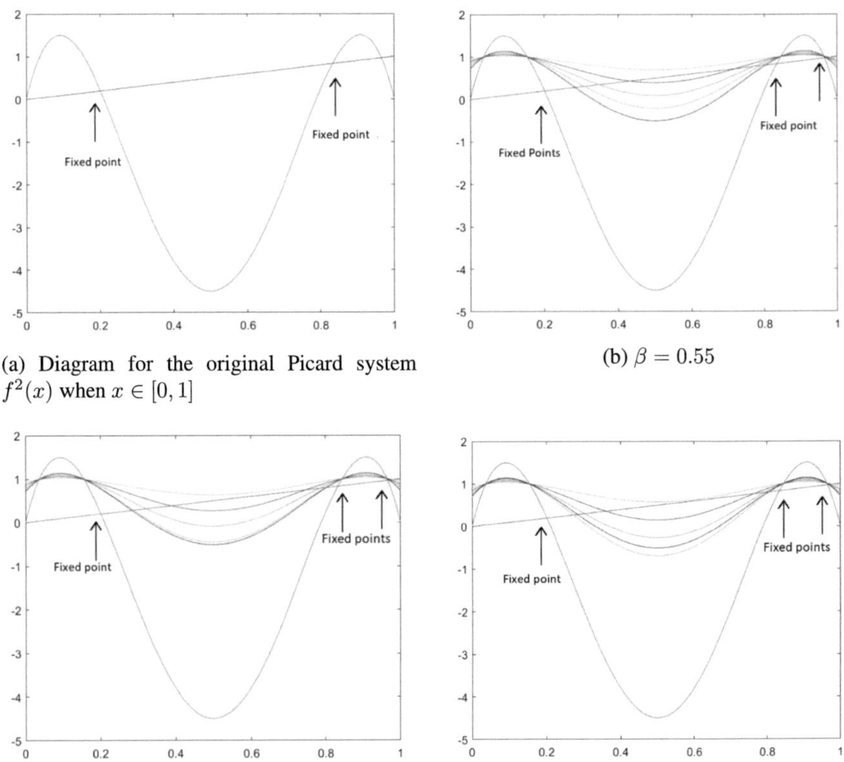

(a) Diagram for the original Picard system $f^2(x)$ when $x \in [0, 1]$

(b) $\beta = 0.55$

(c) $\beta = 0.66$

(d) $\beta = 0.77$

Fig. 1.3 Period-2 Picard dynamics under Ishikawa control

Table 1.3 The values of parameters utilized in visualizing Fig. 1.3

Sr. no.	α_1	α_2	α_3	α_4	α_5	β
a	–	–	–	–	–	–
b	0.1	0.2	0.3	0.4	0.5	0.55
c	0.1	0.2	0.3	0.4	0.5	0.66
d	0.1	0.2	0.3	0.4	0.5	0.77

$$
\begin{aligned}
f^{2'}(0) &= 36, \\
f^{2'}(0.201452) &= -20.00, \\
f^{2'}(0.833333) &= 16.00, \\
f^{2'}(0.965215) &= -20.00.
\end{aligned}
$$

Here we provide a proper graphical and analytical solution to the proposed problem. We have generated Fig. 1.2e for the original Picard system $f^2(x)$ and the chaotic behavior is shown in Fig. 1.2f. The impact of the values of α and β for the values in Table 1.3 for S. No's. b, c, and d are plotted graphically in Figs. 1.3b, 1.3c, and 1.3d, respectively. The yellow, purple, green, blue, and brownish red lines correspond to $\alpha = 0.5$, 0.4, 0.3, 0.2, and 0.1, respectively, for the preferred Ishikawa system. So that we get $\overline{f}'(y) = -0.8353$ for $\beta = 0.99$, then $\alpha \in (-1.0897, 0)$ which cannot be considered since α has a restricted range between 0 and 1. □

Example 1.9 Let $f(x) = 5x(1-x)^2$, where $x \in [0, 1]$. Next, we check the stability of unstable period-2 fixed points of f^2 for an effectual range (potential range) of parameters in the Ishikawa control system (1.8).

Proof Let $f^2(x) = 25x(1-x)^2(1-5x(1-x)^2)^2$ be an original system, which has five fixed points $x_1 = 0$, $x_2 = 0.2764$, $x_3 = 0.5528$, $x_4 = 0.7236$, and $x_5 = 1.4472$, where x_3 and x_4 are stable and the x_1, x_2 and x_5 are unstable fixed points for period 2. Figure 1.4a is a graph for the function of period 2. As we have

$$
\begin{aligned}
f^{2'}(0) &= 25, \\
f^{2'}(0.2764) &= 13.4167, \\
f^{2'}(0.5528) &= -5.9394, \\
f^{2'}(0.7236) &= -13.4165, \\
f^{2'}(1.4772) &= 853.3881.
\end{aligned}
$$

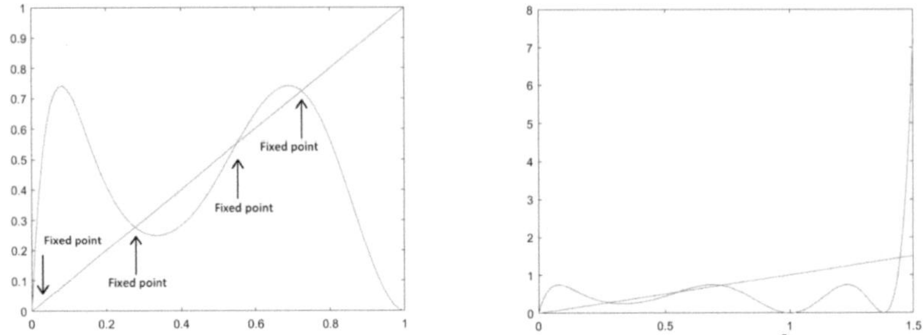

(a) The Ishikawa system $f^2(x)$ when $x \in [0, 1]$ (b) The Ishikawa system $f^2(x)$ when $x \in [0, 1.5]$

Fig. 1.4 Period-2 Picard dynamics with Ishikawa control

Table 1.4 The values of parameters utilized in visualizing Figs. 1.4 and 1.5

Sr. no.	α_1	α_2	α_3	α_4	β
4.a	–	–	–	–	–
4.b	–	–	–	–	–
5.a	–	–	–	–	–
5.b	0.1	0.2	0.3	0.4	0.99
5.c	0.1	0.2	0.3	0.4	0.77
5.d	0.1	0.2	0.3	0.4	0.55

So that we get $\overline{f}'(y) = -0.8353$ for $\beta = 0.99$ then $\alpha \in (-1.0897, 0)$, which cannot be considered since α has a restricted range between 0 and 1. In Table 1.4, $\alpha = 0.4$, 0.3, 0.2, and 0.1 represents the yellow, purple, green, and blue lines, respectively, for the preferred Ishikawa system (see, Figs. 1.4a–d and 1.5a–d). □

Similarly, in Table 1.5, we use the values of α in Fig. 1.6e, we compare the behavior of $f_\lambda(x)$ for Ishikawa and Picard Iterations. Note that as we decrease the value of β then other graphs behave like $y = x$.

Stability Examination Using Lyapunov Exponent

The Lyapunov exponent plays a crucial role in determining the sensitivity of orbits. This subsection examines the behavior of the Lyapunov exponent for both fixed and periodic point states within the preferred Ishikawa control system. The system under consideration is

$$\overline{f}(y) = x + \beta(f(x) - x),$$
$$\overline{\overline{f}}(x) = x + \alpha(\overline{f}(y) - x),$$

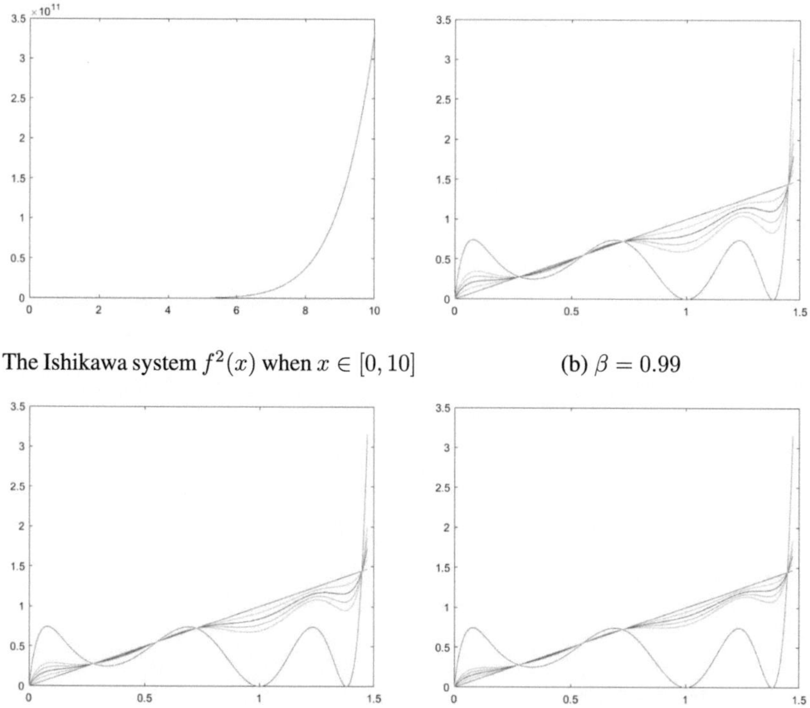

(a) The Ishikawa system $f^2(x)$ when $x \in [0, 10]$

(b) $\beta = 0.99$

(c) $\beta = 0.77$

(d) $\beta = 0.55$

Fig. 1.5 Period-2 Ishikawa dynamics for varying β

Table 1.5 The values of parameters utilized in visualizing Fig. 1.6

Sr. no.	α_1	α_2	α_3	α_4	β
a	0.6	0.7	0.8	0.9	0.33
b	0.6	0.7	0.8	0.9	0.66
c	0.6	0.7	0.8	0.9	0.88
d	0.1	0.2	0.3	0.4	0.22
e	0.1	0.2	0.3	0.4	0.88

where $f(x)$ corresponds to the original Picard system. We consider two arbitrary points x_0 and y_0 from the interval $[a, b]$ having a minor difference $\epsilon \in (0, 1)$. In this case, the kth iteration for the system $\overline{\overline{f}}$ is given by

$$|y_p - x_p| = |\overline{\overline{f^k}}(y_0) - \overline{\overline{f^k}}(x_0)| = P(y_0, x_0).$$

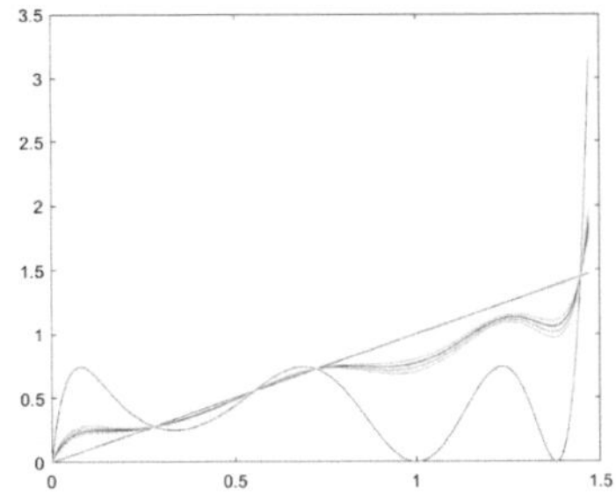

(a) $\beta = 0.33$

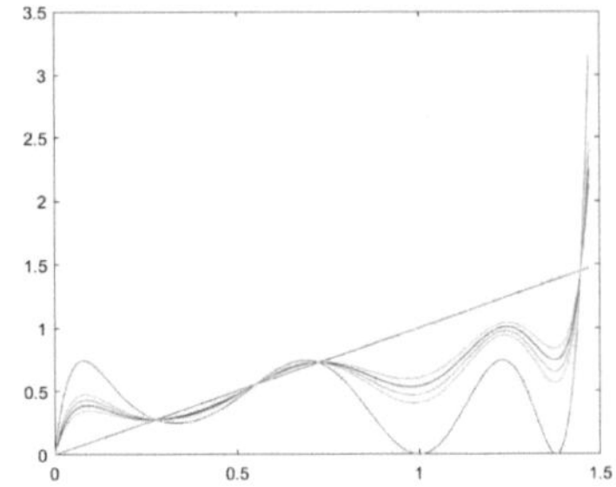

(b) $\beta = 0.66$

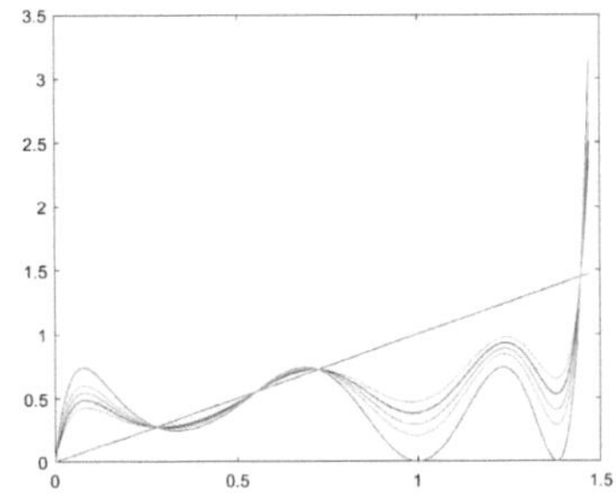

(c) $\beta = 0.88$

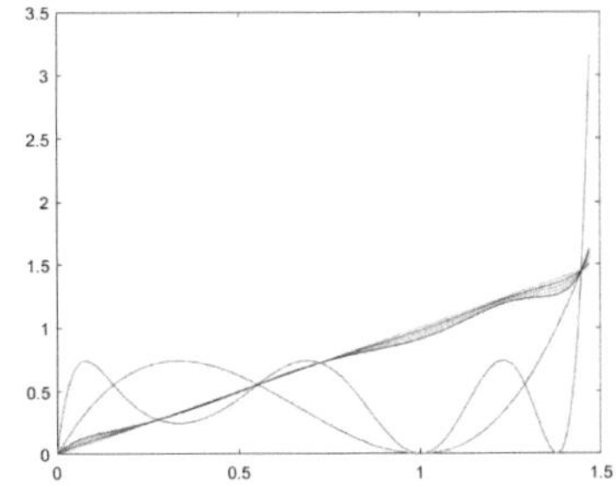

(d) $\beta = 0.22$

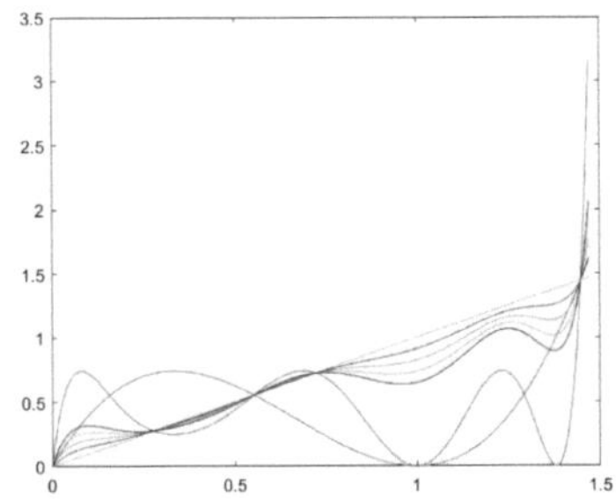

(e) $\beta = 0.88$

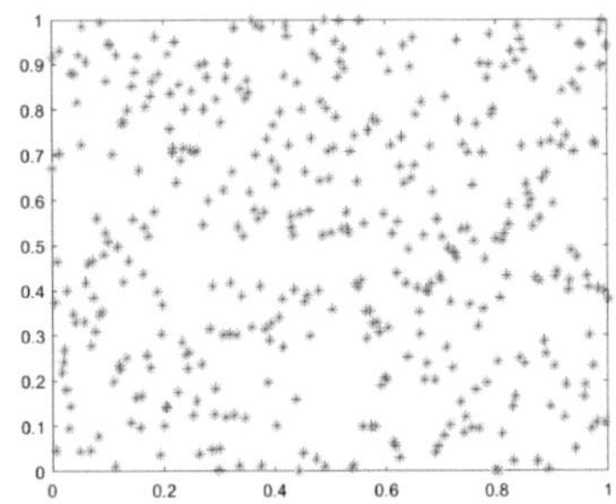

(f) Unstablity of an original Picard control system f(x) for $x_0 = 0.2345$ and N = 500

Fig. 1.6 Ishikawa dynamics for varying values of α and β, and instability of the original Picard system

Now, using Taylor's series expansion and considering only terms of first-order derivatives in it, we attain

$$\overline{\overline{f^k}}(y_0) = \overline{\overline{f^k}}(x_0) + \overline{\overline{f^k}}'(x_0)(y_0 - x_0)$$
$$\overline{\overline{f^k}}(y_0) - \overline{\overline{f^k}}(x_0) = \overline{\overline{f^k}}'(x_0)(y_0 - x_0)$$
$$\frac{\overline{\overline{f^k}}(y_0) - \overline{\overline{f^k}}(x_0)}{(y_0 - x_0)} = \overline{\overline{f^k}}'(x_0).$$

Now, in view of Devaney's chain rule

$$\frac{\overline{\overline{f^k}}(y_0) - \overline{\overline{f^k}}(x_0)}{(y_0 - x_0)} = (\overline{\overline{f}}'(x_0).\overline{\overline{f}}'(x_1).\overline{\overline{f}}'(x_2)...\overline{\overline{f}}'(x_{k-1})].$$

Letting $k \to \infty$, we get

$$\lim_{k\to\infty} \frac{\overline{\overline{f^k}}(y_0) - \overline{\overline{f^k}}(x_0)}{(y_0 - x_0)} = \lim_{k\to\infty} (\overline{\overline{f}}'(x_0).\overline{\overline{f}}'(x_1).\overline{\overline{f}}'(x_2)...\overline{\overline{f}}'(x_{k-1})]. \quad (1.11)$$

Also, it is a well-known fact that the rate of growth follows an exponential asymptotic pattern in periodic orbits, it follows that

$$\lim_{k\to\infty} \frac{\overline{\overline{f^k}}(y_0) - \overline{\overline{f^k}}(x_0)}{(y_0 - x_0)} = \lim_{p\to\infty} e^{k\lambda}. \quad (1.12)$$

Using Eqs. (1.9) and (1.10), we can write

$$\lim_{k\to\infty} (\overline{\overline{f}}'(x_0).\overline{\overline{f}}'(x_1).\overline{\overline{f}}'(x_2)...\overline{\overline{f}}'(x_{k-1})] = \lim_{k\to\infty} e^{k\lambda}. \quad (1.13)$$

Taking Logarithms on both sides, we get

$$\lambda = \lim_{k\to\infty} \frac{1}{k} \log(\overline{\overline{f}}'(x_0).\overline{\overline{f}}'(x_1).\overline{\overline{f}}'(x_2)...\overline{\overline{f}}'(x_{k-1})] \quad (1.14)$$
$$= \lim_{k\to\infty} \frac{1}{k} \sum_{i=1}^{k} \log|\overline{\overline{f}}'(x_i)|, \quad (1.15)$$

where k is the period of function and λ is the Lyapunov exponent for the preferred Ishikawa-type control system. Particularly, on assigning k to be 1, only a single fixed point is present and the Lyapunov exponent takes the form

$$\lambda = \log|\overline{\overline{f}}'(x)|. \quad (1.16)$$

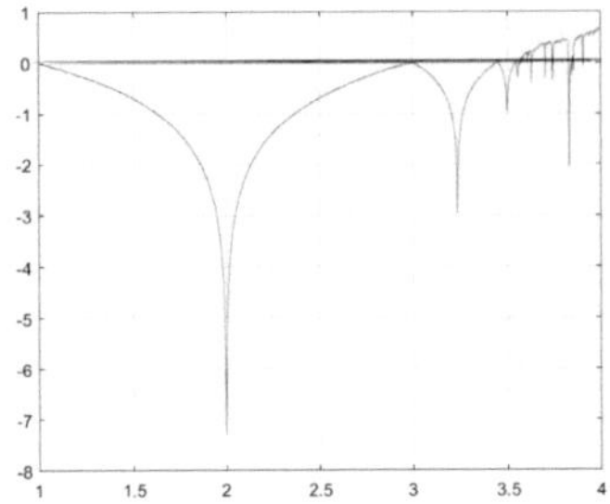

(a) Lyapunov exponent for $r \in [0, 4]$ for Example 3.3

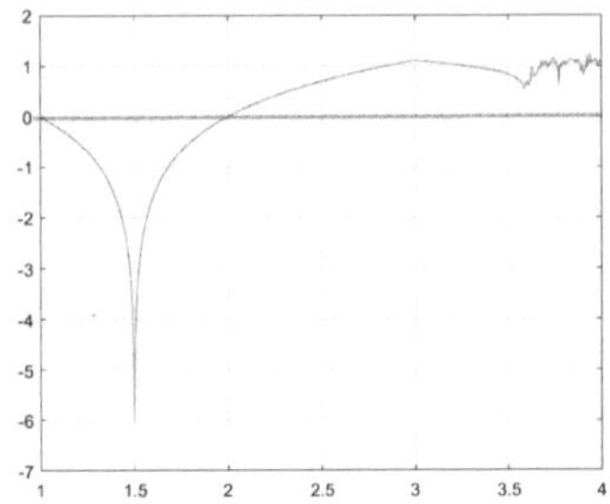

(b) Lyapunov exponent for $r \in [0, 4]$ for Example 3.5

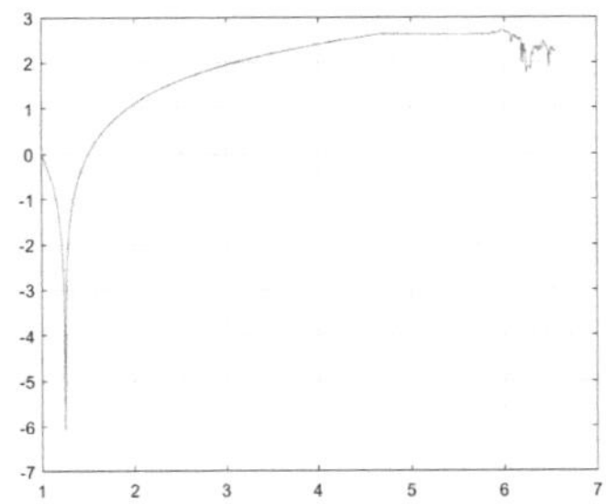

(c) Lyapunov exponent for $\alpha = 0.6, \beta = 0.9$ and $r \in [0, 7]$ for Example 3.3

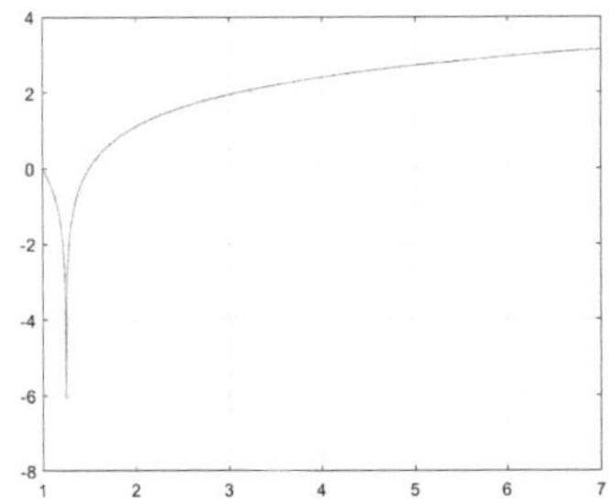

(d) Lyapunov exponent for $\alpha = 0.5, \beta = 0.5$ and $r \in [0, 7]$ for Example 3.3

Fig. 1.7 Lyapunov exponent spectra for the Picard and Ishikawa systems under varying parameters

Remarkably, it is to be noted that the fixed and periodic conditions are stabilized for $\lambda < 0$, whereas, for $\lambda > 0$ these states get unstabilized. See Fig. 1.7.

Emergence of Chaos in Ishikawa Orbit of the Logistic Map

Theorem 1.10 *Let $f_\lambda(x) = \lambda x(1 - x)$, be the logistic map for $\lambda \in [4.143, 0.4.157]$ and $x \in [0, 1]$. Then, $M^3_{\alpha\beta\lambda}(x)$ has eight fixed point for $\alpha = 0.9$.*

Proof Let $M_{\alpha,\beta,\lambda} = (x + \alpha(\beta(f_\lambda(x) - x)))$, where $\overline{f}_\lambda(x) = \lambda x(1 - x)$. Then

$$\begin{aligned} M^3_{\alpha,\beta,\lambda} &= M_{\alpha,\beta,\lambda}(M_{\alpha,\beta,\lambda}(M_{\alpha,\beta,\lambda})) \\ &= M^2_{\alpha,\beta,\lambda}(x + \alpha(\beta(f_\lambda(x) - x))) \\ &= M^2_{\alpha,\beta,\lambda}((1 - \alpha\beta)x + \alpha\beta\lambda(x - x^2)) \end{aligned}$$

$$= M_{\alpha,\beta,\lambda}[(1-\alpha\beta)[(1-\alpha\beta)x+\alpha\beta\lambda(x-x^2)]$$
$$+\alpha\beta\lambda([(1-\alpha\beta)x+\alpha\beta\lambda(x-x^2)]-[(1-\alpha\beta)x+\alpha\beta\lambda(x-x^2)]^2)]\ (A)$$

$$= M_{\alpha\beta\lambda}(x[(1-\alpha\beta)^2+2\alpha\beta\lambda(1-\alpha\beta)+\alpha^2\beta^2\lambda^2]$$
$$-x^2[\alpha\beta\lambda(1-\alpha\beta)+\alpha^2\beta^2\lambda^2+\alpha\beta\lambda(1-\alpha\beta)^2+\alpha^3\beta^3\lambda^3+2\alpha^2\beta^2\lambda^2(1-\alpha\beta)]$$
$$+x^3[2\alpha^3\beta^3\lambda^3+2\alpha^2\beta^2\lambda^2(1-\alpha\beta)]-x^4[\alpha^3\beta^3\lambda^3])\ (B)$$

$$M^3_{\alpha\beta\lambda}(x)=((1-\alpha\beta)P+\alpha\beta\lambda(P-P^2))$$

where

$$P=(x[(1-\alpha\beta)^2+2\alpha\beta\lambda(1-\alpha\beta)+\alpha^2\beta^2\lambda^2]-x^2[\alpha\beta\lambda(1-\alpha\beta)+\alpha^2\beta^2\lambda^2$$
$$+\alpha\beta\lambda(1-\alpha\beta)^2+\alpha^3\beta^3\lambda^3+2\alpha^2\beta^2\lambda^2(1-\alpha\beta)]+x^3[2\alpha^3\beta^3\lambda^3+2\alpha^2\beta^2\lambda^2(1-\alpha\beta)]$$
$$-x^4[\alpha^3\beta^3\lambda^3])$$

$$(1-\alpha\beta)P=-x^4[\alpha^3\beta^3\lambda^3(1-\alpha\beta)]+x^3[2\alpha^3\beta^3\lambda^3(1-\alpha\beta)+2\alpha^2\beta^2\lambda^2(1-\alpha\beta)^2]$$
$$-x^2[\alpha\beta\lambda(1-\alpha\beta)^2+\alpha^2\beta^2\lambda^2(1-\alpha\beta)+\alpha\beta\lambda(1-\alpha\beta)^3+\alpha^3\beta^3\lambda^3(1-\alpha\beta)$$
$$+2\alpha^2\beta^2\lambda^2(1-\alpha\beta)^2]++x[(1-\alpha\beta)^3+2\alpha\beta\lambda(1-\alpha\beta)^2+\alpha^2\beta^2\lambda^2(1-\alpha\beta)]$$

$$P^2=x^8[\alpha^6\beta^6\lambda^6]-4x^7[\alpha^6\beta^6\lambda^6+\alpha^5\beta^5\lambda^5(1-\alpha\beta)]$$
$$+2x^6[\alpha^6\beta^6\lambda^6+\alpha^5\beta^5\lambda^5+2\alpha^5\beta^5\lambda^5(1-\alpha\beta)+\alpha^4\beta^4\lambda^4(1-\alpha\beta)+\alpha^4\beta^4\lambda^4(1-\alpha\beta)^2]$$
$$+4x^6[\alpha^6\beta^6\lambda^6+2\alpha^5\beta^5\lambda^5(1-\alpha\beta)+\alpha^4\beta^4\lambda^4(1-\alpha\beta)^2]$$
$$-2x^5[\alpha^5\beta^5\lambda^5+2\alpha^4\beta^4\lambda^4(1-\alpha\beta)+\alpha^3\beta^3\lambda^3(1-\alpha\beta)^2]$$
$$-2x^5[\alpha^6\beta^6\lambda^6+\alpha^5\beta^5\lambda^5+3\alpha^5\beta^5\lambda^5(1-\alpha\beta)+2\alpha^4\beta^4\lambda^4(1-\alpha\beta)+3\alpha^4\beta^4\lambda^4(1-\alpha\beta)^2$$
$$+\alpha^3\beta^3\lambda^3(1-\alpha\beta)^2+\alpha^3\beta^3\lambda^3(1-\alpha\beta)^3]$$
$$+4x^4[3\alpha^4\beta^4\lambda^4(1-\alpha\beta)+\alpha^5\beta^5\lambda^5+\alpha^2\beta^2\lambda^2(1-\alpha\beta)^3+3\alpha^3\beta^3\lambda^3(1-\alpha\beta)^2]$$
$$+x^4[\alpha^2\beta^2\lambda^2(1-\alpha\beta)^2+2\alpha^2\beta^2\lambda^2(1-\alpha\beta)^3+\alpha^2\beta^2\lambda^2(1-\alpha\beta)^4+2\alpha^3\beta^3\lambda^3(1-\alpha\beta)$$
$$+6\alpha^3\beta^3\lambda^3(1-\alpha\beta)^2+4\alpha^3\beta^3\lambda^3(1-\alpha\beta)^3+\alpha^4\beta^4\lambda^4+6\alpha^4\beta^4\lambda^4(1-\alpha\beta)$$
$$+6\alpha^4\beta^4\lambda^4(1-\alpha\beta)^2+2\alpha^5\beta^5\lambda^5+4\alpha^5\beta^5\lambda^5(1-\alpha\beta)+\alpha^6\beta^6\lambda^6]$$
$$-2x^3[\alpha\beta\lambda(1-\alpha\beta)^4+\alpha\beta\lambda(1-\alpha\beta)^5+3\alpha^2\beta^2\lambda^2(1-\alpha\beta)^3+4\alpha^2\beta^2\lambda^2(1-\alpha\beta)^4$$
$$+3\alpha^3\beta^3\lambda^3(1-\alpha\beta)^2+6\alpha^3\beta^3\lambda^3(1-\alpha\beta)^3+\alpha^4\beta^4\lambda^4(1-\alpha\beta)$$
$$+4\alpha^4\beta^4\lambda^4(1-\alpha\beta)^2+\alpha^5\beta^5\lambda^5(1-\alpha\beta)]$$
$$+x^2[(1-\alpha\beta)^4+4\alpha\beta\lambda(1-\alpha\beta)^3+6\alpha^2\beta^2\lambda^2(1-\alpha\beta)^2$$
$$+4\alpha^3\beta^3\lambda^3(1-\alpha\beta)+\alpha^4\beta^4\lambda^4]\ (C)$$

$$
\begin{aligned}
P - P^2 = {} & [-x^4[\alpha^3\beta^3\lambda^3] \\
& + x^3[2\alpha^3\beta^3\lambda^3 + 2\alpha^2\beta^2\lambda^2(1-\alpha\beta)] \\
& - x^2[\alpha\beta\lambda(1-\alpha\beta) + \alpha^2\beta^2\lambda^2 + \alpha\beta\lambda(1-\alpha\beta)^2 + \alpha^3\beta^3\lambda^3 + 2\alpha^2\beta^2\lambda^2(1-\alpha\beta)] \\
& + x[(1-\alpha\beta)^2 + 2\alpha\beta\lambda(1-\alpha\beta) + \alpha^2\beta^2\lambda^2]] \\
& - [x^8[\alpha^6\beta^6\lambda^6] - 4x^7[\alpha^6\beta^6\lambda^6 + \alpha^5\beta^5\lambda^5(1-\alpha\beta)] \\
& + 2x^6[\alpha^6\beta^6\lambda^6 + \alpha^5\beta^5\lambda^5 + 2\alpha^5\beta^5\lambda^5(1-\alpha\beta) + \alpha^4\beta^4\lambda^4(1-\alpha\beta) + \alpha^4\beta^4\lambda^4(1-\alpha\beta)^2] \\
& + 4x^6[\alpha^6\beta^6\lambda^6 + 2\alpha^5\beta^5\lambda^5(1-\alpha\beta) + \alpha^4\beta^4\lambda^4(1-\alpha\beta)^2] \\
& - 2x^5[\alpha^5\beta^5\lambda^5 + 2\alpha^4\beta^4\lambda^4(1-\alpha\beta) + \alpha^3\beta^3\lambda^3(1-\alpha\beta)^2] \\
& - 2x^5[\alpha^6\beta^6\lambda^6 + \alpha^5\beta^5\lambda^5 + 3\alpha^5\beta^5\lambda^5(1-\alpha\beta) + 2\alpha^4\beta^4\lambda^4(1-\alpha\beta) + 3\alpha^4\beta^4\lambda^4(1-\alpha\beta)^2 \\
& + \alpha^3\beta^3\lambda^3(1-\alpha\beta)^2 + \alpha^3\beta^3\lambda^3(1-\alpha\beta)^3] \\
& + 4x^4[3\alpha^4\beta^4\lambda^4(1-\alpha\beta) + \alpha^5\beta^5\lambda^5 + \alpha^2\beta^2\lambda^2(1-\alpha\beta)^3 + 3\alpha^3\beta^3\lambda^3(1-\alpha\beta)^2] \\
& + x^4[\alpha^2\beta^2\lambda^2(1-\alpha\beta)^2 + 2\alpha^2\beta^2\lambda^2(1-\alpha\beta)^3 + \alpha^2\beta^2\lambda^2(1-\alpha\beta)^4 + 2\alpha^3\beta^3\lambda^3(1-\alpha\beta) \\
& + 6\alpha^3\beta^3\lambda^3(1-\alpha\beta)^2 + 4\alpha^3\beta^3\lambda^3(1-\alpha\beta)^3 + \alpha^4\beta^4\lambda^4 + 6\alpha^4\beta^4\lambda^4(1-\alpha\beta) \\
& + 6\alpha^4\beta^4\lambda^4(1-\alpha\beta)^2 + 2\alpha^5\beta^5\lambda^5 + 4\alpha^5\beta^5\lambda^5(1-\alpha\beta) + \alpha^6\beta^6\lambda^6] \\
& - 2x^3[\alpha\beta\lambda(1-\alpha\beta)^4 + \alpha\beta\lambda(1-\alpha\beta)^5 + 3\alpha^2\beta^2\lambda^2(1-\alpha\beta)^3 + 4\alpha^2\beta^2\lambda^2(1-\alpha\beta)^4 \\
& + 3\alpha^3\beta^3\lambda^3(1-\alpha\beta)^2 + 6\alpha^3\beta^3\lambda^3(1-\alpha\beta)^3 + \alpha^4\beta^4\lambda^4(1-\alpha\beta) \\
& + 4\alpha^4\beta^4\lambda^4(1-\alpha\beta)^2 + \alpha^5\beta^5\lambda^5(1-\alpha\beta)] \\
& + x^2[(1-\alpha\beta)^4 + 4\alpha\beta\lambda(1-\alpha\beta)^3 + 6\alpha^2\beta^2\lambda^2(1-\alpha\beta)^2 + 4\alpha^3\beta^3\lambda^3(1-\alpha\beta) + \alpha^4\beta^4\lambda^4]]\ (D)
\end{aligned}
$$

$$
\begin{aligned}
M_{\alpha\beta\lambda}(P) = {} & -x^4[\alpha^3\beta^3\lambda^3(1-\alpha\beta)] + x^3[2\alpha^3\beta^3\lambda^3(1-\alpha\beta) + 2\alpha^2\beta^2\lambda^2(1-\alpha\beta)^2] \\
& - x^2[\alpha\beta\lambda(1-\alpha\beta)^2 + \alpha^2\beta^2\lambda^2(1-\alpha\beta) + \alpha\beta\lambda(1-\alpha\beta)^3 + \alpha^3\beta^3\lambda^3(1-\alpha\beta) \\
& + 2\alpha^2\beta^2\lambda^2(1-\alpha\beta)^2] + x[(1-\alpha\beta)^3 + 2\alpha\beta\lambda(1-\alpha\beta)^2 + \alpha^2\beta^2\lambda^2(1-\alpha\beta)] \\
& + \alpha\beta\lambda[-x^8[\alpha^6\beta^6\lambda^6] + 4x^7[\alpha^6\beta^6\lambda^6 + \alpha^5\beta^5\lambda^5(1-\alpha\beta)] \\
& - 2x^6[\alpha^6\beta^6\lambda^6 + \alpha^5\beta^5\lambda^5 + 2\alpha^5\beta^5\lambda^5(1-\alpha\beta) + \alpha^4\beta^4\lambda^4(1-\alpha\beta) + \alpha^4\beta^4\lambda^4(1-\alpha\beta)^2] \\
& - 4x^6[\alpha^6\beta^6\lambda^6 + 2\alpha^5\beta^5\lambda^5(1-\alpha\beta) + \alpha^4\beta^4\lambda^4(1-\alpha\beta)^2] \\
& + 2x^5[\alpha^5\beta^5\lambda^5 + 2\alpha^4\beta^4\lambda^4(1-\alpha\beta) + \alpha^3\beta^3\lambda^3(1-\alpha\beta)^2] \\
& + 2x^5[\alpha^6\beta^6\lambda^6 + \alpha^5\beta^5\lambda^5 + 3\alpha^5\beta^5\lambda^5(1-\alpha\beta) + 2\alpha^4\beta^4\lambda^4(1-\alpha\beta) + 3\alpha^4\beta^4\lambda^4(1-\alpha\beta)^2 \\
& + \alpha^3\beta^3\lambda^3(1-\alpha\beta)^2 + \alpha^3\beta^3\lambda^3(1-\alpha\beta)^3] \\
& - x^4[\alpha^3\beta^3\lambda^3(1-\alpha\beta)] \\
& - 4x^4[3\alpha^4\beta^4\lambda^4(1-\alpha\beta) + \alpha^5\beta^5\lambda^5 + \alpha^2\beta^2\lambda^2(1-\alpha\beta)^3 + 3\alpha^3\beta^3\lambda^3(1-\alpha\beta)^2] \\
& - x^4[\alpha^2\beta^2\lambda^2(1-\alpha\beta)^2 + 2\alpha^2\beta^2\lambda^2(1-\alpha\beta)^3 + \alpha^2\beta^2\lambda^2(1-\alpha\beta)^4 + 2\alpha^3\beta^3\lambda^3(1-\alpha\beta) \\
& + 6\alpha^3\beta^3\lambda^3(1-\alpha\beta)^2 + 4\alpha^3\beta^3\lambda^3(1-\alpha\beta)^3 + \alpha^4\beta^4\lambda^4 + 6\alpha^4\beta^4\lambda^4(1-\alpha\beta) \\
& + 6\alpha^4\beta^4\lambda^4(1-\alpha\beta)^2 + 2\alpha^5\beta^5\lambda^5 + 4\alpha^5\beta^5\lambda^5(1-\alpha\beta) + \alpha^6\beta^6\lambda^6] \\
& + x^3[2\alpha^3\beta^3\lambda^3(1-\alpha\beta) + 2\alpha^2\beta^2\lambda^2(1-\alpha\beta)^2] \\
& + 2x^3[\alpha\beta\lambda(1-\alpha\beta)^4 + \alpha\beta\lambda(1-\alpha\beta)^5 + 3\alpha^2\beta^2\lambda^2(1-\alpha\beta)^3 + 4\alpha^2\beta^2\lambda^2(1-\alpha\beta)^4 \\
& + 3\alpha^3\beta^3\lambda^3(1-\alpha\beta)^2 + 6\alpha^3\beta^3\lambda^3(1-\alpha\beta)^3 + \alpha^4\beta^4\lambda^4(1-\alpha\beta) \\
& + 4\alpha^4\beta^4\lambda^4(1-\alpha\beta)^2 + \alpha^5\beta^5\lambda^5(1-\alpha\beta)]
\end{aligned}
$$

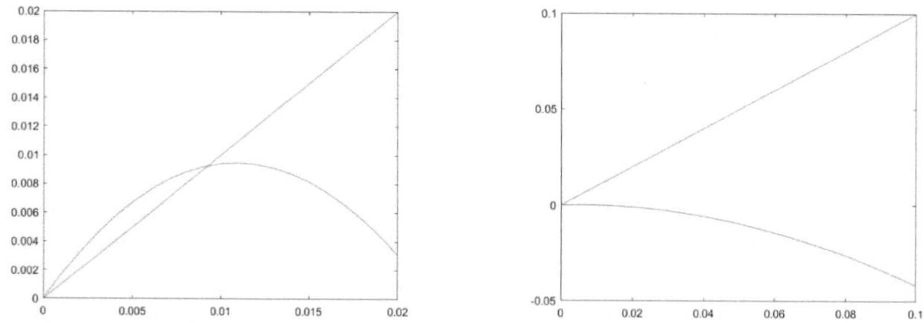

(a) Graphical plot of $M^3_{\alpha,\beta\lambda}(x)$ for $\alpha = 0.8, \beta = 0.5$ and $\lambda = 4.5$

(b) Graphical plot of $M^3_{\alpha,\beta\lambda}(x)$ for $\alpha = 0.1, \beta = 0.5$ and $\lambda = 4.5$

Fig. 1.8 Fixed-point structure of the 3-periodic mapping $M^3_{\alpha,\beta\lambda}$

$$
\begin{aligned}
&- x^2[(1-\alpha\beta)^4 + 4\alpha\beta\lambda(1-\alpha\beta)^3 + 6\alpha^2\beta^2\lambda^2(1-\alpha\beta)^2 + 4\alpha^3\beta^3\lambda^3(1-\alpha\beta) + \alpha^4\beta^4\lambda^4] \\
&- x^2[\alpha\beta\lambda(1-\alpha\beta)^2 + \alpha^2\beta^2\lambda^2(1-\alpha\beta) + \alpha\beta\lambda(1-\alpha\beta)^3 + \alpha^3\beta^3\lambda^3(1-\alpha\beta) \\
&+ 2\alpha^2\beta^2\lambda^2(1-\alpha\beta)^2] \\
&+ x[(1-\alpha\beta)^3 + 2\alpha\beta\lambda(1-\alpha\beta)^2 + \alpha^2\beta^2\lambda^2(1-\alpha\beta)]]\ (E)
\end{aligned}
$$

solving equation (E), we get unique and non-unique fixed point solutions for $M^3_{\alpha\beta\lambda}(x)$ for some value of α, β and λ, which can be shown by a graphical approach. □

Example 1.11 We examine the stability of 3-periodic mapping $M^3_{\alpha\beta\lambda}$ for $\alpha = 0.1$, $\beta = 0.5$, and $\lambda = 4.5$.

Proof By using Theorem 1.10, we show that $M^3_{\alpha\beta\lambda}(x)$ have both unique and non-unique fixed points for some values of α, β, and λ. Clearly, it is not easy to get a solution for an eight-degree polynomial algebraically. So we use MATLAB to determine for $(\alpha, \beta, \lambda) = (0.1, 0.5, 4.5)$ and $(0.8, 0.5, 4.5)$. The absolute derivative method is used to evaluate stability, that is, the fixed point is said to be stable (attracting) for $M^n_{\alpha,\beta\lambda}(x)$ if $|M^n_{\alpha,\beta\lambda}(x)| < 1$ and unstable (repelling) if $|M^n_{\alpha,\beta\lambda}(x)| > 1$ and also it is neutral or weak attracting if $|M^n_{\alpha,\beta\lambda}(x)| = 1$. Then for $n = 3$, we get Fig. 1.8a, b, in which we show that $f_\lambda(x)$ has unique and non-unique fixed points for different values α, β, and γ.

□

1.3.2 Traffic Flow Model

A dynamical system can be used to describe the traffic network. In traffic control models (see Marusin & Danilov 2018; Seliverstov & Seliverstov 2018; Xiao & Lo 2015), chaos or chaotic flow implies that traffic is unstable and unexpected, which is hazardous for drivers. A bifurcation interpreted by Stoop and Radons (1997) is an alteration in an external parameter that results in a system's solution coming to a crossroads and is known as a control parameter because it can revise the system's qualitative behavior. Experimentally, it has been observed that increasing the value of the control parameter causes period doubling, which transforms simple systems into complex ones and eventually, degenerates into a state of chaos. But, period doubling is not only the cause of instigating chaotic behavior. The bifurcation sequences are mathematically described, and it has been observed that basic mechanisms are adequate to produce complicated behavior. Jarrett and Zhang (1993) and Zhang and Jarrett (Zhang & Jarrett 1997) examined chaos in a traffic control system by adopting the car-following model, whereas Kirby and Smith (1993) unfortunately failed to determine chaos in car-following models.

Basically, traffic flow problems can be characterized mainly by macroscopic and microscopic categories. The model defined by Lo and Cho (2005) is a macroscopic traffic flow model and is described by the logistic map

$$x_{n+1} = \lambda x_n(1 - x_n), \tag{1.17}$$

where x_i denotes the population density with time i and λ is the control parameter. Here λ varies from 1 to 4 and a fixed point is present when $\lambda \in [1, 3)$. The bifurcation point is at $\lambda = 3$. When $\lambda \in (3, 4]$ behavior gets complex and chaos occurs. The bifurcation figure initiates the flow–density–speed fundamental model

$$p = h.v, \tag{1.18}$$

where p is the flow, v is speed, and h is density. Since it is well known that speed completely rely on density so p is the function of density and we get

$$p(h) = hv(h). \tag{1.19}$$

If flow satisfies Greenshield's model, then, the formula can be written as

$$v(h) = v_f(1 - \frac{h}{h_j}), \tag{1.20}$$

where v_f is free flow speed and h_j is jam density. Using Eqs. (1.19) and (1.20), we obtain

$$p(h) = h.v_f(1 - \frac{h}{h_j}). \tag{1.21}$$

It is well known that traffic flow in the next span depends on current traffic conditions. So Eq. (1.18) can be rewritten as

$$p_{i+1} = h_i.v_f(1 - \frac{h_i}{h_j}). \tag{1.22}$$

Also, the occupancy variable converts Eq. (1.19) into a single variable function. The mathematically occupancy variable σ is

$$\sigma = \frac{p\overline{L}}{\overline{v}} = h\overline{L}, \tag{1.23}$$

where $\overline{L}$ is the average vehicle length and $\overline{v}$ is the average speed. Occupancy is the fraction of the time taken and available time for specific spaces and lies in [0, 1]

$$p = \frac{\sigma\overline{v}}{\overline{L}} \tag{1.24}$$

$$h = \frac{\sigma}{\overline{L}}. \tag{1.25}$$

Using values of p and h in Eq. (1.19), we obtain

$$\frac{\sigma_{i+1}\overline{v}}{\overline{L}} = v_f \frac{\sigma_i}{\overline{L}}(1 - \frac{\sigma_i}{h_j\overline{L}}). \tag{1.26}$$

Let us consider $\check{\sigma} = \frac{\sigma}{h_j\overline{L}}$ so that

$$\check{\sigma}_{i+1} = \frac{v_f}{\overline{v}}\check{\sigma}_i(1 - \check{\sigma}_i). \tag{1.27}$$

Equation (1.24) is a developed discrete dynamic traffic flow model which is the same as Eq. (1.17). So the fraction of free flow speed and average speed equals the control parameter $\lambda \in [0, 4]$. Here we have some observations

(i). When the control parameter $\lambda \in (0, 1]$, traffic flow is minor and there is nothing to control in a system.
(ii). When the control parameter $\lambda \in (1, 3]$, traffic flow is in a stable state.
(iii). When the control parameter $\lambda \in (3, 3.544]$, the system has 2-stable, 4-stable fixed points, and so on, which means that traffic flow is admitted to be in a stable state.
(iv). When the control parameter $\lambda \in (3.544, 4]$, the system tends to chaos and traffic flow is assumed to be in an unstable state.
(v). It may be seen in Fig. 1.6a that 3 is the point of bifurcation.

Figure 1.9 unveils the full dynamical class of traffic flow on the road and the traffic model (1.24) is an original Picard system. Using an Ishikawa control system (1.2) and (1.3), we obtain

$$\overline{\delta}_i = \sigma_i + \beta[\frac{v_f}{\overline{v}}\check{\sigma}_i(1 - \check{\sigma}_i) - \sigma_i] \tag{1.28}$$

$$\overline{\sigma}_i = \sigma_i + \alpha[\frac{v_f}{\overline{v}}\overline{\delta}_i(1 - \overline{\delta}_i) - \sigma_i]. \tag{1.29}$$

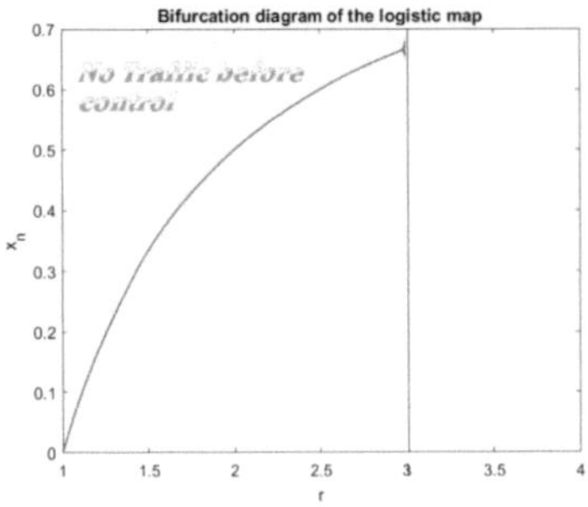

(a) Bifurcation diagram for model (8.3.24) and $r \in [1, 3]$

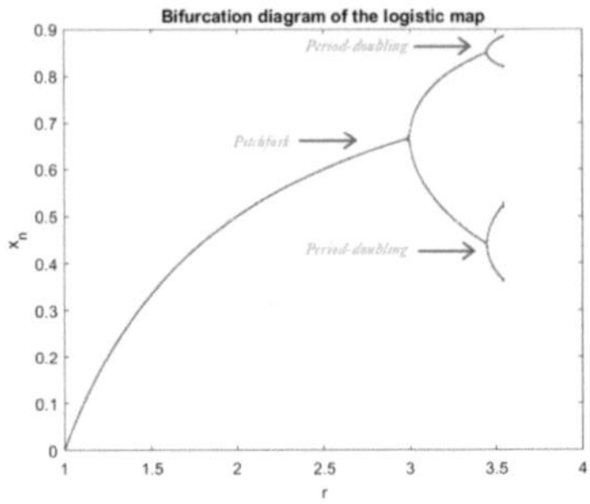

(b) Bifurcation diagram for model (8.3.24) and $r \in [1, 3.544]$

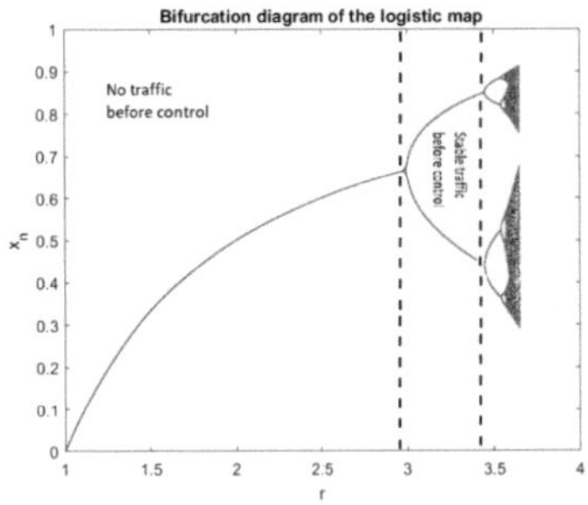

(c) Bifurcation diagram for model (8.3.24) and $r \in [1, 3.6]$

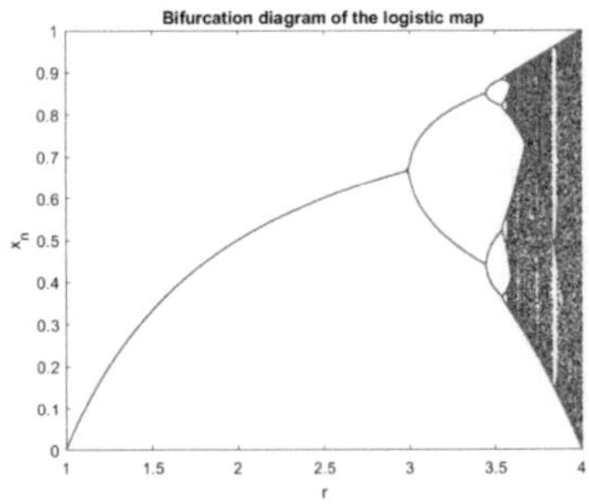

(d) Bifurcation diagram for model (8.3.24) and $r \in [1, 4]$

Fig. 1.9 Bifurcation diagrams of model (8.3.24) for varying ranges of the control parameter r

Here $\frac{v_f}{v} \in [0, 4]$ is the parameter that changes the dynamical system , σ_i is the variable of occupancy, $\alpha, \beta \in (0, 1)$ is the control parameter, and (1.26) is the superior traffic control system.

Figure 1.9a illustrates the absence of traffic control in the original Picard system, while Fig. 1.9b depicts a variational class of traffic systems. This indicates that stable traffic flow is achieved through the superior traffic flow systems (1.28) and (1.29).

1.3.3 Remark

We have observed that a macroscopic chaotic traffic flow system responds to the control parameter. The chaotic behavior becomes highly complex at the bifurcation point, making it difficult to predict the emergence of a chaotic system (see, Fig. 1.9). When traffic flow becomes chaotic and unstable, drivers tend to accelerate or decelerate, which causes them to feel pressured and uncomfortable. As a result, they tend to decrease the volume of total traffic. Also, unstable traffic leads to accidents. So the traffic flow should be controlled and traffic operators and planners should think of stable traffic conditions to work on the safety and level of service of the road. A road should be designed so that chaotic traffic can be easily controlled, with the level of service that is desired in mind. Managing the traffic on a highway speed limit control is also a nice idea because the control parameter is the ratio of the free flow and the average speed. We have noticed the bifurcation when the ratio is greater than 3. So if the average speed is greater than one-third of the free flow speed on the highway, the traffic flow will be in equilibrium. This is an operational strategy to control chaotic traffic.

Data Availability Data sharing does not apply to this article as no datasets were generated or analyzed during the current study.

Competing Interests The authors have no competing interests to declare.

Conflict of Interest The authors declare that they have no conflict of interest.

References

Banach, S. (1922). Sur les opérations dans les ensembles abstraits et leur application aux équations íntégrales. *Fundamenta Mathematicae, 3*(1), 133–181.

Braverman, E., & Liz, E. (2012). Global stabilization of periodic orbits using a proportional feedback control with pulses. *Nonlinear Dynamics, 67*, 2467–2475.

Cao, J., & Chugh, R. (2019). Controlling chaos using superior feedback technique with applications in discrete traffic models. *International Journal of Fuzzy Systems, 21*(5), 1467–1479.

Chaudhary, H., Khan, A., & Sajid, M. (2022a). An investigation on microscopic chaos controlling of identical chemical reactor system via adaptive controlled hybrid projective synchronization. *The European Physical Journal Special Topics, 231*(3), 453–463.

Chaudhary, H., Khan, A., Nigar, U., Kaushik, S., & Sajid, M. (2022b). An effective synchronization approach to stability analysis for chaotic generalized lotka-volterra biological models using active and parameter identification methods. *Entropy, 24*(4), 529.

Chaudhary, H., Sajid, M., Kaushik, S., & Allahem, A. (2023). Stability analysis of chaotic generalized Lotka-Volterra system via active compound difference anti-synchronization method. *Mathematical Biosciences and Engineering, 20*(5), 9410–9422.

Devaney, R. L. (1992). *A first course in chaotic dynamical systems, theory and experiment*. Addison-Wesley.

Feketa, P., Schaum, A., & Meurer, T. (2020). Synchronization and multicluster capabilities of oscillatory networks with adaptive coupling. *IEEE Transactions on Automatic Control, 66*, 3084–3096.

Guo, R., & Qi, Y. (2021). Partial anti-synchronization in a class of chaotic and hyper-chaotic systems. *IEEE Access, 9*, 46303–46312.

Holmgren, R. A. (1994). *A first course in discrete dynamical systems*. Springer.

Ishikawa, S. (1974). Fixed points by a new iteration method. *Proceedings of the American Mathematical Society, 44*, 147–150.

Jahanzaib, L. S., Trikha, P., Chaudhary, H., & Haider, S. (2020). Compound synchronization using disturbance observer based adaptive sliding mode control technique. *Journal of Mathematics and Computer Science, 10*, 1463–1480.

Jarrett, D., & Zhang, Y. (1993). The dynamic behaviour of road traffic flow, stability or chaos? In *Applications of fractals and chaos*. The Shape of Things.

Kirby, H. R., & Smith, M. J. (1993). Can chaos theories have transport application? In *23rd annual conference of the universities transport studies group*. University of Nottingham.
Kumar, S., Matouk, A. E., Chaudhary, H., & Kant, S. (2020) Control and synchronization of fractional-order chaotic satellite systems using feedback and adaptive control techniques. *The International Journal of Adaptive Control and Signal Processing, 35*, 484–497.
Kurganov, A., & Polizzi, A. (2009). Non-oscillatory central schemes for traffic flow models with Arrhenius look-ahead dynamics. *Networks and Heterogeneous Media, 4*(3), 431–451.
Li, T. (2013). Qualitative analysis of some PDE models of traffic flow. *Networks and Heterogeneous Media, 8*(3), 773–781.
Liz, E. (2010). How to control chaotic behaviour and population size with proportional feedback. *Physics Letters A, 374*(5), 725–728.
Lo, Shih-Ching., & Cho, Hsun-Jung. (2005). Chaos and control of discrete dynamic traffic model. *Journal of the Franklin Institute, 342*(7), 839–851.
Ma, X., Bai, Z., & Sun, S. (2023). Stability and bifurcation control for a fractional-order chemostat model with time delays and incommensurate orders. *Mathematical Biosciences and Engineering, 20*(1), 437-455.
Mann, W. R. (1953). Mean value methods in iteration. *Proceedings of the American Mathematical Society, 4*, 506–510.
Marusin, A., & Danilov, I. (2018). A method for assessing the influence of automated traffic enforcement system parameters on traffic safety. *Transportation Research Procedia, 36*, 500–506.
Ott, E., Grebogi, C., & Yorke, J. A. (1990). Controlling chaos. *Physical Review Letters, 64*(11), 1196.
Picard, E. (1890). Memoire sur la theorie des equations aux derivees partielles et la methode des approximations successives. *Journal de Mathématiques Pures et Appliquées, 6*, 145–210.
Pyragas, K. (1992). Continuous control of chaos by self-controlling feedback. *Physics Letters A, 170*(6), 421–428.
Seliverstov, S., & Seliverstov, Y. (2018). Developing principles for building transport networks of conflict-free continuous traffic. *Transportation Research Procedia, 36*, 689–699.
Stoop, R., & Radons, G. (1997). Fromthe routes to chaos to extended systems. *Physics Reports, 290*, 3–9.
Xiao, L., & Lo, H. K. (2015). Combined route choice and adaptive traffic control in a day-to-day dynamical system. *Networks and Spatial Economics, 15*, 697–717.
Yi, X., Guo, R., & Qi, Y. (2020). Stabilization of chaotic systems with both uncertainty and disturbance by the UDE-based control method. *IEEE Access, 8*, 62471–62477.
Zhang, Y., & Jarrett, D. (1997). Stability analysis of classical car-following model. *Transportation Research: Part B, 31*(6), 441–462.

Chapter 2
Unlocking Stock Market Dynamics: An Application of Fixed-Point Theory with Moving Averages and Technical Analysis

Vipul Kumar, Anita Tomar, and U. S. Rana

Abstract Fixed-point theory, which serves as a cornerstone in mathematical analysis, discovers compelling applications extending beyond its theoretical domain. This paper explores the application of fixed-point theory within the realm of stock market analysis, specifically focusing on its practical utilization in conjunction with moving averages and technical analysis. In traditional stock market analysis, moving averages serve as vital tools to discern trends and forecast future price movements. Utilizing fixed-point theory improves this methodology by introducing a profound mathematical framework to interpret and extrapolate market dynamics. By identifying fixed points, where a function maintains its value despite iterative transformations, we establish pivotal moments within stock price movements where equilibrium or stability is reached. This paper underscores how fixed points, when applied to moving averages, provide deeper insights into market behavior and aid in making informed investment decisions. This paper emphasizes the practical relevance of fixed-point theory in navigating the complexities of stock market fluctuations, offering a systematic framework for investors and analysts to enhance their decision-making processes.

Keywords Equilibrium · Fixed point · Moving average · Stock price

AMS Subject Classification 37C25 · 54H25

V. Kumar · U. S. Rana
D.A.V. College Dehradun, Uttarakhand, India

A. Tomar (✉)
Pt.L.M.S. Campus, Sri Dev Suman Uttarakhand University, Rishikesh, India
e-mail: anitatmr@yahoo.co.in

© The Author(s), under exclusive license to Springer Nature Singapore Pte Ltd. 2026

A. Tomar and M. Jain (eds.), *Applications of Fixed-Point Theorem*, Industrial and Applied Mathematics, https://doi.org/10.1007/978-981-95-5160-6_2

2.1 Introduction

Fixed-point theory is a fundamental concept in mathematics with profound applications across various disciplines, from pure mathematics to applied sciences and engineering. The notion of fixed points has a rich history dating back centuries, with roots in classical mathematics and significant advancements in the modern era.

In its simplest form, a fixed point of a function f is a point x in its domain such that $f(x) = x$. In other words, the function f maps the point x to itself. This seemingly straightforward concept holds immense significance due to its applicability in diverse mathematical contexts. Fixed-point theory finds applications in numerous branches of mathematics, including analysis, topology, algebra, and optimization. In analysis, fixed-point theorems provide powerful tools for proving the existence and uniqueness of solutions to differential and integral equations. In topology, fixed points play a crucial role in understanding continuous mappings and homotopy theory. Algebraic structures, such as groups and rings, often exhibit fixed points under certain operations, shedding light on their properties and symmetries. Moreover, in optimization and mathematical economics, fixed-point theorems are essential for establishing convergence properties of iterative algorithms and equilibrium points in dynamic systems. The development of fixed-point theory has been driven by the pioneering work of mathematicians such as Banach (1922), Brouwer (1911), Kakutani (1941), and many others. The fixed-point theorem of Banach (1922), commonly known as the contraction mapping principle, provides a powerful criterion for the existence and uniqueness of fixed points in complete metric spaces. On the other hand, Brouwer (1911) fixed-point theorem, asserts the existence of fixed points for continuous mappings on compact convex sets, laying the foundation for topological methods in fixed-point theory. In recent decades, fixed-point theory has witnessed significant advancements fueled by interdisciplinary collaborations and applications in emerging fields such as computer science, quantum mechanics, and mathematical biology. The exploration of fixed points in dynamical systems (Bacaër, 2011), fractals, (see, Antal et al. 2021, 2022, Tomar et al. 2022, 2023, and so on) and chaos theory (Cao & Chugh, 2019; Devaney, 1992; Lo & Cho, 2005, and so on) have led to profound insights into the behavior of complex systems and nonlinear phenomena.

This research paper aims to elucidate its central role in modern mathematics by navigating the complexities of stock market fluctuations, offering a systematic framework for investors and analysts to enhance their decision-making processes. (see, Cai and Lai 2022, Chong et al. 2017, Jiang et al. 2020, Kar and Pathak 2023, Nabipour et al. 2020, Nayak et al. 2019, Picasso 2019, Sezer et al. 2020, and so on).

A stock candle (Murphy, 1999) often referred to as a candlestick, is a graphical representation of price movements for a specific time interval in financial markets, commonly used in technical analysis. Each candlestick provides information about the opening, closing, high, and low prices of security within the designated time period, typically ranging from minutes to weeks, depending on the time frame of the chart.

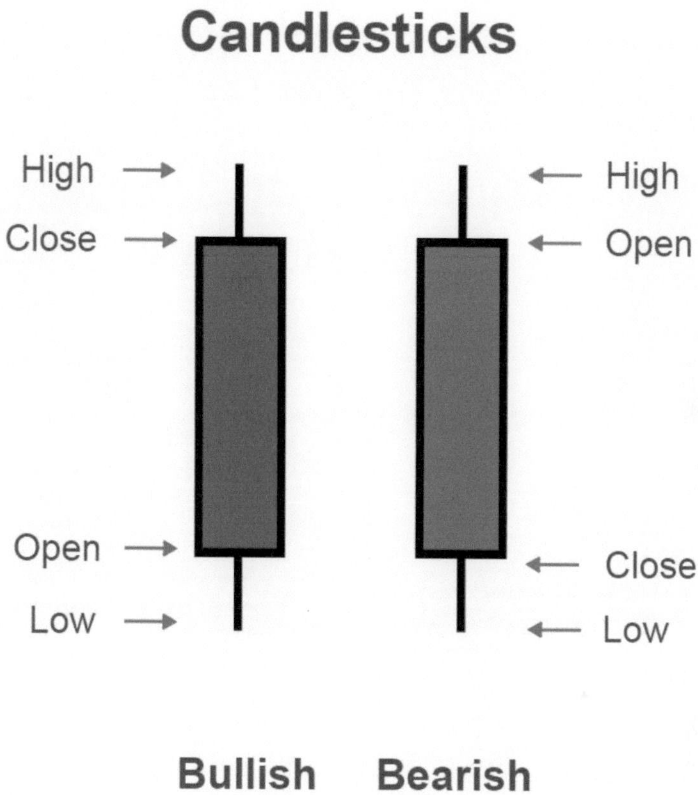

Fig. 2.1 Candlestick

In Fig. 2.1, it is clear that if the open price is less than the closing price then the color of the candle is green and if the open price is greater than the closing price then the color of the candle is red.

Next, we make a moving average line (Fig. 2.2.) with the opening price or closing price.

A moving average (Murphy, 1999) represents the average price of a stock over a predefined time frame. At the end of each period, typically a day or week, the average price of the recent closing trades of stock is calculated. This process is repeated for each period, creating a series of average prices plotted on a price chart.

$$SMA = \frac{P_1 + P_2 + ... + P_n}{n},$$

where P_i is the opening or closing prices for each of the n periods, n is the number of periods, and SMA means Simple Moving Average.

An exponential moving average (Hunter, 1986) places more weight on the most recent prices by adding a weighting multiplier. The exponential moving average (EMA) formula is

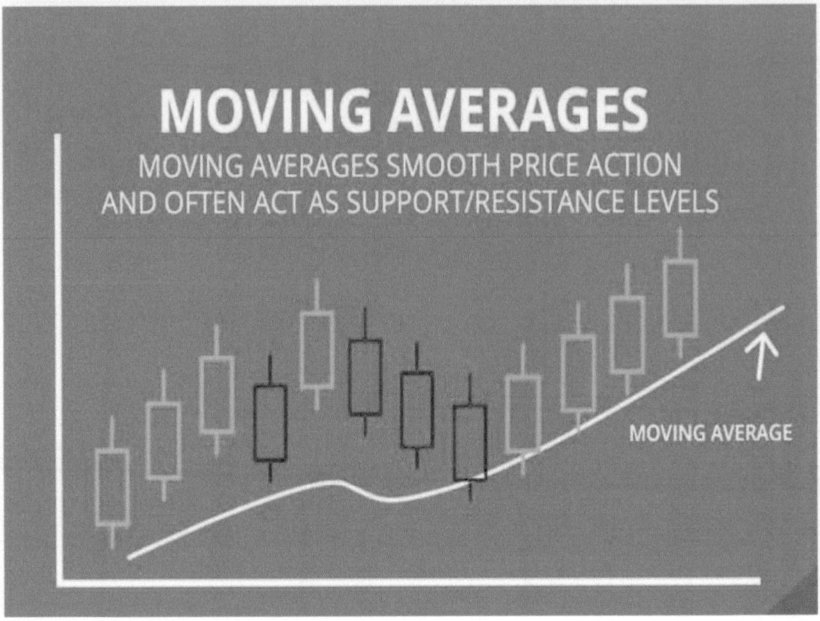

Fig. 2.2 Moving average line

$$EMA = (Close_t - EMA_{t-1}) \times \alpha + EMA_{t-1},$$

where EMA stands for Exponential Moving Average at time t and $close_t$ is the closing price for the current period t. EMA_{t-1} is the exponential moving average for the current period $t-1$. The initial exponential moving average is usually taken as the simple moving average for the first period. α is the smoothing factor calculated as $\frac{2}{n+1}$, where n is the number of periods.

In Fig. 2.2, we illustrates the concept of moving averages as applied in financial time series analysis. A moving average is a statistical tool used to smooth out short-term price fluctuations and highlight longer-term trends or cycles.

Let us define a function $f : \mathbb{X}_{200} \rightarrow \mathbb{X}_{22}$, where $\mathbb{X}_{22}$ and $\mathbb{X}_{200}$ are sets of simple moving averages of closing prices for 22 and 200 candles, respectively, and denoted by $\mathbb{X}_{22} = \{a_1, a_2, a_3, ...a_i\}$ and $\mathbb{X}_{200} = \{b_1, b_2, b_3, ...b_i\}$, where:

- a_1 is the average of closing prices for the first 22 candles.
- b_1 is the average of closing prices for the first 200 candles.
- a_2 is the average of closing prices for 22 candles after removing the first candle.
- b_2 is the average of closing prices for 200 candles after removing the first candle.
- a_i and b_i are averages of closing prices for 22 and 200 candles, respectively, after removing the first i candles.

2.2 Main Results

A point x is a fixed point if $f(x) = x$, which occurs when the simple moving average lines of 22 and 200 candles intersect. In our analysis:

1. The first fixed point indicates a potential buy (or sell) signal.
2. The second fixed point suggests a potential sell (or buy) price.
3. This approach demonstrates how identifying a fixed point between the 22 and 200 candles, moving averages can validate an ideal entry or exit price for a stock position.

This fixed point is the optimal buying point as shown in Fig. 2.3.

2.2.1 Case Studies

Figure 2.4 illustrates a key fixed point at Rs. 34.60. Based on this analysis, we initiated a position in the stock of Punjab National Bank (P.N.B.) on September 7, 2022, at Rs. 37.95, slightly above the identified fixed point. The performance of stock following this entry was as follows (Table 2.1):

1. On December 15, 2022, it reached its first peak at Rs. 62.00, yielding a 63.37% profit.
2. On October 3, 2023, it achieved a second high of Rs. 83.50, representing a 120.00% profit from the entry point.
3. On March11, 2024, the stock touched an extreme high of Rs. 132.95, resulting in a 250.33% profit from the initial purchase.

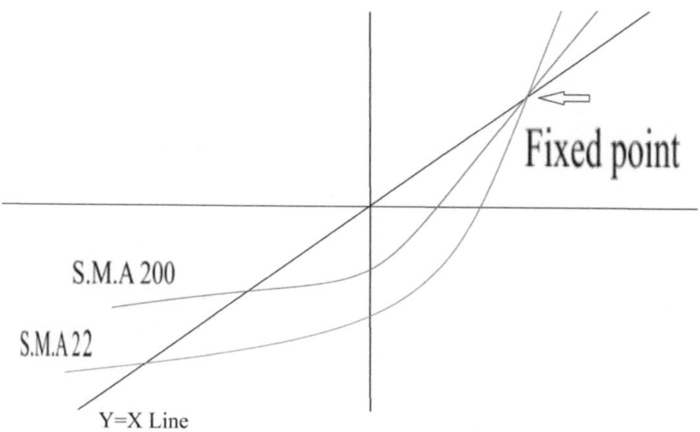

Fig. 2.3 Intersection of moving average lines

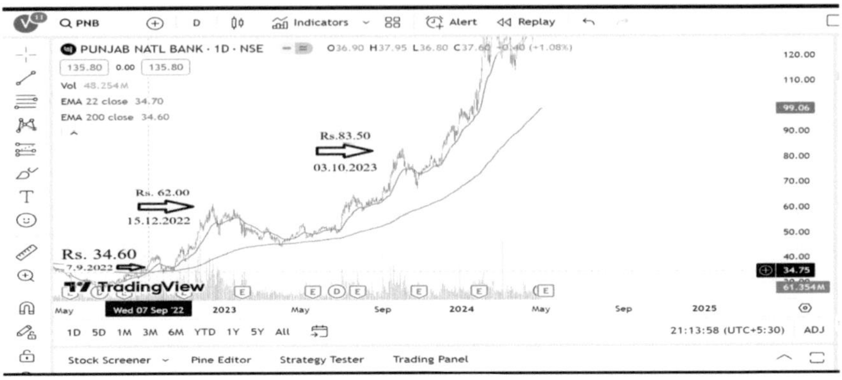

Fig. 2.4 Day candle chart of P.N.B.

Table 2.1 Performance of stock in Fig. 2.4

Sr.no.	Date	Stock price	Profit
1.	07-Sep-23	37.95	
2.	15-Dec-22	62.00	63.37%
3.	03-Oct-23	83.50	120%
4.	11-Mar-24	132.95	250.33%

Fig. 2.5 Weekly candle chart of P.N.B.

In Fig. 2.5, we identify a fixed point at Rs. 51.60. We purchased the stock at a price above Rs. 62.35 after July 10, 2023. By April 29, 2024, the share price had risen to Rs. 142.90, representing a profit of approximately 129.20% (Table 2.2).

In Fig. 2.6, the share price on April 27, 2009, was Rs. 108.94. It increased to Rs. 280.00 within a year, yielding a profit of 157.00% (Table 2.3).

Figure 2.7, illustrates a significant price movement pattern in Voltas Ltd. shares based on the interaction of the 22-candle and 200-candle moving averages. On May

Table 2.2 Performance of stock in Fig. 2.5

Sr.no.	Date	Stock price	Profit
1.	10-Jul-23	62.35	
2.	29-Apr-24	142.90	129.20%

Fig. 2.6 Weekly candle chart of P.N.B.

Table 2.3 Performance of stock in Fig. 2.6

Sr.no.	Date	Stock price	Profit
1.	27-Apr-09	108.94	
2.	08-Nov-10	280.00	157.00%

Fig. 2.7 Day candle chart of Voltas Ltd.

Table 2.4 Performance of stock in Fig. 2.7

Sr.no.	Date	Stock price	Profit
1.	11-May-22	935.00	
2.	30-Jan-23	744.90	20.33%

Table 2.5 Performance of stock in Fig. 2.8

Sr.no.	Date	Stock price	Profit
1.	15-Dec-23	869.90	
2.	04-Mar-24	1127.00	29.56%
3.	06-May-24	1502.30	72.70%

11, 2022, the declining 22-candle moving average intersected the rising 200-candle moving average at approximately Rs. 1172.15. This crossover preceded a substantial price decline, with the share price reaching its lowest point of Rs. 744.90 on January 30, 2023. This movement presented a shorting opportunity with a potential profit of 20.33% (Table 2.4).

Conversely, on December 15, 2023, the ascending 22-candle moving average crossed above the 200-candle moving average at Rs. 843.42. Following this bullish crossover, the share price surged to a peak of Rs. 1502.30 on May 6, 2024, offering a potential profit of 72.70% for long positions (Table 2.5). These observations suggest that the crossover of these moving averages may serve as a valuable indicator for potential trend reversals and trading opportunities in Voltas Ltd. share (Figs. 2.7 and 2.8).

Remark 2.2.1 A shorting opportunity refers to a situation in the financial markets where an investor anticipates that the price of a security (such as a stock or commodity) will decline. By taking advantage of this prediction, the investor can potentially profit from the decrease in the price of the asset. For example, if an investor shorts a stock at Rs. 100 and buys it back at Rs. 80, they make a Rs. 20 profit per share (excluding any costs associated with the transaction). Shorting carries significant risks, as the price of the security could rise instead of falling, potentially leading to unlimited losses since the price of an asset can theoretically increase indefinitely.

Remark 2.2.2 A bullish crossover is a technical analysis term referring to a situation where a shorter-term moving average crosses above a longer-term moving average. This event is often interpreted by traders and investors as a signal that a potential upward trend or bullish movement in the price of the asset is likely to occur. Bullish crossovers are often seen as buying signals because they suggest that the momentum of the asset is shifting upward, and it may continue to rise in price. However, like all technical indicators, they are not foolproof and should be used in conjunction with other forms of analysis and risk management strategies.

Fig. 2.8 Week candle chart of Voltas Ltd.

Table 2.6 Performance of stock in Fig. 2.8

Sr.no.	Date	Stock price	Profit
1.	10-Mar-14	147.75	
2.	29-Apr-24	1501.00	916.00%

Figure 2.8, identifies a fixed point at Rs. 113.00, indicating a buying opportunity. We entered the position above the high price of March 10, 2014, at Rs. 147.75. The performance of stock from this entry point was as follows (Table 2.6):

- On June 22, 2015, it reached a new high of Rs. 359.90, yielding a 143.58% profit.
- On December 26, 2017, it hit Rs. 675.30, representing a 357.00% profit over 3 years.
- By February 8, 2021, it climbed to Rs. 1132.00, achieving a 666.16% profit in 7 years.
- On April 29, 2024, it touched Rs. 1501.00, resulting in a 916.00% profit over 10 years.

This long-term analysis demonstrates the potential of using fixed-point theory to identify entry point.

Figure 2.9, identifies a fixed point at Rs. 563.91, indicating a buying opportunity. We entered the position above the high price of January 11, 2021, at Rs. 691.20. The stock's performance from this entry point was as follows (Table 2.7):

- On October 25, 2021, it reached its first peak of Rs. 866.90, yielding a 25.41% profit in approximately 9 months.

Fig. 2.9 Day candle chart of Axis Bank

Table 2.7 Performance of stock in Fig. 2.9

Sr.no.	Date	Stock price	Profit
1.	11-Jan-21	691.20	
2.	04-Dec-23	1151.85	66.64%

- By January 2, 2023, the price rose to Rs. 970.00, resulting in a 40.33% profit over about 2 years.
- On December 4, 2023, the price climbed further to Rs. 1151.85, reflecting a 66.64% profit in nearly 3 years.

These analyses demonstrate the potential of using fixed-point theory to identify entry points for medium-term investments, capturing significant price appreciation over a period of a few years.

Figure 2.10, identifies a fixed point at Rs. 100.50, indicating a buying opportunity. We entered the position above the high price of May 4, 2009, at Rs. 121.00. The performance of stock from this entry point was as follows (Table 2.8):

- On October 18, 2010, it reached its first peak of Rs. 254.00, yielding a 109.92% profit.
- In Fig. 2.11, by July 16, 2018, the share price had risen to Rs. 1110.00, resulting in an 817.36% profit over approximately 9 years.
- On December 16, 2019, the price climbed further to Rs. 1305.50, reflecting a 978.93% profit over about ten and a half years.

Fig. 2.10 Week candle chart of HDFC Bank

Fig. 2.11 Week candle chart of HDFC Bank

Table 2.8 Performance of stock in Fig. 2.11

Sr.no.	Date	Stock price	Profit
1.	04-May-09	121.00	
2.	18-Oct-10	254.00	109.92%
3.	16-Jul-18	1110.00	817.36%
4.	16-Dec-19	1305.50	978.93%

This analysis of HDFC Bank demonstrates the potential of using the fixed-point theory to identify entry points for long-term investments, capturing substantial price appreciation over extended periods.

2.3 Risk Management

Given the inherent unpredictability of the stock market, effective risk management is crucial. We propose the following risk management strategy:

Step 1. Divide your total investment amount into four parts:

1. 10%
2. 20%
3. 30%
4. 40%

Step 2. Invest 10% of your total capital to open a position (buy or sell) at the identified price. Set a stop-loss order at 10% from the entry price.
Step 3. As the stock price increases, add to your position as follows:

5. At 10% gain: Invest the next 20% portion.
6. At 20% gain: Invest the 30% portion.
7. At 30% gain: Invest the final 40% portion.
8. [Step 4.] This strategy, known as scaled entry or pyramiding, aims to minimize risk by starting with a small position and gradually increasing investment as the stock performs well.

Step 5. It allows for capitalizing on positive momentum while limiting initial exposure.

2.4 Conclusion

This study presents a novel approach to stock market analysis and investment strategy by integrating fixed-point theory with traditional technical analysis tools. Our findings suggest that: Fixed-point analysis can provide valuable insights into optimal timing for stock purchases or sales, potentially enhancing investment decision-making. The application of this fixed-point strategy has demonstrated the potential for substantial profits while maintaining a structured approach to risk management. Our research establishes a framework for minimizing risk exposure in stock market investments, addressing a critical concern for investors. This work represents a significant step in bridging the gap between theoretical fixed-point concepts and their practical implementation in real-world financial markets. By combining mathematical rigor with empirical market data, our approach offers a new perspective on stock market dynamics. It provides investors and analysts with an additional tool to navigate market complexities and make more informed decisions. While promising, we acknowledge that further research is needed to validate these findings across diverse market conditions and longer time horizons. Future studies could explore the integration of this approach with other predictive models or its application to different asset classes. This research contributes to the evolving field of quantitative finance by demonstrating the practical relevance of fixed-point theory in stock market analysis and decision-making processes.

Data Availability Data utilized in this paper are available on www.tradeview.com.

Competing Interests The authors have no competing interests to declare.

Conflict of interest The authors declare that they have no conflict of interest.

References

Antal, S., Tomar, Anita, Prajapati, D. J., & Sajid, M. (2021). Fractals as Julia sets of complex sine function via fixed point iterations. *Fractal and Fractional,5*(4), 272.

Antal S., Tomar Anita, Prajapati D. J., & Sajid M. (2022). Variants of Julia and Mandelbrot sets as fractals via Jungck-Ishikawa fixed point iteration system with sconvexity. *AIMS Mathematics, 7*(6), 10939–10957.

Bacaër, N. (2011). Verhulst and the logistic equation (1838). In *A short history of mathematical population dynamics* (pp. 35–39). Springer

Banach, S. (1922). Sur les opérations dans les ensembles abstraits et leur application aux équations íntégrales. *Fundamenta Mathematicae,3*(1), 133–181.

Brouwer, L. E. J. (1911). Über abbildung von mannigfaltigkeiten. *Mathematische Annalen,71*(1), 97–115.

Cai, X., & Lai, Y. (2022). A novel stock trend prediction method using technical analysis and machine learning. *Expert Systems with Applications,187*, Article 115889.

Cao, J., & Chugh, R. (2019). Controlling chaos using superior feedback technique with applications in discrete traffic models. *International Journal of Fuzzy Systems,21*(5), 1467–1479.

Chong, E., Han, C., & Park, F. C. (2017). Deep learning networks for stock market analysis and prediction: Methodology, data representations, and case studies. *Expert Systems with Applications,83*, 187–205.

Devaney, R. L. (1992). *A first course in chaotic dynamical systems, theory and experiment*. Addison-Wesley.

Hunter, J. S. (1986). The exponentially weighted moving average. *Journal of Quality Technology,18*(4), 203–210.

Jiang, M., Liu, J., Zhang, L., & Liu, C. (2020). An improved Stacking framework for stock index prediction by leveraging tree-based ensemble models and deep learning algorithms. *Physica A: Statistical Mechanics and its Applications,541*, Article 122272.

Kar, A. K., & Pathak, S. (2023). Fixed point theory and its applications in economics and finance: A systematic literature review. *Fixed Point Theory and Applications,1*, 1–25.

Kakutani, S. (1941). A generalization of Brouwer's fixed point theorem. *Duke Mathematical Journal,8*(3), 457–459.

Lo, Shih-Ching., & Cho, Hsun-Jung. (2005). Chaos and control of discrete dynamic traffic model. *Journal of the Franklin Institute,342*(7), 839–851.

Nabipour, M., Nayyeri, P., Jabani, H., & Mosavi, A. (2020). Deep learning for stock market prediction. *Entropy,22*(8), 840.

Nayak, S. C., Misra, B. B., & Behera, H. S. (2019). Efficient forecasting of financial time-series data with virtual adaptive neuro-fuzzy inference system. *International Journal of Business Forecasting and Marketing Intelligence,5*(2), 142–164.

Murphy, J. J. (1999). *Technical analysis of the financial markets: A comprehensive guide to trading methods and applications*. New York Institute of Finance

Picasso, A., Merello, S., Ma, Y., Oneto, L., & Cambria, E. (2019). Technical analysis and sentiment embeddings for market trend prediction. *Expert Systems with Applications,135*, 60–70.

Sezer, O. B., Gudelek, M. U., & Ozbayoglu, A. M. (2020). Financial time series forecasting with deep learning: A systematic literature review: 2005–2019. *Applied Soft Computing,90*, Article 106181.

Tomar, A., Prajapati, D. J., Antal, S., & Rawat, S. (2022). Variants of Mandelbrot and Julia fractals for higher-order complex polynomials. *Mathematical Methods in the Applied Sciences*, 1–13.

Tomar, A., Kumar, V., Rana, U. S., & Sajid, M. (2023). Fractals as Julia and Mandelbrot sets of complex cosine functions via fixed point iterations. *Symmetry,15*, 478.

Chapter 3
A Symmetric Contraction in $\mathcal{GV} - \mathcal{FMS}$ and Coupled Coincidence Points

Manish Jain

Abstract The purpose of this chapter is to present the occurrence of coupled coincidence points for a pair of mixed monotone operators subjected to a symmetric contraction in the setting of partially ordered $\mathcal{GV}$-fuzzy metric spaces having Hadžić type t-norm. The corresponding result in the metric spaces constitutes the application part. Furthermore, a suitable example is incorporated to validate the work.

Keywords Partially ordered set · Hadžić type t-norm · Fuzzy metric spaces · Coupled coincidence points

AMS Subject Classification Primary: 47H10 · Secondary: 54H25

3.1 Introduction and Preliminaries

An important branch of mathematics is fixed point theory, which has made significant advances in the study of nonlinear phenomena. The theory itself consists of many powerful tools that provide valuable insights into nonlinear analysis. Among these tools, one of the most celebrated contributions, introduced by Banach (1922) and known as "Banach Contraction Principle" (shortly, BCP), has applications in wide range of fields, including artificial intelligence, image processing, soft computing, physical and chemical sciences, game theory, economics, and much more. The BCP suggests that "Every contraction mapping defined on a complete metric space has a unique fixed point in the space." This principle has been extended and generalized over time by many distinguished researchers in various abstract spaces.

In general, a set consists of elements that share a particular property. A universal set includes all possible elements involved in any experiment. A crisp set is a collection of distinct, unordered elements derived from a universal set. An element of a set is referred to as a "member" of that set. If any element resides within the set, it is

M. Jain (✉)
Department of Mathematics, Ahir College, Rewari, India
e-mail: jainmanish26128301@gmail.com

© The Author(s), under exclusive license to Springer Nature Singapore Pte Ltd. 2026

A. Tomar and M. Jain (eds.), *Applications of Fixed-Point Theorem*, Industrial and Applied Mathematics, https://doi.org/10.1007/978-981-95-5160-6_3

considered a member; otherwise, it is not. Crisp sets do not allow for the concept of partial membership–an element is either fully present in the set or entirely absent. The major limitation of crisp sets is their use of bi-valued membership; essentially, the membership values are restricted to 0 or 1. A membership value of 0 indicates the absence of the element from the set, while a value of 1 indicates its presence in the set. However, a natural question arises regarding the behavior of elements with membership values between 0 and 1. This gave rise to the theory of fuzzy sets, which addresses ambiguity. Fuzzy sets deal with transitional membership values.

The groundbreaking theory of fuzzy sets was introduced by Zadeh (1965), marking the beginning of a new era. Subsequently, significant advances have been made in this area of research. In mathematical analysis, the concept of distance plays a crucial role, with applications in nearly every scientific field. The introduction of the fuzzy concept spurred the authors to consider the notion of distance in a fuzzy setting. The initial step in this direction was taken by Kramosil and Michalek (1975), who introduced the concept of fuzzy metric spaces, commonly referred to as $\mathcal{KM}-$ fuzzy metric spaces. Subsequently, several non-equivalent definitions of fuzzy metrics were developed (Deng, 1982; Erceg, 1979; Kaleva & Seikkala, 1984). Metrical fixed point theory, an important area of research, also experienced a surge in the use of fuzzy terminology.

Grabiec (1988) noted that the limit in the above definition of convergence is uniquely determined since $*$ is continuous. Further, the following fuzzy version of BCP was also suggested in the same paper by Grabiec (1988)

"Let $(\mathcal{Y}, \mathcal{M}, *)$ be a complete fuzzy metric space with $\lim_{t\to\infty} \mathcal{M}(x, y, t) = 1$ for x, y in $\mathcal{Y}$. Let h be a self-mapping on x satisfying

$$\mathcal{M}(hx, hy, kt) \geq \mathcal{M}(x, y, t), \tag{3.1}$$

for all x, y in $\mathcal{Y}$, $0 < k < 1$ and $t > 0$. Then, h has a unique fixed point in x."

Subsequently, important contractions were formulated in fuzzy metric spaces. The results given by Fang (1992) improved, generalized and unified the works of Edelstein (1962), Istratescu (1981), Sehgal and Bharucha-Reid (1972).

In a related development, George and Veeramani (1994, 1997) introduced modifications to the concept of fuzzy KM metric spaces and derived the Hausdorff topology in such spaces, commonly referred to as $\mathcal{GV}$ fuzzy metric spaces. Subsequently, several authors have demonstrated the existence of fixed points for various mappings in these spaces (George & Veeramani, 1994, 1997). In this study, we focus on $\mathcal{GV}$-fuzzy metric spaces, which will be abbreviated as $\mathcal{GV} - \mathcal{FMS}$.

The following definition of fuzzy metric spaces is due to George and Veeramani (1994):

"The 3-tuple $(\mathcal{Y}, \mathcal{M}, *)$ is called a fuzzy metric space in the sense of George and Veeramani (shortly, $\mathcal{GV}$-fuzzy metric space or $\mathcal{GV} - \mathcal{FMS}$), if $\mathcal{Y}$ is an arbitrary set, $*$ is a continuous t-norm and $\mathcal{M}$ is a fuzzy set on $\mathcal{Y}^2 \times (0, \infty)$ satisfying the following conditions:

(F$\mathcal{M}$-1) $\mathcal{M}(x, y, 0) > 0$,
(F$\mathcal{M}$-2) $\mathcal{M}(x, y, t) = 1$ iff $x = y$,
(F$\mathcal{M}$-3) $\mathcal{M}(x, y, t) = \mathcal{M}(y, x, t)$,
(F$\mathcal{M}$-4) $\mathcal{M}(x, y, t) * \mathcal{M}(y, z, s) \leq \mathcal{M}(x, z, t + s)$,
(F$\mathcal{M}$-85) $\mathcal{M}(x, y, \cdot) : (0, \infty) \to [0, 1]$ is continuous for all $x, y, z \in \mathcal{Y}$ and $s, t > 0$."

In the current work, $(\mathcal{Y}, \mathcal{M}, *)$ denote the $\mathcal{GV} - \mathcal{FMS}$ having the Hadžić type t-norm $*$, $\mathcal{M}(x, y, t) \to 1$ as $t \to \infty$ for all $x, y \in \mathcal{Y}$ *and* $\leq$ be a partial order defined on $\mathcal{Y}$, then, we write it as $(\mathcal{Y}, \mathcal{M}, *, \leq)$ and in short, we call it $\mathcal{PO} - \mathcal{GV} - \mathcal{FMS}$.

Analogously to the notions of Cauchyness, Convergence, and Completeness in metric spaces, these notions were also fabricated in the fuzzy setting as follows: Let $(\mathcal{Y}, \mathcal{M}, *)$ be $\mathcal{GV} - \mathcal{FMS}$,

"a sequence $\{x_n\}$ in $\mathcal{Y}$ is said to be Convergent to a point $x \in \mathcal{Y}$, if $\lim_{n\to\infty} \mathcal{M}(x_n, x, t) = 1$, for all $t > 0$";

"a sequence $\{x_n\}$ in $\mathcal{Y}$ is called a Cauchy sequence if for each $0 < \varepsilon < 1$ and $t > 0$, there exists a positive integer n_0 such that $\mathcal{M}(x_n, x_m, t) > 1 - \varepsilon$ for each $n, m \geq n_0$";

"a fuzzy metric space in which every Cauchy sequence is convergent is said to be complete."

In recent times, the concept of fixed points has gained attention not only for self-mappings but also for mappings with a two-dimensional domain and a one-dimensional range. This particular branch of fixed point theory is known as coupled fixed point theory, with initial results being (Guo & Lakshmikantham, 1987). Subsequently, researchers delved into this area with the influential work of Bhaskar and Lakshmikantham (2006), where the authors established the existence of coupled fixed points (CFP) for mixed monotone mappings in partially ordered metric spaces (POMS). Building on this, Lakshmikantham and Ćirić (2009) extended the concept of mixed monotone mappings to mixed $\mathcal{T}$-monotone mappings and proved results regarding coupled common fixed points for a pair of commuting mappings. This development has significantly advanced the theory of coupled fixed points, leading to the establishment of several important results.

Unless otherwise stated, we consider the mappings $\mathcal{A} : \mathcal{Y} \times \mathcal{Y} \to \mathcal{Y}, \mathcal{T} : \mathcal{Y} \to \mathcal{Y}$.

Next, we shall give some concepts related to coupled fixed point theory which will be useful in formulating our main results. The first definition in this direction was formulated by Lakshmikantham and Ćirić (2009) in the form of ensuing elucidation:

"Let $(\mathcal{Y}, \leq)$ be a partially ordered set ($\mathcal{POS}$) and $\mathcal{A}, \mathcal{T}$ be the mappings. The mapping $\mathcal{A}$ is said to have the mixed $\mathcal{T}$-monotone property (shortly, M$\mathcal{T}$MP) if $\mathcal{A}(x, y)$ is monotone $\mathcal{T}$-non-decreasing in its first argument and is monotone $\mathcal{T}$-non-increasing in its second argument."

Taking $\mathcal{T}$ to be identity mapping in the above definition, we can obtain the definition of mixed monotone property (shortly, MMP).

The following notions are due to Lakshmikantham and Ćirić (2009):

"An element$(x, y) \in \mathcal{Y} \times \mathcal{Y}$, is called a coupled coincidence point (CCP) for the pair $(\mathcal{A}, \mathcal{T})$ of mappings if $\mathcal{A}(x, y) = \mathcal{T}x$ and $\mathcal{A}(y, x) = \mathcal{T}y$."

By taking $\mathcal{T}$ as the identity mapping in the above definition, we can the definition of a coupled fixed point (CFP).

"An element $(x, y) \in \mathcal{Y} \times \mathcal{Y}$, is called a coupled common fixed point (CCFP) of the mappings $\mathcal{A}$ and $\mathcal{T}$ if $x = \mathcal{T}x = \mathcal{A}(x, y)$ and $y = \mathcal{T}y = \mathcal{A}(y, x)$."

Next, we have the concept of regular space as follows:

Let $(\mathcal{Y}, d)$ be a metric space endowed with a partial ordering $(\leq)$. We say that $\mathcal{Y}$ is regular if it satisfies the following property:

(i) "if a non-decreasing sequence $\{x_n\} \to x$, then $x_n \leq x$ for all $n \geq 0$";
(ii) "if a non-decreasing sequence $\{y_n\} \to y$, then $y \leq y_n$ for all $n \geq 0$."

Note that if $(\mathcal{Y}, \mathcal{M}, *)$ is a $\mathcal{GV} - \mathcal{FMS}$ endowed with a partial ordering $(\leq)$. We say that $\mathcal{Y}$ is regular if it satisfies the last two properties mentioned above with respect to the convergence in $\mathcal{GV} - \mathcal{FMS}$.

For more details, the reader may consult (Choudhary & Kundu, 2010; Gordji et al., 2012; Jain et al., 2013, 2014, 2017; Shatanawi et al., 2011). Nowadays, the problematic situation of computing coupled fixed points has received fuzzy treatment; see Choudhury et al. (2013), Hu (2011), Jain et al. (2012, 2013, 2019, 2023), Sedghi et al. (2010).

The current study focuses on calculating coupled coincidence and coupled fixed points for mappings that are subjected to a symmetric contraction in the context of $\mathcal{PO} - \mathcal{GV} - \mathcal{FMS}$ with a Hadžić type t-norm. The corresponding result in the metric space supports the application aspect. Notably, our results do not necessitate any commutative or compatible hypotheses of the involved mappings. The completeness of the space has been substituted by the completeness of the sub-spaces corresponding to any of the mappings involved. Thus, the technique utilized in our study can be applied to generalize many existing works in the literature.

3.2 Main Results

This section focuses on the main coupled coincidence point results in $\mathcal{GV}$-fuzzy metric spaces.

Theorem 3.2.1 *Let $(\mathcal{Y}, \mathcal{M}, *, \leq)$ be a $\mathcal{PO} - \mathcal{GV} - \mathcal{FMS}$. Let $(\mathcal{A}, \mathcal{T})$ be a pair of mappings such that $\mathcal{A}$ has $M\mathcal{T}MP$ and satisfies the following conditions:*

(T-1) $\mathcal{A}(\mathcal{Y} \times \mathcal{Y}) \subseteq \mathcal{T}(\mathcal{Y})$;
(T-2) *one of the range subspaces $\mathcal{A}(\mathcal{Y} \times \mathcal{Y})$ or $\mathcal{T}(\mathcal{Y})$ is complete;*
(T-3) *there exists k in* $(0, 1)$ *such that*

$$\mathcal{M}(\mathcal{A}(x, y), \mathcal{A}(\mu, v), kt) * \mathcal{M}(\mathcal{A}(y, x), \mathcal{A}(v, \mu), kt)$$
$$\geq \gamma(\mathcal{M}(\mathcal{T}(x), \mathcal{T}(\mu), t) * \mathcal{M}(\mathcal{T}(y), \mathcal{T}(v), t)),$$

for all x, y, μ, v in $\mathcal{Y}$ and $t > 0$ with $\mathcal{T}(x) \leq \mathcal{T}(\mu)$ and $\mathcal{T}(y) \geq \mathcal{T}(v)$ (or, $\mathcal{T}(x) \geq \mathcal{T}(\mu)$ and $\mathcal{T}(y) \leq \mathcal{T}(v)$) and γ is a continuous function on $[0, 1]$ and $\gamma(t) \geq t$, for t in $[0, 1]$;

(T-4) there exist two elements x_0, y_0 in $\mathcal{Y}$ such that the following satisfies:

$$\mathcal{T}(x_0) \leq \mathcal{A}(x_0, y_0) \text{ and } \mathcal{T}(y_0) \geq \mathcal{A}(y_0, x_0)$$
$$(\text{or, } \mathcal{T}(x_0) \geq \mathcal{A}(x_0, y_0) \text{ and } \mathcal{T}(y_0) \leq \mathcal{A}(y_0, x_0)).$$

Further, suppose either

(a) both the mappings $\mathcal{A}$ and $\mathcal{T}$ are continuous, or
(b) $\mathcal{Y}$ is regular;

then, the $(\mathcal{A}, \mathcal{T})$ has a CCP in $\mathcal{Y}$.

Proof Without loss of generality, by (T-4) suppose that x_0, y_0 in X be such that $\mathcal{T}(x_0) \leq \mathcal{A}(x_0, y_0)$ and $\mathcal{T}(y_0) \geq \mathcal{A}(y_0, x_0)$. Since $\mathcal{A}(X \times X) \subseteq \mathcal{T}(\mathcal{Y})$ and $\mathcal{A}$ has M$\mathcal{T}$MP, so inductively, the sequences $\{x_n\}$ and $\{y_n\}$ can be constructed in $\mathcal{Y}$ such that

$$\mathcal{T}(x_{n+1}) = \mathcal{A}(x_n, y_n) \text{ and } \mathcal{T}(y_{n+1}) = \mathcal{A}(y_n, x_n), \quad \text{for all } n \geq 0. \tag{3.2}$$

Further, by using (3.2) and the fact that P has M$\mathcal{T}$MP, inductively, it can be shown that

$$\mathcal{T}(x_n) \leq \mathcal{T}(x_{n+1}) \text{ and } \mathcal{T}(y_n) \geq g(y_{n+1}), \quad \text{for all } n \geq 0. \tag{3.3}$$

We suppose either $\mathcal{T}(x_{n+1}) = \mathcal{A}(x_n, y_n) \neq \mathcal{T}(x_n)$ or $\mathcal{T}(y_{n+1}) = \mathcal{A}(y_n, x_n) \neq \mathcal{T}(y_n)$, for all $n \geq 0$, otherwise, we can obtain directly the CCP of the pair $(\mathcal{A}, \mathcal{T})$.

Using (3.2) and (3.3), from (T-3), for $t > 0$, $n > 0$, we have

$$\begin{aligned} &\mathcal{M}(\mathcal{T}(x_n), \mathcal{T}(x_{n+1}), kt) * \mathcal{M}(\mathcal{T}(y_n), \mathcal{T}(y_{n+1}), kt) \\ &\quad = \mathcal{M}(\mathcal{A}(x_{n-1}, y_{n-1}), \mathcal{A}(x_n, y_n), kt) * \mathcal{M}(\mathcal{A}(y_{n-1}, x_{n-1}), \mathcal{A}(y_n, x_n), kt) \\ &\quad \geq \gamma(\mathcal{M}(\mathcal{T}(x_{n-1}), \mathcal{T}(x_n), t) * \mathcal{M}(\mathcal{T}(y_{n-1}), \mathcal{T}(y_n), t)) \\ &\quad \geq \mathcal{M}(\mathcal{T}(x_{n-1}), \mathcal{T}(x_n), t) * \mathcal{M}(\mathcal{T}(y_{n-1}), \mathcal{T}(y_n), t). \end{aligned} \tag{3.4}$$

Using (3.4) and applying the result of Choudhury et al. (2013) which states that "Let $(\mathcal{Y}, \mathcal{M}, *)$ be a fuzzy metric space with a Hadžić type t-norm $*$ such that $\mathcal{M}(x, y, t) \to 1$ as $t \to \infty$ for all $x, y \in \mathcal{Y}$. If the sequences $\{x_n\}$ and $\{y_n\}$ in $\mathcal{Y}$ are such that, for all $n \geq 1$, $t > 0$, $\mathcal{M}(x_n, x_{n+1}, t) * \mathcal{M}(y_n, y_{n+1}, t) \geq \mathcal{M}(x_{n-1}, x_n, t/k) * \mathcal{M}(y_{n-1}, y_n, t/k)$ where $0 < k < 1$, then the sequences $\{x_n\}$ and $\{y_n\}$ are Cauchy sequences." The sequences $\{\mathcal{T}(x_n)\}$ and $\{\mathcal{T}(y_n)\}$ behave like Cauchy sequences. Without loss of generality, suppose $\mathcal{T}(\mathcal{Y})$ is complete, so there exist x, y in $\mathcal{Y}$ such that, for all $t > 0$, we have that

$$\lim_{n\to\infty} \mathcal{M}(\mathcal{T}(x), \mathcal{T}(x_n), t) = \lim_{n\to\infty} \mathcal{M}(\mathcal{T}(x), \mathcal{A}(x_n, y_n), t) = 1,$$
$$\lim_{n\to\infty} \mathcal{M}(\mathcal{T}(y), \mathcal{T}(y_n), t) = \lim_{n\to\infty} \mathcal{M}(\mathcal{T}(y), \mathcal{A}(y_n, x_n), t) = 1. \tag{3.5}$$

Suppose, condition (a) holds.

Define a multifunction $G : \mathcal{T}(\mathcal{Y}) \to 2^{\mathcal{Y}}$ by $G(y) = \{x \in X : \mathcal{T}(x) = y\}$. Making use of Axiom of Choice, a function $h : \mathcal{T}(\mathcal{Y}) \to \mathcal{Y}$, so that $h(y)$ belongs to $G(y)$ for all y in $\mathcal{T}(\mathcal{Y})$ can be constructed, yielding $\mathcal{T}(h(y)) = y$ for all y in $\mathcal{T}(\mathcal{Y})$. Consider $E = \{h(y) : y \in \mathcal{T}(\mathcal{Y})\} \subseteq \mathcal{Y}$. Clearly, the mapping $\mathcal{T} : E \to \mathcal{Y}$ is injective with $\mathcal{T}(E) = G(\mathcal{Y})$.

Define another mapping $H : \mathcal{T}(E) \times \mathcal{T}(E) \to \mathcal{Y}$ by

$$H(\mathcal{T}(e), \mathcal{T}(f)) = \mathcal{A}(e, f) \text{ for } \mathcal{T}(e), \mathcal{T}(f) \text{ in } \mathcal{T}(E)(= \mathcal{T}(\mathcal{Y})). \tag{3.6}$$

Since $\mathcal{T} : E \to \mathcal{Y}$ is injective, the mapping H is well-defined. By (3.5) and (3.6), we have

$$\lim_{n\to\infty} H(\mathcal{T}(x_n), \mathcal{T}(y_n)) = \lim_{n\to\infty} \mathcal{A}(x_n, y_n) = \lim_{n\to\infty} \mathcal{T}(x_n) = \mathcal{T}(x),$$
$$\lim_{n\to\infty} H(\mathcal{T}(y_n), \mathcal{T}(x_n)) = \lim_{n\to\infty} \mathcal{A}(y_n, x_n) = \lim_{n\to\infty} \mathcal{T}(y_n) = \mathcal{T}(y). \tag{3.7}$$

Using the continuity of $\mathcal{A}$ and $\mathcal{T}$, the continuity of the mapping H can be achieved, so by (3.7), we have

$$H(\mathcal{T}(x), \mathcal{T}(y)) = \mathcal{T}(x) \text{ and } H(\mathcal{T}(y), \mathcal{T}(x)) = \mathcal{T}(y). \tag{3.8}$$

On combining (3.6) with (3.8), we see that $\mathcal{A}(x, y) = \mathcal{T}(x)$ and $\mathcal{A}(y, x) = \mathcal{T}(y)$.

Next, we show the result for the condition (b).

By (3.3) and (3.5), using the fact that $\mathcal{Y}$ is regular, we have

$$\mathcal{T}(x_n) \le \mathcal{T}(x) \text{ and } \mathcal{T}(y) \le \mathcal{T}(y_n), \quad \text{for } n \ge 0. \tag{3.9}$$

Suppose that $(\mathcal{T}(x_n), \mathcal{T}(y_n)) \neq (\mathcal{T}(x), \mathcal{T}(y))$ for $n \ge 0$ otherwise (x_n, y_n) is a CCP of the pair $(\mathcal{A}, \mathcal{T})$. Now, for $t > 0$, $n \ge 0$, we obtain

$$\mathcal{M}(\mathcal{A}(x, y), \mathcal{T}(x), t) \ge \mathcal{M}(\mathcal{A}(x, y), \mathcal{A}(x_n, y_n), kt) * \mathcal{M}(\mathcal{A}(x_n, y_n), \mathcal{T}(x), t - kt), \tag{3.10}$$

$$\mathcal{M}(\mathcal{A}(y, x), \mathcal{T}(y), t) \ge \mathcal{M}(\mathcal{A}(y, x), \mathcal{A}(y_n, x_n), kt) * \mathcal{M}(\mathcal{A}(y_n, x_n), \mathcal{T}(y), t - kt). \tag{3.11}$$

From (3.10) and (3.11), for all $t > 0$, we obtain that

$$\begin{aligned}
&\mathcal{M}(\mathcal{A}(x, y), \mathcal{T}(x), t) * \mathcal{M}(\mathcal{A}(y, x), \mathcal{T}(y), t) \\
&\quad \geq [\mathcal{M}(\mathcal{A}(x, y), \mathcal{A}(x_n, y_n), kt) * \mathcal{M}(\mathcal{A}(x_n, y_n), \mathcal{T}(x), t - kt)] \\
&\quad\quad * [\mathcal{M}(\mathcal{A}(y, x), \mathcal{A}(y_n, x_n), kt) * \mathcal{M}(\mathcal{A}(y_n, x_n), \mathcal{T}(y), t - kt)] \\
&\quad = [\mathcal{M}(\mathcal{A}(x, y), \mathcal{A}(x_n, y_n), kt) * \mathcal{M}(\mathcal{A}(y, x), \mathcal{A}(y_n, x_n), kt)] \\
&\quad\quad * [\mathcal{M}(\mathcal{A}(x_n, y_n), \mathcal{T}(x), t - kt) * \mathcal{M}(\mathcal{A}(y_n, x_n), \mathcal{T}(y), t - kt)]. \qquad (3.12)
\end{aligned}$$

Since $\mathcal{T}(x_n) \leq \mathcal{T}(x)$ and $\mathcal{T}(y) \leq \mathcal{T}(y_n)$ for all $n \geq 0$, using (T-3), we obtain that

$$\begin{aligned}
&\mathcal{M}(\mathcal{A}(x, y), \mathcal{A}(x_n, y_n), kt) * \mathcal{M}(\mathcal{A}(y, x), \mathcal{A}(y_n, x_n), kt) \\
&\quad \geq \gamma(\mathcal{M}(\mathcal{T}(x), \mathcal{T}(x_n), t) * \mathcal{M}(\mathcal{T}(y), \mathcal{T}(y_n), t). \qquad (3.13)
\end{aligned}$$

By (3.12) and (3.13), we obtain that

$$\begin{aligned}
&\mathcal{M}(\mathcal{A}(x, y), \mathcal{T}(x), t) * \mathcal{M}(\mathcal{A}(y, x), \mathcal{T}(y), t) \\
&\quad \geq [\gamma(\mathcal{M}(\mathcal{T}(x), \mathcal{T}(x_n), t) * \mathcal{M}(\mathcal{A}(\mathcal{T}(y), \mathcal{T}(y_n), t))] \\
&\quad\quad * [\mathcal{M}(\mathcal{A}(x_n, y_n), \mathcal{T}(x), t - kt) * \mathcal{M}(\mathcal{A}(y_n, x_n), \mathcal{T}(y), t - kt)]. \qquad (3.14)
\end{aligned}$$

Letting the limit as $n \to \infty$ in (3.14), for $t > 0$, we obtain that

$$\begin{aligned}
&\mathcal{M}(\mathcal{A}(x, y), \mathcal{T}(x), t) * \mathcal{M}(\mathcal{A}(y, x), \mathcal{T}(y), t) \\
&\quad \geq \lim_{n\to\infty} [\gamma(\mathcal{M}(\mathcal{T}(x), \mathcal{T}(x_n), t) * \mathcal{M}(E(\mathcal{T}(y), \mathcal{T}(y_n), t))] \\
&\quad\quad * \lim_{n\to\infty} [\mathcal{M}(\mathcal{A}(x_n, y_n), \mathcal{T}(x), t - kt) * \mathcal{M}(\mathcal{A}(y_n, x_n), \mathcal{T}(y), t - kt)].
\end{aligned}$$

By (3.5),

$$\begin{aligned}
&\mathcal{M}(\mathcal{A}(x, y), \mathcal{T}(x), t) * \mathcal{M}(\mathcal{A}(y, x), \mathcal{T}(y), t) \\
&\quad \geq \gamma(1 * 1) * 1 * 1 = 1.
\end{aligned}$$

That is,

$$\mathcal{M}(\mathcal{A}(x, y), \mathcal{T}(x), t) * \mathcal{M}(\mathcal{A}(y, x), \mathcal{T}(y), t) \geq 1.$$

Therefore, $\mathcal{M}(\mathcal{A}(x, y), \mathcal{T}(x), t) = 1$ and $\mathcal{M}(\mathcal{A}(y, x), \mathcal{T}(y), t) = 1$, which implies $\mathcal{T}(x) = \mathcal{A}(x, y)$ and $\mathcal{T}(y) = \mathcal{A}(y, x)$.

Hence, we obtained the result. □

Next, we give the following example in support of Theorem 3.2.1:

Example 3.2.2 Let $(\mathcal{Y}, \leq)$ be the $\mathcal{POS}$ with $X = [0, 1)$ and the natural ordering of real numbers $\leq$ considered as PO $\leq$. Define $\mathcal{M}(x, y, t) = \frac{t}{t+|x-y|}$ for x, y in $\mathcal{Y}$ and $t > 0$ and $x * y = \min\{x, y\}$ for all x, y in $[0, 1]$. Then $(\mathcal{Y}, \mathcal{M}, *, \leq)$ be a $\mathcal{PO} - \mathcal{GV} - \mathcal{FMS}$, which is not complete. Also, $\mathcal{Y}$ is regular. Define the mappings $\mathcal{A} : \mathcal{Y} \times \mathcal{Y} \to \mathcal{Y}$ and $\mathcal{T} : \mathcal{Y} \to \mathcal{Y}$, respectively, by

$$\mathcal{A}(x, y) = 0.5, \quad \text{for all}(x, y) \text{ in } \mathcal{Y} \times \mathcal{Y}$$

and

$$\mathcal{T}(x) = \begin{cases} 0.8 & \text{if } 0 \leq x < 0.6, \\ x - 0.2 & \text{if } 0.6 \leq x < 1. \end{cases}$$

Since $\mathcal{T}(\mathcal{A}(x, y)) = \mathcal{T}(0.5) = 0.8 \neq 0.5 = \mathcal{A}(\mathcal{T}(x), \mathcal{T}(y))$ for all x, y in $\mathcal{Y}$, the mappings $\mathcal{A}$ and $\mathcal{T}$ are neither commutative nor compatible.

The following observations are immediate:

(i) $\mathcal{A}(\mathcal{Y} \times \mathcal{Y}) \subseteq \mathcal{T}(\mathcal{Y})$;
(ii) $\mathcal{T}(\mathcal{Y})$ is complete;
(iii) the mapping $\mathcal{T}$ is not continuous;
(iv) $\mathcal{A}$ has M$\mathcal{T}$MP;
(v) the mapping $\mathcal{T}$ is not monotonic.

Further, there exist $x_0 = 0.2$ and $y_0 = 0.6$ such that $\mathcal{T}(x_0) = \mathcal{T}(0.2) = 0.8 \geq 0.5 = \mathcal{A}(0.2, 0.6) = \mathcal{A}(x_0, y_0)$ and $\mathcal{T}(y_0) = \mathcal{T}(0.6) = 0.4 \leq 0.5 = \mathcal{A}(0.6, 0.2) = \mathcal{A}(y_0, x_0)$.

Also, the inequality (T-3) holds, since, for any k in $(0, 1)$, $t > 0$

$$\begin{aligned} \mathcal{M}(\mathcal{A}(x, y), \mathcal{A}(\mu, v), kt) * \mathcal{M}(\mathcal{A}(y, x), \mathcal{A}(v, \mu), kt) &= 1 * 1 \\ &\geq \gamma(\mathcal{M}(\mathcal{T}(x), \mathcal{T}(\mu), t) * \mathcal{M}(\mathcal{T}(y), \mathcal{T}(v), t)), \end{aligned}$$

for all x, y, μ, v in $\mathcal{Y}$, with $\mathcal{T}(x) \geq \mathcal{T}(\mu)$ and $\mathcal{T}(y) \geq \mathcal{T}(v)$ and $\gamma(t) = t$ in $[0, 1]$.

Hence, all hypotheses of Theorem 3.2.1 hold, thus, the pair $(\mathcal{A}, \mathcal{T})$ has a CCP in $\mathcal{Y} \times \mathcal{Y}$. Thus, on applying Theorem 3.2.1, we get that the point $(0.7, 0.7)$ is a CCP of the pair $(\mathcal{A}, \mathcal{T})$.

Corollary 3.2.3 *Let* $(\mathcal{Y}, \mathcal{M}, *, \leq)$ *be a* $\mathcal{PO} - \mathcal{GV} - \mathcal{FMS}$ *which is complete. Let the mapping* $\mathcal{A}$ *has MMP and be subjected to the following:*

(T-1) there exists k in $(0, 1)$ *such that*

$$\mathcal{M}(\mathcal{A}(x, y), \mathcal{A}(\mu, v), kt) * \mathcal{M}(\mathcal{A}(y, x), \mathcal{A}(v, \mu), kt) \geq \gamma(\mathcal{M}(x, \mu, t) * \mathcal{M}(y, v, t)),$$

for all x, y, μ, v *in* $\mathcal{Y}$, $t > 0$ *with* $x \leq \mu$ *and* $y \geq v$ *(or,* $x \geq \mu$ *and* $y \leq v$*);*

(T-2) *there exist* x_0, y_0 *in* $\mathcal{Y}$ *such that* $x_0 \leq \mathcal{A}(x_0, y_0)$ *and* $y_0 \geq \mathcal{A}(y_0, x_0)$ *(or,* $x_0 \geq \mathcal{A}(x_0, y_0)$ *and* $y_0 \leq \mathcal{A}(y_0, x_0)$*).*

Further, suppose either

(a) *the mapping* $\mathcal{A}$ *is continuous, or*
(b) $\mathcal{Y}$ *is regular;*

then, the mapping $\mathcal{A}$ *has a CFP in* $\mathcal{Y}$.

Proof Proof follows by considering $\mathcal{T} = I_{\mathcal{Y}}$ (identity mapping on $\mathcal{Y}$), in Theorem 3.2.1.

□

3.3 Existence and Uniqueness of CCFP

Recently, to obtain CCFP for nonlinear contractive mappings in cone metric spaces, Abbas et al. (2010) gave the notion of w-compatible mappings. In this section, utilizing the notion of w-compatible mappings, we extend Theorem 3.2.1 to ensure the existence and uniqueness of the CCFP in fuzzy metric space.

Definition 3.3.1 (Abbas et al. 2010) The mappings $\mathcal{A}$ and $\mathcal{T}$ are called w-compatible if $\mathcal{T}(\mathcal{A}(x, y)) = \mathcal{A}(\mathcal{T}x, \mathcal{T}y)$ whenever $\mathcal{T}(x) = \mathcal{A}(x, y)$ and $\mathcal{T}(y) = \mathcal{A}(y, x)$ for $x, y \in \mathcal{Y}$.

Now, we give our CCFP result as follows:

Theorem 3.3.2 *Suppose all the assumptions of Theorem 3.2.1 hold along with an additional hypothesis that: "for every* (x, y), (x', y') *in* $\mathcal{Y} \times \mathcal{Y}$, *there exists a* (μ, v) *in* $\mathcal{Y} \times \mathcal{Y}$ *such that* $(\mathcal{A}(\mu, v), \mathcal{A}(v, u))$ *is comparable to* $(\mathcal{A}(x, y), \mathcal{A}(y, x))$ *and* $(\mathcal{A}(x', y'), \mathcal{A}(y', x'))$. *If the pair* $(\mathcal{A}, \mathcal{T})$ *is* w*-compatible, then it has a unique CCFP.*

Proof Using Theorem 3.2.1, the set of coupled coincidences of pair $(\mathcal{A}, \mathcal{T})$ is not empty. For obtaining the result, we first show: if (x, y) and (x', y') are CCP, then

$$\mathcal{T}(x) = \mathcal{T}(x') \text{ and } \mathcal{T}(y) = \mathcal{T}(y'). \tag{3.15}$$

By assumption there is(μ, v) in $\mathcal{Y} \times \mathcal{Y}$ such that $(\mathcal{A}(\mu, v), \mathcal{A}(v, v))$ is comparable with $(\mathcal{A}(x, y), \mathcal{A}(y, x))$ and $(\mathcal{A}(x', y'), \mathcal{A}(y', x'))$. Put $\mu_0 = \mu$, $v_0 = v$ and choose μ_1, v_1 in X so that

$$\mathcal{T}(\mu_1) = \mathcal{A}(\mu_0, v_0), \quad \mathcal{T}(v_1) = \mathcal{A}(v_0, \mu_0).$$

Then, like in Theorem 3.2.1, sequences $\{\mathcal{T}(\mu_n)\}$ and $\{\mathcal{T}(v_n)\}$ can be defined such that

$$\mathcal{T}(\mu_{n+1}) = \mathcal{A}(\mu_n, v_n) \text{ and } \mathcal{T}(v_{n+1}) = \mathcal{A}(v_n, \mu_n).$$

Further, set $x_0 = x,\ y_0 = y,\ x_0' = x',\ y_0' = y'$, we construct sequences $\{\mathcal{T}(x_n)\}$, $\{\mathcal{T}(y_n)\}$ and $\{\mathcal{T}(x_n')\}$, $\{\mathcal{T}(y_n')\}$ following the same track. Then, it can be easily obtained that

$$\mathcal{T}(x_{n+1}) = \mathcal{A}(x_n, y_n), \mathcal{T}(y_{n+1}) = \mathcal{A}(y_n, x_n)$$

and

$$\mathcal{T}(x_{n+1}') = \mathcal{A}(x_n', y_n'),\ \mathcal{T}(y_{n+1}') = \mathcal{A}(y_n', x_n'), \quad \text{for } n \geq 0.$$

Since $(\mathcal{A}(\mu, v), \mathcal{A}(v, \mu)) = (\mathcal{T}(\mu_1), \mathcal{T}(v_1))$ and $(\mathcal{T}(x), \mathcal{T}(y)) = (\mathcal{A}(x, y), \mathcal{A}(y, x)) = (\mathcal{T}(x_1), \mathcal{T}(y_1))$ are comparable, then $\mathcal{T}(\mu_1) \geq \mathcal{T}(x)$ and $\mathcal{T}(v_1) \leq \mathcal{T}(y)$. Easily, it can be shown that $(\mathcal{T}(\mu_n), \mathcal{T}(v_n))$ and $(\mathcal{T}(x), \mathcal{T}(y))$ are comparable; that is, $\mathcal{T}(\mu_n) \geq \mathcal{T}(x)$ and $\mathcal{T}(v_n) \leq \mathcal{T}(y)$ for all $n > 0$. Thus, from (T-3), for $n > 0$,

$$\begin{aligned}
&\mathcal{M}(\mathcal{T}(\mu_{n+1}), \mathcal{T}(x), kt) * \mathcal{M}(\mathcal{T}(v_{n+1}), \mathcal{T}(y), kt) \\
&\quad = \mathcal{M}(\mathcal{A}(\mu_n, v_n), \mathcal{A}(x, y), kt) * \mathcal{M}(\mathcal{A}(v_n, \mu_n), \mathcal{A}(y, x), kt) \\
&\quad \geq \gamma(\mathcal{M}(\mathcal{T}(\mu_n), \mathcal{T}(x), t) * \mathcal{M}(\mathcal{T}(v_n), \mathcal{T}(y), t)),
\end{aligned}$$

that is, for all $n \in N$, we have that

$$\begin{aligned}
&\mathcal{M}(\mathcal{T}(\mu_{n+1}), \mathcal{T}(x), kt) * \mathcal{M}(\mathcal{T}(v_{n+1}), \mathcal{T}(y), kt) \\
&\quad \geq \gamma(\mathcal{M}(\mathcal{T}(\mu_n), \mathcal{T}(x), t) * \mathcal{M}(\mathcal{T}(v_n), \mathcal{T}(y), t)).
\end{aligned} \tag{3.16}$$

We claim that

$$\lim_{n\to\infty} \mathcal{T}(\mu_n) = \mathcal{T}(x) and \lim_{n\to\infty} \mathcal{T}(v_n) = \mathcal{T}(y). \tag{3.17}$$

If $\lim_{n\to\infty} \mathcal{T}(\mu_n) = \mu$ and $\lim_{n\to\infty} \mathcal{T}(v_n) = v$ for some μ, v in $\mathcal{Y}$, it suffices to obtain that $\mu = \mathcal{T}(x)$ and $v = \mathcal{T}(y)$. Letting the limit as $n \to \infty$ in (3.16) and using the properties of γ, we obtain that

$$\begin{aligned}
\mathcal{M}(\mu, \mathcal{T}(x), kt) * \mathcal{M}(v, \mathcal{T}(y), kt) &\geq \gamma(\mathcal{M}(\mu, \mathcal{T}(x), t) * \mathcal{M}(v, \mathcal{T}(y), t)) \\
&\geq \mathcal{M}(\mu, \mathcal{T}(x), t) * \mathcal{M}(v, \mathcal{T}(y), t).
\end{aligned} \tag{3.18}$$

Thus,

$$\mathcal{M}(\mu, \mathcal{T}(x), t) * \mathcal{M}(v, \mathcal{T}(y), t) \geq \mathcal{M}(\mu, \mathcal{T}(x), t/k^n) * \mathcal{M}(v, \mathcal{T}(y), t/k^n),$$

for all $n \in N$, implying that $\mathcal{M}(\mu, \mathcal{T}(x), t) * \mathcal{M}(v, \mathcal{T}(y), t) = 1$ for $t > 0$. It follows that $\mathcal{M}(\mu, \mathcal{T}(x), t) = \mathcal{M}(v, \mathcal{T}(y), t) = 1$ for all $t > 0$, therefore we can obtain that $\mu = \mathcal{T}(x)$ and $v = \mathcal{T}(y)$, and hence (3.17) holds.

Similarly, we obtain that

$$\lim_{n\to\infty} \mathcal{T}(\mu_n) = \mathcal{T}(x') \text{ and } \lim_{n\to\infty} \mathcal{T}(v_n) = \mathcal{T}(y). \tag{3.19}$$

Using the uniqueness of limit, it is obtainable that $\mathcal{T}(x) = \mathcal{T}(x')$ and $\mathcal{T}(y) = \mathcal{T}(y')$. Thus, we have proved (3.15).

As $\mathcal{T}(x) = \mathcal{A}(x, y)$, $\mathcal{T}(y) = \mathcal{A}(y, x)$ and the pair $(\mathcal{A}, \mathcal{T})$ is w-compatible, it is obtainable that

$$\mathcal{T}(\mathcal{T}(x)) = \mathcal{T}(\mathcal{A}(x, y)) = \mathcal{A}(\mathcal{T}(x), \mathcal{T}(y)) \text{ and } \mathcal{T}(\mathcal{T}(y)) = \mathcal{T}(\mathcal{A}(y, x)) = \mathcal{A}(\mathcal{T}(y), \mathcal{T}(x)). \tag{3.20}$$

Denote $\mathcal{T}(x) = z$, $\mathcal{T}(y) = w$. Then, using (3.20),

$$\mathcal{T}(z) = \mathcal{A}(z, w) \text{ and } \mathcal{T}(z) = \mathcal{A}(w, z). \tag{3.21}$$

Thus, (z, w) is a CCP.

Then, by (3.15) with $x' = z$ and $y' = w$, we get $\mathcal{T}(z) = \mathcal{T}(x)$ and $\mathcal{T}(w) = \mathcal{T}(y)$; that is,

$$\mathcal{T}(z) = z, \ \mathcal{T}(w) = w. \tag{3.22}$$

Using (3.21) and (3.22), we obtain

$$z = \mathcal{T}(z) = \mathcal{A}(z, w) \text{ and } w = \mathcal{T}(w) = \mathcal{A}(w, z).$$

Hence, (z, w) is the CCFP of $\mathcal{A}$ and q.

For uniqueness, let (p, q) be any CCFP. Then, by (3.15), we get

$$p = \mathcal{T}(p) = \mathcal{T}(z) = z \text{ and } q = \mathcal{T}(p) = \mathcal{T}(w) = w.$$

□

3.4 Application in Metric Spaces

In this section, as an application of the results proved in the earlier sections of this paper, we obtain CCP results in the framework of ordered metric spaces.

Theorem 3.4.1 *Let* $(\mathcal{Y}, \preceq)$ *be* $\mathcal{POS}$ *and* d *a metric on* $\mathcal{Y}$ *so that* $(\mathcal{Y}, d)$ *is a metric space, the pair* $(\mathcal{A}, \mathcal{T})$ *of mappings be such that* $\mathcal{A}$ *has* $M\mathcal{T}MP$ *and be subjected to the following:*

(T-1) there exists some k *in* $(0, 1)$ *such that*

$$max\{d(\mathcal{A}(x, y), \mathcal{A}(\mu, v)), d(\mathcal{A}(y, x), \mathcal{A}(v, \mu))\}$$
$$\frac{k}{2}[d(\mathcal{T}(x), \mathcal{T}(\mu)) + d(\mathcal{T}(y), \mathcal{T}(v))],$$

for all x, y, μ, v *in* $\mathcal{Y}$ *for which* $\mathcal{T}(x) \preceq \mathcal{T}(\mu)$, $\mathcal{T}(y) \succeq \mathcal{T}(v)$ *(or,* $\mathcal{T}(x) \succeq \mathcal{T}(\mu)$ *and* $\mathcal{T}(y) \preceq \mathcal{T}(v)$*).*

Suppose that (T-1), (T-2), and (T-4) hold. Also, assume that both the mappings $\mathcal{A}$ *and* $\mathcal{T}$ *are continuous, or*

(a) $\mathcal{Y}$ *is regular,*
(b) then the pair $(\mathcal{A}, \mathcal{T})$ *has a CCP.*

Proof For all x, y in $\mathcal{Y}$ and $t > 0$, define $\mathcal{M}(x, y, t) = \frac{t}{t+d(x,y)}$ and $p * q = \min\{p, q\}$ with p, q in $[0, 1]$. Then, $(\mathcal{Y}, \mathcal{M}, *)$ is F$\mathcal{M}$S, $*$ as the Hadžić type t-norm. Also, it can be easily seen that $\mathcal{M}(x, y, t) = \frac{t}{t+d(x,y)} \to 1$ as $t \to \infty$, for all x, y in $\mathcal{Y}$. We next verify that the inequality (T-6) implies (T-3). If otherwise, for $\gamma(t) = t$, from (T-3), for some $t > 0$ and x, y, μ, v in $\mathcal{Y}$ with $\mathcal{T}(x) \preceq \mathcal{T}(\mu)$, $\mathcal{T}(y) \succeq \mathcal{T}(v)$, we have

$$\min\left\{\frac{t}{t + \frac{1}{k}d(\mathcal{A}(x, y), \mathcal{A}(\mu, v))}, \frac{t}{t + \frac{1}{k}d(\mathcal{A}(y, x), \mathcal{A}(v, \mu))}\right\}$$
$$< \min\left\{\frac{t}{t + d(\mathcal{T}(x), \mathcal{T}(\mu))}, \frac{t}{t + d(\mathcal{T}(y), \mathcal{T}(v))}\right\},$$

then, we have either,

$$\frac{t}{t + \frac{1}{k}d(\mathcal{A}(x, y), \mathcal{A}(\mu, v))} < \min\left\{\frac{t}{t + d(\mathcal{T}(x), \mathcal{T}(\mu))}, \frac{t}{t + d(\mathcal{T}(y), \mathcal{T}(v))}\right\}, \tag{3.23}$$

or,

$$\frac{t}{t+\frac{1}{k}d(\mathcal{A}(y,x),\mathcal{A}(v,\mu))} < \min\left\{\frac{t}{t+d(\mathcal{T}(x),\mathcal{T}(\mu))}, \frac{t}{t+d(\mathcal{T}(y),\mathcal{T}(v))}\right\}. \tag{3.24}$$

From (3.23), it follows

$$t+\frac{1}{k}d(\mathcal{A}(x,y),\mathcal{A}(\mu,v)) > t+d(\mathcal{T}(x),\mathcal{T}(\mu)), \tag{3.25}$$

$$t+\frac{1}{k}d(\mathcal{A}(x,y),\mathcal{A}(\mu,v)) > t+d(\mathcal{T}(y),\mathcal{T}(v)). \tag{3.26}$$

Adding (3.25) and (3.26), we attain

$$d(\mathcal{A}(x,y),\mathcal{A}(\mu,v)) > \frac{k}{2}[d(\mathcal{T}(x),\mathcal{T}(\mu))+d(\mathcal{T}(y),\mathcal{T}(v))]. \tag{3.27}$$

Similarly, from (3.24), we obtain that

$$d(\mathcal{A}(y,x),\mathcal{A}(v,\mu)) > \frac{k}{2}[d(\mathcal{T}(x),\mathcal{T}(\mu))+d(\mathcal{T}(y),\mathcal{T}(v))]. \tag{3.28}$$

Using (3.27) and (3.28), we have that

$$\max\{d(\mathcal{A}(x,y),\mathcal{A}(\mu,v)), d(\mathcal{A}(y,x),\mathcal{A}(v,\mu))\} > \frac{k}{2}[[d(\mathcal{T}(x),\mathcal{T}(\mu))+d(\mathcal{T}(y),\mathcal{T}(v))]],$$

which is a contradiction to (T-7). Now, on applying Theorem 3.2.1, we get the desired result. □

Remark 3.4.2 (i) Theorem 3.4.1 provides an extension and generalization of a result of Berinde (2011, Corollary 1) for a pair of mappings.

(ii) Unique coupled common fixed point for the pair $(\mathcal{A}, g)$ under the hypotheses of Theorem 3.4.1 can be obtained by assuming the additional assumptions as in Theorem 3.3.2.

References

Abbas, M., Ali Khan, M., & Radenović, S. (2010). Common coupled fixed point theorems in cone metric spaces for w-compatible mappings. *Applied Mathematics and Computation, 217*, 195–202.

Banach, S. (1922). Sur les opérations dans les ensembles abstraits et leur application aux équations intégrales. *Fundamenta Mathematicae, 3*(1), 133–181.

Berinde, V. (2011). Generalized coupled fixed point theorems for mixed monotone mappings partially ordered metric spaces. *Nonlinear Analysis, 74*, 7347–7355.

Bhaskar, T. G., & Lakshmikantham, V. (2006). Fixed point theorems in partially ordered metric spaces and applications. *Nonlinear Analysis,65*(7), 1379–1393.

Choudhury, B. S., Das, K., & Das, P. (2013). Coupled coincidence point results for compatible mappings in partially ordered fuzzy metric spaces. *Fuzzy Sets and Systems,222*, 84–97.

Choudhury, B. S., & Kundu, A. (2010). A coupled coincidence point result in partially ordered metric spaces for compatible mappings. *Nonlinear Analysis,73*, 2524–2531.

Deng, Z. K. (1982). Fuzzy pseudo metric spaces. *Journal of Mathematical Analysis and Applications,86*, 74–95.

Edelstein, M. (1962). On fixed and periodic points under contraction mappings. *Journal of the London Mathematical Society,37*, 74–79.

Erceg, M. A. (1979). Metric spaces in fuzzy set theory. *Journal of Mathematical Analysis and Applications,69*, 205–230.

Fang, J. X. (1992). On fixed point theorems in fuzzy metric spaces. *Fuzzy Sets and Systems,46*, 107–113.

George, A., & Veeramani, P. (1994). On some results in fuzzy metric spaces. *Fuzzy Sets and Systems,64*(3), 395–399.

George, A., & Veeramani, P. (1997). On some results of analysis for fuzzy metric spaces. *Fuzzy Sets and Systems,90*, 365–368.

Gordji, M. E., Cho, Y. J., Ghods, S., Ghods, M., & Dehkordi, M. H. (2012). Coupled fixed point theorems for contractions in partially ordered metric spaces and applications. *Mathematical Problems in Engineering, 2012*, Article ID 150363, 20.

Grabiec, M. (1988). Fixed points in fuzzy metric spaces. *Fuzzy Sets and Systems,27*(3), 385–389.

Guo, D., & Lakshmikantham, V. (1987). Coupled fixed points of nonlinear operators with applications. *Nonlinear Analysis,11*, 623–632.

Hu, X.-Q. (2011). Common coupled fixed point theorems for contractive mappings in fuzzy metric spaces. *Fixed Point Theory Application, 2011*, Article ID 363716.

Istratescu, I. (1981). A fixed point theorem for mappings with a probabilistic contractive iterate. *Revue Roumaine de Mathématiques Pures et Appliquées,26*, 431–435.

Jain, M., Gupta, N., & Kumar, S. (2017). A discussion on some recent coupled fixed point results via new generalized nonlinear contractive conditions. *TWMS Journal of Applied and Engineering Mathematics,7*(1), 110–130.

Jain, M., Gupta, N., & Kumar, S. (2019). Common fixed point results for w-compatible mappings along with (CLRST) property in fuzzy metrics spaces. *TWMS Journal of Applied and Engineering, Mathematics,9*(4), 851–863.

Jain, M., Gupta, N., Vetro, C., & Kumar, S. (2013). Coupled fixed point theorems for symmetric (ϕ, ψ)-weakly contractive mappings in ordered partial metric spaces. *The Journal of Mathematics and Computer Sciences,7*(4), 230–304.

Jain, M., Kumar, S., & Chugh, R. (2013). Coupled fixed point theorems for weak compatible mappings in fuzzy metric spaces. *Annals of Fuzzy Mathematics and Informatics,5*(2), 321–336.

Jain, M., Kumar, S., Park, C., Addressing an open problem of Choudhury, et al. (2023). *Fuzzy Sets System, 8*, 84–97. (*Journal of Intelligent and Fuzzy Systems,44*(5), 7211–7223).

Jain, M., Tas, K., Kumar, S., & Gupta, N. (2012). Coupled fixed point theorems for a pair of weakly compatible mappings along with CLRg property in fuzzy metric spaces. *Journal of Applied Mathematics, 2012*, Article ID 961210, 13. https://doi.org/10.1155/2012/961210.

Jain, M., Tas, K., Rhoades, B. E., & Gupta, N. (2014). Coupled fixed point theorems for generalized symmetric contractions in partially ordered metric spaces and applications. *Journal of Computational Analysis and Applications,16*(3), 438–454.

Kaleva, O., & Seikkala, S. (1984). On fuzzy metric spaces. *Fuzzy Sets and Systems,12*, 215–229.

Kramosil, I., & Michalek, J. (1975). Fuzzy metric and statistical metric spaces. *Kybernetica,11*, 326–334.

Lakshmikantham, V., & Ćirić, L. (2009). Coupled fixed point theorems for nonlinear contractions in partially ordered metric spaces. *Nonlinear Analysis,70*, 4341–4349.

Sedghi, S., Altun, I., & Shobe, N. (2010). Coupled fixed point theorems for contractions in fuzzy metric spaces. *Nonlinear Analysis,72*, 1298–1304.
Sehgal, V. M., & Bharucha-Reid, A. T. (1972). Fixed points of contraction mappings on PM-spaces. *Mathematical Systems Theory,6*, 97–100.
Shatanawi, W., Samet, B., & Abbas, M. (2011). Coupled fixed point theorems for mixed monotone mappings in ordered partial metric spaces. *Mathematical and Computer Modelling*. https://doi.org/10.1016/j.mcm.2011.08.042
Zadeh, L. A. (1965). Fuzzy sets. *Information and Control,89*, 338–353.

Chapter 4
On S-Polynomial Type Contractions in S-Metric Spaces and Fixed-Point Results

Deep Chand, Khairul Habib Alam, Yumanm Rohen, and Sanasam Surendra Singh

Abstract In this article, we introduce two new classes of single-valued contraction mappings within the framework of S-metric spaces. For the first class, called S-polynomial type contraction mappings, we establish two fixed-point theorems that demonstrate the existence and uniqueness of fixed points. We begin by considering the case where the mapping is continuous and then relax this continuity requirement. Importantly, we also derive Banach's fixed-point theorem as a special case within the context of S-metric spaces. The second class, termed almost S-polynomial type contraction mappings, builds on the concept of almost contraction mappings introduced by Berinde. We establish two fixed-point theorems that generalize Berinde's results in S-metric spaces. Additionally, we introduce a condition for uniqueness and prove a theorem for a unique fixed point. Several examples are provided to illustrate the significance of our generalizations. Finally, we demonstrate an application of the main result by describing the solution of the integral equation.

Keywords S-polynomial type contractions · Almost S-polynomial type contractions · Picard-continuous · Fixed point · S-metric space

AMS Subject Classification 47H10 · 54H25

4.1 Introduction and Preliminaries

Fixed-point theory is a rich and diverse area of mathematics that explores the properties of fixed points across various types of metric spaces. Understanding these prop-

---

D. Chand · S. S. Singh
Department of Mathematics National Institute of Technology Manipur, Imphal, Manipur, India

Y. Rohen
Department of Mathematics, Manipur University, Canchipur, Imphal, Manipur, India

K. H. Alam (✉)
Mathematical Sciences, Indian Institute of Science Education and Research Berhampur, Berhampur, Odisha, India
e-mail: khairulha@iiserbpr.ac.in; almkrlhb@gmail.com

© The Author(s), under exclusive license to Springer Nature Singapore Pte Ltd. 2026

A. Tomar and M. Jain (eds.), *Applications of Fixed-Point Theorem*, Industrial and Applied Mathematics, https://doi.org/10.1007/978-981-95-5160-6_4

erties provides not only deep theoretical insights but also practical applications in fields such as mathematics, physics, engineering, computer science, and more. Each type of metric space offers distinct results and techniques for analyzing fixed points, contributing to the broader landscape of fixed-point theory. This diversity enhances our understanding and application of mathematical principles, making fixed-point theory a vital and influential area of research with far-reaching implications.

Fixed-point theory examines the existence and properties of fixed points, which are elements that remain unchanged when a function is applied to them. Formally, for a function T mapping a set X to itself, a point $x \in X$ is called a fixed point if $T(x) = x$. One of the most renowned results in this field is the Banach fixed-point theorem (Banach 1992). This theorem states that if (X, d) is a complete metric space and $T : X \to X$ is a mapping that satisfies the condition:

$$d(Tx, Ty) \leq kd(x, y), \text{ for all } x, y \in X, \text{ and some constant } k \in [0, 1), \tag{4.1}$$

then T has a unique fixed point. Moreover, this fixed point can be found by iteratively applying T, starting from any initial point in X.

A mapping $T : X \to X$ that satisfies inequality (1) is known as a contraction. Numerous generalizations and extensions of Banach's fixed-point theorem have been developed. Some of these extensions aim to relax the condition on the right side of inequality (1), as seen in works by Berinde (2004), Rakotch (1962), Boyd (1969), Reich (1972), Ćirić (1974), Rus (2001), and Alam et al. (2023). Additionally, other extensions introduce a generalized distance on the underlying space, which has been explored by researchers such as Dhage (1992), Czerwik (1993), Branciari (2000), Joshi et al. (2021, 2023), and Alam et al. (2024a, d, 2025a, b). These generalizations and extensions provide a broader framework for understanding fixed-point theory in various mathematical contexts.

While Banach's fixed-point theorem focuses on continuous contraction mappings, extensive research has expanded the theorem to encompass broader classes of mappings, including those that are not necessarily continuous. A significant extension by R. Kannan introduces a new class of mappings. A mapping $T : X \to X$ on a metric space (X, d) satisfies the Kannan condition if there exists a constant $k \in [0, \frac{1}{2})$ such that

$$d(Tx, Ty) \leq k[d(x, Tx) + d(y, Ty)],$$

for all $x, y \in X$. Kannan's theorem states that:

Theorem 4.1.1 *If (X, d) is a complete metric space and $T : X \to X$ is a Kannan mapping, then T has a unique fixed point.*

The above theorem expands the scope of fixed-point results by relaxing the continuity requirement and the strict contraction condition of Banach's theorem. Additional generalizations and extensions beyond continuous mappings are documented by Nadler (1969), Chatterjea (1972), Ciric (1971), and Chand et al. (2024) further enriching the field of fixed-point theory.

In particular, Berinde introduced a notable class of mappings known as almost contractions, which encompass Kannan's mappings as well as many other types. Almost contraction mappings are defined as follows:

Definition 4.1.2 (*Berinde* (2004)) If (X, d) is a metric space, a mapping $T : X \to X$ is called an almost contraction if it satisfies the following inequality for every $x, y \in X$:

$$d(Tx, Ty) \leq \lambda d(x, y) + \delta d(y, Ty),$$

where $\lambda \in (0, 1)$ and $\delta > 0$.

The following fixed-point theorem was proved by Berinde for the above class of mappings.

Theorem 4.1.3 (Berinde (2004)) *Let (X, d) be a complete metric space and $T : X \to X$ be an almost contraction. Then the mapping T possesses at least one fixed point.*

It is important to note that almost all contractions can possess more than one fixed point (see Example 1 in Berinde (2004)). On the other hand, another method to advance fixed-point theory and discover new fixed-point results involves generalizing the concept of metric spaces. For example, the concept of an S-metric space was introduced by Sedghi and Shobe (2012) as a generalization of a metric space, defined as follows:

Definition 4.1.4 (*Sedghi et al.* (2012)) Let X be a nonempty set. An S-metric on X is a function $S : X \times X \times X \to [0, \infty)$ that satisfies the following conditions for each $x, y, z, a \in X$:

($\mathcal{S}_1$) $S(x, y, z) \geq 0$,
($\mathcal{S}_2$) $S(x, y, z) = 0$ if and only if $x = y = z$,
($\mathcal{S}_3$) $S(x, y, z) \leq S(x, x, a) + S(y, y, a) + S(z, z, a)$.

The pair (X, S) is called S-metric space.

Note: (2012) For an S-metric space, the symmetry condition can be defined as follows:

$$S(x, x, y) = S(y, y, x) \text{ for all } x, y \in X.$$

The literature provides examples of S-metrics that are not generated by any standard metric (Hieu et al. 2015; Özgür and Taş 2017). Therefore, it is essential to explore new fixed-point theorems within the framework of S-metric spaces. Researchers are actively studying various fixed-point results by employing different techniques to extend and generalize established fixed-point theorems. For further details, refer to references (Joshi et al. 2023; Chand and Rohen 2023 ; Chand et al. 2024; Deep et al. 2024; Tomar et al. 2023).

Motivated by Berinde (2004), we present a novel type of contraction within S-metric spaces, termed S-polynomial contractions. We develop two fixed-point theorems for these mappings, building upon and extending previous findings, particularly those by Jleli. Our initial focus is on cases where the mapping T is continuous. We then extend our investigation to scenarios where the continuity condition is relaxed, thereby expanding the applicability of our results. Furthermore, we introduce the concept of almost S-polynomial contraction mappings in S-metric spaces and establish corresponding fixed-point theorems for these mappings. To demonstrate the validity and practical relevance of our results, we provide several illustrative examples. This work significantly broadens the understanding and utility of fixed-point theory in S-metric spaces. Finally, we illustrate an application of the primary result by explaining the solution to the integral equation.

4.2 Class of S-Polynomial Type Contraction

Now we present the following class of nearly polynomial contractions.

Definition 4.2.1 Let (X, S) be a S-metric space, and $T : X \to X$ be a specified mapping. The mapping T is said to be an S-polynomial type contraction of a first kind if there exists $\lambda \in (0, 1)$, a natural number $k \geq 1$, and a family of mappings $a_i : X \times X \times X \to [0, \infty)$ for $i = 0, ..., k$, such that

$$\sum_{i=0}^{k} a_i(Tx, Ty, Tz)S^i(Tx, Ty, Tz) \leq \lambda \sum_{i=0}^{k} a_i(x, y, z)S^i(x, y, z), \tag{4.2}$$

for every $x, y, z \in X$.

Definition 4.2.2 Let (X, S) be a S-metric space, and $T : X \to X$ be a specified mapping. The mapping T is said to be an S-polynomial type contraction of a second kind if there exists $\lambda \in (0, 1)$, a natural number $k \geq 1$, and a family of mappings $a_i : X \times X \times X \to [0, \infty)$ for $i = 0, ..., k$, such that

$$\sum_{i=0}^{k} a_i(Tx, Tx, Ty)S^i(Tx, Tx, Ty) \leq \lambda \sum_{i=0}^{k} a_i(x, x, y)S^i(x, x, y), \tag{4.3}$$

for every $x, y \in X$.

In this section, we focus on the study of fixed points for the aforementioned class of mappings. We begin by examining the case where T is a continuous mapping.

Theorem 4.2.3 *Let (X, S) be a complete metric space, and $T : X \to X$ be an S-polynomial type contraction of the first kind. If the following conditions hold:*

(a) T is continuous on X,
(b) There exists $j \in \{0, 1, ..., k\}$ and $\alpha_j > 0$ such that

$$a_j(x, y, z) \geq \alpha_j, \quad for\ all\ x, y, z \in X.$$

Then T has only one fixed point $x^* \in X$. Moreover, for any $x_0 \in X$, the Picard sequence $x_{n+1} = Tx_n$, for all $n \geq 0$, converges to x_*.

Proof First of all, we prove the existence of a fixed point. Let us consider a random point $x_0 \in X$ and the Picard sequence

$$x_{n+1} = Tx_n, \quad \text{for all } n \geq 0.$$

Using Eq. (4.2) for $x = y = x_0$ and $z = x_1$, we have

$$\sum_{i=0}^{k} a_i(Tx_0, Tx_0, Tx_1)S^i(Tx_0, Tx_0, Tx_1) \leq \lambda \sum_{i=0}^{k} a_i(x_0, x_0, x_1)S^i(x_0, x_0, x_1),$$

which implies that

$$\sum_{i=0}^{k} a_i(x_1, x_1, x_2)S^i(x_1, x_1, x_2) \leq \lambda \sum_{i=0}^{k} a_i(x_0, x_0, x_1)S^i(x_0, x_0, x_1). \tag{4.4}$$

Again making use of (4.2) for $x = y = x_1$ and $z = x_2$, we have

$$\sum_{i=0}^{k} a_i(Tx_1, Tx_1, Tx_2)S^i(Tx_1, Tx_1, Tx_2) \leq \lambda \sum_{i=0}^{k} a_i(x_1, x_1, x_2)S^i(x_1, x_1, x_2),$$

which implies

$$\sum_{i=0}^{k} a_i(x_2, x_2, x_3)S^i(x_2, x_2, x_3) \leq \lambda \sum_{i=0}^{k} a_i(x_1, x_1, x_2)S^i(x_1, x_1, x_2). \tag{4.5}$$

Therefore, it implies by (4.4) and (4.5) that

$$\sum_{i=0}^{k} a_i(x_2, x_2, x_3)S^i(x_2, x_2, x_3) \leq \lambda^2 \sum_{i=0}^{k} a_i(x_0, x_0, x_1)S^i(x_0, x_0, x_1).$$

Continuing in the same manner, we obtain by induction that

$$\sum_{i=0}^{k} a_i(x_n, x_n, x_{n+1})S^i(x_n, x_n, x_{n+1}) \leq \lambda^n \sum_{i=0}^{k} a_i(x_0, x_0, x_1)S^i(x_0, x_0, x_1). \quad (4.6)$$

It is obvious that

$$a_j(x_n, x_n, x_{n+1})S^j(x_n, x_n, x_{n+1}) \leq \sum_{i=0}^{k} a_i(x_n, x_n, x_{n+1})S^i(x_n, x_n, x_{n+1}).$$

Therefore, by using (b), we obtain

$$\alpha_j S^j(x_n, x_n, x_{n+1}) \leq \sum_{i=0}^{k} a_i(x_n, x_n, x_{n+1})S^i(x_n, x_n, x_{n+1}).$$

It follows by using (4.6) that

$$S^j(x_n, x_n, x_{n+1}) \leq \lambda^n A_{j,0}, \quad \text{for all } n \geq 0, \quad (4.7)$$

where

$$A_{j,0} = \frac{1}{\alpha_j} \sum_{i=0}^{k} a_i(x_0, x_0, x_1)S^i(x_0, x_0, x_1). \quad (4.8)$$

Therefore,

$$S(x_n, x_n, x_{n+1}) \leq \lambda^{\frac{n}{j}} A_{j,0}^{\frac{1}{j}} = \lambda^{\frac{n}{j}} B, \quad \text{for all } n \geq 0, \quad (4.9)$$

where $B = A_{j,0}^{\frac{1}{j}}$. By using the inequality (4.9) and the triangle inequality of the metric, we have

$$\begin{aligned}
&S(x_n, x_n, x_{n+m})\\
&\leq 2S(x_n, x_n, x_{n+1}) + 2S(x_{n+1}, x_{n+1}, x_{n+2}) + ... + S(x_{n+m-1}, x_{n+m-1}, x_{n+m})\\
&\leq 2B(\lambda^{\frac{n}{j}} + \lambda^{\frac{n+1}{j}} + ... + \lambda^{\frac{n+m-1}{j}})\\
&= 2B\lambda^{\frac{n}{j}} \frac{1 - \lambda^{\frac{m}{j}}}{1 - \lambda^{\frac{1}{j}}}\\
&\leq 2B \frac{\lambda^{\frac{n}{j}}}{1 - \lambda^{\frac{1}{j}}},
\end{aligned}$$

which provides $S(x_n, x_n, x_{n+1}) \to 0$, as $n, m \to +\infty$.

This shows that $\{x_n\}$ is a Cauchy sequence. Considering the completeness of the metric space, there exists $x^* \in X$ such that

$$\lim_{n\to\infty} S(x_n, x_n, x^*) = 0.$$

Then, by the continuity of the mapping T, it follows that

$$\lim_{n\to\infty} S(x_{n+1}, x_{n+1}, Tx^*) = \lim_{n\to\infty} S(Tx_n, Tx_n, Tx^*) = 0.$$

Therefore, by the uniqueness of the limit of the sequence $\{x_n\}$, we conclude that $Tx^* = x^*$, that is, x^* is a fixed point of T.

Now, we will prove the uniqueness of the fixed point x^* for T. On the contrary, let us assume that x^{**} is another fixed point for the mapping T that is $Tx^{**} = x^{**}$. Then, by the use of inequality (4.2) for $x = y = x^*$ and $z = x^{**}$, we obtain

$$\sum_{i=0}^{k} a_i(Tx^*, Tx^*, Tx^{**})S^i(Tx^*, Tx^*, Tx^{**}) \leq \lambda \sum_{i=0}^{k} a_i(x^*, x^*, x^{**})S^i(x^*, x^*, x^{**}),$$

that is,

$$\sum_{i=0}^{k} a_i(x^*, x^*, x^{**})S^i(x^*, x^*, x^{**}) \leq \lambda \sum_{i=0}^{k} a_i(x^*, x^*, x^{**})S^i(x^*, x^*, x^{**}). \quad (4.10)$$

Also, from (b), we have

$$\alpha_j S^j(x^*, x^*, x^{**}) \leq \sum_{i=0}^{k} a_i(x^*, x^*, x^{**})S^i(x^*, x^*, x^{**}),$$

since $\alpha_j > 0$ and $S^j(x^*, x^*, x^{**}) \geq 0$, we deduce that

$$\sum_{i=0}^{k} a_i(x^*, x^*, x^{**})S^i(x^*, x^*, x^{**}) > 0.$$

Then, we reached to a contradiction on dividing (4.10) by $\sum_{i=0}^{k} a_i(x^*, x^*, x^{**})$ $S^i(x^*, x^*, x^{**})$, for $\lambda \in (0, 1)$. Therefore, x^* is the only fixed point of the mapping T. □

We will now examine some specific instances of Theorem 4.2.3.

Proposition 4.2.4 *Let (X, S) be an S-metric space and let $T : X \to X$ be an S-polynomial type contraction of the first kind. Assume that the following conditions are satisfied:*

(i) $a_0 = 0$ *that is* $a_0(x, y, z) = 0$, *for all* $x, y, z \in X$;
(ii) *For all* $i \in \{1, ..., k\}$, *there exists* $\beta_i > 0$ *such that* $a_i(x, y, z) \leq \beta_i$, *for all* $x, y, z \in X$;
(iii) *There exist* $j \in \{1, ..., k\}$ *and* $\alpha_j > 0$ *such that* $a_j(x, y, z) \geq \alpha_j$, *for all* $x, y, z \in X$.

Then T is continuous.

Proof Let $\{u_n\} \subset X$ be a sequence such that $\lim_{n\to\infty} u_n = u$ that is $\lim_{n\to\infty} S(u_n, u_n, u) = 0$, for some $u \in X$. To prove that T is continuous, we need to show that $\lim_{n\to\infty} S(Tu_n, Tu_n, Tu) = 0$.

Making use of (i) and relation (4.2) for $x = y = u_n$ and $z = u$, we have

$$\sum_{i=0}^{k} a_i(Tu_n, Tu_n, Tu)S^i(Tu_n, Tu_n, Tu) \leq \lambda \sum_{i=0}^{k} a_i(u_n, u_n, u)S^i(u_n, u_n, u), \text{ for all } n \geq 0,$$

which implies by (ii) and (iii) that

$$\alpha_i S^i(Tu_n, Tu_n, Tu) \leq \lambda \sum_{i=0}^{k} \beta_i S^i(u_n, u_n, u), \text{ for all } n \geq 0. \tag{4.11}$$

Taking the limit as $n \to \infty$ in (4.11), we have

$$\lim_{n\to\infty} S^j(Tu_n, Tu_n, Tu) = 0,$$

which is equivalent to

$$\lim_{n\to\infty} S(Tu_n, Tu_n, Tu) = 0,$$

which proves the assertion. □

Corollary 4.2.5 *Let (X, S) be an S-complete metric space and $T : X \to X$ be an S-polynomial type contraction of the first kind. Assume that the following conditions are satisfied:*

(i) $a_0 \equiv 0$ *that is* $a_0(x, y, z) = 0$, *for all* $x, y, z \in X$;
(ii) *For all* $i \in \{1, ..., k\}$, *there exists* $\beta_i > 0$ *such that* $a_i(x, y, z) \leq \beta_i$, *for all* $x, y, z \in X$;
(iii) *There exist* $j \in \{1, ..., k\}$ *and* $\alpha_j > 0$ *such that* $a_j(x, y, z) \geq \alpha_j$, *for all* $x, y, z \in X$.

Then T admits only one fixed point $x^* \in X$. Moreover, for every $x_0 \in X$, the Picard sequence $\{x_n\} \subset X$ defined by $x_{n+1} = Tx_n$, for all $n \geq 0$, converges to x^*.

Remark 4.2.6 Following the procedure used in the proof of Theorem 4.2.3, we can construct similar results for S-polynomial type contractions of the second kind. These results include Theorem 4.2.3, Proposition 4.2.4, and Corollary 4.2.5, adapted to this new type of contraction.

Corollary 4.2.7 *Let (X, S) be a complete metric space and $T : X \to X$ be a given mapping. Assume that there exist $\lambda \in (0, 1)$, a natural number $k \geq 1$ and a finite sequence $\{a_i\}_{i=1}^{k} \subset (0, \infty)$ such that*

$$\sum_{i=0}^{k} a_i S^i(Tx, Ty, Tz) \leq \lambda \sum_{i=0}^{k} a_i S^i(x, y, z), \tag{4.12}$$

for every $x, y, z \in X$. Then T admits a unique fixed point $x^* \in X$. Moreover, for every $x_0 \in X$, the Picard sequence $\{x_n\} \subset X$ defined by $x_{n+1} = Tx_n$, for all $n \geq 0$, converges to x^*.

Remark 4.2.8 Notice that Corollary 4.2.5 reinstates Banach's fixed-point theorem in the framework of S-metric space. Specifically, by setting $k = 1$ and $a_1 = 1$, Eq. (4.12) simplifies to

$$S(Tx, Ty, Tz) \leq \lambda S(x, y, z), \;\; forall \; x, y, z \in X.$$

The following example illustrates Theorem 4.2.3.

Example 4.2.9 Let $X = \{x_1, x_2, x_3, x_4\}$ and $T : X \to X$ be the mapping defined by

$$Tx_1 = x_1, \;\; Tx_2 = x_3, \;\; Tx_3 = x_4, \;\; Tx_4 = x_1.$$

Let X be equipped with discrete S-metric, that is

$$S(x_i, x_j, x_k) = \begin{cases} 0, \; if \; i = j = k; \\ 1, \; otherwise. \end{cases}$$

Consider the mapping $a_0 : X \times X \times X \to [0, \infty)$ given by

$$a_0(x_i, x_j, x_k) = \begin{cases} 0, \; if \; i = j = k; \\ 1, \; if \; i, j, k \in \{1, 4\} \; except \; i = j = k; \\ 6, \; if \; i, j, k \in \{2, 4\} \; except \; i = j = k; \\ 3, \; if \; at \; least \; one \; of \; i, j, k \; is \; equal \; to \; 2, \; and \; one \; is \; different \; from \; 2 \; and \; 4; \\ 2, \; otherwise. \end{cases}$$

Table 4.1 Values in columns of all different possible variations of (i, j, k) for similar values of i, j, and k are the same. For example, the values for (1, 2, 3), (1, 3, 2), and all other possible variations are the same

(i, j, k)	$a_0(Tx_i, Tx_j, Tx_k) + S(Tx_i, Tx_j, Tx_k)$	$a_0(x_i, x_j, x_k) + S(x_i, x_j, x_k)$
(1 ,1, 2)	3	4
(1, 1, 3)	2	3
(1, 2, 2)	3	4
(1, 3, 3)	2	3
(1, 2, 3)	3	4
(1, 2, 4)	3	4
(1, 3, 4)	2	3
(2, 3, 3)	3	4
(2, 4, 4)	3	7
(2, 3, 4)	3	4
(2, 2, 3)	3	4
(2, 2, 4)	3	7
(3, 4, 4)	2	3
(3, 3, 4)	2	3

We claim that

$$a_0(Tx_i, Tx_j, Tx_k) + S(Tx_i, Tx_j, Tx_k)$$

$$\leq \frac{3}{4}(a_0(Tx_i, Tx_j, Tx_k) + S(Tx_i, Tx_j, Tx_k)), \tag{4.13}$$

for every $i, j, k \in \{1, 2, 3, 4\}$, that is, T is a polynomial contraction in the sense of Definition 4.2 with $k = 1$, $a_1 = 1$ and $\lambda = \frac{3}{4}$. When $i = j = k$ or $i, j, k \in \{1, 4\}$, then (4.13) is obvious. Now, we will show that (4.13) holds for all other possible cases. The following Table 4.1 provides different values of $a_0(Tx_i, Tx_j, Tx_k) + S(Tx_i, Tx_j, Tx_k)$ and $a_0(x_i, x_j, x_k) + S(x_i, x_j, x_k)$, for all possible cases with different values of a_0, which confirms (4.13).

Then, all the conditions of Theorem 4.2.3 are satisfied ((ii) is satisfied with $\alpha_1 = 1$). On the other hand, T admits a unique fixed point $x^* = x_1$, which confirms our obtained result.

Remark 4.2.10 Banach's fixed-point theorem in the context of S-metric space is not applicable in this example. Indeed, we have

$$\frac{S(Tx_1, Tx_1, Tx_2)}{S(x_1, x_1, x_2)} = \frac{S(x_2, x_2, x_3)}{S(x_1, x_1, x_2)} = 1.$$

Definition 4.2.11 Let (X, S) be a S-metric space. A self-map T on X is said to be Picard-continuous, if for all $z, w \in X$, we have

$$\lim_{n\to\infty} S(T^n u, T^n u, v) = 0 \implies \lim_{n\to\infty} S(T(T^n u), T(T^n u), Tv) = 0,$$

where $T^0 u = u$ and $T^{n+1} u = T(T^n u)$, for all $n \geq 0$.

Remark 4.2.12 If T is continuous, then T is Picard-continuous. However, the converse is not true. The following example shows this fact.

Example 4.2.13 Let $X = [0, 2]$ equipped with S-metric defined by $S(x, y, z) = |x - z| + |y - z|$, for all $x, y, z \in X$. Consider the mapping $T : X \to X$ defined by

$$Tx = \begin{cases} 0, \; if \; 0 \leq x < 2, \\ \frac{3}{2}, \; if \; x = 2. \end{cases}$$

Clearly, the mapping T is not continuous at $x = 2$. However, T is Picard-continuous which can be easily verified. Indeed, we observe that for all $u \in [0, 2]$, we have

$$T^n u = 0, \;\; for \; all \; n \geq 2.$$

This implies that, if $u, v \in [0, 2]$, we have

$$\lim_{n\to\infty} S(T^n u, T^n u, v) = 0, \implies v = 0$$

and

$$\lim_{n\to\infty} S(T(T^n u), T(T^n u), Tv) = S(T0, T0, T0) = 0.$$

This completes the proof.

Theorem 4.2.14 *Let (X, S) be a complete S-metric space. If the self-map T on X is a polynomial contraction. Assume that the following conditions hold:*

(i) T is Picard-continuous;
(ii) There exist $j \in \{1, ..., k\}$ and $\alpha_j > 0$ such that

$$a_j(x, y, z) \geq \alpha_j, \;\; x, y, z \in X.$$

Then T admits only one fixed point $x^* \in X$. Moreover, for every $x_0 \in X$, the Picard sequence $\{x_n\} \subset X$ defined by $x_{n+1} = Tx_n$, for all $n \geq 0$, converges to x^*.

Proof First, we will prove the existence of a fixed point of T. Let $x_0 \in X$ be a fixed point and Picard sequence defined by

$$x_{n+1} = Tx_n, \;\; for all \; n \geq 0,$$

that is,

$$x_n = T^n x_0, \text{ for all } n \geq 0.$$

From previous Theorem 4.2.3, we know that $\{x_n\}$ is a Cauchy sequence. The completeness of (X, S) implies that there exists $x^* \in X$ such that

$$\lim_{n\to\infty} S(T^n x_0, T^n x_0, x^*) = 0.$$

Then by Picard's continuity, we can conclude that

$$\lim_{n\to\infty} S(T^{n+1} x_0, T^{n+1} x_0, Tx^*) = \lim_{n\to\infty} S(T(T^n x_0), T(T^n x_0), Tx^*) = 0,$$

which implies by the uniqueness of the limit that $Tx^* = x^*$, that is x^* is a fixed point of T. Next, we proceed as in Theorem 4.2.3. □

Now, we present an example to illustrate Theorem 4.2.14

Example 4.2.15 Let $X = [0, 1]$ equipped with S-metric defined by $S(x, y, z) = |x - z| + |y - z|$, $x, y, z \in X$. The self-map $T : X \to X$ is defined by

$$Tx = \begin{cases} \frac{1}{4}, \ if \ 0 \leq 1, \\ 0, \ if \ x = 1. \end{cases}$$

Notice that the mapping T is not continuous. Therefore, Theorem 4.2.3 is not applicable in this example. Also, we observe that T is Picard-continuous (see Example 4.2.9).

Consider now the mapping $a_0 : X \times X \times X \to [0, \infty)$ defined by

$$a_0(x, y, z) = \frac{5}{6}\left(x\left|x - \frac{1}{4}\right| + y\left|y - \frac{1}{4}\right| + z\left|z - \frac{1}{4}\right|\right), \quad x, y, z \in X.$$

We can easily verify for all possible different cases that

$$a_0(Tx, Ty, Tz) + S(Tx, Ty, Tz) \leq \frac{4}{5}(a_0(x, y, z) + S(x, y, z)), \tag{4.14}$$

for all $x, y, z \in X$, that is, T is an S-polynomial type contraction of the first kind with $k = 1$, $a_1 \equiv 1$ and $\lambda = \frac{4}{5}$. Consequently, Theorem 4.2.14 holds. On the other hand, $x^* = \frac{4}{5}$ is the unique fixed point of T, which confirms the obtained result given by Theorem 4.2.14.

Remark 4.2.16 Notice that in this example Theorem 4.2.3 is not applicable because of the discontinuity of T.

4.3 Class of Almost S-Polynomial Type Contraction

In this section, we define the class of almost S-polynomial type contractions in the framework of S-metric space as below.

Definition 4.3.1 Let (X, S) be an S-metric space and $T : X \to X$ be a self-map. We say that T is an almost S-polynomial type contraction of the first kind, if there exists $\lambda \in (0, 1)$, a natural number $k \geq 1$, two finite sequences $\{L_i\}_{i=0}^k$, $\{M_i\}_{i=0}^k \subset (0, \infty)$ and a family of mappings $a_i : X \times X \times X \to [0, \infty)$, for $i = 0, \ldots, k$, such that

$$\sum_{i=0}^k a_i(Tx, Ty, Tz)S^i(Tx, Ty, Tz)$$

$$\leq \lambda \sum_{i=0}^{k} a_i(x, y, z)[S^i(x, y, z) + L_i S^i(z, z, Tx) + M_i S^i(z, z, Ty)], \quad (4.15)$$

for every $x, y, z \in X$.

Definition 4.3.2 Let (X, S) be an S-metric space and $T : X \to X$ be a self-map. We say that T is an almost S-polynomial type contraction of the second kind, if there exists $\lambda \in (0, 1)$, a natural number $k \geq 1$, a finite sequences $\{L_i\}_{i=0}^k \subset (0, \infty)$, and a family of mappings $a_i : X \times X \times X \to [0, \infty)$, for $i = 0, \ldots, k$, such that

$$\sum_{i=0}^k a_i(Tx, Tx, Ty)S^i(Tx, Tx, Ty)$$

$$\leq \lambda \sum_{i=0}^{k} a_i(x, x, y)[S^i(x, x, y) + L_i S^i(y, y, Tx)], \quad (4.16)$$

for every $x, y \in X$.

Next, we will prove our main theorem in this section.

Theorem 4.3.3 *Let (X, S) be a complete S-Metric space, and $T : X \to X$ is an almost S-polynomial type contraction of the first kind. If the following conditions are satisfied:*

(i) T is Picard-continuous;
(ii) There exist $j \in \{1, \ldots, k\}$ and $\alpha_j > 0$ such that $a_j(x, y, z) \geq \alpha_j$, for all $n \geq 0$.

Then T has at least one fixed point.

Proof Let $x_0 \in X$ be a random fixed point and the Picard sequence $\{x_n\} \subset X$ defined by

$$x_{n+1} = Tx_n, \text{ for all } n \geq 0.$$

Putting $x = y = x_0$ and $z = x_1$ in (4.15), we obtain

$\sum_{i=0}^{k} a_i(Tx_0, Tx_0, Tx_1)S^i(Tx_0, Tx_0, Tx_1)$

$$\leq \lambda \sum_{i=0}^{k} a_i(x_0, x_0, x_1)[S^i(x_0, x_0, x_1) + L_i S^i(x_1, x_1, Tx_0) + M_i S^i(x_1, x_1, Tx_0)],$$

that is,

$$\sum_{i=0}^{k} a_i(x_1, x_1, x_2)S^i(x_1, x_1, x_2) \leq \lambda \sum_{i=0}^{k} a_i(x_0, x_0, x_1)S^i(x_0, x_0, x_1). \quad (4.17)$$

Again putting $x = y = x_1$ and $z = x_2$ is (4.15), we obtain

$\sum_{i=0}^{k} a_i(Tx_1, Tx_1, Tx_2)S^i(Tx_1, Tx_1, Tx_2)$

$$\leq \lambda \sum_{i=0}^{k} a_i(x_1, x_1, x_2)[S^i(x_1, x_1, x_2) + L_i S^i(x_2, x_2, Tx_1) + M_i S^i(x_2, x_2, Tx_1)],$$

that is,

$$\sum_{i=0}^{k} a_i(x_2, x_2, x_3)S^i(x_2, x_2, x_3) \leq \lambda \sum_{i=0}^{k} a_i(x_1, x_1, x_2)S^i(x_1, x_1, x_2). \quad (4.18)$$

Hence, by using (4.17) in (4.18), we have

$$\sum_{i=0}^{k} a_i(x_2, x_2, x_3)S^i(x_2, x_2, x_3) \leq \lambda^2 \sum_{i=0}^{k} a_i(x_0, x_0, x_1)S^i(x_0, x_0, x_1).$$

Continuing this process, we get by induction that

$$\sum_{i=0}^{k} a_i(x_n, x_n, x_{n+1})S^i(x_n, x_n, x_{n+1}) \leq \lambda^n \sum_{i=0}^{k} a_i(x_0, x_0, x_1)S^i(x_0, x_0, x_1),$$

which implies from (ii) that

$$S^i(x_n, x_n, x_{n+1}) \leq \lambda^n A_{j,0}, \text{ for all } n \geq 0,$$

where $A_{j,0}$ is given by (4.8). Next, proceeding as in the proof of Theorem 4.2.3, we obtain that $\{x_n\}$ is a Cauchy sequence, which guaranteed the existence of $x^* \in X$ by completeness of (X, S) such that

$$\lim_{n\to\infty} S(x_n, x_n, x^*) = 0.$$

Finally, taking into consideration that the T is Picard-continuous, we get

$$\lim_{n\to\infty} S(Tx_n, Tx_n, Tx^*) = 0,$$

which implies the uniqueness of the limit that $x^* = Tx^*$. This completes the proof. □

Now, we will prove the uniqueness of the fixed point of the self-map T in the following theorem.

Theorem 4.3.4 *Let (X, S) be a complete S-metric space and $T : X \to X$ is an almost S-Contraction type Mapping of the first kind. Then the mapping T has only one fixed point in Theorem 4.3.3, if an additional condition:*

$$\lambda < \frac{1}{\mathcal{N}}, \ where\ \mathcal{N} = \max_i (1 + L_i + M_i),$$

holds with (i) and (ii).

Proof Following the proof of Theorem 4.3.3, let x^* be a fixed point of T. Indeed if it is possible let us consider x^{**} as another fixed point of self-map T.

Now, making use of (4.16) for $x = y = x^*$ and $z = x^{**}$, we obtain

$$\begin{aligned}
&\textstyle\sum_{i=0}^k a_i(Tx^*, Tx^*, Tx^{**})S^i(Tx^*, Tx^*, Tx^{**})\\
&\quad \le \lambda \sum_{i=0}^k a_i(x^*, x^*, x^{**})[S^i(x^*, x^*, x^{**}) + L_i S^i(x^{**}, x^{**}, Tx^*)\\
&\quad + M_i S^i(x^{**}, x^{**}, Tx^*)],
\end{aligned}$$

that is,

$$\begin{aligned}
&\textstyle\sum_{i=0}^k a_i(x^*, x^*, x^{**})S^i(x^*, x^*, x^{**})\\
&\quad \le \lambda \sum_{i=0}^k a_i(x^*, x^*, x^{**})[S^i(x^*, x^*, x^{**}) + L_i S^i(x^{**}, x^{**}, x^*)\\
&\quad + M_i S^i(x^{**}, x^{**}, x^*)]\\
&\quad = \lambda \sum_{i=0}^k a_i(x^*, x^*, x^{**})[1 + L_i + M_i]S^i(x^*, x^*, x^{**})\\
&\quad \le \lambda\mathcal{N} \sum_{i=0}^k a_i(x^*, x^*, x^{**})S^i(x^*, x^*, x^{**}).
\end{aligned} \tag{4.19}$$

Also from (ii), we have

$$\alpha_j S^j(x^*, x^*, x^{**}) \le \sum_{i=0}^{k} a_i(x^*, x^*, x^{**}) S^i(x^*, x^*, x^{**}),$$

since $\alpha_j > 0$ and $S^j(x^*, x^*, x^{**}) \ge 0$, we deduce that

$$\sum_{i=0}^{k} a_i(x^*, x^*, x^{**}) S^i(x^*, x^*, x^{**}) > 0.$$

Then, on dividing (4.19) by $\sum_{i=0}^{k} a_i(x^*, x^*, x^{**}) S^i(x^*, x^*, x^{**})$, we have $\lambda \ge \frac{1}{\mathcal{N}}$, which is a contradiction. Therefore, x^* is the only fixed point of the mapping T. □

Remark 4.3.5 Following the procedure used in the proof of Theorem 4.3.3, we can construct similar results for almost S-polynomial type contractions of the second kind. These results include Theorems 4.3.3, and 4.3.4, adapted to this new type of contraction.

We now investigate some special cases of Theorems 4.3.3 and 4.3.4.

Proposition 4.3.6 *Let (X, S) be an S-metric space and $T : X \to X$ be a self-mapping. Assume that there exist $\lambda \in (0, 1)$, a natural number $k \ge 1$, and three finite sequences $\{a_i\}_{i=1}^{k}$, $\{L_i\}_{i=1}^{k}$, $\{M_i\}_{i=1}^{k} \subset (0, \infty)$ such that*

$$\sum_{i=1}^{k} a_i S^i(Tx, Ty, Tz) \le \lambda \sum_{i=1}^{k} a_i[S^i(x, y, z) + L_i S^i(z, z, Tx) + M_i S^i(z, z, Ty)], \tag{4.20}$$

for every $x, y, z \in X$. Then T is Picard-continuous.

Proof Let $u, v \in X$ be such that

$$\lim_{n\to\infty} S(T^n u, T^n u, v) = 0 \tag{4.21}$$

Making use of (4.20) for $x = y = T^n u$ and $z = v$, we have

$$\sum_{i=1}^{k} a_i S^i(T(T^n u), T(T^n u), Tv)$$

$$\le \lambda \sum_{i=1}^{k} a_i[S^i(T^n u, T^n u, v) + L_i S^i(v, v, T(T^n u)) + M_i S^i(v, v, T(T^n u))],$$

that is,

$\sum_{i=1}^{k} a_i S^i(T(T^n u), T(T^n u), Tv)$

$$\leq \lambda \sum_{i=1}^{k} a_i [S^i(T^n u, T^n u, v) + L_i S^i(v, v, T^{n+1}u) + M_i S^i(v, v, T^{n+1}u)],$$

which implies that

$$S(T(T^n u), T(T^n u), Tv) \leq \frac{\lambda}{a_1} \sum_{i=1}^{k} a_i [S^i(T^n u, T^n u, v) + (L_i + M_i) S^i(v, v, T^{n+1}u)].$$

Then, passing to the limit as $n \to \infty$ in the above inequality and making use of (4.21), we obtain

$$\lim_{n\to\infty} S(T(T^n u), T(T^n u), Tv) = 0,$$

which proves that T is Picard-continuous. □

From Theorems 4.3.3, 4.3.4, and Proposition 4.3.6, we deduce the following result.

Corollary 4.3.7 *Let (X, S) be a complete S-metric space and $T : X \to X$ is a given mapping. If there exist $\lambda \in (0, 1)$, a natural number $k \geq 1$ and three finite sequences $\{a_i\}_{i=1}^k, \{L_i\}_{i=1}^k, \{M_i\}_{i=1}^k \subset (0, \infty)$ such that (4.20) holds for every $x, y, z \in X$. Then T has a fixed point.*

Proof Notice that (4.20) is a special case of (4.15) with $a_0 \equiv 0$ and each a_i is constant. Then, by Proposition 4.3.6, Theorem 4.3.3 applies. □

Corollary 4.3.8 *Let (X, S) be a complete S-metric space, self-map $T : X \to X$ with all assumptions in Corollary 4.3.7 with an extra condition $\lambda < \frac{1}{\mathcal{N}}$, where $\mathcal{N} = \max_i[1 + L_i + M_i]$ have only one fixed point.*

Proof The proof is obvious by the use of Proposition 4.3.6 and Theorem 4.3.4. □

Example 4.3.9 Let $X = \{1, 2, 3\}$ and $T : X \to X$ be the mapping defined by

$$T1 = 1, \;\; T2 = 2, \;\; T3 = 1.$$

The set X is equipped with discrete S-metric defined as

$$S(x, y, z) = \begin{cases} 0, \; if \; x = y = z, \\ 1, \; otherwise. \end{cases}$$

We assert that the mapping T meets condition (4.20) with $k = 2, a_1 = a_2 = 1, \lambda = \frac{2}{3}$ and $L_1 = L_2 = \frac{1}{2} = M_1 = M_2$, that is,

Table 4.2 Values in different variations for (x, y, z)

(x, y, z)	A	B
(1, 2, 2)	2	3
(2, 1, 2)	2	3
(2, 2, 1)	2	4
(1, 1, 2)	2	4
(1, 2, 1)	2	3
(2, 1, 1)	2	3
(2, 3, 3)	2	4
(3, 2, 3)	2	4
(3, 3, 2)	2	4
(2, 2, 3)	2	4
(2, 3, 2)	2	3
(3, 2, 2)	2	3
(1, 2, 3)	2	4
(1, 3, 2)	2	4
(2, 3, 1)	2	3
(3, 2, 1)	2	3
(2, 1, 3)	2	4
(3, 1, 2)	2	4

$$S(Tx, Ty, Tz) + S^2(Tx, Ty, Tz) \leq \frac{2}{3} \begin{Bmatrix} S(x, y, z) + \frac{S(z,z,Tx)+S(z,z,Ty)}{2} \\ +S^2(x, y, z) + \frac{S^2(z,z,Tx)+S^2(z,z,Ty)}{2} \end{Bmatrix}, \tag{4.22}$$

for all $x, y, z \in X$. If $x = y = z$ or $x, y, z \in \{1, 3\}$, then (4.22) is obvious. For all other possible cases the following Table 4.2 gives values of $A = S(Tx, Ty, Tz) + S^2(Tx, Ty, Tz)$ and $B = S(x, y, z) + \frac{1}{2}(S(z, z, Tx) + S(z, z, Ty)) + S^2(x, y, z) + \frac{1}{2}(S^2(z, z, Tx) + S^2(z, z, Ty))$, for all $x, y, z \in X$. Then Corollary 4.3.7 applies. On the other hand, the fixed point set of T is nonempty, which confirms the obtained result provided by Corollary 4.3.7. Remark that the set of points of T is equal to $\{1, 2\}$.

Here is another example that demonstrates Theorem 4.3.3.

Example 4.3.10 Let $X = [0, 2]$ and $T : X \to X$ be the mapping defined by

$$Tx = \begin{cases} 1, & if\ 0 \leq x < 2, \\ 0, & if\ x = 2. \end{cases}$$

Let S-metric be defined on X as

$$S(x, y, z) = |x - z| + |y - z|, \quad x, y, z \in X.$$

Notice that T is Picard-continuous (see Example 4.2.13).

Let $a_0 : X \times X \times X \to [0, \infty)$ be defined as

$$a_0(x, y, z) = |2x^2 - 3x + 1| + |2y^2 - 3x + 1| + z|z - 1|, \; forall\; x, y \in X.$$

Now we will show that,

$$a_0(Tx, Ty, Tz) + S(Tx, Ty, Tz)$$

$$\leq \frac{1}{2}[3a_0(x, y, z) + S(x, y, z) + S(z, z, Tx) + S(z, z, Ty)], \tag{4.23}$$

for all $x, y, z \in X$, that T is almost S-polynomial type contraction of the first kind with $k = 1,\ a_1 \equiv 1,\ L_0 = L_1 = 1 = M_0 = M_1$ and $\lambda = \frac{1}{2}$.

Now we will discuss all possible cases. For the cases $0 \leq x, y, z < 2$ and $x = y = z = 2$, it is obvious to (4.23). All other possible cases are discussed below:

- Case 1.
  $0 \leq x, y < 2$, and $z = 2$;
  $a_0(Tx, Ty, T2) + S(Tx, Ty, T2)$
  $= a_0(1, 1, 0) + S(1, 1, 0)$
  $= 2$
  $= \frac{1}{2}[S(2, 2, 1) + S(2, 2, 1)].$
  $= \frac{1}{2}[S(2, 2, Tx) + S(2, 2, Ty)]$
  $\leq \frac{1}{2}[3a_0(x, y, 2) + S(x, y, 2)$
  $+S(2, 2, Tx) + S(2, 2, Ty)].$

- Case 2.
  $0 \leq x, z < 2$, and $y = 2$;
  $a_0(Tx, T2, Tz) + S(Tx, T2, Tz)$
  $= a_0(1, 0, 1) + S(1, 0, 1)$
  $= 2$
  $\leq 3 \leq \frac{3\times 3}{2} \leq \frac{1}{2}[3a_0(x, 2, z)]$
  $\leq \frac{1}{2}[3a_0(x, 2, z) + S(x, 2, z)$
  $+S(z, z, Tx) + S(z, z, T2)].$

- Case 3.
  $0 \leq y, z < 2$, and $x = 2$;
  $a_0(T2, Ty, Tz) + S(T2, Ty, Tz)$
  $= a_0(0, 1, 1) + S(0, 1, 1)$
  $= 2$
  $\leq 3 \leq \frac{3\times 3}{2} \leq \frac{1}{2}[3a_0(2, y, z)]$
  $\leq \frac{1}{2}[3a_0(2, y, z) + S(2, y, z)$
  $+S(z, z, T2) + S(z, z, Ty)].$

- Case 4.
  $0 \leq x < 2$, and $y = z = 2$;
  $a_0(Tx, T2, T2) + S(Tx, T2, T2)$
  $= a_0(1, 0, 0) + S(1, 0, 0)$
  $= 2$
  $\leq \frac{6}{2} \leq \frac{1}{2}[S(2, 2, Tx) +$
  $S(2, 2, T2)]$
  $\leq \frac{1}{2}[3a_0(x, 2, 2) + S(x, 2, 2)$
  $+S(2, 2, Tx) + S(2, 2, T2)].$

- Case 5.
  $0 \leq y < 2$, and $x = z = 2$;
  $a_0(T2, Ty, T2) + S(T2, Ty, T2)$
  $= a_0(0, 1, 0) + S(0, 1, 0)$
  $= 2$
  $\leq \frac{6}{2} \leq \frac{1}{2}[S(2, 2, T2) +$
  $S(2, 2, Ty)]$
  $\leq \frac{1}{2}[3a_0(2, y, 2) + S(2, y, 2)$
  $+S(2, 2, T2) + S(2, 2, Ty)].$

- Case 6.
  $0 \leq z < 2$, and $x = y = 2$;
  $a_0(T2, T2, Tz) + S(T2, T2, Tz)$
  $= a_0(0, 0, 1) + S(0, 0, 1)$
  $= 4$
  $\leq 6 \leq \frac{3\times 6}{2} \leq \frac{1}{2}[3a_0(2, 2, z)]$
  $\leq \frac{1}{2}[3a_0(2, 2, z) + S(2, 2, z)$
  $+S(z, z, T2) + S(z, z, T2)].$

Consequently, (4.23) is satisfied for all $x, y \in X$. Notice also that condition (ii) of Theorem 4.3.3 holds for $j = 1$ and $\alpha_1 = 1$. Then Theorem 4.3.3 applies. On the other hand, $x^* = 1$ is a fixed point of T, which verifies our obtained result Theorem 4.3.3.

4.4 An Application to The Integral Equation

Integral equations with two parameters are fundamental in various scientific and engineering disciplines. These equations typically involve an unknown function that appears within an integral and is parameterized by two variables. They are particularly useful in solving complex boundary value problems and are employed in fields such as fluid dynamics, heat conduction, and electromagnetics. The versatility of these equations allows for the modeling of diverse phenomena, from wave propagation to potential theory.

The study of integral equations with two parameters often requires sophisticated mathematical techniques for their solution (Singh et al. 2023, 2024). Methods such as numerical approximation (Alam and Rohen 2024b; Alam et al. 2024c), series expansion, and transform methods are commonly used to tackle these equations. Understanding their properties and behavior is crucial for accurate modeling and simulation in applied sciences. The continued development of analytical and computational tools for solving integral equations ensures their ongoing relevance and applicability in both theoretical research and practical applications. In this section, we demonstrate an application of the main result by describing the solution of the integral equation.

Consider $\mathcal{M} = C(I, \mathbb{R})$, which represents the set of all continuous functions on $I = [0, 1]$, equipped with the metric $S(x, y, z) = \sup\{|x(s) - z(s)| + |y(s) - z(s)|,\ s \in I\}$, for all $x, y, z \in \mathcal{M}$. It is important to note that $(\mathcal{M}, S)$ constitutes a complete S-metric space. Within this framework, we explore the integral equation:

$$x(s) = \chi(s) + \int_0^1 \mathcal{K}(s, u)\eta(u, x(u))du,\ s \in I, \tag{4.24}$$

where, the functions $\eta : I \times \mathbb{R} \to \mathbb{R}$ and $\chi : I \to \mathbb{R}$ are continuous, and $\mathcal{K} : I \times I \to \mathbb{R}_0^+$ is a function satisfying $\mathcal{K}(s, \cdot) \in L^1(I)$, for all $s \in I$. We examine the mapping $T : \mathcal{M} \to \mathcal{M}$, which is defined as follows:

$$T(x)(s) = \chi(s) + \int_0^1 \mathcal{K}(s, u)\eta(u, x(u))du,\ s \in I. \tag{4.25}$$

Theorem 4.4.1 *Equation (4.25) has at least one solution in $\mathcal{M}$, if for all $u \in I$ and $a, b \in \mathcal{R}$, we have*

$$|\eta(u, x(u)) - \eta(u, y(u))| \leq |x(u) - y(u)|, \ \textit{for all } u \in I,$$

followed by the assumed inequality

$$\sup_{s \in I} \int_0^1 \mathcal{K}(s, u) du = \lambda,$$

where $\lambda \in (0, 1)$ *is a fixed constant.*

Proof Note that finding the solution of (4.24) is equivalent to finding $x^* \in \mathcal{M}$ that is a fixed point of T. Now, let $x, y \in \mathcal{M}$, on account of the above inequalities, we find that

$$\begin{aligned}
|T(x)(s) - T(z)(s)| &+ |T(y)(s) - T(z)(s)| \\
&= \left| \int_0^1 \mathcal{K}(s, u)[\eta(u, x(u)) - \eta(u, z(u))] du \right| \\
&+ \left| \int_0^1 \mathcal{K}(s, u)[\eta(u, y(u)) - \eta(u, z(u))] du \right| \\
&\leq \int_0^1 \mathcal{K}(s, u) du.[|\eta(u, x(u)) - \eta(u, z(u))| + |\eta(u, y(u)) - \eta(u, z(u))|] \\
&\leq \int_0^1 \mathcal{K}(s, u) du.[|x(u) - z(u)| + |y(u) - z(u)|] \\
&\leq \lambda[|x(u) - z(u)| + |y(u) - z(u)|],
\end{aligned}$$

which implies that

$$|T(x)(s) - T(z)(s)| + |T(y)(s) - T(z)(s)| \leq \lambda[|x(u) - z(u)| + |y(u) - z(u)|].$$

Taking supremum both sides, we get

$$S(Tx, Ty, Tz) \leq \lambda S(x, y, z).$$

Now

$$\begin{aligned}
S(Tx, Ty, Tz) + 2S^2(Tx, Ty, Tz) &\leq \lambda S(x, y, z) + 2\lambda^2 S^2(x, y, z) \\
&\leq \lambda[S(x, y, z) + 2 S^2(x, y, z)],
\end{aligned}$$

implies

$$\sum_{i=0}^{2} a_i(Tx, Ty, Tz) S(Tx, Ty, Tz) \leq \lambda \sum_{i=0}^{2} a_i(x, y, z) S(x, y, z),$$

for all $u \in I$ and $x, y \in \mathcal{M}$.

This shows that T is S-Polynomial type contraction mapping for $k = 2$ with $a_0 = \lambda, a_1 = 1$ and $a_2 = 2$. Since T is a continuous mapping, it satisfies all the assumptions of Theorem 4.2.3 with $\alpha_0 = \lambda > 0$. Therefore, we can conclude that there exists an $x^* \in \mathcal{M}$ such that $x^* = Tx^*$. In other words, x^* is a solution to the integral equation (4.24). □

4.5 Conclusion

In this study, we introduced a novel type of contraction within S-metric spaces, termed S-polynomial contractions. We developed two fixed-point theorems for these mappings, building upon and extending previous findings, particularly those by Jleli. Initially, we focused on cases where the mapping T is continuous. We then extended our investigation to scenarios where the continuity condition was relaxed, thereby broadening the applicability of our results. Furthermore, we introduced the concept of almost S-polynomial contraction mappings in S-metric spaces and established corresponding fixed-point theorems for these mappings. To demonstrate the validity and practical relevance of our results, we provided several illustrative examples. This work significantly broadened the understanding and utility of fixed-point theory in S-metric spaces. Finally, we demonstrated an application of the main result by describing the solution of the integral equation.

Conflict of Interest Not applicable.

Acknowledgements The first author is thankful to the University Grant Commission (UGC), India, for financial support to this work under the Senior Research Fellowship through the award ID 1174/(CSIRNETJUNE2019) and the second author is thankful to IISER Berhampur for financial support to this work under the Post-Doctoral Research Fellowship through the employment ID C.B0160.

References

Alam, K. H., Rohen, Y., & Tomar, A. (in press). $(\alpha, F)-$Geraghty type generalized $F-$contractions on non-Archimedean fuzzy metric-unlike spaces. *Demonstratio Mathematica*.

Alam, K. H., Rohen, Y., Kallel, I. A., & Ahmad, J. (2024). Solution of an algebraic linear system of equations using fixed point results in C^*-algebra valued extended Branciari S_b-metric spaces. *International Journal of Analysis and Applications, 22*, 139. https://doi.org/10.28924/2291-8639-22-2024-139

Alam, K. H., Rohen, Y., Tomar, A., & Sajid, M. (in press). On fixed points in M_v^b-metric space and solutions to nonlinear matrix equations and differential equations related to beam theory. *Journal of Nonlinear and Convex Analysis*.

Alam, K. H., & Rohen, Y. (2024). An efficient iterative procedure in hyperbolic space and application to non-linear delay integral equation. *Journal of Applied Mathematics and Computing*. https://doi.org/10.1007/s12190-024-02134-z

Alam, K. H., Rohen, Y., & Saleem, N. (2023). Fixed points of (α, β, F^*) and (α, β, F^{**})-weak Geraghty contractions with an application. *Symmetry,15*(1), 243. https://doi.org/10.3390/sym15010243

Alam, K. H., Rohen, Y., Saleem, N., Aphane, M., & Razzaque, A. (2024). Convergence of Fibonacci-Ishikawa iteration procedure for monotone asymptotically non-expansive mappings. *Journal of Inequalities and Applications,2024*, 81. https://doi.org/10.1186/s13660-024-03156-8

Alam, K. H., Rohen, Y., & Tomar, A. (2024). On fixed point and its application to the spread of infectious diseases model in M_v^b−metric space. *Mathematical Methods in the Applied Science,47*(7), 6489–6503. https://doi.org/10.1002/mma.9933

Banach, S. (1992). Sur les opérations dans les ensembles abstraits et leur applications auxéquations intégrales. *Fundamenta Mathematicae,3*, 133–181.

Berinde, V. (2004). Approximating fixed points of weak contractions using the Picard iteration. *Nonlinear Analysis Forum,9*(1), 43–53.

Boyd, D. W., & Wong, J. S. W. (1969). On nonlinear contractions. *Proceedings of the American Mathematical Society,20*, 458–464.

Branciari, A. (2000). A fixed point theorem of Banach-Caccioppoli type on a class of generalized metric spaces. *Publicationes Mathematicae Debrecen,57*, 31–37.

Chand, D., & Rohen, Y. (2023, May). Fixed points of $\alpha_s - \beta_s - \psi$-contractive mappings in S-metric spaces. *Nonlinear Functional Analysis and Applications,28*, 571–587.

Chand, D., Rohen, Y., Saleem, N., Aphane, M., & Razzaque, A. (2024, April 18). S-Pata-type contraction: a new approach to fixed-point theory with an application. *Journal of Inequalities and Applications, 2024*(1), 59.

Chand, D., & Rohen, Y. (2024). Paired contractive mappings and fixed point results. *AIMS Mathematics,9*(1), 1959–1968.

Chatterjea, S. K. (1972, January 1). Fixed-point theorems. *Dokladi na Bolgarskata Akademiya na Naukite, 25*(6), 727–730.

Ciric, L. B. (1971). Generalized contractions and fixed-point theorems. *Publications de l'Institut Mathematique,12*(26), 19–26.

Ćirić, L. B. (1974). A generalization of Banach's contraction principle. *Proceedings of the American Mathematical Society,45*(2), 267–273.

Czerwik, S. (1993). Contraction mappings in b-metric spaces. *Acta Mathematica et Informatica Universitatis Ostraviensis,1*(1), 5–11.

Deep, C., Yumnam, R., & Nicola, F. (2024). Zamfirescu mappings under Pata-type condition: results and application to an integral equation. *Vojnotehnički glasnik,72*(2), 583–604.

Dhage, B. C. (1992). Generalized metric space and mapping with fixed point. *Bulletin of the Calcutta Mathematical Society,84*, 329–336.

Hieu, N. T., Ly, N. T., & Dung, N. V. (2015). A generalization of Ciric quasi contractions for maps on S-metric spaces. *Thai Journal of Mathematics,13*(2), 369–380.

Joshi, M., Tomar, A., Nabwey, H. A., & George, R. (2021). On unique and nonunique fixed points and fixed circles in M_v^b-metric space and application to cantilever beam problem. *Journal of Function Spaces, 2021*, 15. Art. ID 6681044. https://doi.org/10.1155/2021/6681044

Joshi, M., Tomar, A., & Abdeljawad, T. (2023). On Fixed Point, its geometry and application to satellite web coupling problem in S-Metric Spaces. *AIMS Mathematics,8*(2), 4407–4441. https://doi.org/10.3934/math.20232202023501

Joshi, M., Tomar, A., & Uddin, I. (2023). Fixed point in M_v^b-Metric space and applications. *Acta Universitatis Sapientiae, Mathematica,15*(2), 272–287. https://doi.org/10.2478/ausm-2023-00015

Nadler, S. B., Jr. (1969). Multi-valued contraction mappings. *Pacific Journal of Mathematics,30*, 475–488.

Özgür, N. Y., & Taş, N. (2017). Some new contractive mappings on S-metric spaces and their relationships with the mapping (S25). *Mathematical Sciences,11*(1), 7–16.

Rakotch, E. (1962). A note on contractive mappings. *Proceedings of the American Mathematical Society,13*, 459–465.

Reich, S. (1972). Fixed points of contractive functions. *Bollettino dell'Unione Matematica Italiana,5*, 26–42.

Rus, I. A. (2001). *Generalized contractions and applications*. Cluj-Napoca: Cluj University Press.

Sedghi, S., Shobe, N., & Aliouche, A. (2012). A generalization of fixed point theorems in S-metric spaces. *Matematički Vesnik,64*(3), 258–266.

Singh, M. P., Rohen, Y., Alam, K. H., Ahmad, J., & Emam, W. (2024). On fixed point and an application of C^*-algebra valued $(\alpha, \beta)-$Bianchini-Grandolfi gauge contractions. *AIMS Mathematics,9*(6), 15172–15189. https://doi.org/10.3934/math.2024736

Singh, M. P., Rohen, Y., Saleem, N., Alam, K. H., Singh, K. A., & Razzaque, A. (2023). On fixed point equations involving Geraghty type contractions with solution to integral equation. *Mathematics,11*(24), 4882. https://doi.org/10.3390/math11244882

Tomar, A., Tas, N., & Joshi, M. (2023). On interpolative type non-unique fixed points, their geometry and applications on S-metric spaces. *Applied Mathematics E-Notes,23*, 243–249.

Chapter 5
Common Fixed Points for Asymptotically Regular Kannan ν-Lipschitzian Semigroup in the Variable Exponent Sequence Spaces

Amit Gangwar and Shivam Rawat

Abstract In this article, we introduce the concept of Kannan ν-Lipschitzian semigroup in a variable exponent sequence space $l_{p(.)}$ and prove some common fixed point results for these newly defined mappings. These results generalize various results in the existing literature.

Keywords Variable exponent sequence space $l_{p(.)}$ · Kannan ν-Lipschitzian semigroup · Asymptotically regular

AMS Subject Classification 47H09 · 47H10

5.1 Introduction and Preliminaries

In 1922, the Banach contraction principle (BCP) was introduced by Banach (1922), which is a very powerful result in the field of metric fixed point theory because of its applications in several applied mathematical analysis and nonlinear analysis domains. This method has undergone extensive development and extension in the last few decades. It makes sense to wonder whether a discontinuous map satisfying a contractive type condition has a fixed point or not because the map that follows the BCP is continuous. In response to this query on a metric space $(\mathfrak{X}, \mathfrak{d}^*)$, Kannan (1969) defined a contractive condition for a discontinuous map T as follows:

$$\mathfrak{d}^*(Ta, Tb) \leq c[\mathfrak{d}^*(a, Ta) + \mathfrak{d}^*(b, Tb)],$$

A. Gangwar
Department of Mathematics, Bharat Institute of Engineering and Technology, Mangalpally, Hyderabad, Telangana, India

S. Rawat (✉)
Department of Mathematics, Graphic Era (Deemed to be University), Dehradun, Uttarakhand, India
e-mail: rawat.shivam09@gmail.com

© The Author(s), under exclusive license to Springer Nature Singapore Pte Ltd. 2026
A. Tomar and M. Jain (eds.), *Applications of Fixed-Point Theorem*, Industrial and Applied Mathematics, https://doi.org/10.1007/978-981-95-5160-6_5

for all $a, b \in \mathfrak{X}$ and $c \in [0, \frac{1}{2})$. Moreover, some existence and uniqueness results in the framework of a complete metric space were also proved by Kannan. The "variable exponent sequence spaces" was introduced in 1931 by Orlicz (1931). "Variable exponent sequence spaces" extend the concept of classical sequence spaces by allowing the exponents to vary in the norm. These spaces find applications in functional analysis, particularly in the study of sequences with varying growth conditions. The norm in these spaces is defined in terms of variable exponents associated with the sequence elements. The vector space

$$l_{p(.)} = \{\{a_n\} \in \mathbb{R}^{\mathbb{N}}; \sum_{n=0}^{\infty} \frac{1}{p(n)} |\lambda a_n|^{p(n)} < \infty \text{ for some } \lambda > 0\},$$

where $p : \mathbb{N} \to \infty$ was defined by Orlicz. Many authors have studied these spaces so far (see, for instance, Diening et al. 2011, El Amri and Khamsi 2022, Klee 1965, Sundaresan 1971, Waterman et al. 1969). In 1950, Nakano (1950) gave the formal definition of a modular. Modular vector spaces have also been explored by several authors (see, for instance, Khamsi and Kozlowski 2015, Kozlowski 1988, Nakano 1951).

Definition 5.1.1 (Orlicz 1931) A sequence vector space is defined as

$$l_{p(.)} = \{\{a_n\} \in \mathbb{R}^{\mathbb{N}}; \sum_{n=0}^{\infty} \frac{1}{p(n)} |\lambda a_n|^{p(n)} < \infty \ for\ some\ \lambda > 0\},$$

where $p : \mathbb{N} \to \infty$.

The term "variable exponent sequence spaces" was not employed by Orlicz (1931) for $l_{p(.)}$. It was in a subsequent phase that these types of spaces became integral to the emerging theory of variable exponent spaces.

Definition 5.1.2 (Klee 1965, Sundaresan 1971, Nakano 1950) Let $l_{p(.)}$ be a vector space and define a function $\nu : l_{p(.)} \to [0, \infty)$ by

$$\nu(a) = \nu(\{a_n\}) = \sum_{n=0}^{\infty} \frac{1}{p(n)} |a_n|^{p(n)}.$$

Then the below conditions hold:
1. $\nu(a) = 0$ if and only if $a = 0$;
2. $\nu(\gamma a) = \nu(a)$, if $|\gamma| = 1$;
3. $\nu(sa + (1 - s)b) \leq s\nu(a) + (1 - s)\nu(b)$, for all $s \in [0, 1]$, and $a, b \in l_{p(.)}$.

In this situation, it is stated that ν is a convex modular.

Definition 5.1.3 (Khamsi et al. 1990) Let $l_{p(.)}$ be a vector space.
1. Any sequence $\{a_n\} \subseteq l_{p(.)}$ is said to be ν-convergent to a if and only if $\nu(a_n - a) = 0$.
It is clear that if the ν-limit exists, then it is unique.
2. Any sequence $\{a_n\} \subseteq l_{p(.)}$ is said to be ν-Cauchy if $\nu(a_n - a_m) = 0$ for all $m, n \in \mathbb{N}$.
3. Any set K in $l_{p(.)}$ is said to be ν-closed, if for any ν-convergent sequence $\{a_n\} \subseteq K$, which converges to a, we have a in K.

Observe that the property Fatou is satisfied by ν. It states that for any $a \in l_{p(.)}$ and any ν-convergent sequence $\{a_n\} \subseteq l_{p(.)}$ which converges to a^*, we obtain

$$\nu(a - a^*) \leq \liminf_{n\to\infty} \nu(a - a_n).$$

In 2020, Abdou and Khamsi (2020) introduced the notion of Kannan ν-Lipschitzian mappings.

Definition 5.1.4 (Abdou and Khamsi 2020) Let K be a nonempty subset of $l_{p(.)}$. A self-mapping $T : K \to K$ is known as Kannan ν-Lipschitzian if for any $a, b \in K$, some $L > 0$ exists for which

$$\nu(Ta - Tb) \leq L(\nu(a - Ta) + \nu(b - Tb)).$$

We state that T is:

(i) Kannan ν-nonexpansive if $L = \frac{1}{2}$;
(ii) Kannan ν-contraction if $L < \frac{1}{2}$.

The most important result of Abdou and Khamsi (2020) is given next.

Theorem 5.1.5 (Abdou and Khamsi 2020) *Consider a self-mapping T on K to be a Kannan ν-contraction mapping, where K is a nonempty ν-closed subset of $l_{p(.)}$. Let $r \in K$ be such that $\nu(r - Tr) < +\infty$. Then, $\{T^n r\}$ is ν-convergent to some $w \in K$. Moreover, we get $\nu(r - Tr) < +\infty$ or $\nu(r - Tr) = 0$.*

5.2 Main Results

Abdou and Khamsi (2020) proved Theorem 5.1.5 for a Kannan ν-Lipschitzian self-map on a nonempty ν-closed subset of $l_{p(.)}$. Now, we will prove the results for a class of mappings known as Kannan ν-Lipschitzian semigroups. We start by giving the following definition.

Definition 5.2.1 Let K be a nonempty subset of a variable exponent sequence space $l_p(.)$. A one-parameter family of mappings $S_K = \{T_t : t \geq 0\}$ from K to K is called Kannan ν-Lipschitzian semigroup on K if it fulfills the given assumptions:

(i) $T_0(a) = a$ for all $a \in K$;
(ii) $T_{s+t}(a) = T_s T_t(a)$ for all $s, t \geq 0$;
(iii) $\nu(T_t(a) - T_t(b)) \leq L(\nu(a - T_t(a)) + \nu(b - T_t(b))$ for any $a, b \in K, t \geq 0$ and $0 \leq L < \frac{1}{2}$.

Theorem 5.2.2 *Let S_K be Kannan ν-Lipschitzian semigroup on a nonempty ν-closed subset K of a variable exponent sequence space $l_{p(.)}$. Let for all $r \in K$, $\nu(r - T_{t_0}(r)) < +\infty$ for some $t_0 \geq 0$. Then, a common fixed point exists for S_K.*

Proof Let $r \in K$. Then there exists $t_0 \geq 0$ such that $\nu(r - T_{t_0}(r)) < +\infty$.

Claim: $\{T_{t_0}^n r\}$ is a ν-Cauchy sequence.

Since S_K is Kannan ν-Lipschitzian semigroup, this implies that there exists $L \in [0, \frac{1}{2})$ in such a way that

$$\nu(T_{t_0}a - T_{t_0}b) \leq L(\nu(a - T_{t_0}a) + \nu(b - T_{t_0}b)),$$

for any $a, b \in K$. Take $\theta = \frac{L}{1-L}$. Observe that $\theta < 1$, and taking $a = T_{t_0}^{n-1}r$ and $b = T_{t_0}^n r$, we get

$$\nu(T_{t_0}^n r - T_{t_0}^{n+1} r) \leq L(\nu(T_{t_0}^{n-1} r - T_{t_0}^n r) + \nu(T_{t_0}^n r - T_{t_0}^{n+1} r)).$$

This implies

$$\begin{aligned}\nu(T_{t_0}^n r - T_{t_0}^{n+1} r) &\leq \frac{L}{1-L}\nu(T_{t_0}^{n-1} r - T_{t_0}^n r) \\ &= \theta\nu(T_{t_0}^{n-1} r - T_{t_0}^n r),\end{aligned}$$

for any $n \geq 1$. Hence, we have

$$\nu(T_{t_0}^n r - T_{t_0}^{n+1} r) \leq \theta^n \nu(r - T_{t_0} r),$$

for all $n \in \mathbb{N}$. As S_K is Kannan ν-Lipschitzian semigroup, we obtain

$$\nu(T_{t_0}^n r - T_{t_0}^{n+h} r) \leq L(\nu(T_{t_0}^{n-1} r - T_{t_0}^n r) + \nu(T_{t_0}^{n+h-1} r - T_{t_0}^{n+h} r)). \tag{5.1}$$

This implies

$$\nu(T_{t_0}^n r - T_{t_0}^{n+h} r) \leq L(\theta^{n-1}\nu(r - T_{t_0} r) + \theta^{n+h-1}\nu(r - T_{t_0} r)), \tag{5.2}$$

for $h \in \mathbb{N}$ and any $n \geq 1$. As $k < 1$ and $\nu(r - T_{t_0}(r)) < \infty$, we infer that $\{T_{t_0}^n r\}$ is a ν-Cauchy sequence.

Let the sequence $\{T_{t_0}^n r\}$ has a limit $w \in K$.

$$\begin{aligned}\nu(T_{t_0}^n r - T_{t_0} w) &\leq L(\nu(T_{t_0}^{n-1} r - T_{t_0}^n r) + \nu(w - T_{t_0} w)) \\ &\leq L(\theta^{n-1}\nu(r - T_{t_0} r) + \nu(w - T_{t_0} w)),\end{aligned}$$

for every $n \geq 1$. From Fatou's property, we obtain

$$\begin{aligned}\nu(w - T_{t_0}w) &\leq \liminf_{n\to\infty} \nu(T_{t_0}^n r - T_{t_0}w) \\ &\leq L\nu(w - T_{t_0}w).\end{aligned}$$

This implies that w is a fixed point of T_{t_0}.

As $L < \frac{1}{2}$, we deduce that $\nu(w - T_{t_0}w) = 0$.

Let $t \geq 0$, then we have $t = nt_0 + \epsilon$, for $0 \leq \epsilon < t_0$ and $n \geq 1$. Then

$$\begin{aligned}\nu(T_t w - w) &= \nu(T_{nt_0+\epsilon} w - w) \\ &= \nu((T_\epsilon . T_{nt_0})w - w) \\ &= \nu(T_\epsilon . T_{t_0}^n w - T_{t_0}^n w) \\ &\leq L(\nu(T_\epsilon . T_{t_0}^n w - T_{t_0}^n w) + \nu(T_{t_0}^n w - w)) \\ &\leq L(\theta^n \nu(T_\epsilon w - w) + \nu(T_{t_0}^n w - w)),\end{aligned}$$

By Fatou's property, we obtain

$$\begin{aligned}\nu(T_t w - w) &\leq L \liminf_{n\to\infty} \nu(T_{t_0}^n w - w) \\ &= L\nu(w - w).\end{aligned}$$

Thus, $\nu(T_t w - w) = 0$. Since $t \geq 0$ was arbitrary, we conclude that w is a common fixed point of S_k. □

Note that, if we expand the condition $\nu(r - T_{t_0}(r)) < +\infty$ for some $t_0 \geq 0$ to $\nu(r - T_t(r)) < +\infty$ for all $t \geq 0$, then we have the following corollary.

Corollary 5.2.3 *Let S_K be Kannan ν-Lipschitzian semigroup on a nonempty ν-closed subset K of a variable exponent sequence space $l_{p(.)}$. Let $r \in K$ in such a way that $\nu(r - T_t(r)) < +\infty$, $\forall\, t \geq 0$. Also, if w is a common fixed point of S_K and L is a Kannan constant related to the semigroup S_K, we obtain*

$$\nu(T_t^n r - w) \leq L\left(\frac{L}{1-L}\right)^n \nu(r - T_t r),$$

for all $t \geq 0$, $n \geq 1$ and $r \in K$.

Proof As $t_0 \geq 0$, from the inequality (5.2) and using Fatou's property, we have

$$\nu(T_t^n r - w) \leq L\theta^n \nu(r - T_t r) = L\left(\frac{L}{1-L}\right)^n \nu(r - T_t r),$$

for all $t \geq 0$, $n \geq 1$ and $r \in K$. □

Definition 5.2.4 Let S_K be the Kannan ν-Lipschitzian semigroup on a nonempty, ν-closed subset of $l_{p(.)}$. Then the semigroup S_K is called asymptotically regular if

$$\lim_{t\to\infty} \nu(T_t r - T_{t+s} r) = 0,$$

for all $s \geq 0$ and $r \in K$.

Theorem 5.2.5 *Let S_K be Kannan ν-Lipschitzian asymptotically regular semigroup on a nonempty ν-closed subset K of a variable exponent sequence space $l_p(.)$. Let there exists $w \in K$ such that $\nu(w - T_t(w)) < +\infty$, for all $t \geq 0$. Then, a common fixed point w exists for S_K.*

Proof Let $t_0 \geq 0$ be an arbitrary element. If $t_0 = 0$, then nothing to prove, i.e., every point in K is the fixed point for T_0.

Let us assume that $t_0 > 0$. From the inequality (5.1), we have

$$\nu(T_{t_0}^n r - T_{t_0}^{n+h} r) \leq L(\nu(T_{t_0}^{n-1} r - T_{t_0}^n r) + \nu(T_{t_0}^{n+h-1} r - T_{t_0}^{n+h} r), \tag{5.3}$$

for all $n \geq 1$, $h \in \mathbb{N}$ and $r \in K$. This is equivalent to

$$\nu(T_{nt_0} r - T_{nt_0+ht_0} r) \leq L(\nu(T_{(n-1)t_0} r - T_{nt_0} r) + \nu(T_{(nt_0+(h-1)t_0} r - T_{nt_0+ht_0} r), \tag{5.4}$$

for all $nt_0 > 0$ and $(h-1)t_0 \geq 0$. As the semigroup S_K is asymptotically regular, so if $n \to \infty$, then $nt_0 \to \infty$ and hence we conclude that $\{T_{t_0}^n r\}$ is a ν-Cauchy sequence.

Let the sequence $\{T_{t_0}^n r\}$ has the limit $w \in K$.

$$\begin{aligned} \nu(T_{t_0}^n r - T_{t_0} w) &\leq L(\nu(T_{t_0}^{n-1} r - T_{t_0}^n r) + \nu(w - T_{t_0} w) \\ &\leq L(\theta^{n-1} \nu(r - T_{t_0} r) + \nu(w - T_{t_0} w), \end{aligned}$$

for every $n \geq 1$. By Fatou's property, we obtain

$$\begin{aligned} \nu(w - T_{t_0} w) &\leq \liminf_{n\to\infty} \nu(T_{t_0}^n r - T_{t_0} w) \\ &\leq L\nu(w - T_{t_0} w). \end{aligned}$$

As $L < \frac{1}{2}$, we deduce that $\nu(w - T_{t_0} w) = 0$. Thus, w is a fixed point of T_{t_0}.

As $t_0 \geq 0$ is arbitrary, therefore w is a common fixed point of S_K. □

Conflict of interest The authors declare that they have no conflict of interest.

Acknowledgements The first author would like to thank UGC, New Delhi, India, for financial support.

References

Abdou, A. A. N., & Khamsi, M. A. (2020). Fixed points of Kannan maps in the variable exponent sequence spaces $l_{p(.)}$. *Mathematics*, *8*, 375.

Banach, S. (1922). Sur les opérations dans les ensembles abstraits et leurs applications. *Fundamenta Mathematicae,3*, 133–181.

Diening, L., Harjulehto, P., Hästö, P., & Ruzicka, M. (2011). Lebesgue and Sobolev spaces with variable exponents. Lecture Note in Mathematics 2017. Springer.

El Amri, A., & Khamsi, M. A. (2022). New modular fixed point theorem in the variable exponent spaces $l_{p(.)}$. *Mathematics*, *10*, 869.

Kannan, R. (1969). Some results on fixed points- II. *The American Mathematical Monthly,76*, 405–408.

Khamsi, M.A., & Kozlowski, W. M. (2015). *Fixed point theory in modular function spaces*. Birkhauser.

Khamsi, M. A., Kozlowski, W. K., & Reich, S. (1990). Fixed point theory in modular function spaces. *Nonlinear Analysis,14*, 935–953.

Klee, V. (1965). Summability in $l_{(p_1,p_2,...)}$ Spaces. *Studia Mathematica,25*, 277–280.

Kozlowski, W. M. (1988). Modular function spaces. In *Series of monographs and textbooks in pure and applied mathematics* (Vol. 122). Dekker

Nakano, H. (1950). *Modulared semi-ordered linear spaces*. Maruzen Co.

Nakano, H. (1951). Modulared Sequence Spaces. *Proceedings of the Japan Academy,27*, 508–512.

Orlicz, W. (1931). Über konjugierte exponentenfolgen. *Studia Mathematica,3*(1), 200–211.

Sundaresan, K. (1971). Uniform convexity of Banach spaces $l(p_i)$. *Studia Mathematica,39*(3), 227–231.

Waterman, D., Ito, T., Barber, F., & Ratti, J. (1969). Reflexivity and summability: The Nakano $l(p_i)$ spaces. *Studia Mathematica,33*(2), 141–146.

Chapter 6
General Common Fixed Point Results in *b*-Metric Space and Applications

Surjeet Singh Chauhan Gonder and Manju Devi

Abstract For a pair of multivalued mappings on a b-metric space, common fixed point results, for a new Suzuki contractive condition, are proved. An applied illustration accords the validity of the work. Further, applications in dynamic programming, regarding the existence of common solutions for a class of functional equations, are also discussed.

Keywords Common fixed point · b-metric space · Hausdorff metric · Functional equations · Dynamic programming

AMS Subject Classification 47H10 · 54H25

Abbreviations and Acronyms

FP–Fixed Point
CFP–Common Fixed Point
MS–Metric Space
bMS–b-Metric Space
CMS–Complete Metric Space
CbMS–Complete b-Metric Space
BCT–Banach Contraction Theorem

S. S. Chauhan Gonder · M. Devi (✉)
Department of Mathematics, UIS, Chandigarh University, Gharuan, Mohali, Punjab, India
e-mail: manjuchahal1991m@gmail.com

M. Devi
Department of Mathematics, Dr. B. R. Ambedkar Government College,
Jagdishpura, Kaithal, Haryana, India

© The Author(s), under exclusive license to Springer Nature Singapore Pte Ltd. 2026

A. Tomar and M. Jain (eds.), *Applications of Fixed-Point Theorem*, Industrial and Applied Mathematics, https://doi.org/10.1007/978-981-95-5160-6_6

6.1 Introduction and Preliminaries

The notion of fixed point theory is an exciting area of inquiry that started formally at the beginning of the 20th century as a crucial component of nonlinear analysis. Although basic ideas of fixed points are found in the work of Augustin-Louis Cauchy (1789–1857) (see also, Granas and Dugundji (2003), and Zeidler (1986)).

Polish mathematician Stefan Banach (1922) obtained the result popularly known as the BCT as follows.

Theorem 6.1.1 *Let (Z, d) be a CMS and $E : Z \to Z$. Assume $r \in [0, 1)$ satisfying for each $l, m \in Z$,*

$$d(El, Em) \leq rd(l, m).$$

Then uniquely, $q \in Z$ exists satisfying $q = Eq$.

The following notations are required for the sequel.
Let (Z, d) be an MS and for $A, B \subseteq Z$, we signify by:

$$CL(Z) \neq \emptyset \text{ the set of all closed subsets of } Z.$$
$$CB(Z) \neq \emptyset \text{ the set of all closed and bounded subsets of } Z.$$
$$D(A, B) = \inf\{d(l, m) : l \in A, m \in B\},$$
$$\delta(A, B) = \sup\{d(l, m) : l \in A, m \in B\},$$
$$D(a, B) = \inf\{d(l, m) : m \in B\},$$
$$\delta(a, B) = \sup\{d(l, m) : m \in B\}.$$

Nadler (1969) introduced multivalued contractions in 1969. He established the result, which includes the Banach contraction theorem as a specific example, using the idea of the Hausdorff metric. Keep in mind that a metric d on Z induces a Hausdorff metric H, which is provided by

$$H(A, B) = \max\left\{\sup_{l\in A} d(l, B), \sup_{m\in B} d(m, A)\right\},$$

for $A, B \in CB(Z)$ and $d(l, B) = \inf_{m\in B} d(l, m)$.

Numerous mathematicians, physicists, and economists have applied the BCT to hybrid maps under various contractive conditions in diverse spaces (see, for instance, Aliouche and Hamaizia (2022), Ćirić (1972), Khojasteh et al. (2014), Rhoades (2015), Rouhani and Moradi (2010), Rus (2001), Tomar et al. (2017), Tomar et al. (2020)). For the study of diverse spaces, one may refer to Chauhan et al. (2022) and for a fundamental overview of numerous contractive conditions for one and two mappings, see Rhoades (1977).

The most important theorem also referred to as the Nadler theorem and a multi-valued version of Theorem 6.1.1, is given by Nadler (1969).

Theorem 6.1.2 *Let* (Z, d) *be a CMS and* $E : Z \to CB(Z)$. *Assume* $r \in [0, 1)$ *satisfying for each* $l, m \in Z$,

$$H(El, Em) \leq rd(l, m).$$

Then, $q \in Z$ *exists satisfying* $q \in Eq$.

Later on, Nadler Jr. replaced the condition "$E : Z \to CB(Z)$" in Theorem 6.1.2 by "$E : Z \to CL(Z)$" (see Nadler Jr. (1978) for details). *Nadler multi-valued contraction theorem* garnered much focus in applicable mathematics, especially in solving inclusions and multivalued differential & functional equations, and was expanded upon and applied in numerous contexts.

Khojasteh et al. (2014) obtained an interesting generalization of the result of Nadler (1969) in the manner described below:

Theorem 6.1.3 *Let* (Z, d) *be a CMS and* $E : Z \to CB(Z)$ *satisfying for each* $l, m \in Z$,

$$H(El, Em) \leq \frac{D(m, El) + D(l, Em)}{\delta(l, El) + \delta(m, Em) + 1} d(l, m).$$

Then, $q \in Z$ *exists satisfying* $q \in Eq$.

By putting this theorem for two multi-valued mappings into practice, Rhoades (2015) extended the aforementioned theorem.

Theorem 6.1.4 *Let* (Z, d) *be a CMS and* $E, F : Z \to CL(Z)$ *satisfying for each* $l, m \in Z$,

$$H(El, Fm) \leq U(l, m) \cdot V(l, m),$$

where

$$U(l, m) = \frac{\max\{d(l, m), D(l, El) + D(m, Fm), D(l, Fm) + D(m, El)\}}{\delta(l, El) + \delta(m, Fm) + 1},$$

and

$$V(l, m) = \max\left\{d(l, m), D(l, El), D(m, Fm), \frac{D(l, Fm) + D(m, El)}{2}\right\}.$$

Then, at least one, $q \in Z$ *exists satisfying* $q \in Eq$ *and* $q \in Fq$.

An innovative distance, on a set that is not empty, known as a b-metric was introduced in 1989 by Bakhtin (1989) (see also Bourbaki (1974)) and in 1993 by Czerwik (1993). A bMS aims to expand the MS by removing all but the triangle inequality and adding only one real constant. It is intriguing to note that a few properties that hold true in metric space do not hold true in b-metric space.

The definition of bMS is unmistakably an expansion of MS. Furthermore, if we use the idea of bMS and assume that the real constant is 1, we get the definition of MS (see for instance, Dung et al. 2015). Our solutions are therefore more generic than the equivalent ones in MS.

For further studies, we will use some preliminaries.

Let (Z, d) be a bMS. Assume that A is a subset of Z and that $\{l_n\}$ is a sequence in Z. Then it can be argued that

(i) If $\lim d(l, l_n) = 0$ as $n \to \infty$, then $\{l_n\}$ converges to l.
(ii) If the condition $\lim \sup\{d(l_n, l_m) : m > n\} = 0$ is true, $\{l_n\}$ is then said to be Cauchy.
(iii) Z is considered complete if all of the Cauchy sequences converge.

Lemma 6.1.5 *Let $\{l_n\}$ represent a sequence in Z and (Z, d) serve as the b-metric space. Assume that $r \in [0, 1)$ exists and that $d(l_{n+1}, l_{n+2}) \leq rd(l_n, l_{n+1})$ for every $n \in N$. Then $\{l_n\}$ is Cauchy.*

In order to generalize the BCT, Czerwik (1993) formalized a fixed point theorem as follows.

Theorem 6.1.6 *Assume that (Z, d) is the CbMS and $E : Z \to CB(Z)$. Assume that $r \in [0, 1/K)$ exists and that it meets $H(El, Em) \leq rd(l, m)$ for all $l, m \in Z$. Then, $q \in Z$ exists satisfying $q \in Eq$.*

The forthcoming fixed point theorem was established by using Jungck type common fixed point result.

Theorem 6.1.7 (Radenović and Došenović 2017) *If (Z, d, A) is a bMS, then $E, F : Z \to Z$ with $E(Z) \subseteq F(Z)$ in such a way that for all $l, m \in Z$*

$$d(El, Em) \leq \mu d(Fl, Fm),$$

where $\mu \in (0, 1)$.

Also, consider that at least one of the following is true:

(i) $(E(Z), d)$ or $(F(Z), d)$ is b-complete,
(ii) (Z, d, A) is b-complete, F is b-continuous, and (E, F) are commuting,

Then E and F discover a unique coincidence point. Furthermore, $q \in Z$ exists satisfying $q \in Eq$ and $q \in Fq$, if they are weakly compatible.

Following is a typical Zamfirescu type FP theorem in bMS.

Theorem 6.1.8 (Jovanović 2016) *Let (Z, d, A) be a b-complete bMS, and $E : Z \to Z$, suppose nonnegative values x, y, and z exist such that for all $l, m \in Z$, a minimum of one of these circumstances exists:*

(i) $d(El, Em) \leq xd(l, m)$
(ii) $d(El, Em) \leq y[d(l, El) + d(m, Em)]$
(iii) $d(El, Em) \leq Z[d(l, Em) + d(m, El)]$ *holds.*

Then, $q \in Z$ *exists satisfying* $q \in Eq$*, if* $x < 1/A$, $y < 1/2A^2$, $Z < 1/2A^2$.

The following theorem is given by Aliouche and Hamaizia (2022).

Theorem 6.1.9 *Let* (Z, d) *be a CbMS and* $E, F : Z \to CB(Z)$ *satisfying for each* $l, m \in Z$,

$$s^{\alpha} H(El, Fm) \leq U(l, m) \cdot V(l, m),$$

with $\alpha \geq 1$ *is a constant, where*

$$U(l, m) = \frac{\max\{d(l, m), D(l, El) + D(m, Fm), D(l, Fm) + D(m, El) + \beta\}}{\delta(l, El) + \delta(m, Fm) + \gamma}, \quad \gamma > \beta \geq 0,$$

and $V(l, m) = \max\left\{d(l, m), D(l, El), D(m, Fm), \frac{D(l, Fm)+D(m, El)}{2s}\right\}$.
If one of the following is true

(i) *with* $\alpha \geq 2$ *and* $s > 1$, d *is discontinuous,*
(ii) *with* $\alpha \geq 1$ *and* $s \geq 1$, d *is continuous,*
(iii) *with* $1 \leq \alpha < 2$, $s > 1$, d *is discontinuous and either* E *or* F *is continuous in* Z.

Then E *and* F *have at least one CFP* $q \in Z$.

But in each of these fixed point theorems, the contractive conditions or contractive requirements must hold for each of the points l and m of the domain. One should, of course, anticipate the day when this stipulation is significantly eased without impairing a conclusion of the theorem.

Consider a function θ from $[0, 1)$ onto $\left(\frac{1}{2}, 1\right]$ that is non-increasing with

$$\theta(r) = \begin{cases} 1 & \text{if } 0 \leq r \leq \frac{1}{2}(\sqrt{5} - 1), \\ \frac{1-r}{r^2} & \text{if } \frac{1}{2}(\sqrt{5} - 1) \leq r \leq \frac{1}{\sqrt{2}}, \\ \frac{1}{1+r} & \text{if } \frac{1}{\sqrt{2}} \leq r < 1. \end{cases}$$

In 2007, Suzuki (2007) presented a novel kind of outcome by restricting the domain for a map, which is a generalisation of the BCP, as in the following manners:

Theorem 6.1.10 *Let* (Z, d) *be a CMS and* $E : Z \to Z$*. Assume* $r \in [0, 1)$ *such that for each* $l, m \in Z$,

$$\theta(r)d(l, El) \leq d(l, m) \text{ implies } d(El, Em) \leq rd(l, m).$$

Then E *has an* FP.

Theorem 6.1.2 is generalized by the following result attributed to Kikkawa and Suzuki (2008).

Theorem 6.1.11 *Let* (Z, d) *be a CMS and* $E : Z \to CB(Z)$. *Assume* $r \in [0, 1)$ *such that for each* $l, m \in Z$,

$$d(l, El) \leq (1 + r)d(l, m) \ \textit{implies} \ H(El, Em) \leq rd(l, m).$$

Then E *has an* FP.

The forceful nature of Suzuki contraction theorem has prompted numerous researchers to offer numerous elegant and intriguing additions and generalizations along these lines over a short period of the past one and a half decades (see, for instance, Dhompongsa and Yingtaweesittikul (2009), Dorić and Lazović (2011), Kikkawa and Suzuki (2008), Mot and Petrusel (2009), Singh et al. (2013), Singh and Mishra (2010a), Singh and Mishra (2010b), Singh and Mishra (2011), Singh et al. (2012), Suzuki (2007), Suzuki (2008), Tomar et al. (2024) and others).

In order to derive a new result for a pair of multivalued maps in a bMS, we integrate the concepts of Theorems 6.1.9 and 6.1.11 in this study. Our findings broaden and generalize various results of Aliouche and Hamaizia (2022), Khojasteh et al. (2014), Rhoades (2015), Suzuki (2007) and others. As applications, we extend and enhance the findings of Bellman (1973) and Bellman and Lee (1978) by using our findings to explore the common solutions for functional equations emerging in dynamic programming (see also Bhakta and Mitra 1984, Baskaran and Subrahmanyam 1986, Ćirić 1972, Zeidler 1986).

The work embodied in this chapter is organized into three sections depending on the diversity of topics contained therein.

6.2 Main Results

Essentially we require the following lemma (see Ćirić 1972, Nadler 1969, Rouhani and Moradi 2010, p. 4, Rus 2001, p. 76).

Lemma 6.2.1 *If* $A, B \in CL(Z)$, $l \in A$, *then there is* $m \in B$ *such that* $d(l, m) \leq \frac{1}{h_1} H(A, B)$ *for any* $0 < h_1 < 1$.

Theorem 6.2.2 *Let* (Z, d) *be a CbMS and* $E, F : Z \to CB(Z)$ *satisfying for each* $l, m \in Z$,

$$\frac{1}{2s} \min\{d(l, El), d(m, Fm)\} \leq d(l, m) \ \textit{implies} \ s^{\alpha} H(El, Fm) \leq U(l, m) \cdot V(l, m), \tag{6.1}$$

with $\alpha \geq 1$ *is a constant, where*

$$U(l,m) = \frac{\max\{d(l,m),\, D(l,El) + D(m,Fm),\, D(l,Fm) + D(m,El) + \beta\}}{\delta(l,El) + \delta(m,Fm) + \gamma},\ \gamma > \beta \geq 0, \tag{6.2}$$

and

$$V(l,m) = \max\left\{d(l,m),\, D(l,El),\, D(m,Fm),\, \frac{D(l,Fm) + D(m,El)}{2s}\right\}. \tag{6.3}$$

If one of the circumstances listed below is true

(i) *with* $\alpha \geq 2$ *and* $s > 1$, d *is discontinuous,*
(ii) *with* $\alpha \geq 1$ *and* $s \geq 1$, d *is continuous,*
(iii) *with* $1 \leq \alpha < 2$, $s > 1$, d *is discontinuous and either* E *or* F *is continuous in* Z.

Then E *and* F *have at least one CFP* $q \in Z$.

Its proof is followed by the subsequent lemma.

Lemma 6.2.3 *Let* (Z, d) *be a CbMS and* $E, F : Z \to CB(Z)$ *satisfying* (6.1), (6.2) *and* (6.3). *Then each FP of* E *is an FP of* F *and vice versa.*

Proof. Suppose $q \in Eq$, then

$$\frac{1}{2s}\min\{d(q,Eq), d(q,Fq)\} = 0 = d(q,q).$$

So

$$s^{\alpha} D(q,Fq) \leq s^{\alpha} H(Eq,Fq) \leq U(q,q)V(q,q), \tag{6.4}$$

where

$$U(q,q) = \frac{D(q,Fq) + \beta}{\delta(q,Eq) + \delta(q,Fq) + \gamma} \leq \frac{D(q,Fq) + \beta}{D(q,Fq) + \gamma} < 1, \tag{6.5}$$

and

$$V(q,q) = D(q,Fq). \tag{6.6}$$

So using inequality (6.5) and Eq. (6.6) in inequality (6.4), we have

$$s^{\alpha} D(q,Fq) < D(q,Fq), \tag{6.7}$$

which implies that $q \in Fq$.

It can be demonstrated similarly that if $q \in Fq$, then $q \in Eq$. □

Proof of Theorem 6.2.2 Let $l_0 \in X$ be an arbitrary and $l_1 \in El_0$. Using Lemma 6.2.1, $\exists\, l_2 \in Fl_1$ so that $d(l_1, l_2) \leq \frac{1}{h_1} H(El_0, Fl_1)$ for any $0 < h_1 < 1$.

Similar to this, there exists $l_3 \in El_2$ such that $d(l, l_3) \leq \frac{1}{h_2} H(El_1, Fl_2)$ for any $0 < h_2 < 1$.

Inductively, define $\{l_n\}$ in Z so that for each $n \in N$,

$$l_{2n+1} \in El_{2n}, l_{2n+2} \in Fl_{2n+1},$$

$$d(l_{2n+1}, l_{2n+2}) \leq \frac{1}{h_{2n+1}} H(El_{2n}, Fl_{2n+1}) \quad \text{for any} 0 < h_{2n+1} < 1,$$

$$d(l_{2n}, l_{2n+1}) \leq \frac{1}{h_{2n}} H(El_{2n}, Fl_{2n-1}) \quad \text{for any } 0 < h_{2n} < 1.$$

If there exists n for which $H(El_{2n}, Fl_{2n+1}) = 0$, therefore $l_{2n+1} \in Fl_{2n+1}$. By Lemma 6.2.3, $l_{2n+1} \in El_{2n+1}$ and hence the result. Similarly, if there exists n for which $H(El_{2n}, Fl_{2n-1}) = 0$, then $l_{2n} \in El_{2n}$, $l_{2n} \in Fl_{2n}$, nothing to prove. Hence we suppose that

$$H(El_{2n}, Fl_{2n+1}) \neq 0 \text{ and } H(El_{2n}, Fl_{2n-1}) \neq 0, \quad \text{for all } n \in n.$$

Put $l = l_{2n}$, $m = l_{2n+1}$ in (1), we have

$$\begin{aligned}
\frac{1}{2s} \min\{d(l, El), d(m, Fm)\} &= \frac{1}{2s} \min\{d(l_{2n}, El_{2n}), d(l_{2n+1}, Fl_{2n+1})\} \\
&= \frac{1}{2s} \min\{d(l_{2n}, l_{2n+1}), d(l_{2n+1}, l_{2n+2})\} \\
&\leq d(l_{2n}, l_{2n+1}),
\end{aligned}$$

implies

$$\begin{aligned}
d(l_{2n+1}, l_{2n+2}) &\leq \frac{1}{h_{2n+1}} H(El_{2n}, Fl_{2n+1}) \\
&\leq \frac{1}{s^{\alpha} h_{2n+1}} U(l_{2n}, l_{2n+1}) \cdot V(l_{2n}, l_{2n+1}),
\end{aligned} \tag{6.8}$$

where h_n is established using the formula $h_n = \sqrt{\beta_n}$ and β_n is specified below. Using inequalities (6.2) and (6.3), we find

$$U(l_{2n}, l_{2n+1}) = \frac{\begin{pmatrix} \max\{d(l_{2n}, l_{2n+1}), D(l_{2n}, El_{2n}) + D(l_{2n+1}, Fl_{2n+1}), \\ D(l_{2n}, Fl_{2n+1}) + D(l_{2n+1}, El_{2n}) + \beta\} \end{pmatrix}}{\delta(l_{2n}, El_{2n}) + \delta(l_{2n+1}, Fl_{2n+1}) + \gamma}$$
$$\leq \frac{\max\{d(l_{2n}, l_{2n+1}), d(l_{2n}, l_{2n+1}) + d(l_{2n+1}, l_{2n+2}), d(l_{2n}, l_{2n+2}) + \beta\}}{d(l_{2n}, l_{2n+1}) + d(l_{2n+1}, l_{2n+2}) + \gamma}$$
$$\leq \frac{\begin{pmatrix} \max\{d(l_{2n}, l_{2n+1}), d(l_{2n}, l_{2n+1}) + d(l_{2n+1}, l_{2n+2}), s(d(l_{2n}, l_{2n+1}) \\ +d(l_{2n+1}, l_{2n+2})) + \beta\} \end{pmatrix}}{d(l_{2n}, l_{2n+1}) + d(l_{2n+1}, l_{2n+2}) + \gamma},$$

which leads to

$$\begin{aligned} U(l_{2n}, l_{2n+1}) &\leq \frac{s(d(l_{2n}, l_{2n+1}) + d(l_{2n+1}, l_{2n+2})) + \beta}{d(l_{2n}, l_{2n+1}) + d(l_{2n+1}, l_{2n+2}) + \gamma} \\ &= s\beta_{2n+1}, \\ 0 < \beta_n &= \frac{d(l_{n-1}, l_n) + d(l_n, l_{n+1}) + \beta}{d(l_{n-1}, l_n) + d(l_n, l_{n+1}) + \gamma} < 1, \ n \geq 1, \end{aligned} \tag{6.9}$$

and

$$\begin{aligned} V(l_{2n}, l_{2n+1}) &= \max\left\{d(l_{2n}, l_{2n+1}), D(l_{2n}, El_{2n}), D(l_{2n+1}, Fl_{2n+1}), \right. \\ &\quad \left. \frac{D(l_{2n}, Fl_{2n+1}) + D(l_{2n+1}, El_{2n})}{2s}\right\} \\ &\leq \max\left\{d(l_{2n}, l_{2n+1}), d(l_{2n+1}, l_{2n+2}), \frac{d(l_{2n}, l_{2n+2})}{2s}\right\} \\ &\leq \max\left\{d(l_{2n}, l_{2n+1}), d(l_{2n+1}, l_{2n+2}), \frac{d(l_{2n}, l_{2n+1}) + d(l_{2n+1}, l_{2n+2})}{2}\right\}, \end{aligned}$$

which gives

$$V(l_{2n}, l_{2n+1}) \leq \max\{d(l_{2n}, l_{2n+1}), d(l_{2n+1}, l_{2n+2})\}. \tag{6.10}$$

Substituting inequalities (6.9) and (6.10) into (6.8),

$$\begin{aligned} d(l_{2n+1}, l_{2n+2}) &\leq \frac{1}{s^\alpha h_{2n+1}} U(l_{2n}, l_{2n+1}) \cdot V(l_{2n}, l_{2n+1}) \\ &\leq \frac{1}{s^\alpha \sqrt{\beta_{2n+1}}} s\beta_{2n+1} \max\{d(l_{2n}, l_{2n+1}), d(l_{2n+1}, l_{2n+2})\} \\ &\cdot \frac{\sqrt{\beta_{2n+1}}}{s^{\alpha-1}} d(l_{2n}, l_{2n+1}). \end{aligned} \tag{6.11}$$

Similarly, we obtain

$$d(l_{2n}, l_{2n+1}) \leq \frac{\sqrt{\beta_{2n}}}{s^{\alpha-1}} d(l_{2n-1}, l_{2n}). \tag{6.12}$$

Thus from inequalities (6.11) and (6.12), we get for all $n \geq 1$,

$$d(l_n, l_{n+1}) \leq \theta_n \cdot d(l_{n-1}, l_n),$$

where $\theta_n = \frac{\sqrt{\beta_n}}{s^{\alpha-1}}$.

Accordingly, $\{l_n\}$ is a Cauchy, so $\exists\, q \in Z$ such that $\lim_{n\to\infty} l_n = q$.

Case 1: If $\alpha \in [2, \infty)$, $s \in (1, \infty)$ and d is discontinuous. Suppose q is not an FP of F.
Since

$$d(q, l_n) \leq \frac{1}{3s} d(q, m), \quad \text{for } m \neq q \text{ and all } n \geq n_0.$$

Afterwards as in Suzuki (2007, p. 1862)

$$\begin{aligned} d(l_{2n-1}, Fl_{2n-1}) &\leq s[d(l_{2n-1}, q) + d(q, l_{2n})] \\ &\leq s\frac{2}{3s} d(m, q) \\ &= d(m, q) - \frac{1}{3} d(m, q) \\ &= d(m, q) - \frac{s}{3s} d(m, q) \\ &\leq d(m, q) - sd(q, l_{2n-1}) \\ &\leq sd(q, l_{2n-1}) + sd(l_{2n-1}, m) - sd(l_{2n-1}, q) \\ &= sd(l_{2n-1}, m). \end{aligned}$$

That is,

$$d(l_{2n-1}, Fl_{2n-1} \leq sd(l_{2n-1}, m).$$

That is,

$$s^{-1} d(l_{2n-1}, Fl_{2n-1}) \leq d(l_{2n-1}, m).$$

Hence

$$\frac{1}{2} \min\{d(m, Em), d(l_{2n-1}, Fl_{2n-1})\} \leq sd(m, l_{2n-1}).$$

Therefore, by the assumption

$$s^{\alpha} H(Em, Fl_{2n-1}) \le U(m, l_{2n-1}) \cdot V(m, l_{2n-1}),$$

where

$$\begin{aligned}
&U(m, l_{2n-1}) \\
&= \frac{\max\{d(m, l_{2n-1}), D(m, Em) + D(l_{2n-1}, Fl_{2n-1}), D(m, Fl_{2n-1}) + D(l_{2n-1}, Em) + \beta\}}{\delta(m, Em) + \delta(l_{2n-1}, Fl_{2n-1}) + \gamma} \\
&\le \frac{\max\{d(m, l_{2n-1}), D(m, Em) + d(l_{2n-1}, l_{2n}), d(m, l_{2n}) + D(l_{2n-1}, Em) + \beta\}}{\delta(m, Em) + d(l_{2n-1}, l_{2n}) + \gamma} \\
&\le \frac{\max\{d(m, l_{2n-1}), D(m, Em) + sd(m, l_{2n-1}), d(m, l_{2n}) + D(l_{2n-1}, Em) + \beta\}}{\delta(m, Em) + d(l_{2n-1}, l_{2n}) + \gamma},
\end{aligned}$$

and

$$V(m, l_{2n-1}) = \max\left\{d(m, l_{2n-1}), D(m, Em), d(l_{2n-1}, l_{2n}), \frac{d(m, l_{2n}) + D(l_{2n-1}, Em)}{2s}\right\}.$$

Making $n \to \infty$,

$$\begin{aligned}
s^{\alpha} d(Em, q) &= \frac{\max\{d(m, q), s[d(m, q) + d(q, Em)], d(m, q) + D(q, Em) + \beta\}}{\delta(m, Em) + \gamma} \\
&\quad \times \max\left\{d(m, q), D(m, Em), \frac{d(m, q) + D(q, Em)}{2s}\right\} \\
&= \frac{s[d(m, q) + D(q, Em)] + \beta}{\delta(m, Em) + \gamma} \times \max\{d(m, q), s[d(m, q) + D(q, Em)]\} \\
&\le s[d(m, q) + D(q, Em)].
\end{aligned}$$

Thus,

$$(s^{\alpha} - s)d(Em, q) \le s[d(m, q)],$$

that is,

$$(s^{\alpha-1} - 1)d(Em, q) \le d(m, q).$$

Therefore

$$\begin{aligned}
d(m, Em) &\le s[d(Em, q) + d(q, m)] \\
&\le \left(\frac{s}{s^{\alpha-1} - 1}\right) d(m, q) + sd(m, q) \\
&\le sd(m, q) + sd(m, q) \\
&= 2sd(m, q).
\end{aligned}$$

Hence $\min\{d(m, Em), d(q, Fq)\} \leq 2sd(m, q)$, so by the assumption

$$s^{\alpha} H(Em, Fq) \leq U(m, q) \cdot V(m, q),$$

where

$$U(m, q) = \frac{\max\{d(m, q), D(m, Em) + D(q, Fq), D(m, Fq) + D(q, Em) + \beta\}}{\delta(m, Em) + \delta(q, Fq) + \gamma},$$

and

$$V(m, q) = \max\left\{d(m, q), D(m, Em), d(q, Fq), \frac{d(m, Fq) + D(q, Em)}{2s}\right\}.$$

Put $m = l_{2n}$, we have

$$s^{\alpha} H(El_{2n}, Fq) \leq U(l_{2n}, q) \cdot V(l_{2n}, q),$$

where

$$\begin{aligned} U(l_{2n}, q) &= \frac{\max\{d(l_{2n}, q), D(l_{2n}, El_{2n}) + D(q, Fq), D(l_{2n}, Fq) + D(q, El_{2n}) + \beta\}}{\delta(l_{2n}, El_{2n}) + \delta(q, Fq) + \gamma} \\ &\leq \frac{\max\{d(l_{2n}, q), d(l_{2n}, l_{2n+1}) + d(q, Fq), sd(l_{2n}, q) + sd(q, Fq) + d(q, l_{2n+1}) + \beta\}}{d(l_{2n}, l_{2n+1}) + d(q, Fq) + \gamma}, \end{aligned}$$

and

$$V(l_{2n}, q) = \max\left\{d(l_{2n}, q), d(l_{2n}, l_{2n+1}), d(q, Fq), \frac{sd(l_{2n}, q) + sd(q, Fq) + d(q, l_{2n+1})}{2s}\right\}.$$

Letting $n \to \infty$, we have

$$D(q, Fq) \leq \frac{1}{s^{\alpha-1}} \frac{sd(q, Fq) + \beta}{d(q, Fq) + \gamma} D(q, Fq) < D(q, Fq),$$

which implies that $q \in Fq$, by Lemma 6.2.3, $q \in Eq$ and hence the result.

Case 2: If $\alpha, s \in [1, \infty)$ and d is continuous. Put $l = l_{2n}$ and $m = q$ in the inequality (6.1) and take $n \to \infty$, we have

$$D(q, Fq) \leq \frac{1}{s^{\alpha}} \frac{d(q, Fq) + \beta}{d(q, Fq) + \gamma} D(q, Fq) < D(q, Fq),$$

which implies that $q \in Fq$, by Lemma 6.2.3, $q \in Eq$ and hence the result.

Case 3: Either E or F is continuous on $Z, \alpha \in [1, 2)$, $s \in (1, \infty)$ and d is discontinuous. If E is continuous on Z. So $q \in Eq$, by Lemma 6.2.3, $q \in Fq$ and hence nothing to prove. Again F is continuous on Z. So $q \in Fq$, by Lemma 6.2.3, $q \in Eq$ and hence the conclusion. □

Corollary 6.2.4 *Let (Z, d) be a CbMS and $E, F : Z \to Z$ satisfying for each $l, m \in Z$,*

$$\frac{1}{2s} \min\{d(l, El), d(m, Fm)\} \le d(l, m) \text{ implies } s^{\alpha} d(El, Fm) \le U(l, m)\, V(l, m),$$

with $\alpha \ge 1$ is a constant, where

$$U(l, m) = \frac{\max\{d(l, m), d(l, El) + d(m, Fm), d(l, Fm) + d(m, El) + \beta\}}{d(l, El) + d(m, Fm) + \gamma}, \quad \gamma > \beta \ge 0,$$

and

$$V(l, m) = \max\left\{d(l, m), d(l, El), d(m, Fm), \frac{d(l, Fm) + d(m, El)}{2s}\right\}.$$

If one of the conditions listed below is true

(i) with $\alpha \ge 2$ and $s > 1$, d is discontinuous,
(ii) with $\alpha \ge 1$ and $s \ge 1$, d is continuous,
(iii) with $1 \le \alpha < 2$, $s > 1$, d is discontinuous and either E or F is continuous in Z.

Then E and F have at least one CFP $q \in Z$.

Proof. It originates from Theorem 6.2.2 when E and F are single valued maps. □

Corollary 6.2.5 *Let (Z, d) be a CbMS and $E : Z \to CB(Z)$ satisfying for each $l, m \in Z$,*

$$\frac{1}{2s} \min\{d(l, El), d(m, Em)\} \le d(l, m) \text{ implies } s^{\alpha} H(El, Em) \le U(l, m) \cdot V(l, m),$$

with $\alpha \ge 1$ is a constant, where

$$U(l, m) = \frac{\max\{d(l, m), D(l, El) + D(m, Em), D(l, Em) + D(m, El) + \beta\}}{\delta(l, El) + \delta(m, Em) + \gamma}, \quad \gamma > \beta \ge 0,$$

and

$$V(l, m) = \max\left\{d(l, m), D(l, El), D(m, Em), \frac{D(l, Em) + D(m, El)}{2s}\right\}.$$

If one of the conditions listed below is true

(i) *with* $\alpha \geq 2$ *and* $s > 1$, d *is discontinuous,*
(ii) *with* $\alpha \geq 1$ *and* $s \geq 1$, d *is continuous,*
(iii) *with* $1 \leq \alpha < 2$, $s > 1$, d *is discontinuous and either* E *or* F *is continuous in* Z.

Then E *has atleast one FP* $q \in Z$.

Proof. It originates from Theorem 6.2.2 when $F = E$. □

Corollary 6.2.6 *Let* (Z, d) *be a CbMS and* $E : Z \to Z$ *satisfying for each* $l, m \in Z$,

$$\frac{1}{2s}\min\{d(l, El), d(m, Em)\} \leq d(l, m) \text{ implies } s^{\alpha} d(El, Em) \leq U(l, m) \cdot V(l, m),$$

with $\alpha \geq 1$ *is a constant, where*

$$U(l, m) = \frac{\max\{d(l, m), d(l, El) + d(m, Em), d(l, Em) + d(m, El) + \beta\}}{d(l, El) + d(m, Em) + \gamma}, \quad \gamma > \beta \geq 0$$

and

$$V(l, m) = \max\left\{d(l, m), d(l, El), d(m, Em), \frac{d(l, Em) + d(m, El)}{2s}\right\}.$$

If one of the circumstances listed below is true

(i) *with* $\alpha \geq 2$ *and* $s > 1$, d *is discontinuous,*
(ii) *with* $\alpha \geq 1$ *and* $s \geq 1$, d *is continuous,*
(iii) *with* $1 \leq \alpha < 2$, $s > 1$, d *is discontinuous and either* E *or* F *is continuous in* Z.

Then E *has atleast one FP* $q \in Z$.

Proof. It originates from Corollary 6.2.4 when $F = E$. □

The generality of Theorem 6.2.2 over Theorem 6.1.9 is demonstrated by the subsequent example.

Example 6.2.7 Let $Z = \left\{0, 1, \frac{1}{2}, \frac{1}{3}\right\}$ and $d : Z \times Z \to R^{+}$ defined as

$$d(l, l) = 0, \quad \text{for all } l \in Z,$$

and for $l \neq m$

$$d(l,m) = \begin{cases} \frac{2}{3} & \text{if } l,m = 0, \frac{1}{2} \\ \frac{3}{2} & \text{if } l,m = 0, \frac{1}{3} \\ \frac{4}{3} & \text{if } l,m = \frac{1}{2}, \frac{1}{3} \\ 1 & \text{if } l,m = 0, 1 \\ 2 & \text{if } l,m = 1, \frac{1}{2} \\ 2 & \text{if } l,m = 1, \frac{1}{3} \end{cases}$$

Now to check that d is a b-metric, the only requirement is to check the triangular inequality condition, i.e.,

$$d(l,m) \leq s[d(l,q) + d(q,m)].$$

For $s = \frac{4}{3} > 1$, the above inequality holds for every $l, m, q \in Z$.
Hence, with $s = \frac{4}{3}$, d is b-metric.
Define $E, F : Z \to CB(Z)$ by

$$E(Z) = \begin{cases} \{0\} & \text{if } l = 0, \frac{1}{2} \\ \{1\} & \text{if } l = 1 \\ \left\{\frac{1}{2}\right\} & \text{if } l = \frac{1}{3} \end{cases} \quad \text{and} \quad F(Z) = \begin{cases} \{0\} & \text{if } l = 0, \frac{1}{2} \\ \{1\} & \text{if } l = 1, \frac{1}{3} \end{cases}.$$

We need to show that for any $l, m \in Z$ and $\alpha = 1, \beta = \frac{1}{2}, \gamma = 1$,

$$\frac{1}{2s} \min\{d(l, El), d(m, Fm)\} \leq d(l,m) \text{ implies } s^{\alpha} H(El, Fm) \leq U(l,m)V(l,m). \tag{6.13}$$

Note that $H(El, Fm) = 0$ if $(l,m) \in \left\{\left(0, \frac{1}{2}\right), \left(1, \frac{1}{3}\right), \left(\frac{1}{2}, 0\right), (0,0), (1,1), \left(\frac{1}{2}, \frac{1}{2}\right)\right\}$ and the inequality (6.13) holds.

(1) At $(l,m) = (0,1)$, we obtain

$$s^{\alpha} H(E0, F1) = \frac{4}{3} H(E0, F1) = \frac{4}{3} \cdot 1 \leq U(0,1) \cdot V(0,1) = 2.1$$

(2) $At(l,m) = \left(0, \frac{1}{3}\right)$, we obtain

$$s^{\alpha} H\left(E0, F\frac{1}{3}\right) = \frac{4}{3} H\left(E0, F\frac{1}{3}\right) = \frac{4}{3} \cdot 1 \leq U\left(0, \frac{1}{3}\right) \cdot V\left(0, \frac{1}{3}\right) = 1.2$$

(3) At $(l, m) = \left(1, \frac{1}{2}\right)$, we obtain

$$s^{\propto} H\left(E1, F\frac{1}{2}\right) = \frac{4}{3} H\left(E1, F\frac{1}{2}\right) = \frac{4}{3} \cdot 1 \leq U\left(1, \frac{1}{2}\right) \cdot V\left(1, \frac{1}{2}\right) = \frac{9}{5} \cdot 2$$

(4) At $(l, m) = \left(\frac{1}{2}, \frac{1}{3}\right)$, we obtain

$$s^{\propto} H\left(E\frac{1}{2}, F\frac{1}{3}\right) = \frac{4}{3} H\left(E\frac{1}{2}, F\frac{1}{3}\right) = \frac{4}{3} \cdot 1 \leq U\left(\frac{1}{2}, \frac{1}{3}\right) \cdot V\left(\frac{1}{2}, \frac{1}{3}\right) = \frac{12}{11} \cdot 2$$

(5) At $(l, m) = (1, 0)$, we obtain

$$s^{\propto} H\,(E1, F0) = \frac{4}{3} H(E1, F0) = \frac{4}{3} \cdot 1 \leq U(1, 0) \cdot V(1, 0) = 2 \cdot 1$$

(6) At $(l, m) = \left(\frac{1}{3}, 0\right)$, we obtain

$$s^{\propto} H\left(E\frac{1}{3}, F0\right) = \frac{4}{3} H\left(E\frac{1}{3}, F0\right) = \frac{4}{3} \cdot \frac{2}{3} \leq U\left(\frac{1}{3}, 0\right) \cdot V\left(\frac{1}{3}, 0\right) = \frac{48}{42} \cdot \frac{3}{2}$$

(7) At $(l, m) = \left(\frac{1}{2}, 1\right)$, we obtain

$$s^{\propto} H\left(E\frac{1}{2}, F1\right) = \frac{4}{3} H\left(E\frac{1}{2}, F1\right) = \frac{4}{3} \cdot 1 \leq U\left(\frac{1}{2}, 1\right) \cdot V\left(\frac{1}{2}, 1\right) = \frac{9}{5} \cdot 2$$

(8) At $(l, m) = \left(\frac{1}{3}, 1\right)$, we obtain

$$s^{\propto} H\left(E\frac{1}{3}, F1\right) = \frac{4}{3} H\left(E\frac{1}{3}, F1\right) = \frac{4}{3} \cdot 2 \leq U\left(\frac{1}{3}, 1\right) \cdot V\left(\frac{1}{3}, 1\right) = \frac{27}{14} \cdot \frac{12}{8}$$

(9) At $(l, m) = \left(\frac{1}{3}, \frac{1}{2}\right)$, we obtain

$$s^{\propto} H\left(E\frac{1}{3}, F\frac{1}{2}\right) = \frac{4}{3} H\left(E\frac{1}{3}, F\frac{1}{2}\right) = \frac{4}{3} \cdot \frac{2}{3} \leq U\left(\frac{1}{3}, \frac{1}{2}\right) \cdot V\left(\frac{1}{3}, \frac{1}{2}\right) = \frac{2}{3} \cdot \frac{4}{3}$$

(10) At $(l, m) = \left(\frac{1}{3}, \frac{1}{3}\right)$, we obtain

$$s^{\propto} H\left(E\frac{1}{3}, F\frac{1}{3}\right) = \frac{4}{3} H\left(E\frac{1}{3}, F\frac{1}{3}\right) = \frac{4}{3} \cdot 2 > U\left(\frac{1}{3}, \frac{1}{3}\right) \cdot V\left(\frac{1}{3}, \frac{1}{3}\right) = \frac{23}{26} \cdot 2$$

Thus we note that $s^{\alpha} H(El, Fm) \leq U(l,m) \cdot V(l,m)$ if $(l,m) \neq \left(\frac{1}{3}, \frac{1}{3}\right)$, since

$$\frac{1}{2s} \min\left\{d\left(\frac{1}{3}, E\frac{1}{3}\right), d\left(\frac{1}{3}, F\frac{1}{3}\right)\right\} = \frac{3}{8} \cdot \frac{4}{3} = \frac{1}{2} > 0 = d\left(\frac{1}{3}, \frac{1}{3}\right).$$

Thus E and F satisfy Theorem 6.2.2 but not Theorem 6.1.9.

6.3 Application

This section focuses on the application of the work done in the previous section to the dynamic programming problem. Here, we take Z as a Banach space such that $W, D \subseteq Z$. Let R represents the field of real numbers and consider the mappings $H_1, H_2 : W \times D \times R \to R$. Considering W and D as the state space and decision space respectively, dynamic programming is an issue of solving

$$p_i = \sup_{m \in D} H_i(l, m, p_i(l,m)), \quad l \in W, \ i = 1, 2. \tag{6.14}$$

For more details, one may refer to Bellman (1973), Bellman and Lee (1978), Bhakta and Mitra (1984), Baskaran and Subrahmanyam (1986), Ćirić (1972) and Zeidler (1986). Here, we investigate the common solutions of (6.14) for $i = 1, 2$.

Assign the term $B(W)$ to the collection of all bounded real-valued functions on W. For any $k \in B(W)$, define $\|k\| = \sup_{l \in W} |k(l)|^2$. Then $(B(W), \|\cdot\|)$ is a Banach space. Considering the aforementioned conditions to be true:

(Dρ-1) H_1 and H_2 are bounded.

(Dρ-2) Suppose $\exists\, s > 1$ so that for all $(l,m) \in W \times D, h, k \in B(W)$ and $t \in W$,

$$\frac{1}{2s} \min\{|h(t) - A_1 h(t)|^2, |k(t) - A_2 k(t)|^2\} \leq |h(t) - k(t)|^2,$$

implies

$$s^{\alpha} |H_1(l, m, h(t)) - H_2(l, m, k(t))|^2$$
$$\leq \frac{\left(\max\left\{|h(t) - k(t)|^2, |h(t) - A_1 h(t)|^2 + |k(t) - A_2 k(t)|^2, |h(t) - A_2 k(t)|^2 + |k(t) - A_1 h(t)|^2 + \beta\right\}\right)}{|h(t) - A_1 h(t)|^2 + |k(t) - A_2 k(t)|^2 + \gamma}$$
$$\times \max\left\{|h(t) - k(t)|^2, |h(t) - A_1 h(t)|^2, |k(t) - A_2 k(t)|^2, \frac{|h(t) - A_2 k(t)|^2 + |k(t) - A_1 h(t)|^2}{2s}\right\},$$

where

$$A_i h(l) = \sup_{m \in D} H_i(l, m, h(l, m)), \quad l \in W, h \in B(W), \ i = 1, 2.$$

Theorem 6.3.1 *Assume that (Dρ-1) and (Dρ-2) are true. Then the functional equations* (6.14), $i = 1, 2$ *have atleast one solution in* $B(W)$.

Proof. For $h, k \in B(W)$, let $d(h, k) = \sup\{|h(l) - k(l)|^2 : l \in W\}$. Then $(B(W), d)$ is a complete b-metric space. Let $\varepsilon > 0$ be a given and $h_1, h_2 \in B(W)$. Take $l \in W$, and choose $m_1, m_2 \in B$ such that

$$A_i h_j < H_i(l, m, h_j(l_j)) + \varepsilon, \quad i = 1, 2, \tag{6.15}$$

where $l_j = (l, m_j)$.
Further,

$$A_1 h_1 \geq H_1(l, m_1, h_1(l_1)), \tag{6.16}$$
$$A_2 h_2 \geq H_2(l, m_1, h_2(l_1)). \tag{6.17}$$

Now, as in Singh et al. (2013)

$$\frac{1}{2s} \min\left\{|h_1 - A_1 h_1|^2, |h_2 - A_2 h_2|^2\right\} \leq |h_1 - h_2|^2$$

and this along with (6.15), (6.16) and (6.17) implies

$$A_1 h_1 - A_2 h_2 < H_1(l, m_1, h_1(l_1)) - H_2(l, m_1, h_2(l_1)) + \varepsilon.$$
$$|A_1 h_1 - A_2 h_2|^2 < |H_1(l, m_1, h_1(l_1)) - H_2(l, m_1, h_2(l_1)) + \varepsilon|^2.$$

Since $\varepsilon > 0$ is an arbitrary positive real number, so

$$\begin{aligned} |A_1 h_1 - A_2 h_2|^2 &< |H_1(l, m_1, h_1(l_1)) - H_2(l, m_1, h_2(l_1))|^2 \\ &\leq s^\alpha |H_1(l, m_1, h_1(l_1)) - H_2(l, m_1, h_2(l_1))|^2 \\ &\leq U(h_1, h_2) \cdot V(h_1, h_2). \end{aligned}$$

Therefore by Corollary 6.2.4, A_1 and A_2 have at least one CFP w^*, that is, $w^*(x)$ is at least one bounded common solution of (6.14), $i = 1, 2$. □

Corollary 6.3.2 *Assume* H *is bounded and* $s > 1$ *so that for all* $l, m \in W \times D, h, k \in B(W)$ *and* $t \in W$,

$$\frac{1}{2s} \min\{|h(t) - Ah(t)|^2, |k(t) - Ak(t)|^2\} \leq |h(t) - k(t)|^2$$

implies

$$s^{\alpha}|H(l,m,h(t)) - H(l,m,k(t))|^2$$
$$\leq \frac{\begin{pmatrix} \max\{|h(t)-k(t)|^2, |h(t)-Ah(t)|^2+|k(t)-Ak(t)|^2, \\ |h(t)-Ak(t)|^2+|k(t)-Ah(t)|^2+\beta\} \end{pmatrix}}{|h(t)-Ah(t)|^2+|k(t)-Ak(t)|^2+\gamma}$$
$$\times \max\left\{|h(t)) - k(t))|^2, |h(t)-Ah(t)|^2, |k(t)-Ak(t)|^2, \right.$$
$$\left. \frac{|h(t)-Ak(t)|^2+|k(t)-Ah(t)|^2}{2s}\right\},$$

where $Ah(l) = \sup_{m \in D} H(l,m,h(l,m))$, $l \in W$, $h \in B(W)$.

Then the Eq. (6.14) *with* $H_i = H$ *has at least one solution in* $B(W)$.

Proof. Take $A_1 = A_1 = A$ in Theorem 6.3.1. □

6.4 Conclusion

Common fixed point results for multi-valued mappings subjected to a Suzuki type contraction in b-metric space have been established. A practical example confirms the accuracy of the finding. We also apply the findings to the existence of a common solution in dynamic programming. The present work will inspire further development in fixed point theory on diverse spaces, with different contractive conditions, and for hybrid maps.

Acknowledgements The Editors' insightful recommendations to enhance the original typescript are much appreciated by the Authors.

References

Aliouche, A., & Hamaizia, T. (2022). Common fixed point theorems for multivalued mappings in b-metric spaces with an application to integral inclusions. *Journal of Analysis,30*(1), 43–62.

Bakhtin, I. A. (1989). The contraction mapping principle in almost metric spaces. *Functional Analysis,30*, 26–37.

Banach, S. (1922). Sur les operations dans les ensembles abstraits et leur application aur equations integrals. *Fundamenta Mathematicae*, *3*, 133–181.

Baskaran, R., & Subrahmanyam, P. V. (1986). A note on the solution of a class of functional equations. *Applicable Analysis,22*(3–4), 235–241.

Bellman, R. (1973). *Methods of nonlinear analysis* (Vol. II). New York, USA: Academic Press.

Bellman, R., & Lee, E. S. (1978). Functional equations in dynamic programming. *Aequationes Mathematicae,17*(1), 1–18.

Bhakta, P. C., & Mitra, S. (1984). Some existence theorems for functional equations arising in dynamic programming. *Journal of Mathematical Analysis and Applications,98*(2), 348–362.

Bourbaki, N. (1974). *Topologie Generale*. Herman.

Chauhan (Gonder), S. S., Garg, P., & Thakur, K. (2022). Study of metric space and its variants. *Journal of Mathematics, 2022*, 1–22.

Ćirić, L. (1972). Fixed points for generalized multivalued contractions. *Matematički Vesnik, 9*(24), 265–272.

Czerwik, S. (1993). Contraction mappings in *b*-metric spaces. *Acta Mathematica et Informatica Universitatis Ostraviensis,1*, 5–11.

Dhompongsa, S., & Yingtaweesittikul, H. (2009). Fixed points for multivalued mappings and the metric completeness. *Fixed Point Theory Applications, 2009*, Art. ID 972395, 15.

Dorić, D., & Lazović, R. (2011). Some Suzuki-type fixed point theorems for generalized multivalued mappings and applications. *Fixed Point Theory Applications, 2011*(40), 13.

Dung, N. V., Hang, V. T. L., & Sedghi, S. (2015). Remarks on metric-type spaces and applications. *Southeast Asian Bulletin of Mathematics,39*(6), 755–768.

Granas, A., & Dugundji, J. (2003). *Fixed point theory*. Springer.

Jovanović, M. (2016). Contribution to the theory of abstract metric spaces. Ph.D. thesis, University of Belgrade, Faculty of Mathematics, Belgrade.

Khojasteh, F., Abbas, M., & Costache, S. (2014). Two new fixed point theorems in complete metric spaces. *Abstract and Applied Analysis, 2014*, Article ID 325840, 5.

Kikkawa, M., & Suzuki, T. (2008). Three fixed point theorems for generalized contractions with constants in complete metric spaces. *Nonlinear Analysis,69*(9), 2942–2949.

Mot, G., & Petrusel, A. (2009). Fixed point theory for a new type of contractive multivalued operators. *Nonlinear Analysis,70*, 3371–3377.

Nadler, S. B., Jr. (1969). Multi-valued contraction mappings. *Pacific Journal of Mathematics,30*, 475–488.

Nadler, S. B., Jr. (1978). *Hyperspaces of sets*. Marcel Dekker.

Radenović, S., & Došenović, T. (2017). Contractive conditions in *b*-metric spaces. *Vojnotehnicki Glasnik*. https://doi.org/10.5937/vojtehg65-14817

Rhoades, B. E. (1977). A comparison of various definitions of contractive mappings. *Transactions of the American Mathematical Society,226*, 257–290.

Rhoades, B. E. (2015). Two new fixed point theorems. *General Mathematics Notes,27*(2), 123–132.

Rouhani, B.D., & Moradi, S. (2010). Common fixed point of multivalued generalized φ-weak contractive mappings. *Fixed Point Theory Applications, 2010*, Art. ID 708984, 13.

Rus, I. A. (2001). Generalized Contractions nd Applications. Cluj-Napoca.

Singh, S. L., Chugh, R., & Kamal, R. (2013). Suzuki type common fixed point theorems and applications. *Fixed Point Theory,14*(2), 497–506.

Singh, S. L., & Mishra, S. N. (2010). Coincidence theorems for certain classes of hybrid contractions. *Fixed Point Theory Applications, 2010*, Art. ID 898109, 14.

Singh, S. L., & Mishra, S. N. (2010). Remarks on recent fixed point theorems. *Fixed Point Theory Applications, 2010*, Art. ID 452905, 18.

Singh, S. L., & Mishra, S. N. (2011). Fixed point theorems for single-valued and multi-valued maps. *Nonlinear Analysis,74*(6), 2243–2248.

Singh, S. L., Mishra, S. N., Chugh, R., & Kamal, R. (2012). General common fixed point theorems and applications. *Journal of Applied Mathematics, 2012*, Art. ID 902312, 14.

Suzuki, T. (2007). A generalized Banach contraction principle that characterizes metric completeness. *Proceedings of the American Mathematical Society,136*(5), 1861–1869.

Suzuki, T. (2008). Fixed point theorems and convergence theorems for some generalized nonexpansive mappings. *Journal of Mathematical Analysis and Applications,340*, 1088–1095.

Tomar, A., Upadhyay, S., & Sharma, R. (2017). On existence of strict coincidence and common strict fixed point of a faintly compatible hybrid pair of maps. *Electronic Journal of Mathematical Analysis and Applications*, *5*(2), 298–305.

Tomar, A., Sharma, R., & Ansari, A. H. (2020). Strict coincidence and common strict fixed point of a faintly compatible hybrid pair of maps via C-class function and applications. *Palestine Journal of Mathematics*, *9*(1), 274–288.

Tomar, A., Rana, U. S., & Kumar, V. (2024). Fixed point, its geometry and application via w-interpolative contraction of Suzuki type mapping. *Mathematical Methods in the Applied Sciences*, *47*(5), 3507–3528.

Zeidler, E. (1986). *Nonlinear functional analysis and its applications, Part I: Fixed point theorems*. Springer.

Chapter 7
SPK Iterative Algorithm for Monotone Nonexpansive Mappings: Convergence and Applications

Samir Dashputre and Kavita Sakure

Abstract This chapter focuses on the SPK iterative algorithm, which is used to approximate the fixed points of monotone Reich-Suzuki type nonexpansive mappings in partially ordered hyperbolic spaces. The main objective is to present convergence results for this algorithm when applied to such mappings. Additionally, the chapter includes an illustrative example to demonstrate its main findings. Finally, the application of the results to nonlinear integral equations is discussed.

Keywords Partial ordered hyperbolic spaces · Monotone nonexpansive mappings · SPK iterative algorithm · Δ-convergence

AMS Subject Classification 47H10 · 54H25

7.1 Introduction

Fixed point theory is a fundamental and significant area of study in both pure and applied mathematics, offering versatile applications across various mathematical disciplines. The impact of fixed point results goes beyond mathematics and reaches into diverse fields such as statistics, computer science, chemistry, physics, economics, biology, medicine, engineering, and game theory (see, e.g., Li et al. (2021), Tian and Guo (2022)). This broad applicability underscores the importance and relevance of fixed point theory in tackling real-world problems and advancing knowledge in a wide range of scientific and practical domains.

Fixed point theory thrives through two primary avenues of research. The first direction involves expanding our understanding by identifying broader classes of

S. Dashputre
Department of Mathematics, Government College, Arjunda, Balod, Chhattisgarh, India

K. Sakure (✉)
Department of Mathematics, Government Didvijay Autonomous P.G. College, Rajnandgaon, Chhattisgarh, India
e-mail: kavitaaage@gmail.com

© The Author(s), under exclusive license to Springer Nature Singapore Pte Ltd. 2026

A. Tomar and M. Jain (eds.), *Applications of Fixed-Point Theorem*, Industrial and Applied Mathematics, https://doi.org/10.1007/978-981-95-5160-6_7

functions and establishing conditions that ensure the existence of fixed points. This pursuit enriches our knowledge of fixed-point theorems and unveils essential principles governing them. The second research direction centers on developing iterative algorithms capable of effectively approximating the fixed points of these functions. As direct methods may not always suffice, these algorithms provide an alternative approach. Researchers in this field continually work on defining and improving these algorithms, taking into account factors such as convergence, stability, computational complexity, and error estimates. With its wide-ranging applications and ongoing research efforts, fixed point theory remains a captivating area of study, driving progress and breakthroughs not only in mathematics but also in numerous other disciplines. Its importance lies in the valuable insights it offers and the practical solutions it provides, making it a vital and dynamic field in the realm of scientific exploration and problem-solving.

The Fixed Point Problem (FPP) found a significant application in 1890 when French mathematician Picard (1890) used it to establish the existence result for differential equations. Later in 1922, Banach (1922) made a crucial advancement in this area with the introduction of the Banach Contraction Principle. This principle states that *if* ψ *is a function from a complete metric space* (Y, d) *into itself satisfying* $d(\psi(\tau_1), \psi(\tau_2)) \leq \lambda d(\tau_1, \tau_2)$ *for all* $\tau_1, \tau_2 \in Y$ *and for some* λ, $0 \leq \lambda < 1$, *then function* ψ *possesses a unique fixed point*. The Banach Contraction Principle not only guarantees the existence of fixed points but also provides a systematic approach to computing them using iterative algorithms. Over time, numerous iterative algorithms have been developed and can be found in literature. This concept has evolved to become essential not only in nonlinear analysis but also in various other mathematical fields. Moreover, its applications extend to science and technology. As a result of its significance, the Banach Contraction Principle has been subject to many extensions and generalizations, further broadening its scope and applicability.

A nonexpansive mapping is one whose Lipschitz constant is equal to one. The existence results for non-expansive mappings were independently established by Browder (1965), Göhde (1965), and Kirk (1965). Since then, numerous researchers have extended and generalized their findings in various ways. In 1970, Takahashi (1970) pioneered the investigation of fixed point theory for nonexpansive mappings in hyperbolic metric spaces. He referred to members of this class as convex metric spaces. Goebel and Kirk (1983) extended this concept by using hyperbolic type spaces, which encompass spaces with hyperbolic metrics (also see Gromov et al. (1999)). In 1990, Reich and Shafrir (1990) further developed the theory by considering hyperbolic metric spaces. Kohlenbach (2005) proposed a more comprehensive definition of hyperbolic metric spaces in 2004, which accommodates previous definitions of hyperbolic metric spaces. Busemann spaces, which were introduced by Busemann (1948) in 1948, are widely recognized as prominent instances of hyperbolic metric spaces. In a more contemporary context, Abdullatif et al. (2016) established a fixed point theorem concerning a monotone nonexpansive mapping within partially ordered hyperbolic metric spaces. Overall, these developments have seen contributions from various researchers and have led to the exploration of different types of hyperbolic metric spaces, opening up new avenues for studying fixed point properties in such settings.

Banach (1922) introduced the Picard iteration method in his Banach contraction principle—the first fixed point approximation method. Subsequently, numerous researchers have continued to delve into this field and have developed various fixed point approximation techniques tailored to different types of mappings. In the modern era, there is a significant amount of research focused on studying the convergence of fixed point approximation methods in hyperbolic spaces. Many authors have explored various extensions and generalizations of nonexpansive mappings Dashputre et al. (2021), Dashputre et al. (2020), Dashputre and Sakure (2024), Kim et al. (2021), Kim et al. (2022), Suzuki (2008). This continued research highlights the ongoing importance and relevance of fixed point approximation methods, particularly in the context of hyperbolic spaces, and their applications in various mathematical scenarios. The investigation of extensions and generalizations of nonexpansive mappings further enriches the field and opens up new avenues for exploring fixed point properties in different mathematical settings. In the available literature, numerous iterative algorithms have been put forward to approximate fixed points of nonexpansive mappings. These algorithms include Mann's iteration (Mann, 1953), Ishikawa's iteration (Ishikawa, 1974), Noor's iteration (Noor, 2000), S-iteration process (Agarwal et al., 2007), normal S-iteration (Sahu, 2011) and Picard normal S-iteration process (Kadioglu & Yildrim, 2015).

Recently, Kim et al. (2022) introduced a novel iterative algorithm termed as SPK iterative algorithm, which is presented in their paper. The algorithm operates in the context of a complete uniformly convex hyperbolic space (UCHS) denoted as (H, d, W), where H is a set equipped with a metric d and a gauge function W. Within this space, they consider a nonempty, convex, and closed subset D of Y, and $\psi : D \to P(D)$ represents a multivalued mapping. The SPK iterative algorithm is designed to generate a sequence $\{\tau_m\}$ as follows:

$$\begin{cases} \tau_1 \in D \\ \upsilon_a = \psi((1-\beta_a)\tau_a + \beta_a \psi(\tau_a)), \\ u_a = \psi(\upsilon_a), \\ \tau_{a+1} = \psi((1-\alpha_a)\psi(\upsilon_a) + \alpha_a \psi(u_a)),\ \forall a \in \mathbb{N}, \end{cases} \tag{7.1}$$

where $\{\alpha_a\}$ and $\{\beta_a\}$ are real sequences with values in the open interval $(0, 1)$.

In this chapter, we explore the SPK iterative algorithm, which was introduced by Kim et al. (2022) in 2022. The algorithm is employed to approximate the fixed point of a monotone RS-Type nonexpansive mapping within the context of partially ordered hyperbolic spaces (POHS). The chapter is structured into six sections. We present fundamental definitions and results that form the basis for our main findings in Sect. 7.2. Section 7.3 provides the strong and $\Delta-$convergence results for the SPK iterative algorithm applied to the monotone RS-Type nonexpansive mapping. A supporting example is demonstrated in Sect. 7.4 to validate our key results. We show the application of our conclusions to solving nonlinear integral equations in Sect. 7.5. The last section encapsulates the essence of the chapter, serving as a conclusion.

7.2 Preliminaries

Recall the following definition presented by Kohlenbach (2005).

Definition 7.2.1 Assuming we have a metric space (H, d), the triplet (H, d, W) is regarded as a hyperbolic metric space if there exists a mapping $W : H \times H \times [0, 1] \to H$ with the following properties for all $s, t, u, v \in H$ and $\beta, \gamma \in [0, 1]$:

(i) $d(u, W(s, t, \beta)) \leq \beta d(u, s) + (1 - \beta)\, d(u, t)$
(ii) $d(W(s, t, \beta), W(s, t, \gamma)) = |\beta - \gamma|\, d(s, t)$
(iii) $W(s, t, \beta) = W(t, s, (1 - \beta))$
(iv) $d(W(s, t, \beta), W(u, v, \beta)) \leq \beta d(s, t) + (1 - \beta)\, d(u, v)$.

Definition 7.2.2 (Leuştean (2007)) Suppose we have a hyperbolic metric space (H, d, W). Then H is considered uniformly convex if there exists a value δ in the range $(0, 1]$ such that the following inequality holds for any $s, \tau, \theta \in H$, $\varepsilon \in (0, 2]$, and $\rho > 0$:

$$d\left(\frac{1}{2}s \oplus \frac{1}{2}\tau, \theta\right) \leq (1 - \delta)\rho,$$

such that $d(s, \theta) \leq \rho$, $d(\tau, \theta) \leq \rho$, and $d(s, \tau) \geq \varepsilon\rho$.

Considering a hyperbolic metric space (H, d, W), and D being a nonempty subset of H, if $\{\theta_m\}$ is a bounded sequence in H, and θ belongs to D. Then

1. $r(\{\theta_m\}, \theta) = \limsup_{m\to\infty} d(\theta_m, \theta)$ is referred to as the asymptotic radius of the sequence $\{\theta_m\}$ at θ;
2. $$r(\{\theta_m\}, M) = \inf\{r(\{\theta_m\}, \theta) : \theta \in Y\}$$

   is referred to as the asymptotic radius of the sequence $\{\theta_m\}$ with respect to M; and
3. $$A(\{\theta_m\}, M) = \{\theta \in Y : r(\{\theta_m\}, M) = r(\{\theta_m\}, \theta)\}$$

   is referred to as the asymptotic center of the sequence $\{\theta_m\}$ with respect to M.

Lim (1981) introduced the concept of Δ-convergence within a metric space. Kirk and Panyanak (2008) extended this idea to CAT(0) spaces.

Definition 7.2.3 (Kirk and Panyanak (2008)) Suppose we have a hyperbolic metric space (H, d, W), and $\{\theta_m\} \in H$ is a bounded sequence. Then, $\{\theta_m\}$ is said to Δ-converge to a point $\theta \in H$ if θ serves as the unique asymptotic center for every subsequence $\{\theta_{m_k}\}$ of $\{\theta_m\}$.

Definition 7.2.4 (Bin Dehaish and Khamsi (2016)) Assume that (H, d, W) represents a metric space with the partial order $\preceq$ and $\tau_1, \tau_2 \in H$. A self-mapping ψ on H is considered as

1. monotone, if for all $\tau_1, \tau_2 \in H$,

$$\psi(\tau_1) \preceq \psi(\tau_2), \quad \text{whenever } \tau_1 \preceq \tau_2.$$

2. monotone nonexpansive, if ψ is monotone and $\tau_1, \tau_2 \in H$,

$$d(\psi(\tau_1), \psi(\tau_2)) \leq d(\tau_1, \tau_2),$$

where τ_1 and τ_2 are comparable.

In 2018, Pandey et al. (2018) introduced the notion of a monotone Reich-Suzuki type (RS-Type) nonexpansive mapping in the following manner:

Definition 7.2.5 Let (H, d, W) represent a metric space with the partial order $\preceq$. A self mapping ψ on H is referred to monotone Reich-Suzuki type (RS-Type) nonexpansive mapping, if ψ is monotone and

$$\frac{1}{2}d(\tau_1, \psi(\tau_1)) \leq d(\tau_1, \tau_2) \implies d(\psi(\tau_1), \psi(\tau_2)) \leq \alpha d(\tau_1, \psi(\tau_1)) + \alpha d(\psi(\tau_2), \tau_2) + (1 - 2\alpha)d(\tau_1, \tau_2)$$

for all $\tau_1, \tau_2 \in H$, where τ_1 and τ_2 are comparable.

Next, we provide several results that will be utilized in the proof of our main findings.

Proposition 7.2.6 (Pandey (2018)) *Suppose we have a nonempty subset D of a POHS $(H, d, \preceq)$, and $\psi : D \to D$ is a monotone RS-Type nonexpansive mapping with $F(\psi) \neq \emptyset$. Then, ψ is quasi-nonexpansive mapping.*

Lemma 7.2.7 (Khan et al. (2012)) *Assuming that (H, d, W) is a UCHS with a monotone modulus of uniform convexity η, and $\tau^* \in H$, consider the sequence α_a, where $0 < p \leq \alpha_a \leq q < 1$ for all $a \in \mathbb{N}$. If $\{u_a\}$ and $\{v_a\}$ are sequences in H satisfying $\limsup_{a\to\infty} d(u_a, \tau) \leq \rho$, $\limsup_{a\to\infty} d(v_a, \tau) \leq r\rho$, and $\limsup_{a\to\infty} d(\alpha_a v_a \oplus (1 - \alpha_a)u_a, \tau^*) = \rho)$, where $\rho \geq 0$, then it follows that $\lim_{a\to\infty} d(u_a, v_a) = 0$.*

Lemma 7.2.8 *Assume that D is a nonempty subset of a POHS $(H, d, \preceq)$, and $\psi : D \to D$ is a monotone RS-Type nonexpansive mapping. Then, for all $\theta_1, \theta_2 \in D$ such that θ_1 and θ_2 are comparable, we have:*

$$d(\theta_1, \psi(\theta_2)) \leq \left(\frac{3 + \alpha}{1 - \alpha}\right)d(\theta_1, \psi(\theta_1)) + d(\theta_1, \theta_2).$$

The following outcome follows as a consequence of Pandey et al.'s work in Pandey (2018).

Lemma 7.2.9 *Assume that D is a nonempty, closed, and convex subset of a uniformly convex POHS* $(H, d, \preceq)$, *and* $\psi : D \to D$ *is a monotone mapping. Let* $\theta_1 \in D$ *be such that* $\theta_1 \preceq \psi(\theta_1)$ *(or* $\psi(\theta_1) \preceq \theta_1$*). Then, for the sequence* $\{\theta_a\}$ *defined by the SPK iterative algorithm* (7.1), *the following hold:*

1. $\theta_a \preceq \psi(\theta_a) \preceq \theta_{a+1}$ *(or* $\theta_{a+1} \preceq \psi(\theta_a) \preceq \theta_a$*).*
2. $\theta_a \leq \theta^*$ *(or* $\theta^* \preceq \theta_a$*), provided* $\{\theta_a\}$ Δ*-converges to a point* $\theta^* \in D$*, for all* $a \subset \mathbb{N}$.

7.3 Main Results

In this section, we will investigate the strong convergence and Δ-convergence results of the SPK iterative algorithm (7.1) in partially ordered hyperbolic spaces (POHS).

Theorem 7.3.1 *Consider a nonempty, closed, and convex subset D of a uniformly convex POHS* $(H, d, \preceq)$, *where* $\psi : D \to D$ *is a monotone RS-Type nonexpansive mapping. Assume that* $F(\psi)$ *is nonempty, and for every* $\theta^* \in F(\psi)$, θ_1 *and* θ^* *are comparable. Let* $\theta_1 \in D$ *be such that* θ_1 *and* $\psi(\theta_1)$ *are comparable. We define the sequence* $\{\theta_a\}$ *using the SPK iterative algorithm* (7.1). *The following statements hold:*

1. *The sequence* $\{\theta_a\}$ *is bounded, and* $\lim_{a\to\infty} d(\theta_a, \theta^*)$ *exists.*
2. *For all* $a \in \mathbb{N}$, $\max\{d(\theta_{a+1}, \theta^*), d(u_a, \theta^*), d(v_a, \theta^*)\} \leq d(\theta_a, \theta^*)$.

Proof Without loss of generality, we assume that $\theta_1 \preceq \theta^*$. We have $\psi(\theta_1) \preceq \psi(\theta^*) = \theta^*$, because of monotonocity of ψ. Using the definition of the SPK iterative algorithm (7.1), one gets

$$\theta_1 \preceq v_1 \preceq \psi(\theta_1) \preceq u_1 \leq \psi(v_1). \tag{7.2}$$

So, $v_1 \leq \tau^*$. Now, we have

$$\psi(v_1) \preceq \psi(\theta^*) = \theta^*,$$

and also, from (7.2), we have $w_1 \preceq \theta^*$. Since ψ is monotone,

$$\psi(u_1) \preceq \psi(\theta^*) = \theta^*, \tag{7.3}$$

By (7.1)

$$\psi(v_1) \preceq \theta_2 \preceq \psi(u_1) \tag{7.4}$$

From (7.3) and (7.4), we have

$$\psi(\theta_2) \preceq \psi(\theta^*) = \theta^*. \tag{7.5}$$

Again from Lemma 7.2.9 and (7.5), for $a = 2$, we have

$$\theta_2 \preceq \psi(\theta_2) \preceq \theta^*.$$

Applying the same argument, we get

$$\theta_a \preceq \psi(\theta_a) \preceq \theta^*.$$

Now, by (7.1) and Proposition 7.2.6, one gets

$$\begin{aligned} d(v_a, \theta^*) &= d(\psi((1-\beta_a)\theta_a \oplus \beta_a\psi(\theta_a)), \theta^*) \\ &\leq (1-\beta_a)d(\theta_a, \theta^*) + \beta_a d(\theta_a, \theta^*) \\ &= d(\theta_a, \theta^*). \end{aligned} \tag{7.6}$$

$$\begin{aligned} d(u_a, \theta^*) &= d(\psi(v_a), \theta^*) \\ &\leq d(\theta_a, \theta^*) \end{aligned}$$

and

$$\begin{aligned} d(\theta_{a+1}, \theta^*) &= d((1-\alpha_a)\psi(u_a) \oplus \alpha_a\psi(v_a), \theta^*) \\ &\leq (1-\alpha_a)d(v_a, \theta^*) + \alpha_a d(v_a, \theta^*) \\ &\leq d(v_a, \theta^*) \\ &= d(\theta_a, \theta^*). \end{aligned} \tag{7.7}$$

Hence, the sequence $d(\theta_a, \theta^*)$ is both bounded and decreasing. $\lim_{a\to\infty} d(\theta_a, \theta^*)$ exists and

$$\max\{d(\theta_{a+1}, \theta^*), d(u_a, \theta^*), d(v_a, \theta^*) \leq d(\theta_a, \theta^*)\}$$

for all $a \in \mathbb{N}$.

□

Theorem 7.3.2 *Consider a nonempty, closed, and convex subset D of a uniformly convex POHS $(H, d, \preceq)$, where $\psi : D \to D$ is a monotone RS-Type nonexpansive mapping. Assume that $F(\psi)$ is nonempty, and for every $\theta^* \in F(\psi)$, θ_1 and θ^* are comparable. Let $\theta_1 \in D$ be such that θ_1 and $\psi(\theta_1)$ are comparable. We define the sequence $\{\theta_a\}$ using the SPK iterative algorithm (7.1). The following statements hold:*

1. *$\lim_{a\to\infty} dist(\theta_a, F(\psi))$ exists, where $dist(\theta, F(\psi))$ denotes the distance from θ to $F(\psi)$.*
2. $\lim_{a\to\infty} d(\theta_a, \psi(\theta_a)) = 0$.

Proof For every $\theta^* \in F(\psi)$ and $a \in \mathbb{N}$, one gets

$$d(\theta_{a+1}, \theta^*) \le d(\theta_a, \theta^*).$$

For $\tau^* \in F(\psi)$, one gets

$$\text{dist}(\theta_{a+1}, F(\psi)) \le \text{dist}(\theta_a, F(\psi))$$

for all $a \in \mathbb{N}$. Therefore the sequence $\text{dist}(\theta_a, F(\psi))$ is bounded and decreasing. Hence $\lim_{a\to\infty} \text{dist}(\theta_a, F(\psi))$ exists. Let

$$\lim_{a\to\infty} d(\theta_a, \theta^*) = l. \tag{7.8}$$

From (7.8) and Proposition 7.2.6, one gets

$$\limsup_{a\to\infty} d(\psi(\theta_a), \theta^*) \le l. \tag{7.9}$$

By (7.6) and (7.8), one gets

$$\limsup_{a\to\infty} d(v_a, \theta^*) \le l. \tag{7.10}$$

From (7.10) and Proposition 7.2.6, one gets

$$\limsup_{a\to\infty} d(\psi(v_a), \theta^*) \le l. \tag{7.11}$$

By (7.7) and (7.8), one gets

$$\limsup_{a\to\infty} d(u_a, \theta^*) \le l. \tag{7.12}$$

From (7.12) and Proposition 7.2.6, one gets

$$\limsup_{a\to\infty} d(\psi(u_a), \theta^*) \le l. \tag{7.13}$$

From (7.8) and (7.11), one get

$$\limsup_{a\to\infty} d(\theta_{a+1}, \theta^*) \le l. \tag{7.14}$$

Now,

$$\begin{aligned} l &= \limsup_{a\to\infty}\{\gamma_a d(\theta_a, \theta^*) + (1-\gamma_a) d(\psi(\theta_a), \theta^*)\} \\ &\le \limsup_{a\to\infty} d(\theta_a, \theta^*). \end{aligned} \tag{7.15}$$

From (7.15) and Lemma 7.2.7, one gets

$$\lim_{a\to\infty} d(\psi(\theta_a), \theta_a) = 0.$$

□

Now, we present the Δ-convergence result as follows.

Theorem 7.3.3 *Assume that D is a nonempty, closed, and convex subset of a uniformly convex POHS $(H, d, \preceq)$, where $\psi : D \to D$ is a monotone RS-Type nonexpansive mapping. Let θ_1 and θ^* be comparable for every $\theta^* \in F(\psi)$, where $F(\psi)$ is a nonempty and totally ordered set. Assume that there exists $\theta_1 \in D$ such that θ_1 and $\psi(\theta_1)$ are comparable. Then, the sequence θ_a defined by the SPK iterative algorithm (7.1) Δ-converges to the fixed point of ψ.*

Proof Due to Theorem 7.3.2, $\{\theta_a\}$ is a bounded sequence. Thus, there is a subsequence $\{\theta_{a_d}\}$ of $\{\theta_a\}$ such that $\{\theta_{a_d}\}$ Δ-converges to point $\theta^* \in M$ such that

$$\theta_1 \preceq \theta_{a_d} \preceq \theta^* \quad \forall d \in \mathbb{N}.$$

We will now demonstrate that for each Δ-convergent subsequence $\{\theta_{a_d}\}$ of $\{\theta_a\}$, there exists a unique Δ-limit in $F(\psi)$. Let us consider two subsequences $\{\theta_{a_d}\}$ and $\{\theta_{a_e}\}$ of $\{\theta_a\}$ that Δ-converge to l_1 and l_2, respectively. Since $\{\theta_{a_d}\}$ is bounded, according to Theorem 7.3.2, we can infer that

$$\lim_{a\to\infty} d(\psi(\theta_{a_d}), \theta_{a_d}) = 0 \tag{7.16}$$

which concludes that $l_1 \in F(\psi)$. Suppose that $\Gamma : D \to [0, \infty)$ is a type mapping generated by $\{\theta_{a_d}\}$. Then,

$$\Gamma(t) = \limsup_{a\to infty} d(\theta_{a_d}, l_1). \tag{7.17}$$

By Lemma 7.2.8,

$$\begin{aligned}\Gamma(\psi(l_1)) &= \limsup_{a\to\infty} d(\theta_{a_d}, \psi(l_1)) \\ &\le \frac{3+\alpha}{1-\alpha} \limsup_{a\to\infty} d(\theta_{a_d}, \psi((\theta_{a_d}))) + \limsup_{a\to\infty} d(\theta_{a_d}, l_1) \\ &\le \Gamma(l_1).\end{aligned} \tag{7.18}$$

Since l_1 is unique, one gets

$$\psi(l_1) = l_1.$$

Similarly,

$$\psi(l_2) = l_2.$$

Since Γ is continuous, one gets

$$\begin{aligned}
\limsup_{a\to\infty} d(\theta_a, l_1) &= \limsup_{d\to\infty} d(\theta_{a_d}, l_1) \\
&< \limsup_{d\to\infty} d(\theta_{a_d}, l_2) \\
&= \limsup_{a\to\infty} d(\theta_a, l_2) \\
&= \limsup_{e\to\infty} d(\theta_{a_e}, l_2) \\
&< \limsup_{e\to\infty} d(\theta_{a_e}, l_1) \\
&= \limsup_{a\to\infty} d(\theta_a, l_1),
\end{aligned}$$

which is a contradiction, unless $l_1 = l_2$. □

Theorem 7.3.4 *Assuming that D is a nonempty, closed, and convex subset of a uniformly convex POHS* $(H, d, \preceq)$, *and* $\psi : D \to D$ *is a monotone RS-Type nonexpansive mapping. Let* θ_1 *and* θ^* *be comparable for every* $\theta^* \in F(\psi)$, *and suppose there exists* $\theta_1 \in D$ *such that* θ_1 *and* $\psi(\theta_1)$ *are comparable. Also, let* $F(\psi)$ *be a nonempty set. The sequence* $\{\theta_a\}$, *defined by the SPK iterative algorithm* (7.1), *converges strongly to a fixed point of* ψ *if and only if, when* $F(\psi)$ *is totally ordered,* $\liminf_{a\to\infty}$ *dist* $(\theta_a, F(\psi)) = 0$.

Proof Assume that $\liminf_{a\to\infty} \text{dist}\,(\theta_a, F(\psi)) = 0$. From Theorem 7.3.2, $\lim_{a\to\infty} \text{dist}\,(\theta_a, F(\psi)) = 0$ exists, so

$$\lim_{a\to\infty} \text{dist}\,(\theta_a, F(\psi)) = 0.$$

First, we will prove that $F(\psi)$ is closed. For this, assume that $\{f_a\}$ is sequence in $F(\psi)$ converging to point $f^* \in D$. Since

$$\frac{1}{2} d(f_a, \psi(f_a)) = 0 \le d(f_a, f^*),$$

for all $a \in \mathbb{N}$, we have

$$\begin{aligned}
d(\psi(f_a), \psi(f^*)) &= d(f_a, \psi(f^*)) \\
&\le \alpha d(\psi(f_a), f_a) + \alpha d(\psi(f^*), f^*) + (1 - 2\alpha) d(f_a, f^*) \\
&\le \alpha d(\psi(f^*), f_a) + \alpha d(\psi(f_a), f^*) + (1 - 2\alpha) d(f_a, f^*).
\end{aligned}$$

As $1 - \alpha > 0$,

$$\limsup_{a \to \infty} d(f_a, f^*) \le \limsup_{a \to \infty} d(f_a, f^*) = 0,$$

therefore, $\{f_a\}$ converges strongly to $\psi(f^*)$, which implies that $\psi(f^*) = f^*$. Hence, $F(\psi)$ is closed. Since $\lim_{a \to \infty} \text{dist}\,(\theta_a, F(\psi)) = 0$, assume that $\{\theta_{a_n}\}$ is a subsequence of $\{\theta_a\}$ such that $d(\theta_{a_n}, f_r) \le \frac{1}{2^r}$ for all $r \ge 1$, where $\{f_r\}$ is a sequence in $F(\psi)$. Due to Theorem 7.3.2, we get

$$d(\theta_{a_{r+1}}, f_r) \le d(\theta_{a_r}, f_r) \le \frac{1}{2^r},$$

hence by triangular inequality,

$$\begin{aligned} d(f_{r+1}, f_r) &\le d(f_{r+1}, u_{a_{r+1}}) + d(u_{a_{r+1}}, f_r) \\ &< \frac{1}{2^{r-1}}, \end{aligned}$$

which shows that $\{f_r\}$ is a Cauchy sequence. Since $F(\psi)$ is closed, so that $\{f_r\}$ converges to some point $f^* \in F(S)$. Now,

$$d(\theta_{a_r}, f^*) \le d(\theta_{a_r}, f_r) + d(f_r, f^*),$$

as $r \to \infty$, $\{\theta_{a_n}\}$ converges strongly to f^*. Due to Theorem 7.3.2, $\lim_{a \to \infty} d(\theta_a, f^*)$ exists, so that $\{\theta_a\}$ converges strongly to f^*. The converse part of the Theorem is trivial, hence the proof is completed.

□

In the following result, we demonstrate the strong convergence of the sequence $\{\theta_a\}$ generated by the SPK iterative algorithm (7.1) using Condition (I).

Definition 7.3.5 (Senter and Dotson (1974)) Assuming that D is a nonempty, convex, and closed subset of the hyperbolic space H, the mapping $\psi : D \to D$ is said to satisfy *Condition (I)* if there exists a non-decreasing function $\phi : [0, \infty) \to [0, \infty)$ with $\phi(0) = 0$ and $\phi(r) > 0$ for $r > 0$ such that for all $v \in D$,

$$d(v, \psi(v)) \ge \phi(d(v, F(\psi)) \;\; \forall v \in D.$$

Theorem 7.3.6 *In the context of a uniformly convex POHS $(H, d, \preceq)$, consider a nonempty, closed, and convex subset D. Let $\psi : D \to D$ be a monotone RS-Type nonexpansive mapping. We make the assumption that θ_1 and θ^* are comparable for every $\theta^* \in F(\psi)$. Additionally, suppose there exists a point $\theta_1 \in D$ such that θ_1 and $\psi(\theta_1)$ are comparable, and $F(\psi)$ is nonempty. Under the condition of Condition (I) being satisfied by ψ, the SPK iterative algorithm* (7.1) *generates a sequence $\{\theta_a\}$ that converges strongly to a fixed point of ψ.*

Proof Due to Theorem 7.3.2,

$$\liminf_{a\to\infty} d(\psi(\theta_a), \theta_a) = 0.$$

Since ψ satisfies Condition (I), we get

$$d(\psi(\theta_a), \theta_a) \geq \phi(\text{dist}\,(\theta_a, F(\psi)),$$

so, we have

$$\liminf_{a\to\infty} \phi(\text{dist}\,(\theta_a, F(\psi)) = 0.$$

Therefore, by assumption, one gets

$$\liminf_{a\to\infty} \text{dist}\,(\theta_a, F(\psi)) = 0.$$

As all the assumptions of Theorem 7.3.4 are satisfied, so $\{\theta_a\}$ converges to a fixed point of ψ. □

7.4 Example

We demonstrate a monotone RS-Type nonexpansive mapping in this section.

Example 7.4.1 Assume that $H = \mathbb{R}$ is equipped with the usual norm $\|\theta\| = |\theta|$ and usual ordering, $D = [0, 1] \subset \mathbb{R}$, $\psi : D \to D$ is mapping defined by

$$\psi(\theta) = \begin{cases} 1 - \frac{\theta}{3},\ if\ \theta \in [0, \frac{1}{7}) \\ \frac{\theta}{7} + \frac{1}{5},\ if\ \theta \in [\frac{1}{7}, 1]. \end{cases}$$

Then ψ is monotone RS-Type nonexpansive mapping while ψ is not nonexpansive mapping. Indeed, for $\theta_1 = 0$ and $\theta_2 = \frac{1}{7}$, we have

$$\|\psi(\theta_1) - \psi(\theta_2)\| = \frac{11}{14} > \frac{1}{7} = \|\theta_1 - \theta_2\|.$$

Now, we will demonstrate that ψ is a monotone RS-Type nonexpansive mapping with $\alpha = \frac{1}{2}$. We will consider the following cases:

Case I: Assume that $\theta_1, \theta_2 \in [0, \frac{1}{7})$, we get

$$\begin{aligned}\frac{1}{2}\Big(\|\psi(\theta_1)-\theta_1\|+\|\psi(\theta_2)-\theta_2\|\Big) &= \frac{1}{2}\Big(\Big|\frac{4\theta_1}{3}-1\Big|+\Big|\frac{4\theta_2}{3}-1\Big|\Big)\\ &\geq \frac{1}{2}\Big|\Big(\frac{4\theta_1}{3}-1\Big|-\Big|\Big(\frac{4\theta_2}{3}-1\Big|\Big)\\ &= \frac{2}{3}\big|\theta_1-\theta_2\big|\\ &\geq \frac{1}{3}\big|\theta_1-\theta_2\big|\\ &= \big\|\psi(\theta_1)-\psi(\theta_2)\big\|.\end{aligned}$$

Case II: Assume that $\theta_1, \theta_2 \in [\frac{1}{7}, 1]$, we get

$$\begin{aligned}\frac{1}{2}\Big(\|\psi(\theta_1)-\theta_1\|+\|\psi(\theta_2)-\theta_2\|\Big) &= \frac{1}{2}\Big(\Big|\frac{6\theta_1}{7}-\frac{1}{5}\Big|+\Big|\frac{6\theta_2}{7}-\frac{1}{5}\Big|\Big)\\ &\geq \frac{1}{2}\Big|\Big(\frac{6\theta_1}{7}-\frac{1}{5}\Big|-\Big|\Big(\frac{6\theta_2}{7}-\frac{1}{5}\Big|\Big)\\ &= \frac{3}{7}\big|\theta_1-\theta_2\big|\\ &\geq \frac{1}{7}\big|\theta_1-\theta_2\big|\\ &= \big\|\psi(\theta_1)-\psi(\theta_2)\big\|.\end{aligned}$$

Case III: Assume that $\theta_1 \in [0, \frac{1}{7})$ and $\theta_2 \in [\frac{1}{7}, 1]$, we get

$$\begin{aligned}\frac{1}{2}\Big(\|\psi(\theta_1)-\theta_1\|+\|\psi(\theta_2)-\theta_2\|\Big) &= \frac{1}{2}\Big(\Big|\frac{4\theta_1}{3}-1\Big|+\Big|\frac{6\theta_2}{7}-\frac{1}{5}\Big|\Big)\\ &\geq \frac{1}{2}\Big|\Big(\frac{3\theta_1}{4}-1\Big|-\Big|\Big(\frac{6\theta_2}{7}-\frac{1}{5}\Big|\Big)\\ &= \frac{1}{105}\Big|70-70\theta_1+\frac{9}{2}\theta_2\Big|\\ &\geq \frac{1}{105}\big|84-35\theta_1-15\theta_2\big|\\ &= \big\|\psi(\theta_1)-\psi(\theta_2)\big\|.\end{aligned}$$

As a result, ψ is a monotone RS-Type nonexpansive mapping with a unique fixed point $\frac{7}{30}$.

7.5 Application

In this section, we will explore the application of our discoveries to nonlinear integral equations within the context of an ordered Banach space (OBS). It's essential to note that an OBS is a specific type of Partially Ordered Hilbert Space (POHS). Within this framework, we will present a theorem that establishes the existence of solutions to these equations.

Theorem 7.5.1 *Consider a nonempty compact subset D of an ordered Banach space H, and let $\psi : D \to D$ be a monotone RS-Type nonexpansive mapping. If an AFPS $\{\theta_m\}$ exists, then it possesses the existence of at least one fixed point for ψ within the set D.*

Proof Due to Theorem 7.3.2 and Suzuki (2008, Theorem 4), the proof is completed. □

Take the following nonlinear functional integral equation (FIE) as an example.

$$\zeta(\theta) = f(\theta) + \lambda_1 \int_0^\theta \kappa(\theta, \varrho) g(s, \zeta(\varrho)) d\varrho + \lambda_2 \int_0^1 \overline{\kappa}(\theta, \varrho) \overline{g}(\varrho, \zeta(\varrho)) d\varrho \tag{7.19}$$

for all $\theta, \varrho \in [0, 1]$, where λ_1, λ_2 are positive constants and $f : [0, 1] \longrightarrow \mathbb{R}_+, g, \overline{g} : [0, 1] \times \mathbb{R} \longrightarrow \mathbb{R}_+, \kappa, \overline{\kappa} : [0, 1] \times [0, 1] \longrightarrow \mathbb{R}_+$. Consider the following assumptions.

(K_1) The function $f : [0, 1] \longrightarrow [0, 1]$ is continuous.
(K_2) The functions $\kappa, \overline{\kappa} : [0, 1] \times [0, 1] \longrightarrow \mathbb{R}_+$ are continuous such that for all $\theta, \varrho \in [0, 1]$

$$\int_0^\theta \kappa(\theta, \varrho) d\varrho \leq K_1 \quad \text{and} \quad \int_0^1 \overline{\kappa}(\theta, \varrho) d\varrho \leq K_2.$$

(K_3) The functions $g, \overline{g} : [0, 1] \times \mathbb{R} \longrightarrow \mathbb{R}$ are continuous and there are two constants L_1, L_2 such that for all $t \in [0, 1]$; $\zeta_1, \zeta_2 \in C$,

$$|g(\theta, \zeta_1) - g(\theta, \zeta_2)| \leq L_1 |\zeta_1(\theta) - \zeta_2(\theta)|$$

$$|\overline{g}(\theta, \zeta_1) - \overline{g}(\theta, \zeta_2)| \leq L_2 |\zeta_1(\theta) - \zeta_2(\theta)|$$

(K_4) $\lambda_1 L_1 K_1 + \lambda_2 L_2 K_2 = 1$.

Theorem 7.5.2 *Let us consider the space $H = C[0, 1]$, which comprises continuous functions defined on the interval $[0, 1]$. In this context, let C represent a compact subset of H, and the norm in H is established using the supremum norm denoted* *as* $\|\zeta_1 - \zeta_2\| = \sup_{\theta\in[0,1]} |\zeta_1(\theta) - \zeta_2(\theta)|$. *Under the assumptions* $(\mathbf{K_1})$ *to* $(\mathbf{K_4})$*, we have a mapping* $\psi : C \to C$. $(\mathbf{K_1})$ *to* $(\mathbf{K_4})$ *are the specified conditions that hold in this context such that*

$$\psi(\zeta(\theta)) = f(\theta) + \lambda_1 \int_0^\theta \kappa(\theta, \varrho) g(\varrho, \zeta(\varrho)) d\varrho + \lambda_2 \int_0^1 \overline{\kappa}(\theta, \varrho)\overline{g}(\varrho, \zeta(\varrho)) d\varrho.$$

Then the nonlinear FIE (7.19) has solution in $H = C[0, 1]$.

Proof Assume that $\zeta_1, \zeta_2 \in C$. Then

$$\begin{aligned}
&|\psi(\zeta_1(\theta)) - \psi(\zeta_2(\theta))| \\
&= \Big|f(\theta) + \lambda_1 \textstyle\int_0^\theta \kappa(\theta, \varrho) g(\varrho, \zeta_1(\varrho)) d\varrho + \lambda_2 \int_0^1 \overline{\kappa}(\theta, \varrho)\overline{g}(\varrho, \zeta_1(\varrho)) d\varrho \\
&\quad - f(\theta) - \lambda_1 \textstyle\int_0^\theta \kappa(\theta, \varrho) g(\varrho, \zeta_2(\varrho)) d\varrho - \lambda_2 \int_0^1 \overline{\kappa}(\theta, \varrho)\overline{g}(\varrho, \zeta_2(\varrho)) d\varrho\Big| \\
&\le \lambda_1 \textstyle\int_0^\theta \kappa(\theta, \varrho) \big|g(\varrho, \zeta_1(\varrho)) - g(\varrho, \zeta_2(\varrho))\big| d\varrho \\
&\quad + \lambda_2 \textstyle\int_0^1 \overline{\kappa}(\theta, \varrho) \big|\overline{g}(\varrho, \zeta_1(\varrho)) - \overline{g}(\varrho, \zeta_2(\varrho))\big| du \\
&\le \lambda_1 \textstyle\int_0^\theta \kappa(\theta, \varrho) L_1 \big|\zeta_1(\varrho) - \zeta_2(\varrho)\big| d\varrho + \lambda_2 \int_0^1 \overline{\kappa}(\theta, \varrho) L_2 \big|\zeta_1(\varrho) - \zeta_2(\varrho)\big| d\varrho.
\end{aligned}$$

On taking supremum of both sides, we get

$$\begin{aligned}
\|\psi(\zeta_1) - \psi(\zeta_2)\| &\le (\lambda_1 K_1 L_1 + \lambda_2 K_2 L_2)\|\zeta_1 - \zeta_2\| \\
&<= \|\zeta_1 - \zeta_2\|.
\end{aligned}$$

This demonstrates that the mapping ψ satisfies the properties of being a monotone RS-Type nonexpansive mapping with $\alpha = 0$. All the conditions required by Theorem 7.5.1 are met, implying that ψ possesses a fixed point. Consequently, the nonlinear Functional Integral Equation (FIE) represented by (7.19) has a solution within the compact subset C of $H = C[0, 1]$. □

Corollary 7.5.3 *Let us consider the space* $H = C[0, 1]$*, which comprises continuous functions defined on the interval* $[0, 1]$. *In this context, let C represent a compact subset of* H*, and the norm in* H *is established using the supremum norm denoted as* $\|\zeta_1 - \zeta_2\| = \sup_{\theta\in[0,1]} |\zeta_1(\theta) - \zeta_2(\theta)|$. *Under the assumptions* $(\mathbf{K_1})$ *to* $(\mathbf{K_4})$*, we have a mapping* $\psi : C \to C$. $(\mathbf{K_1})$ *to* $(\mathbf{K_4})$ *are the specified conditions that hold in this context such that*

$$\psi(\zeta(\theta)) = f(\theta) + \lambda_1 \int_0^\theta \kappa(\theta, \varrho) g(\theta, \zeta(\varrho)) d\varrho.$$

Then the nonlinear FIE

$$\zeta(\theta) = f(\theta) + \lambda_1 \int_0^\theta \kappa(\theta, \varrho) g(\varrho, \zeta(\varrho)) d\varrho.$$

has solution in $H = C[0, 1]$.

Proof On setting $\overline{\kappa}(\theta, \varrho) \equiv 0$, we get the desired result. □

Corollary 7.5.4 *Let us consider the space* $H = C[0, 1]$, *which comprises continuous functions defined on the interval* $[0, 1]$. *In this context, let C represent a compact subset of H, and the norm in H is established using the supremum norm denoted as* $\|\zeta_1 - \zeta_2\| = \sup_{\theta \in [0,1]} |\zeta_1(\theta) - \zeta_2(\theta)|$. *Under the assumptions* ($\mathbf{K_1}$) *to* ($\mathbf{K_4}$), *we have a mapping* $\psi : C \to C$. ($\mathbf{K_1}$) *to* ($\mathbf{K_4}$) *are the specified conditions that hold true in this context such that*

$$\psi(\zeta(\theta)) = f(\theta) + \lambda_2 \int_0^1 \overline{\kappa}(\theta, \varrho) \overline{g}(\varrho, \zeta(\varrho)) d\varrho.$$

Then the nonlinear FIE

$$\zeta(\theta) = f(\theta) + \lambda_2 \int_0^1 \overline{\kappa}(\theta, \varrho) \overline{g}(\varrho, \zeta(\varrho)) d\varrho.$$

has solution in $H = C[0, 1]$ *provided that* ψ *has fixed point.*

Proof On setting $\kappa(\theta, \varrho) \equiv 0$, we get the desired result. □

Example 7.5.5 Let us consider the following nonlinear FIE : For $\theta, s \in [0, 1]$,

$$\zeta(\theta) = (\theta^2 + 4) + \frac{3}{20} \int_0^\theta [(\theta^3 + 1)(2\varrho + 4)] \frac{|\zeta(\varrho)|}{3} d\varrho + \frac{5}{12} \int_0^1 [\theta^2 (2\varrho + 5)] \frac{|\zeta(\varrho)|}{5} d\varrho. \tag{7.20}$$

If we take

$$\lambda_1 = \frac{3}{20}, \ \lambda_2 = \frac{5}{12}, \ f(\theta) = \theta^2 + 4;$$

$$\kappa(\theta, \varrho) = (\theta^3 + 1)(2\varrho + 4), \qquad \overline{\kappa}(\theta, \varrho) = \theta^2 (2\varrho + 5);$$

$$g(\varrho, \zeta(\varrho)) = \frac{|\zeta(\varrho)|}{3} \qquad \overline{g}(\varrho, \zeta(\varrho)) = \frac{|\zeta(\varrho)|}{5}.$$

Then (7.20) will be in the form of (7.19).

It is clear that function $f(\theta) = \theta^2 + 4$, $\forall \theta \in [0, 1]$ is continuous.

For each $\theta, \varrho \in [0, 1]$,

$$\int_0^1 \kappa(\theta, \varrho) = \int_0^1 (\theta^3 + 1)(2\varrho + 4)d\varrho \leq 10;$$

$$\int_0^1 \overline{\kappa}(\theta, \varrho) = \int_0^\theta \theta^2(2\varrho + 5)d\varrho \leq 6.$$

For $\zeta_1, \zeta_2 \in C$; $\varrho \in [0, 1]$,

$$\begin{aligned}\left|g(\varrho, \zeta_1(\varrho)) - g(\varrho, \zeta_2(\varrho))\right| &= \left|\tfrac{|\zeta_1(\varrho)|}{3} - \tfrac{|\zeta_2(\varrho)|}{3}\right| \\ &= \tfrac{1}{3}\left||\zeta_1(\varrho)| - |\zeta_2(\varrho)|\right| \\ &\leq \tfrac{1}{3}\left|\zeta_1(\varrho) - \zeta_2(\varrho)\right|\end{aligned}$$

$$\begin{aligned}\left|\overline{g}(\varrho, \zeta_1(\varrho)) - \overline{g}(\varrho, \zeta_2(\varrho))\right| &= \left|\tfrac{|\zeta_1(\varrho)|}{5} - \tfrac{|\zeta_2(\varrho)|}{5}\right| \\ &= \tfrac{1}{5}\left||\zeta_1(\varrho)| - |\zeta_2(\varrho)|\right| \\ &\leq \tfrac{1}{5}\left|\zeta_1(\varrho) - \zeta_2(\varrho)\right|\end{aligned}$$

Since all assumptions of Theorem 7.5.2 are satisfied with $K_1 = 10$, $K_2 = 6$ and $\lambda_1 K_1 L_1 + \lambda_2 K_2 L_2 = 1$. Therefore nonlinear FIE (7.20) has a solution.

7.6 Conclusion

This chapter focused on the utilization of the SPK iterative algorithm to estimate fixed points of monotone Reich-Suzuki type nonexpansive mappings within partially ordered hyperbolic spaces. The primary aim was to show the convergence outcomes of applying this algorithm to such mappings. Moreover, an example was provided to illustrate and reinforce the main findings. Lastly, the chapter explored the implications of these results in the context of nonlinear integral equations.

References

Abdullatif, B., Dehaish, B., & Khamsi, M. A. (2016). Browder and Göhde fixed point theorem for monotone nonexpansive mappings. *Fixed Point Theory Application, 1*, 1–9.

Agarwal, R. P., O'Regan, D., & Sahu, D. R. (2007). Iterative construction of fixed points of nearly asymptotically nonexpansive mappings. *Journal of Nonlinear and Convex Analysis, 8*(1), 61–79.

Banach, S. (1922). Sur les opérations dans les ensembles abstraits et leur applications aux équations intégrals. *Fundamenta Mathematicae, 3*, 133–181.

Bin Dehaish, B. A., & Khamsi, M. A. (2016). Browder and Göhde fixed point theorem for monotone nonexpansive mappings. *Fixed Point Theory Application, 2016*(1), 1–9.

Busemann, H. (1948). Spaces with non-positive curvature. *Acta Mathematica, 80*, 259–310.

Browder, F. E. (1965) Nonexpansive nonlinear operators in a Banach space. *Proceedings of the National Academy of Sciences of the USA, 54*(4), 1041–1044.

Dashputre, S., Padmavati, & Sakure, K. (2021). Fixed point approximation of countably infinite family of nonexpansive mappings. *South East Asian Journal of Mathematics and Mathematical Sciences, 17*(1), 361–382.

Dashputre, S., Padmavati, & Sakure, K. (2020). Strong and Δ-convergence results for generalized nonexpansive mapping in hyperbolic space. *Communications in Mathematics and Applications, 11*(3), 389–401.

Dashputre, S., & Sakure, K. (2024). Monotone generalized (α, β)-nonexpansive mappings in ordered hyperbolic metric spaces. *Journal of Advanced Mathematical Studies, 17*(2), 250–263.

Goebel, K., & Kirk, W. A. (1983). Iteration processes for nonexpansive mappings. *Contemporary Mathematics, 21*, 115–123.

Göhde, D. (1965). Zum Prinzip der kontraktiven Abbildung. *Mathematische Nachrichten, 30*, 251–258.

Gromov, M., Katz, M., Pansu, P., & Semmes, S. (1999). *Metric structures for Riemannian and non-Riemannian spaces*. Boston: Birkhäuser Boston, Inc.

Ishikawa, S. (1974). Fixed points by a new iteration method. *Proceedings of the American Mathematical Society, 44*(1), 147–150.

Kadioglu, N., & Yildrim, I. (2015). Approximating fixed points of nonexpansive mappings by faster iteration process. *Journal of Advanced Mathematical Studies, 8*(2), 257–264.

Khan, A. R., Fukhar-ud-din, H., & Khan, M. (2012). An implicit algorithm for two finite families of nonexpansive maps in hyperbolic spaces. *Fixed Point Theory Application, 54*(12), 1–12.

Kirk, W. A. (1965). A fixed point theorem for mappings which do not increase distances. *Am. Math. Mon., 72*(9), 1004–1006.

Kirk, W. A., & Panyanak, B. (2008). A concept of convergence in geodesic spaces. *Nonlinear Anal., 68*, 3689–3969.

Kim, J. K., Dashputre, S., Padmavati, & Sakure, K. (2021). Convergence results for generalized nonexpansive multivalued mapping in hyperbolic spaces via SP-iteration process. *Journal of Nonlinear and Convex Analysis,22*(4), 839–854.

Kim, J. K., Dashputre, S., Padmavati, & Sakure, K. (2022). Generalized α-nonexpansive mappings in hyperbolic spaces. *Nonlinear Functional Analysis and Applications, 27*(3), 449–469.

Kohlenbach, U. (2005). Some logical metatheorems with applications in functional analysis. *Transactions of the American Mathematical Society,357*(1), 89–128.

Leuştean, L. (2007). A quadratic rate of asymptotic regularity for CAT(0) spaces. *Journal of Mathematical Analysis and Applications, 325*(1), 386–399.

Li, Z., Shu, X. B., & Miao, T. (2021). The existence of solutions for Sturm-Liouville differential equation with random impulses and boundary value problems. *Boundary Value Problem, 2021*, 1–23.

Lim, T. C. (1981). On the center of convex set. *Proceedings of the American Mathematical Society, 81*(2), 345–346.

Mann, W. R. (1953). Mean value methods in iteration. *Proceedings of the American Mathematical Society, 4*, 506–610.

Noor, M. A. (2000). New approximation schemes for general variational inequalities. *Journal of Mathematical Analysis and Applications, 251*(1), 217–229.

Pandey, R., Pant, R., & Al-Rawashdeh, A. (2018). Fixed point results for a class of monotone nonexpansive type mappings in hyperbolic spaces. *Journal of Function Spaces, 2018*.

Picard, E. (1890). Mémoire sur la théorie des équations aux dérivées partielles et la méthode des approximations successive. *Journal de Mathématiques Pures et Appliquées, 6*, 145–210.

Reich, S., & Shafrir, I. (1990). Nonexpansive iterations in hyperbolic spaces. *Nonlinear Anal., 15*(6), 537–558.

Sahu, D. R. (2011). Application of the s-iteration process to constrained minimization problems and feasibility problems. *Fixed Point Theory, 12*, 187–204.

Senter, H. F., & Dotson, W. G. (1974). Approximating fixed points of nonexpansive mappings. *Proceedings of the American Mathematical Society, 44*, 375–380.

Suzuki, T. (2008). Fixed point theorems and convergence theorems for some generalized nonexpansive mappings. *Journal of Mathematical Analysis and Applications, 340*(2), 1088–1095.

Takahashi, W. (1970). A convexity in metric space and nonexpansive mappings I. *Kodai Mathematical Seminar Reports, 22*(2), 142–149.

Tian, X., & Guo, S. (2022). Traveling wave solutions for nonlocal dispersal Fisher–KPP model with age structure. *Applied Mathematics Letters, 123*, Article 107593.

Chapter 8
Probabilistic Contraction Result Involving Functionals in Menger Space and Heat Equation

Samir Kumar Bhandari, Pulak Konar, Rabha W. Ibrahim, and Anita Tomar

Abstract The current topic is about an extended probabilistic $C-$contraction. Probable extension of metric spaces is known as Probabilistic metric spaces. We develop a fixed-point result of substituting the constant term by functional. This kind of contraction generalizes several previous fixed-point findings. Finally, we are able to present a unique fixed point solution for such a contraction. An illustrative example of a heat equation with a local internal controller is also provided to verify our primary conclusion.

Keywords C-contraction · Cauchy sequence · Fixed point · t-norm · Menger space

AMS Subject Classification 47H10 · 54H25

8.1 Introduction

Fixed point theory and its application are becoming a major part of mathematical analysis. The applicability of metric fixed point theory is seen in differential calculus, integral calculus, optimization problems, matrix equation and many more. In the present work, we have established generalized C-contraction results on probabilistic metric spaces. The publication of the first-ever fixed point result using probabilistic

---

S. K. Bhandari
Bajkul Milani Mahavidyalaya, P.O- Kismat Bajkul, Bajkul, West Bengal, India

P. Konar
Vellore Institute of Technology, Chennai, India

R. W. Ibrahim
Vellore Institute of Technology, Chennai, India

A. Tomar (✉)
Pt. L.M.S Campus, Sridev Suman Uttarakhand University, Rishikesh, Uttarakhand, India
e-mail: anitatmr@yahoo.com

© The Author(s), under exclusive license to Springer Nature Singapore Pte Ltd. 2026
A. Tomar and M. Jain (eds.), *Applications of Fixed-Point Theorem*, Industrial and Applied Mathematics, https://doi.org/10.1007/978-981-95-5160-6_8

metric spaces by Sehgal and Bharucha-Reid (1972) is known as B-contraction. In 1983, Hicks proved another extension of the contraction mapping principle (Banach 1922) in 1983 which is known as C-contraction and this type of contraction is also discussed by Raidu (1987). Further, these types of results are extensively discussed in scientific literature. A useful development in the book by Hadzic et al. (2001) is also seen prominently. Some important literature in this area of scientific research is noted as Bhandari (2019), Chang et al. (1991), Choudhury et al. (2011), Choudhury et al. (2009), Choudhury et al. (2008), Mihet (2009), O'Regan et al. (2008) and Razani (2007). The works Azhdari et al. (2009), Bhandari et al. (2017), Bhandari (2017), Choudhury et al. (2011), Dutta et al. (2010) and Zikic-Dosenovic (2005) in the direction of C-contraction and their generalization can be mentioned.

The altering distance function is one of the important tools for establishing the fixed point result and Khan et al. (1984) introduced the idea as a control function. The distance between two points in a metric space can be altered by this function. Some references may be noted in Khan et al. (1984), Naidu (2003), Sastry et al. (1999) and Sastry (2000).

The main findings of the paper are as follows:

- Here we establish a generalized probabilistic C-contraction using functional values instead of constants. We get a unique fixed point for such contraction.
- The use of t-norm establishes the result.
- A supportive illustration of a heat equation with a local internal controller is made to justify our result (Fig. 8.1).

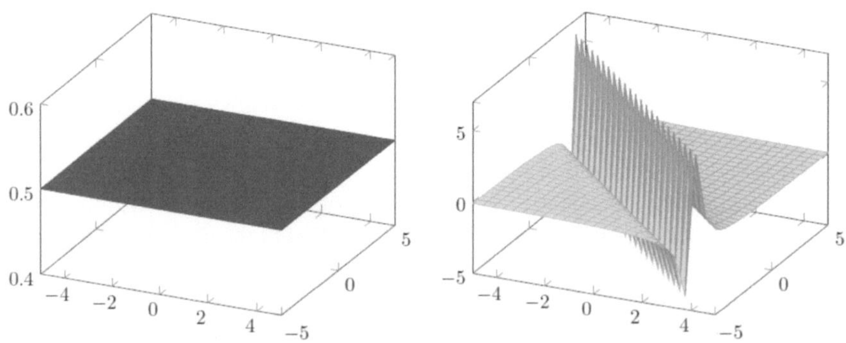

Fig. 8.1 The 3D-plot of ω_1, ω_2 showing the controller of the heat when ω_1 and $\omega_2 \in (0, 1)$

8.2 Preliminaries

The following are the mathematical preliminaries that have been used in our main result.

Definition 8.2.1 (Hadzic et al. (2001) and Schweizer et al. (1983)) A distribution function is a mapping $\mathcal{P} : \mathcal{R} \to \mathcal{R}^+$ such as it is left continuous and non-decreasing with $\inf_{\mathtt{t}\in\mathcal{R}} \mathcal{P}(\mathtt{t}) = 0$ and $\sup_{\mathtt{t}\in\mathcal{R}} \mathcal{P}(\mathtt{t}) = 1$, where $\mathcal{R}^+$ denotes the set of all non-negative real numbers, $\mathcal{R}$ is the set of real numbers.

To denote the probability of the event that the distance between the points α and β is less than $\mathtt{t}$, we use $\mathcal{P}_{\alpha,\beta}(\mathtt{t})$. A metric space becomes a Menger space if we write $\mathcal{P}_{\alpha,\beta}(\mathtt{t}) = \mathcal{H}(\mathtt{t} - \rho(\alpha, \beta))$ where ρ is the metric function and $\mathcal{H}$ is the Heaviside function given by

$$\mathcal{H}(\mathtt{t}) = 1, \text{ if } \mathtt{t} > 0,$$
$$= 0, \text{ if } \mathtt{t} \leq 0.$$

Definition 8.2.2 $\mathtt{t}$**-norm** [Hadzic et al. (2001) and Schweizer et al. (1983)] A mapping $\triangledown : [0, 1] \times [0, 1] \to [0, 1]$ is $\mathtt{t}$-norm if it satisfies the following assertions for all $i, j, k, l \in [0, 1]$

- $\triangledown(1, i) = i$,
- $\triangledown(i, j) = \triangledown(j, i)$,
- $\triangledown(k, l) \geq \triangledown(i, j)$ whenever $k \geq i$ and $l \geq j$,
- $\triangledown(\triangledown(i, j), k) = \triangledown(i, \triangledown(j, k))$.

The examples of $\mathtt{t}$-norm are as follows:

(i) The minimum $\mathtt{t}$-norm, $\triangledown = \mathcal{T}_m$, defined by $\mathcal{T}_m(i, j) = \min\{i, j\}$.
(ii) The product $\mathtt{t}$-norm, $\triangledown = \mathcal{T}_p$, defined by $\mathcal{T}_p(i, j) = i.j$.

Definition 8.2.3 Probabilistic metric space[Hadzic et al. (2001) and Schweizer et al. (1983)] An ordered pair $(\Omega, \mathcal{P})$ is called probabilistic metric space (or PM-space), where $\Omega \neq \emptyset$ and there exists a function $\mathcal{P}$ from $\Omega \times \Omega$ into the set of all distribution functions. For all $\alpha, \beta, \gamma \in \Omega$, $\mathcal{P}_{\alpha, \beta}$ is satisfying the following assertions,

(i) $\mathcal{P}_{\alpha, \beta}(0) = 0$,
(ii) $\mathcal{P}_{\alpha, \beta}(\mathtt{t}) = 1$ for all $\mathtt{t} > 0$ iff $\alpha = \gamma$,
(iii) $\mathcal{P}_{\alpha, \beta}(\mathtt{t}) = \mathcal{P}_{\beta, \alpha}(\mathtt{t})$ for all $\mathtt{t} > 0$,
(iv) if $\mathcal{P}_{\alpha, \beta}(\mathtt{t}_1) = 1$ and $\mathcal{P}_{\beta, \gamma}(\mathtt{t}_2) = 1$, then $\mathcal{P}_{\alpha, \gamma}(\mathtt{t}_1 + \mathtt{t}_2) = 1$ for all $\mathtt{t}_1, \mathtt{t}_2 > 0$.

Definition 8.2.4 Menger space[Hadzic et al. (2001) and Schweizer et al. (1983)]
A triplet $(\Omega,\ \mathcal{P},\ \triangledown)$ is known as a Menger space, where $\Omega \neq \emptyset$, $\triangledown$ is a $\mathtt{t}$-norm and $\mathcal{P}$ is a function defined on $X \times X$ to the set of distribution functions such that for all $\alpha, \beta, \gamma \in \Omega$ and for all $\lambda,\ \mu \geq 0$, the following are satisfied:

- $\mathcal{P}_{\alpha,\ \beta}(0) = 0$ for all $\alpha,\ \beta \in \Omega$,
- $\mathcal{P}_{\alpha,\ \beta}(s) = 1$ for all $s > 0$ and $\alpha,\ \beta \in X$ if and only if $\alpha = \beta$,
- $\mathcal{P}_{\alpha,\ \beta}(s) = \mathcal{P}_{\beta,\ \alpha}(s)$ for all $\alpha,\ \beta \in X, s > 0$ and
- $\mathcal{P}_{\alpha,\ \beta}(\lambda + \mu) \geq \triangledown\,(\mathcal{P}_{\alpha,\ \gamma}(\lambda), \mathcal{P}_{\gamma,\ \beta}(\mu))$.

The definition of τ-convergency, τ-Cauchy sequence and τ-completeness is first seen in Chang et al. (1996) and hugely discussed in the existing literature.

Definition 8.2.5 ***C*-contraction**[Hicks (1983)]
Let $(\Omega,\ \mathcal{P})$ be a probabilistic metric space. A mapping $\chi : \Omega \to \Omega$ is called a C-contraction if there exists $\omega \in (0,\ 1)$ such that for every $p, q \in \Omega$ and $\mathtt{t} > 0$ the following implication holds:

$$\mathcal{P}_{p,q}(t) > 1 - \mathtt{t} \text{ implies } F_{\chi p,\ \chi q}(\omega\mathtt{t}) > 1 - \omega\mathtt{t}.$$

8.3 Main Results

In this section, we establish generalized probabilistic C-contraction results. The main result depends on the following lemma.

Lemma (Chang et al. (1991)) Assume that $(\Omega,\ \mathcal{P},\ \triangledown)$ is a Menger PM-space, $\triangledown$ is a $\mathtt{t}$-norm which satisfies $lim_{\mathtt{t}\to 1-}\ \triangledown\,(\mathtt{t},\ s) = s$ for all $s \in [0, 1]$ and $\{\alpha_n\}, \{\beta_n\}$ be two sequences in Ω with

$\alpha_n \overrightarrow{\tau}\, \alpha,\ \beta_n \overrightarrow{\tau}\, \beta$, respectively.

(i) For any $\mathtt{t} \in \mathcal{R}$, $lim\ \ inf_{n\to\infty}\mathcal{P}_{\alpha_n,\ \beta_n}(\mathtt{t}) \geq \mathcal{P}_{\alpha,\ \beta}(\mathtt{t})$.
(ii) If $\mathtt{t} \in \mathcal{R}$ is a continuous point of $\mathcal{P}_{\alpha,\ \beta}$, then
$lim\ \ inf_{n\to\infty}\mathcal{P}_{\alpha_n,\beta_n}(\mathtt{t}) = \mathcal{P}_{\alpha,\ \beta}(\mathtt{t})$.

Theorem 8.3.1 is divided into three parts namely Section I, Section II and Section III respectively.

Theorem 8.3.1 *Assume that* $(\Omega,\ \mathcal{P},\ \triangledown)$ *is a Menger space with completeness and a* $\mathtt{t}$-*norm* $\triangledown$ *that satisfies* $\triangledown(\mathtt{t},\ \mathtt{t}) \geq \mathtt{t}$ *for all* $\mathtt{t} \in [0, 1]$. *Let* $\chi : \Omega \to \Omega$ *be a mapping satisfying the following condition:*

$$\mathcal{P}_{\alpha,\ \beta}(\mathtt{t}) > 1 - \nu \text{ implies } \mathcal{P}_{\chi\alpha,\chi\beta}(\omega_1(x,\ y)\mathtt{t}) > 1 - \omega_2(x,\ y)\nu, \tag{2}$$

for all $\alpha, \beta \in \Omega$, $\mathtt{t} > 0$ *and* $x, y \in (0, +\infty)$ *with* $\mathcal{P}_{\alpha,\ \beta}(x) > 0$ *and* $\mathcal{P}_{\alpha,\ \beta}(y) < 1$, *where* $\omega_1(x,\ y), \omega_2(x,\ y) : (0, +\infty)^2 \to (0,\ 1)$ *are two functions.*
Then χ *has a unique fixed point in* Ω.

Proof **Section I**:
Let us take $\epsilon > 0$ and $\nu \in (0, 1)$. There exists $\alpha_0 \in \Omega$ and an (ϵ, ν)-nbd of α_0, $N_\nu(\alpha_0, \epsilon) = \{\alpha \in X : \mathcal{P}_{\alpha,\ \alpha_0}(\epsilon) > 1 - \nu\}$, such that

$$\chi : N_\nu(\alpha_0, \epsilon) \to N_\nu(\alpha_0, \epsilon).$$

Suppose it is not true. Then there exists $\epsilon_0 > 0$ and $\nu_0 \in (0, 1)$ such that for each $\alpha \in \Omega$, there is an $\alpha_1 \in \Omega$ with

$$\mathcal{P}_{\alpha,\ \alpha_1}(\epsilon_0) > 1 - \nu_0, \mathcal{P}_{\chi\alpha_1,\ \alpha}(\epsilon_0) \leq 1 - \nu_0. \tag{3}$$

Now, we get two possible cases.

Case-I: If $\mathcal{P}_{\alpha,\ \alpha_1}(\frac{\epsilon_0}{2}) > 1 - \frac{\nu_0}{2}$, then

$$\mathcal{P}_{\alpha,\ \chi\alpha_1}(\epsilon_0) \geq \nabla(\mathcal{P}_{\alpha,\ \chi\alpha}(\tfrac{\epsilon_0}{2}),\ \mathcal{P}_{\chi\alpha,\chi\alpha_1}(\tfrac{\epsilon_0}{2})). \tag{4}$$

If $\alpha_1 = \alpha$, then $\mathcal{P}_{\chi\alpha,\chi\alpha_1}(\frac{\epsilon_0}{2}) = 1$.
By using inequalities (3) and (4), we have

$$\mathcal{P}_{\alpha,\chi\alpha}(\tfrac{\epsilon_0}{2}) \leq \mathcal{P}_{\alpha,\chi\alpha_1}(\epsilon_0) \leq 1 - \nu_0. \tag{5}$$

If $\alpha_1 \neq \alpha$, then there exists $\mathtt{t}_0 > 0$ such that $\mathcal{P}_{\alpha,\ \alpha_1}(t_0) < 1$.
using the inequality (2) and the assumption of Case-I, we have

$$\mathcal{P}_{\alpha,\ \alpha_1}(\epsilon_0) > 1 - \nu_0 \Rightarrow \mathcal{P}_{\chi\alpha,\ \chi\alpha_1}(\omega_1(\tfrac{\epsilon_0}{2}, \mathtt{t}_0)\epsilon_0) > 1 - \omega_2(\tfrac{\epsilon_0}{2}, \mathtt{t}_0)\nu_0 > 1 - \nu_0. \tag{6}$$

$$(\text{since } \omega_2 \in (0, 1)).$$

From inequalities (3) and (6), we have
$\nabla(\mathcal{P}_{\alpha,\ \chi\alpha}(\epsilon_0 - \omega_1(\frac{\epsilon_0}{2}, \mathtt{t}_0)\epsilon_0), 1 - \nu_0) \leq \nabla(\mathcal{P}_{\alpha,\ \chi\alpha}(\epsilon_0 - \omega_1(\frac{\epsilon_0}{2}, \mathtt{t}_0)\epsilon_0),$
$\mathcal{P}_{\chi\alpha,\ \chi\alpha_1}(\omega_1(\frac{\epsilon_0}{2}, \mathtt{t}_0)\epsilon_0))$
$\leq \mathcal{P}_{\alpha,\ \chi\alpha_1}(\epsilon_0) \leq 1 - \nu_0.$

Therefore, we have

$$\mathcal{P}_{\alpha,\ \chi\alpha}(\epsilon_0 - \omega_1(\tfrac{\epsilon_0}{2}, \mathtt{t}_0)\epsilon_0) \leq 1 - \nu_0. \tag{7}$$

Case-II: If $\mathcal{P}_{\alpha,\ \alpha_1}(\frac{\epsilon_0}{2}) \leq 1 - \frac{\nu_0}{2}$, using the triangle inequality, we have

$$\mathcal{P}_{\alpha,\ \chi\alpha_1}(\epsilon_0) \geq \nabla(\mathcal{P}_{\alpha,\ \chi\alpha}((1 - \omega_1(\tfrac{\epsilon_0}{2}, \epsilon_0))\epsilon_0), \mathcal{P}_{\chi\alpha,\ \chi\alpha_1}(\omega_1(\tfrac{\epsilon_0}{2}, \epsilon_0)\epsilon_0)). \tag{8}$$

Now, using inequalities (2) and (3), we have

$$\mathcal{P}_{\alpha,\ \alpha_1}(\epsilon_0) > 1-\nu_0 \Rightarrow \mathcal{P}_{\chi\alpha,\ \chi\alpha_1}(\omega_1(\tfrac{\epsilon_0}{2},\epsilon_0)\epsilon_0) > 1-\omega_2(\tfrac{\epsilon_0}{2},\epsilon_0)\nu_0 > 1-\nu_0. \quad (9)$$
$$\text{(since } \omega_2 \in (0,1)\text{)}.$$

Using inequality (9) in (8), we have

$$\mathcal{P}_{\alpha,\ \chi\alpha_1}(\epsilon_0) \geq \nabla(\mathcal{P}_{\alpha,\ \chi\alpha}((1-\omega_1(\tfrac{\epsilon_0}{2},\epsilon_0))\epsilon_0),\ 1-\nu_0).$$

By the inequality (3), we have

$$1-\nu_0 \geq \mathcal{P}_{\alpha,\ \chi\alpha_1}(\epsilon_0) \geq \nabla(\mathcal{P}_{\alpha,\ \chi\alpha}((1-\omega_1(\tfrac{\epsilon_0}{2},\epsilon_0))\epsilon_0),\ 1-\nu_0),$$

and so we have

$$\mathcal{P}_{\alpha,\ \chi\alpha}((1-\omega_1(\tfrac{\epsilon_0}{2},\epsilon_0))\epsilon_0) \leq 1-\nu_0.$$

Letting $\eta_0 = \min\{\epsilon_0 - \omega_1(\frac{\epsilon_0}{2}, \mathrm{t}_0)\epsilon_0, \epsilon_0 - \omega_1(\frac{\epsilon_0}{2}, \epsilon_0)\epsilon_0\}$.
By substituting the inequality (7), we have

$$\mathcal{P}_{\alpha,\ \chi\alpha}(\eta_0) \leq 1-\nu_0, \text{ for } \alpha \in \Omega. \quad (10)$$

Besides, for any fixed point $\alpha_0 \in \Omega$, since

$$lim_{\mathrm{t}\to\infty}\mathcal{P}_{\alpha_0,\ \chi\alpha_0}(t) = 1, \text{ there exists } t > 0 \text{ such that}$$

$\mathcal{P}_{\alpha_0,\ \chi\alpha_0}(\mathrm{t}) > 1-\nu_0$, which implies $\mathcal{P}_{\chi\alpha_0,\ \chi^2\alpha_0}(\omega_1(\eta_0,\mathrm{t}_0)\mathrm{t}) > 1-\omega_2(\eta_0,\mathrm{t}_0)\nu_0$.
(applying the inequality (2), where $\mathcal{P}_{\alpha_0,\ \chi\alpha_0}(\eta_0) > 0$ and $\mathcal{P}_{\alpha_0,\chi\alpha_0}(\mathrm{t}_0) < 1$)

Now, again by the inequality (2), we have
$\mathcal{P}_{\chi(\chi\alpha_0),\ \chi(\chi^2\alpha_0)}(\omega_1^2(\eta_0,\mathrm{t}_0)\mathrm{t}) > 1-\omega_2^2(\eta_0,\mathrm{t}_0)\nu_0$ that implies
$\mathcal{P}_{\chi^2\alpha_0,\ \chi^3\alpha_0}(\omega_1^2(\eta_0,\mathrm{t}_0)\mathrm{t}) > 1-\omega_2^2(\eta_0,\mathrm{t}_0)\nu_0$.
Successively, we have
$\mathcal{P}_{\chi^n\alpha_0,\ \chi^{n+1}\alpha_0}(\omega_1^n(\eta_0,\mathrm{t}_0)\mathrm{t}) > 1-\omega_2^n(\eta_0,\mathrm{t}_0)\nu_0$, for some $\mathrm{t} > 0, n \geq 1$.
Assuming n as large enough such that taking $\omega_1^n(\eta_0,\mathrm{t}_0)\mathrm{t} = \eta_0$, for some $\mathrm{t} > 0$, we have
$\mathcal{P}_{\chi^n\alpha_0,\ \chi^{n+1}\alpha_0}(\eta_0) > 1-\omega_2^n(\eta_0,\mathrm{t}_0)\nu_0 > 1-\nu_0$, which contradicts inequality (9) for taking $k_2 \in (0,1)$.
Hence the desired result is obtained.
As χ satisfies condition (2) and by the continuity of χ, we have

$$\chi : N_\nu(\alpha_n,\epsilon) \to N_\nu(\alpha_0,\epsilon).$$

The other two sections are proved in the same manner as in Chang et al. (1996).

Section-II:

We can choose $\epsilon_\iota > 0$ and $\nu_\iota \in (0, 1)$, $\iota = 1, 2, 3, ...$, with $\epsilon_\iota > \epsilon_{\iota+1}$, $\nu_\iota > \nu_{\iota+1}$, $\epsilon_\iota \to 0^+$ and $\nu_\iota \to 0^+$, by the result proved in Section-I, there exists $\alpha_\iota \in \Omega$ and a $(\epsilon_\iota,\ \nu_\iota)$-neighborhood $N_\nu(x_i, \epsilon_\iota)$ of α_ι such that

$$\chi : N_{\nu_\iota}(\alpha_\iota, \epsilon_\iota) \to N_{\nu_\iota}(\alpha_\iota, \epsilon_\iota), \iota = 1, 2, 3, ...$$

It is obvious that $N_{\nu_\iota}(\alpha_\iota, \epsilon_\iota) \supseteq N_{\nu_{\iota+1}}(\alpha_{\iota+1}, \epsilon_{\iota+1})$, for all $\iota \geq 1$.
Therefore, we have

$$\chi : \bigcap_{\iota\geq1} N_{\nu_\iota}(\alpha_\iota, \epsilon_\iota) \to \bigcap_{\iota\geq1} N_{\nu_\iota}(\alpha_\iota, \epsilon_\iota).$$

Next, we prove that $\bigcap_{\iota\geq1} N_{\nu_\iota}(\alpha_\iota, \epsilon_\iota) \neq \emptyset$.
Taking $\epsilon > 0$, $\nu > 0$, and ι large enough such that $\epsilon_\iota < \epsilon$, $\nu_\iota < \lambda$, take $t_{\iota,\ \jmath} \in (\epsilon_\iota, \epsilon)$ such that $t_{\iota,\ \jmath}$ is a continuous point of $\mathcal{P}_{\alpha_\iota,\ \alpha_{\iota+\jmath}}(t)$. Since $\alpha_{\iota+\jmath} \in N_{\nu_\iota}(\alpha_\iota,\ \epsilon_\iota)$, by the Lemma 2.1, we have

$$\mathcal{P}_{\alpha_n,\ \alpha_{n+m}}(\epsilon) \geq \mathcal{P}_{\alpha_n,\ \alpha_{n+m}}(\mathtt{t}_{n,m}) \geq 1 - \nu_n > 1 - \nu.$$

Therefore, $\{\alpha_\iota\}$ is a τ-Cauchy sequence $\in \Omega$. Assume that $\alpha_\iota \overrightarrow{\tau} \alpha$. Then,

$$\Omega \in \bigcap_{i\geq1} N_{\lambda_i}(x_i, \epsilon_i).$$

the desired result.

Section-III:

Finally, we prove that $\bigcap_{\iota\geq1} N_{\nu_\iota}(\alpha_\iota, \epsilon_\iota)$ has a unique point.
Suppose that there exist $\alpha, \beta \in \bigcap_{\iota\geq1} N_{\nu_\iota}(\alpha_\iota, \epsilon_\iota)$ with $\alpha \neq \beta$ and as a consequence, there is $\mathtt{t}_0 > 0$ and $\nu_0 \in (0, 1)$ such that $\mathcal{P}_{\alpha,\ \beta}(\mathtt{t}_0) < \nu_0$.
Considering ι be large enough so that $\epsilon_\iota < \frac{\mathtt{t}_0}{2}$ and $1 - \nu_\iota > \nu_0$, then we have

$$\mathcal{P}_{\alpha,\ \beta}(\mathtt{t}_0) \geq \nabla(\mathcal{P}_{\alpha,\ \alpha_\iota}(\tfrac{\mathtt{t}_0}{2}), \mathcal{P}_{\alpha_\iota,\beta}(\tfrac{\mathtt{t}_0}{2})).$$

Taking two continuous points $\mathtt{t}_\iota \in (\epsilon_\iota, \frac{\mathtt{t}_0}{2})$ and $s_\iota \in (\epsilon_\iota, \frac{\mathtt{t}_0}{2})$ respectively, such that $\mathtt{t}_\iota$ being a continuous point, is in $\mathcal{P}_{\alpha,\ \alpha_\iota}(\mathtt{t})$ and s_ι being a continuous point, is in $\mathcal{P}_{\alpha_\iota,\ \beta}(\mathtt{t})$. Then we have

$\mathcal{P}_{\alpha,\ \alpha_\iota}(\mathtt{t}_\iota) \geq 1 - \nu_\iota, \mathcal{P}_{\alpha_\iota,\ \iota}(s_\iota) \geq 1 - \nu_\iota$ and so $\mathcal{P}_{\alpha,\ \beta}(\mathtt{t}_0) > \nu_0$, which is a contradiction.

Therefore $\bigcap_{\iota\geq1} N_{\nu_\iota}(\alpha_\iota, \epsilon_\iota)$ has a unique point, α_*(say), and it is the unique fixed point of χ. □

Remarks: In our main results if we take $\omega_1(x, y) = \omega_2(x, y) = \omega \in (0, 1)$, then for any $\alpha, \beta \in \Omega$ and $\mathtt{t} > 0$, we get C-contraction which is defined in inequality (6). So our results are more generalized than on C-contraction. The results of Chang et al. Chang et al. (1996) help us very much in constructing such a type of contraction. The contraction is newer and more extensive than the contraction earlier mentioned in Raidu (1987).

Example 8.3.2 We suggest the following heat equation with the local internal controller, which is approximately controllable in time $\mathtt{t} > 0$ (see Chuang (2008)). Let $X = \{\alpha_1, \alpha_2, \alpha_3\}$, the $\mathtt{t}$-norm $\triangledown$ is a minimum $\mathtt{t}$-norm and $\mathcal{P}$ be defined as

$$\mathcal{P}_{\alpha_2, \alpha_3}(\mathtt{t}) = \mathcal{P}_{\alpha_3, \alpha_1}(\mathtt{t}) = \begin{cases} 0, & \text{if } \mathtt{t} \leq 0, \\ 0.75, & \text{if } 0 < \mathtt{t} \leq 2, \\ 1, & \text{if } \mathtt{t} > 7, \end{cases}$$

$$\mathcal{P}_{\alpha_1, \alpha_2}(\mathtt{t}) = \begin{cases} 0, & \text{if } \mathtt{t} \leq 0, \\ 1, & \text{if } \mathtt{t} > 1, \end{cases}$$

where $\omega_1(x, y) = \frac{1}{2}, \omega_2(x, y) = \frac{1}{x+y+1}$, $\Omega, y \in (0, \infty)$ and $\omega_1, \omega_2 \in (0, 1)$. Then, $(\Omega, \mathcal{P}, \triangledown)$ is a complete Menger space. Defining $\chi : X \to \Omega$ as follows: $\chi\alpha_1 = \alpha_1, \chi\alpha_2 = \alpha_1, \chi\alpha_3 = \alpha_2$. χ satisfies all the assertions of the Theorem 2.1. Hence, we get α_1 as the unique fixed point of χ in Ω.

Open problem: To consider any existing problem where the solutions of differential equations in probabilistic metric space can be obtained considering this type of $C-$contraction.

8.4 Conclusion

The present article is a special discussion on a new type of C—contraction in Menger spaces which is the functional value dependent instead of constant values. The known C-contraction is extended now. The work is completely different from the work presented in Raidu (1987). The definition of C- contraction is extended through the use of the functional value of ω_1 and ω_2 within the interval of $[0, 1]$.

Conflict of interest The authors declare that they have no conflicts of interest. The authors do not hold any stocks or shares in any organization that may have an interest in the submitted work.

Acknowledgements The authors are grateful to the reviewers for their insightful comments which help us improve the quality of the manuscript.

References

Azhdari, P., & Farnoosh, R. (2009). Fixed point theorems for the generalized C-contractions. *Applied Mathematical Sciences,3*, 1265–1273.

Banach, S. (1922). Sur les Operations dans les Ensembles Abstraits et leur Application aux Equations Integrales. *Fundamenta Mathematicae,3*, 133–181.

Bhandari, S. K. (2017). Unique probabilistic p-cyclic C-contractions using special type product t-norm. *Bulletin of the Calcutta Mathematical Society,109*(1), 55–68.

Bhandari, S. K. (2019). Rational type probabilistic p-cyclic contraction results using some control function. *Bulletin of the Calcutta Mathematical Society,111*(1), 1–12.

Bhandari, S. K., & Choudhury, B. S. (2017). Two unique fixed point results of p-cyclic probabilistic C-contractions using different types of t-norm. *Journal of International Mathematical Virtual Institute,7*, 147–164.

Chang, S. S., Chen, Y. Q., & Guo, J. L. (1991). Ekeland's variational principle and Caristi's fixed point theorem in probabilistic metric spaces. *Acta Mathematicae Applicatae Sinica,7*, 217–228.

Chang, S. S., Lee, B. S., Cho, Y. J., Chen, Y. Q., Kang, S. M., & Jung, J. S. (1996). Generalized contraction mapping principle and differential equations in probabilistic metric spaces. *Proceedings of the American Mathematical Society,124*(8), 2367–2376.

Choudhury, B. S., & Das, K. P. (2008). A new contraction principle in Menger spaces. *Acta Mathematica Sinica, English Series,24*, 1379–1386.

Choudhury, B. S., & Das, K. P. (2009). Fixed points of generalized Kannan type mappings in generalized Menger spaces. *Communications of the Korean Mathematical Society,24*, 529–537.

Choudhury, B. S., Das, K. P., & Bhandari, S. K. (2011). Fixed point theorem for mappings with cyclic contraction in Menger spaces. *International Journal of Applied Science and Technology,4*, 1–9.

Choudhury, B. S., Das, K. P., & Bhandari, S. K. (2011). A generalized cyclic C-contraction principle in Menger spaces using a control function. *International Journal of Applied Mathematics,24*(5), 663–673.

Chuang, Z. (2008). Controllability of the time discrete heat equation. *Asymptotic Analysis,59*(3–4), 139–177.

Dutta, P. N., & Choudhury, B. S. (2010). A generalized contraction principle in Menger spaces using control function. *Analysis in Theory and Applications,26*, 110–121.

Hadzic, O., & Pap, E. (2001). *Fixed point theory in probabilistic metric spaces* (p. 273). Kluwer Academic Publishers.

Hicks, T. L. (1983). Fixed point theory in probabilistic metric spaces. *Zbornik Radova Prirodno-Matematičkog Fakulteta Serija Matematika,13*, 63–72.

Khan, M. S., Swaleh, M., & Sessa, S. (1984). Fixed point theorems by altering distances between the points. *Bulletin of the Australian Mathematical Society,30*, 1–9.

Mihet, D. (2009). Altering distances in probabilistic Menger spaces. *Nonlinear Analysis,71*, 2734–2738.

Naidu, S. V. R. (2003). Some fixed point theorems in metric spaces by altering distances. *Czechoslovak Mathematical Journal,53*, 205–212.

O'Regan, D., & Saadati, R. (2008). Nonlinear contraction theorems in probabilistic spaces. *Applied Mathematics and Computation,195*, 86–93.

Radu, V. (1987). *Some fixed point theorems in probabilistic metric spaces*. Lecture Notes in Mathematics (Vol. 1233, pp. 125–133). Springer, New York, Berlin.

Razani, A., & Fouladgar, K. (2007). Extension of contractive maps in the Menger probabilistic metric space. *Chaos Solitons Fractals,34*, 1724–1731.

Sastry, K. P. R., & Babu, G. V. R. (1999). Some fixed point theorems by altering distances between the points. *Indian Journal of Pure and Applied Mathematics,30*, 641–647.

Sastry, K. P. R., Naidu, S. V. R., Babu, G. V. R., & Naidu, G. A. (2000). Generalisation of common fixed point theorems for weakly commuting maps by altering distances. *Tamkang Journal of Mathematics,31*(3), 243–250.

Schweizer, B., & Sklar A. (1983). *Probabilistic metric spaces*. Elsevier North-Holland.
Sehgal, V. M., & Bharucha-Reid, A. T. (1972). Fixed point of contraction mappings on PM space. *Mathematical Systems Theory,6*(2), 97–100.
Zikic-Dosenovic, T. (2005). A multivalued generalization of Hicks C-Contraction. *Fuzzy Sets and Systems,151*, 549–562.

Chapter 9
Common Fixed Point Theorems of Ćirić-Type Weak Contractions in Cone B-Metric Spaces with an Application

Santosh Kumar and Pauline Shashi

Abstract In this paper, we aim to prove the existence of common and coincidence point theorems for two weakly compatible self-mappings with a Ćirić weak contractions condition in cone b-metric spaces. The uniqueness of a common fixed point in cone b-metric spaces is proved for the pair of weakly compatible mappings. This paper extends and generalizes the fixed point result of Babu and Sailaja (2012), who established a common fixed point theorem for Ćirić-type weak contractions in cone metric spaces. Our results generalize many recent known results in metric spaces, b-metric spaces and cone metric spaces.

Keyword Cone b-metric spaces · Coincidence points · Asymptotically regular pair · Property (E.A) · Weakly compatible mappings · A common fixed point

AMS Subject Classification 47H10 · 54H25 · 46T99

9.1 Introduction

In 2007, Huang and Zhang (2007) developed the concept of cone metric spaces, where real numbers were replaced by an arbitrary ordered set. They developed various fixed point theorems for contractive mappings in cone metric spaces that broadened Banach (1922) contractive principle for metric spaces. Hussain and Shah (2011) introduced the concept of cone b-metric spaces as the generalization of metric spaces, b-metric spaces, and cone metric spaces. Ćirić et al. (2010) introduced a new concept of Ćirić-type f—weak contractive mappings and the existence of common fixed points. In this study, we extend the results of Babu and Sailaja (2012) on common fixed point theorems of Ćirić-type weak contraction in cone metric spaces to cone b-metric

S. Kumar
Department of Mathematics, School of Physical Sciences, North Eastern Hill University, Shillong, Meghalaya, India

P. Shashi (✉)
Department of Management Studies, Tanzania Insititute of Accountancy, Dar es Salaam, Tanzania
e-mail: shashipauline@gmail.com

© The Author(s), under exclusive license to Springer Nature Singapore Pte Ltd. 2026

A. Tomar and M. Jain (eds.), *Applications of Fixed-Point Theorem*, Industrial and Applied Mathematics, https://doi.org/10.1007/978-981-95-5160-6_9

spaces. Some examples and an application are given to support our main results. More references can be found on Ćirić-type contractions, cone metric and cone $b-$metric spaces in Alsulami (2017), Aydi et al. (2012), Karapinar (2020), Karapinar (2013), Vetro (2007), Tomar et al. (2021), Tomar and Joshi (2021), Kumar and Kapinga (2007), Pauline and Kumar (2021), Wangwe and Kumar (2022) and the references therein.

9.2 Preliminaries

The definitions, lemmas, and theorems outlined below are crucial for developing the main results of this paper.

Definition 9.2.1 (Rangamma and Prudhvi (2012)) A subset P of a Banach space E is called a cone if and only if it satisfies the following conditions:

(P1) P is closed, nonempty and $P \neq \{0\}$;
(P2) If $m, n \in \mathbb{R}$ such that $m, n \geq 0$ and $u, v \in P$ then $mu + nv \in P$;
(P3) If $u \in P$ and $-u \in P \Rightarrow u = 0$.

The cone P is called solid if $P \neq \emptyset$.

Definition 9.2.2 (Malhotra and Shukla (2012)) Given a cone $P \subset E$ and $u, v \in P$., then we define a partial ordering "$\preceq$" for P by $u \preceq v$ if and only if $v - u \in P$. We define $u \prec v$ to indicate that $u \preceq v$ but $u \neq v$. While $u \ll v$ if $v - u \in int P$.

Definition 9.2.3 (Huang and Xu (2013)) Let X be a nonempty set, E be a real Banach space and assume that the mapping $\varrho : X \times X \to E$ satisfies the following conditions:

(d1) $0 \preceq \varrho(u, v)$ and $\varrho(u, v) = 0$ if and only if $u = v$ for all u and v in X;
(d2) $\varrho(u, v) = \varrho(v, u)$ for all u and v in X;
(d3) $\varrho(u, v) \preceq \varrho(u, z) + \varrho(z, v)$ for all u, v and z in X. Then ϱ is called cone metric and (X, ϱ) is called a cone metric space.

The following is an example to illustrate the Definition 9.2.3 of a cone metric space.

Example 9.2.4 Let $X = \mathbb{R}$, $E = \mathbb{R}^2$, $P = \{(u, v) \in E : u, v \geq 0\} \subset E$, and $\varrho : X \times X \to E$ such that $\varrho(u, v) = (\mid u - v \mid, \alpha \mid u - v \mid)$, where $\alpha > 0$ is a constant. Then (X, ϱ) is a cone metric space.

Definition 9.2.5 (Rangamma and Prudhvi (2012)) The sequence $\{u_n\}$ in a cone metric space (X, ϱ) is said to be convergent if for every $c \in E$ with $0 \ll c$ there is a natural number N such that $\varrho(u_m, u_n) \ll c$ for all $n \geq N$.

Definition 9.2.6 (Jungcl et al. (2009)) The sequence $\{u_n\}$ in a cone metric space (X, ϱ) is said to be Cauchy if for every $c \in E$ with $0 \ll c$ there is a natural number N such that $\varrho(u_m, u_n) \ll c$ for all $n, m \geq N$.

Definition 9.2.7 (Aage and Salunke (2011)) The cone metric space (X, ϱ) is said to be complete if every Cauchy sequence in X is convergent.

The following is an example that justifies the Definition 9.2.7 of a Cauchy sequence.

Example 9.2.8 Let $X = \mathbb{R}$, $E = \mathbb{R}^2$, $P = \{(u, v) \in E : u, v \geq 0\} \subset E$, and $\varrho : X \times X \to E$ such that $\varrho(u, v) = (\mid u - v \mid, 2 \mid u - v \mid)$. If $\{u_n\} = (1 + \frac{1}{n})$, Then $\{u_n\}$ is a Cauchy sequence in a cone metric space (X, ϱ).

Lemma 9.2.9 (Babu and Sailaja (2012)) *Let P be a cone in a real Banach space E, $a, b, c, x \in E$.*

(i) If $a \preceq b$ and $b \ll c$, then $a \ll c$;
(ii) If $0_E \preceq x \ll c$ for each $c \in int P$, then $x = 0_E$;
(iii) if $0_E \preceq a \ll c$ with $c \in int P$, then $\|a\| \to 0_E$,

where 0_E is the zero vector in E.

Lemma 9.2.10 *If $a \preceq b + c$ for all $c \in int P$ then $a \preceq b$*

Proof Let $c \in int P$. Then $\frac{1}{n}c \in int P$ for each $n \in \mathbb{Z}^+$. Then, by our assumption $a \preceq b + \frac{1}{n}c$ for each $n \in \mathbb{Z}^+$, we have $b + \frac{1}{n}c - a \in P$.
Now, as $n \to \infty$, we get $b - a \in P$, which implies $a \preceq b$. □

Huang and Zhang (2010) introduced the following definition of cone $b-$ metric space as follows:

Definition 9.2.11 (Huang and Zhang (2010)) Let X be a nonempty set and $s \geq 1$ be a given real number. A mapping $\varrho : X \times X \to E$ is said to be cone b-metric if and only if, for all $u, v, z \in X$, the following conditions are satisfied:

(i) $0 \preceq \varrho(u, v)$ with $u \neq v$ and $\varrho(u, v) = 0$ if and only if $u = v$;
(ii) $\varrho(u, v) = \varrho(v, u)$;
(iii) $\varrho(u, v) \preceq s[\varrho(u, z) + \varrho(z, v)]$.

The pair (X, ϱ) is called a cone b-metric space.

Remark 9.2.1 Cone b-metric space is the generalization of cone metric space since every cone metric space is a cone b-metric space but the converse may not be true.

The following is an example that satisfies the conditions of cone $b-$ metric space but not of the cone metric space as the condition (d3) of Definition 9.2.3 fails.

Example 9.2.12 Let $X = \mathbb{R}$, $E = \mathbb{R}^2$, $P = \{(u, v) \in E : u, v \geq 0\} \subset E$, and $\varrho : X \times X \to E$ such that $\varrho(u, v) = (\mid u - v \mid^q, \alpha \mid u - v \mid^q)$, where $q \geq 1$ and $\alpha > 0$ are constants. Then (X, ϱ) is a cone b-metric space but not a cone metric space since the condition (d3) in Definition 9.2.3 fails.

The following definitions are useful in the establishment of our main results.

Definition 9.2.13 (Kaneko and Sessa (1989)) Let X be a non empty set and h and g be self-mappings on X and $u, v \in X$. Then u is called the coincidence point of pair (h, g) if $hu = gu$, and v is called the point of coincidence of pairs (h, g) if $hu = gu = v$.

Aamri and El Moutawakil (2002) introduced $(E.A.)$ property which will be useful in proving the results.

Definition 9.2.14 (Aamri and El Moutawakil (2002)) A pair (h, g) of self-mappings are said to satisfy property $(E.A.)$ if for an arbitrary sequence $\{u_n\} \in X$ and $t \in X$ such that $\lim_{n\to\infty} hu_n = \lim_{n\to\infty} gu_n = t$.

Definition 9.2.15 (Jungck (1986)) Two self-mappings $h, g : u \to X$ are said to be compatible if and only if $\lim_{n\to\infty} d(hg(u_n), gh(u_n)) = 0$ whenever u_n is an arbitrary sequence in X such that $\lim_{n\to\infty} hu_n = \lim_{n\to\infty} gu_n = t$ for some $t \in X$.

Definition 9.2.16 (Jungck and Rhoades (1998)) Two self-mappings $h, g : X \to X$ are said to be weakly compatible if for some $u \in X$, $hu = gu$ implies that $hgu = ghu$.

For details on weaker forms of commutativity, one may refer to Singh and Tomar (2003).

Lemma 9.2.17 *If the pair (h, g) of self-mappings on the cone metric space (X, ϱ) is compatible then it is weakly compatible but the converse does not hold.*

Proof Let $hy = gy$, for some $y \in X$. Then we have to prove that

$$hgy = ghy.$$

Put $u_n = y$ for every $n \in \mathbb{N}$, then we have $hu_n = gu_n = hy = gy$ (since (h, g) is compatible). If $c \in P$ with $c \in int(P)$ then we have

$$\varrho(hgu_n, ghu_n) = \varrho(hgy, ghy) \ll c.$$

By using Lemma 9.2.9 (iii), we get

$$\|\varrho(hgy, ghy)\| \to 0$$

$$\Rightarrow hgy = ghy.$$

□

Definition 9.2.18 (Babu and Sailaja (2012)) Let (X, ϱ) be a cone metric space with a cone P. Assume that h and g are self-mappings of X such that $h(X) \subseteq g(X)$. Suppose $u_0 \in X$. Then, we can construct the sequence $\{v_n\}$ and $\{u_n\}$ in X such that

$$v_n = hu_n = gu_{n+1},\ n = 0, 1, 2, ...$$

We say that the pair (h, g) is asymptotically regular at u_0, if for each $c \in int P$, there exists $n_0 \in \mathbb{Z}^+$ such that

$$\varrho(v_n, v_{n+1}) \ll c\ forall\ n \geq n_0. \tag{9.1}$$

If the pair (h, g) is asymptotically regular at every point of X, then it is asymptotically regular on X.

Remark 9.2.2 The asymptotic regularity and the property (E. A.) are independent of each other. One can see the example below.

Example 9.2.19 Let $(X, \varrho) = [0, +\infty)$, $E = C'_R[0, 1]$ with supremum norm and $P = \{u \in E : u \geq 0\}$, and $\varrho : X \times X \to E$ defined by $\varrho(u, v) = |v - u|\varphi$ with $\varphi(t) = e^t, t \geq 0$. We define $h, g : X \to X$ by $h(u) = \frac{u}{3}$ and $g(u) = \frac{u}{4}$. Then, the pair (h, g) satisfies the property (E. A.) with the sequence defined by $u_n = \frac{1}{n}$ for $n = 1, 2, 3, ...$ On the other hand, let $u_0 = \frac{1}{2}$. Since $h(X) \subseteq g(X)$, we can construct the sequence $\{v_n\}$ and $\{u_n\}$ such that

$$v_n = hu_n = gu_{n+1} = \frac{1}{6}\left(\frac{3}{2}\right)^n,$$
$$\Rightarrow v_{n+1} = \frac{1}{4}\left(\frac{3}{2}\right)^n \quad for\ n = 0, 1, 2, ...$$

Now, it is easy to observe that

$$\begin{aligned}\varrho(v_n, v_{n+1}) &= |\frac{1}{6}\left(\frac{3}{2}\right)^n - \frac{1}{4}\left(\frac{3}{2}\right)^n|e^t,\\ &= \frac{1}{12}\left(\frac{3}{2}\right)^n e^t \to \infty \text{ as } n \to \infty.\end{aligned}$$

Hence the pair (h, g) is not asymptotically regular on X.

The following is the Theorem 1.1 of Aydi et al. (2012) in b-metric spaces.

Theorem 9.2.20 (Aydi et al. (2012)) *Let (X, ϱ) be a complete metric space and g be a mapping of X into itself satisfying:*

$$\varrho(gu, gv) \preceq \alpha\varrho(u, v), \forall u, v \in X,$$

where α is a constant in [0, 1). Then, g has a unique fixed point $u* \in X$.

The following theorem extends the Theorem 1.1 of Aydi et al. (2012).

Theorem 9.2.21 (Babu and Sailaja (2012)) *Let (X, ϱ) be a complete cone metric space with a solid cone P. Suppose that (h, g) is a weakly compatible pair of self-mappings on X such that*

$$\varrho(hu, hv) = \varphi(P(u, v)) \ for \ u, v \in X,$$

where, $P(u, v) = \{\varrho(gu, gv), \varrho(hu, gu), \varrho(hv, gv)\}$ and $\varphi : P \to P$ is a map satisfying the following conditions:

(i) *if $k_1 \preceq k_2 \Rightarrow \varphi(k_1) \preceq \varphi(k_2)$,*
(ii) *$\varphi(0) = 0$ and $\varphi(k) \prec k$ for $k \in P \setminus \{0\}$,*
(iii) *$k \in int P \Rightarrow k - \varphi(k) \in int P$, and*
(iv) *if $k \in P \setminus \{0\}$ and $c \in int P$, then there exists $n_0 \in \mathbb{Z}^+$ such that $\varphi^n(k) \ll c, \forall n \geq n_0$.*

Suppose that $h(X) \subseteq g(X)$ and either $g(X)$ or $h(X)$ is a complete subspace of X. Then, the mappings g and h have a unique common fixed point in X.

Definition 9.2.22 (Babu and Sailaja (2012)) Let (X, ϱ) be a metric space and g, h be self-mappings on X. Then, h, is said to be a Ćirić g-weak contraction if there exists a mapping $\varphi : \mathbb{R}^+ \to \mathbb{R}^+$ which satisfies the following conditions:

(i) $\varphi(t) > 0, \ \forall t > 0$,
(ii) $\lim_{s \to t^+} \varphi(s) > 0, \ \forall t > 0$,
(iii) $t - \varphi(t)$ is non-decreasing $\forall t > 0$,
(iv) $\lim_{t \to \infty} \varphi(t) = \infty, \forall t > 0$

such that

$$\varrho(hu, hv) \leq M(u, v) - \varphi(M(u, v)), \forall u, v \in X,$$

where,

$$M(u, v) = \text{Max}\{\varrho(gu, gv), \varrho(hu, gu), \varrho(hv, gv), \varrho(hv, gu), \varrho(hu, gv)\}.$$

The following is the Theorem 2.4 of Aage and Salunke (2011) on a common fixed point in cone metric spaces.

Theorem 9.2.23 (Aage and Salunke (2011)) *Let (X, ϱ) be a cone metric space, let P be an order cone and let $h, g : X \to X$ be a weakly ϕ-pair. Assume that h and g are weakly compatible with $h(X) \subset g(X)$. If $h(X)$ or $g(X)$ is a complete subspace of X, then the mappings h and g have a unique common fixed point in X. Moreover for any $u_0 \in X$, every h-g-sequence $(h(u_n))$ with initial point u_0 converges to the common fixed point of h and g.*

The following theorem is due to Babu and Sailaja (2012) on a common fixed point theorems of Ćirić-type weak contraction in cone metric spaces.

Theorem 9.2.24 (Babu and Sailaja (2012)) *Let (X, ϱ) be a cone metric space with solid cone P. Suppose $h, g : X \to X$ are such that $h(X) \subseteq g(X)$ and the pair (h, g) is asymptotically regular at some point $u_0 \in X$. Suppose that there exists a continuous mapping $\varphi : P \to P$ satisfying $\varphi(0) = 0$ and $\varphi(t) \succ 0$, for $t \in P \setminus \{0\}$ such that*

$$\varrho(hu, hv) \preceq P(u, v) - \varphi(P(u, v)), \quad \forall u, v \in X,$$

where,

$$P(u, v) \in \{\varrho(gu, gv), \varrho(hu, gu), \varrho(hv, gv), \varrho(hu, gv), \varrho(hv, gu)\}.$$

If either $h(X)$ or $g(X)$ is a complete subspace of X then, h and g have a unique point of coincidence. Moreover, if the pair (h, g) is weakly compatible, then h and g have a unique common fixed point in X.

9.3 Main Results

We now present some coincidence and common fixed point theorems in complete cone b-metric spaces as an extension of Theorem 9.2.24.

Theorem 9.3.1 *Let (X, ϱ) be a cone b-metric space with solid cone P and $s \geq 1$, $s^2\lambda < 1$ where $\lambda \in [0, 1)$. Suppose that $h, g : X \to X$ are such that $h(X) \subseteq g(X)$ and the pair (h, g) is asymptotically regular at some point $u_0 \in X$. Suppose that there exists a continuous mapping $\varphi : P \to P$ satisfying $\varphi(0) = 0$ and $\varphi(t) \succ 0$, for $t \in P \setminus \{0\}$ such that*

$$\varrho(hu, hv) \preceq R(u, v) - \varphi(R(u, v)) \ \forall u, v \in X, \tag{9.2}$$

where,

$$R(u, v) \in \{\varrho(gu, gv), \varrho(hu, gu), \varrho(hv, gv), \varrho(hu, gv), \varrho(hv, gu)\}.$$

If either $h(X)$ or $g(X)$ is a complete subspace of X then, h and g have a unique point of coincidence. Moreover, if the pair (h, g) is weakly compatible, then h and g have a unique common fixed point in X.

Proof Let $u_0 \in X$. Since $h(X) \subseteq g(X)$ there exists $u_1 \in X$ such that $hu_0 = gu_1 = y_0$ (say). After defining v_n, we define $v_{n+1} = hu_{n+1} = gu_{n+2}, n = 0, 1, 2, ...$

Since (h, g) is asymptotically regular at u_0, for each $c \in int P$, there exists $n_0 \in \mathbb{Z}^+$ such that $\varrho(v_n, v_{n+1}) \ll c$, $\forall n \geq n_0$.

We now show that $\{v_n\}$ is a Cauchy sequence in X. Fix a positive integer n such that $n \geq n_0$. First, we show that

$$\varrho(v_n, v_{n+p}) \ll c, \ \forall c \in intP \text{ and } p = 1, 2, 3, ... \tag{9.3}$$

For $p = 1$, it obviously holds. By induction, suppose that the statement is true for some k. Hence, we have $\varrho(v_n, v_{n+k}) \ll c$, $\forall c \in intP$. Now, consider

$$\begin{aligned}
\varrho(v_n, v_{n+k+1}) &\preceq s(\varrho(v_n, v_{n+1}) + \varrho(v_{n+1}, v_{n+k+1})),\\
&= s(\varrho(v_n, v_{n+1}) + \varrho(fu_{n+1}, fu_{n+k+1})),\\
&\preceq s(\varrho(v_n, v_{n+1}) + R(u_{n+1}, u_{n+k+1}) - \varphi(R(u_{n+1}, u_{n+k+1})),
\end{aligned}$$

where,

$$\begin{aligned}
R(u_{n+1}, u_{n+k+1}) &= \{\varrho(gu_{n+1}, gu_{n+k+1}), \varrho(hu_{n+1}, gu_{n+1}), \varrho(hu_{n+k+1}, gu_{n+k+1}),\\
&\quad \varrho(hu_{n+1}, gu_{n+k+1}), \varrho(hu_{n+k+1}, gu_{n+1})\},\\
&= \{\varrho(v_n, v_{n+k}), \varrho(v_{n+1}, v_n), \varrho(v_{n+k+1}, v_{n+k}),\\
&\quad \varrho(v_{n+1}, v_{n+k}), \varrho(v_{n+k+1}, v_n)\}.
\end{aligned}$$

By using the definition of cone b-metric space and inequality (2.1), we have five cases for $n \to \infty$ as follows:

(i) For $R(x_{n+1}, u_{n+k+1}) = \varrho(v_n, v_{n+k})$, we have

$$\begin{aligned}
\varrho(v_n, v_{n+k+1}) &\preceq s\varrho(v_n, v_{n+1}) + s\varrho(v_n, v_{n+k}) - s\varphi(\varrho(v_n, v_{n+k})),\\
&\preceq s\varrho(v_n, v_{n+1}) + s\varrho(v_n, v_{n+k}),\\
&\preceq \frac{c}{2} + \frac{c}{2},\\
&= c.\\
\Rightarrow \varrho(v_n, v_{n+k+1}) &\ll c, \ \text{for } c \in intP.
\end{aligned}$$

(ii) If $P(u_{n+1}, u_{n+k+1}) = \varrho(v_{n+1}, v_n)$ then,

$$\begin{aligned}
\varrho(v_n, v_{n+k+1}) &\preceq s\varrho(v_n, v_{n+1}) + s\varrho(v_{n+1}, v_n) - s\varphi(\varrho(v_{n+1}, v_n)),\\
&\preceq s\varrho(v_n, v_{n+1}) + s\varrho(v_{n+1}, v_n)),\\
&\preceq \frac{c}{2} + \frac{c}{2},\\
&= c.\\
\Rightarrow \varrho(v_n, v_{n+k+1}) &\ll c, \ \text{for } c \in intP.
\end{aligned}$$

(iii) If $R(u_{n+1}, u_{n+k+1}) = \varrho(v_{n+k+1}, v_{n+k})$,

$$\begin{aligned}\Rightarrow \varrho(v_n, v_{n+k+1}) &\preceq s\varrho(v_n, v_{n+1}) + s\varrho(v_{n+k+1}, v_{n+k}) - s\varphi(\varrho(v_{n+k+1}, v_{n+k})),\\ &\preceq s\varrho(v_n, v_{n+1}) + s\varrho(v_{n+k+1}, v_{n+k})),\\ &\preceq \frac{c}{2} + \frac{c}{2},\\ &= c.\\ \Rightarrow \varrho(v_n, v_{n+k+1}) &\ll c, \text{ for } c \in int P.\end{aligned}$$

(iv) If $R(u_{n+1}, u_{n+k+1}) = \varrho(v_{n+1}, v_{n+k})$,

$$\begin{aligned}\Rightarrow \varrho(v_n, v_{n+k+1}) &\preceq s\varrho(v_n, v_{n+1}) + s\varrho(v_{n+1}, v_{n+k}) - s\varphi(\varrho(v_{n+1}, v_{n+k})),\\ &\preceq s\varrho(v_n, v_{n+1}) + s\varrho(v_{n+1}, v_{n+k}),\\ &\preceq s(\varrho(v_n, v_{n+1}) + s(\varrho(v_{n+1}, v_n) + \varrho(v_n, v_{n+k})),\\ &= s\varrho(v_n, v_{n+1}) + s^2\varrho(v_n, v_{n+1}) + s^2\varrho(v_n, v_{n+k}),\\ &\preceq \frac{c}{3} + \frac{c}{3} + \frac{c}{3},\\ &= c.\\ \Rightarrow \varrho(v_n, v_{n+k+1}) &\ll c, \text{ for } c \in int P.\end{aligned}$$

(v) If $R(u_{n+1}, u_{n+k+1}) = \varrho(v_n, v_{n+k+1})$, then

$$\begin{aligned}\varrho(v_n, v_{n+k+1}) &\preceq s\varrho(v_n, v_{n+1}) + s\varrho(v_n, v_{n+k+1}) - s\varphi(\varrho(v_n, v_{n+k+1})),\\ &\preceq s\varrho(v_n, v_{n+1}) + s\varrho(v_n, v_{n+k+1}),\\ &\preceq \frac{c}{2} + \frac{c}{2},\\ &= c.\\ \Rightarrow \varrho(v_n, v_{n+k+1}) &\ll c, \text{ for } c \in int P.\end{aligned}$$

Hence, for all five cases, $\varrho(v_n, v_{n+k+1}) \ll c$, for $c \in int P$. Therefore, condition (2.2) holds for $p = 1, 2, 3, ...$ So, $\{v_n\}$ is a Cauchy sequence in X. Assume that $g(X)$ is a complete subspace of X. Therefore, there exists $r \in g(X)$ such that

$$\lim_{n\to\infty} v_n = \lim_{n\to\infty} hu_n = \lim_{n\to\infty} gu_{n+1} = z. \tag{9.4}$$

Since $z \in g(X)$, there exists $x \in X$ such that $z = gx$. Now, we show that $hx = z$. From the Condition (iii) of Definition 9.2.11, we get

$$\begin{aligned}\varrho(hx, z) &= s\varrho(hx, hu_n) + s\varrho(hu_n, z), \\ &\preceq sR(x, u_n) - s\varphi(P(x, u_n)) + s\varrho(hu_n, z),\end{aligned} \tag{9.5}$$

where

$$\begin{aligned}R(x, u_n) &\in \{\varrho(gx, gu_n), \varrho(hx, gx), \varrho(hx, gu_n), \varrho(hu_n, gu_n), \varrho(hu_n, gx)\}, \\ &= \{\varrho(z, gu_n), \varrho(hx, z), \varrho(hx, gu_n), \varrho(hu_n, gu_n), \varrho(hu_n, z)\}.\end{aligned} \tag{9.6}$$

For $n \to \infty$ and the condition (2.3), we have five cases as follows:

Case I. For $R(x, u_n) = \varrho(z, gu_n)$, we have

$$\begin{aligned}\varrho(hx, z) &\preceq s\varrho(z, gu_n) - s\varphi(\varrho(gu_n, z)) + s\varrho(hu_n, z) \text{ from }, \\ &\preceq s\varrho(z, gu_n) + s\varrho(hu_n, z), \\ &\ll \frac{c}{2} + \frac{c}{2}, \\ &= c,\end{aligned}$$

for each $n \geq n_0 = n_0(c)$

$$\Rightarrow 0 \preceq \varrho(hx, z) \ll c, \text{ for } c \in int P,$$
$$\Rightarrow \varrho(hx, z) = 0.$$

Case II. Take $R(x, u_n) = \varrho(hx, z)$, we have

$$\begin{aligned}\varrho(hx, z) &\preceq s\varrho(hx, z) - s\varphi(\varrho(hx, z)) + s\varrho(hu_n, z), \\ &\preceq s\varrho(hx, z) + s\varrho(hu_n, z), \\ (1 - s)\varrho(hx, z) &\preceq s\varrho(hu_n, z).\end{aligned} \tag{9.7}$$

From (2.3), $\lim_{n\to\infty} hu_n = z$.

$$\Rightarrow (1 - s)\varrho(hx, z) \preceq 0,$$
$$\Rightarrow \varrho(hx, z) = 0.$$

Case III. For $R(x, u_n) = \varrho(hu_n, gu_n)$, we have

$$\begin{aligned}\varrho(hx, z) &\preceq s\varrho(hu_n, gu_n) - s\varphi(\varrho(hu_n, gu_n)) + s\varrho(hu_n, z), \\ &\preceq s\varrho(hu_n, gu_n) + s\varrho(hu_n, z), \\ &\preceq s^2\varrho(hu_n, z) + s^2\varrho(z, gu_n) + s\varrho(hu_n, z), \\ &\preceq \frac{c}{3} + \frac{c}{3} + \frac{c}{3}, \\ &= c.\end{aligned}$$

$$\Rightarrow 0 \preceq \varrho(hx, z) \ll c, \text{ for } c \in int P.$$
$$\Rightarrow \varrho(hx, z) = 0.$$

Case IV. For $R(x, u_n) = \varrho(hx, gu_n)$, we have

$$\begin{aligned}\varrho(hx, z) &\preceq s\varrho(hx, gu_n) - s\varphi(\varrho(hx, gu_n)) + s\varrho(hu_n, z),\\ &\preceq s\varrho(hx, gu_n) + s\varrho(hu_n, z)\end{aligned}$$

Assume that, $hx = y$

$$\begin{aligned}\varrho(hx, z) &\preceq s\varrho(y, gu_n) + s\varrho(hu_n, z),\\ \Rightarrow \varrho(hx, z) &\preceq \frac{c}{2} + \frac{c}{2},\\ &= c.\\ \Rightarrow \varrho(hx, z) &= 0.\end{aligned}$$

Case V. For $R(x, u_n) = \varrho(z, hu_n)$, we have

$$\begin{aligned}\varrho(hx, z) &\preceq s\varrho(z, hu_n) - s\varphi(\varrho(z, hu_n)) + s\varrho(hu_n, z),\\ &\preceq s\varrho(z, hu_n) + s\varrho(hu_n, z),\\ &\preceq \frac{c}{2} + \frac{c}{2},\\ &= c.\\ \Rightarrow \varrho(hx, z) &= 0.\end{aligned}$$

So, from all five cases, $hx = z$.

$$\Rightarrow hx = gx = z.$$

Thus, z is a point of coincidence of h and g.

We now show that z is unique. Assume that there exist $x', z' \in X$ such that $hx' = gx' = z'$, then we have

$$\begin{aligned}\varrho(z, z') &= \varrho(hx, hx') \preceq P(x, x') - \varphi(P(x, x')),\\ \text{where } R(x, x') &= \{\varrho(gx, gx'), \varrho(hx, gx), \varrho(hx', gx'), \varrho(hx, gx'), \varrho(hx', gx)\},\\ &= \{\varrho(z, z'), 0, 0, \varrho(z, z'), \varrho(z', z)\}.,\\ &= \{\varrho(z, z'), 0\}.\end{aligned}$$

$$\text{If } R(x, x') = 0,$$
$$\Rightarrow \varrho(z, z') = 0,$$
$$\Rightarrow z = z'.$$

Again, if $R(x, x') = \varrho(z, z')$,

$$\Rightarrow \varrho(z, z') = \varrho(z, z') - \varphi\varrho(z, z'),$$
$$\Rightarrow \varphi\varrho(z, z') = 0,$$
$$\Rightarrow \varrho(z, z') = 0,$$
$$\Rightarrow z = z'.$$

Then, z is a unique point of coincidence of h and g.
If the pair (h, g) is weakly compatible, then

$$hz = hgx = ghx = gz = t \text{ (say)}. \tag{9.8}$$

Thus, t is another point of coincidence of h and g. But z is unique. Thus, $t = z$.

$$\Rightarrow hz = hgx = ghx = gz = z.$$
$$\Rightarrow hz = gz = z.$$

This implies z is a unique common fixed point of h and g.

If we assume that $h(X)$ is a complete subspace of X, and since $h(X) \subseteq g(X)$, there exists $z \in g(X)$ such that

$$\lim_{n\to\infty} v_n = \lim_{n\to\infty} hu_n = \lim_{n\to\infty} gu_{n+1} = z,$$

and the proof follows as above.

□

If we take $R(u, v) = \varrho(gu, gv)$ in Theorem 9.3.1, we obtain the following corollary as an extension of the Theorem 2.1 of Ćirić et al. (2010) and the Corollary 2.2 of Babu and Sailaja (2012) to cone b-metric spaces.

Corollary 9.3.2 *Let (X, ϱ) be a cone b-metric space with solid cone P and $s \geq 1, s^2\lambda < 1$ where $\lambda \in [0, 1)$. Suppose $h, g : X \to X$ are such that $h(X) \subseteq g(X)$ and the pair (h, g) is asymptotically regular at some point $u_0 \in X$. Suppose that there exists a continuous mapping $\varphi : P \to P$ satisfying $\varphi(0) = 0$ and $\varphi(t) \succ 0$ for $t \in P \setminus \{0\}$ such that*

$$\varrho(hu, hv) \preceq \varrho(gu, gv) - \varphi\varrho(gu, gv)\, \forall u, v \in X.$$

If either $h(X)$ or $g(X)$ is a complete subspace of X then, h and g have a unique point of coincidence. Moreover, if the pair (h, g) is weakly compatible, then h and g have a unique common fixed point in X.

Proof The proof of this corollary follows by taking

$$R(u, v) = \varrho(gu, gv),$$

in the Theorem 9.3.1.

□

Taking $g = I$ (an identity mapping) in Theorem 9.3.1, we get the following corollary as an extension of the Corollary 2.3 of Babu and Sailaja (2012) to cone b-metric spaces.

Corollary 9.3.3 *Let (X, ϱ) be a cone b-metric space with solid cone P and $s \geq 1, s^2\lambda < 1$ where $\lambda \in [0, 1)$. Suppose $h, g : X \to X$ such that $h(X) \subseteq g(X)$ and the pair (h, g) is asymptotically regular at some point $u_0 \in X$. Suppose that there exists a continuous mapping $\varphi : P \to P$ satisfying $\varphi(0) = 0$ and $\varphi(t) \succ 0$, for $t \in P \setminus \{0\}$ such that*

$$\varrho(hu, hv) \preceq R(u, v) - \varphi(R(u, v)) \; \forall u, v \in X, \tag{9.9}$$

where,

$$R(u, v) \in \{\varrho(u, v), \varrho(u, hu), \varrho(v, hv), \varrho(v, hu), \varrho(u, hv)\}.$$

If either $h(X)$ or $g(X)$ is a complete subspace of X then, h and g have a unique point of coincidence. Moreover, if the pair (h, g) is weakly compatible, then h and g have a unique common fixed point in X.

Proof If we take $g = I$ (an identity mapping) in Theorem 9.3.1, the proof follows as in the Theorem 9.3.1.

□

The following example justifies the Theorem 9.3.1 on a common fixed point in cone b-metric spaces.

Example 9.3.4 Let $X = (0, 1]$, $E = \mathbb{R}$ and $P\{(u, v) : u, v > 0\}$ is a solid cone. Define $\varrho : X \times X \to E$ by $\varrho(u, v) = (|u - v|\psi$, for $\psi(t) = e^t$. Define $h, g : X \to X$ and $\varphi : P \to P$ by

$$h(u) = \begin{cases} \frac{2}{5} \; if \; u \in (0, \frac{2}{3}), \\ 1 - \frac{u}{2} \; if \; u \in [\frac{2}{3}, 1], \\ \frac{3}{4} \; if \; u = 1, \end{cases}$$

$$g(u) = \begin{cases} \frac{3}{4} \; if \; u \in (0, \frac{2}{3}), \\ \frac{4}{3} - u \; if \; u \in [\frac{2}{3}, 1], \end{cases}$$

and $\varphi(u(t)) = \frac{u(t)^2}{4\psi(t)}$, for $u(t) \geq 0$. In this case, $h(X) = \{\frac{2}{5}, \frac{3}{4}\} \cup (\frac{1}{2}, \frac{2}{3}]$ and $g(X) = \{\frac{3}{4}\} \cup [\frac{1}{3}, \frac{2}{3}]$. So it is clear that $f(X) \subset g(X)$ and $g(X)$ is a complete subspace of X. We consider the following cases:

Case (i) Let $u, v \in (0, \frac{2}{3}]$. In this case $\varrho(hu, hv) = 0$. So, (2.1) holds.
Case (ii) If $u \in (0, \frac{2}{3})$ and $v \in [\frac{2}{3}, 1)$. So, $\varrho(hu, hv) = (\frac{3}{5} - \frac{v}{2})\psi$ and $\varrho(hu, gu) = |\frac{7}{20}|\psi$. So, $\varrho(hu, hv) = (\frac{3}{5} - \frac{v}{2})\psi < |\frac{7}{20}|\psi - \frac{|(\frac{7}{20}|\psi)^2}{4\psi}$. Then, $\varrho(hu, hv) < \varrho(hu, gu) - \varphi(\varrho(hu, gu))$. So, (2.1) holds.
Case (iii) For $u \in (0, \frac{2}{3})$ and $v = 1$. So, $\varrho(hu, hv) = \frac{1}{5}\psi$ and $\varrho(gu, gv) = \frac{5}{12}\psi$. So, $\varrho(hu, hv) = \frac{1}{5}\psi < |\frac{5}{12}|\psi - \frac{(|\frac{5}{12}\psi)^2}{4\psi}$. Then, $\varrho(hu, hv) < \varrho(gu, gv) - \varphi(\varrho(gu, gv))$. Therefore, (2.1) holds.
Case (iv) For $u, v \in [\frac{2}{3}, 1)$. So, $\varrho(hu, hv) = \frac{1}{2}|u - v|\psi$ and $\varrho(hv, gv) = |v - \frac{1}{3}|\psi$. So, $\varrho(hu, hv) = \frac{1}{2}|u - v|\psi < |v - \frac{1}{3}|\psi - \frac{(|v - \frac{1}{3}|\psi)^2}{4\psi}$. Then, $\varrho(hu, hv) < \varrho(hv, gv) - \varphi(\varrho(hv, gv))$. Hence, (2.1) holds.
Case (v) If $u = 1$ and $v = 1$ then, $\varrho(hu, hv) = 0$. So, (2.1) holds.

Hence, all the conditions of Theorem 2.1 are satisfied. Therefore, h and g have a unique common fixed point. Thus, $\frac{2}{3}$ is a unique common fixed point of h and g.

The following theorem extends the Theorem 2.4 of Aage and Salunke (2011) and the Theorem 2.11 of Babu and Sailaja (2012).

Theorem 9.3.5 *Let (X, ϱ) be a cone b-metric space with $s \geq 1$ and $s^2\lambda < 1$ where $\lambda \in [0, 1)$. Suppose that h and g are self-maps of X and assume that the pair (h, g) satisfies property (E. A.). Suppose that there exists a continuous mapping $\varphi : P \to P$ satisfying $\varphi(0) = 0$ and $\varphi(t) \succ 0$, for $t \in P \setminus \{0\}$ such that*

$$\varrho(hu, hv) \preceq \varphi(R(u, v)), \ \forall u, v \in X, \tag{9.10}$$

$$\text{where, } R(u, v) \in \{\varrho(gu, gv), \varrho(hu, gu), \varrho(hv, gv), \frac{1}{2}[\varrho(hu, gv) + \varrho(hv, gu)]\}.$$

Assume that $g(X)$ is a closed subspace of X and either $h(X)$ or $g(X)$ is a complete subspace of X then, h and g have a unique point of coincidence in X. Moreover, if the pair (h, g) is weakly compatible, then h and g have a unique common fixed point in X.

Proof Let $\{u_n\}$ be a sequence in X and $z \in X$ such that

$$\lim_{n\to\infty} hu_n = \lim_{n\to\infty} gu_n = z.$$

This is possible since the pair (h, g) satisfies the property (E. A.). Since $g(X)$ is a closed subspace of X, there exists $x \in X$ such that $gx = z$. We show that $hx = z$. Now, consider

$$\begin{aligned} \varrho(hx, z) &= s\varrho(hx, hu_n) + s\varrho(hu_n, z) \\ &\preceq s\varphi(R(x, u_n)) + s\varrho(hu_n, z) \text{ (from 2.9)}, \end{aligned} \tag{9.11}$$

$$\text{where, } R(x,u_n) \in \{\varrho(gx,gu_n), \varrho(hx,gx), \varrho(hu_n,gu_n), \frac{1}{2}[\varrho(hx,gu_n)+\varrho(hu_n,gx)]\},$$
$$= \{\varrho(z,gu_n), \varrho(hx,z), \varrho(hu_n,gu_n), \frac{1}{2}[\varrho(hx,gu_n)+\varrho(hu_n,z)]\}.$$

For $n \to \infty$ and using inequality (2.10) we have the following four cases:

Case I. For $R(x,u_n) = \varrho(z,gu_n)$, we get

$$\begin{aligned}
\varrho(hx,z) &\preceq s\varphi(\varrho(z,gu_n)) + s\varrho(hu_n,z),\\
&\prec s\varrho(z,gu_n) + s\varrho(hu_n,z),\\
&\ll \frac{c}{2}+\frac{c}{2}, \text{ for each } n \geq n_0 = n_0(c),\\
&= c.\\
&\Rightarrow 0 \preceq \varrho(hx,z) \ll c.
\end{aligned}$$

From Lemma 9.2.9 (ii), we have

$$\varrho(hx,z) = 0.$$

Case II. If $R(x,u_n) = \varrho(hx,z)$, we have

$$\begin{aligned}
\varrho(hx,z) &\preceq s\varphi\varrho(hx,z)) + s\varrho(hu_n,z),\\
&\prec s\varrho(hx,z)) + s\varrho(hu_n,z),\\
&\prec s^2\varrho(hx,gu_n)) + s^2\varrho(gu_n,z) + s\varrho(hu_n,z).
\end{aligned}$$

From condition (iii) of the Definition 9.2.11, let $hx = a, \forall a \in X$. Then,

$$\begin{aligned}
\varrho(hx,z) &\ll \frac{c}{3}+\frac{c}{3}+\frac{c}{3}, \text{ for each } n \geq n_0 = n_0(c)\\
&= c.\\
&\Rightarrow 0 \preceq \varrho(hx,z) \ll c.\\
&\Rightarrow \varrho(hx,z) = 0.
\end{aligned}$$

Case III. If $R(x,u_n) = \varrho(hu_n,gu_n)$, we have

$$\begin{aligned}
\varrho(hx,z) &\preceq s\varphi(\varrho(hu_n,gu_n)) + s\varrho(hu_n,z),\\
&\prec s\varrho(hu_n,gu_n) + s\varrho(hu_n,z),\\
&\prec s^2\varrho(hu_n,z) + s^2\varrho(z,gu_n) + s\varrho(hu_n,z).
\end{aligned}$$

From condition (iii) of the Definition 9.2.11, we have

$$\begin{aligned}\varrho(hx, z) &\ll \frac{c}{3} + \frac{c}{3} + \frac{c}{3}, \text{ for each } n \geq n_0 = n_0(c)\\ &= c\\ &\Rightarrow 0 \preceq \varrho(hx, z) \ll c.\\ \Rightarrow \varrho(hu, z) &= 0.\end{aligned}$$

Case IV. For $R(u, u_n) = \frac{1}{2}[\varrho(hx, gu_n) + \varrho(z, hu_n)]$, from (2.10), we get

$$\begin{aligned}\varrho(hx, z) &\preceq s\varphi\frac{1}{2}[\varrho(hx, gu_n) + \varrho(z, hu_n)] + s\varrho(hu_n, z),\\ &\prec \frac{1}{2}s[\varrho(hx, gu_n) + \varrho(z, hu_n)] + s\varrho(hu_n, z),\\ &\preceq \frac{1}{2}s\varrho(hx, gu_n) + 2\,s\varrho(hu_n, z),\\ &\ll \frac{1}{2}c + 2\frac{c}{4}, \text{ for each } n \geq n_0 = n_0(c)\\ &= c.\\ &\Rightarrow 0 \preceq \varrho(hx, z) \ll c\\ \Rightarrow \varrho(hx, z) &= 0.\end{aligned}$$

Hence, for all four cases, we have $\varrho(hx, z) = 0$. This implies that $hx = gx = z$. Thus, z is a point of coincidence of h and g. We now show that z is unique.

Suppose that there exists $x', z' \in X$ such that $hx' = gx' = z'$ for $z' \neq z$. Then,

$$\begin{aligned}\varrho(z, z') &= \varrho(hx, hx')\\ &\preceq \varphi(P(x, x')) \text{ from (2.9)},\end{aligned}$$

where,

$$\begin{aligned}R(x, x') &\in \{\varrho(gx, gx'), \varrho(hx, gx), \varrho(hx', gx'), \frac{1}{2}[\varrho(hx, gx') + \varrho(hx', gx)]\},\\ &= \{\varrho(z, z'), 0, 0, \frac{1}{2}[\varrho(z, z') + \varrho(z', z)]\}.\\ &= \{\varrho(z, z'), 0, \varrho(z, z')\}.\\ &= \{\varrho(z, z'), 0\}.\end{aligned}$$

If $R(x, x') = 0$, then $\varrho(z, z') = 0$. This implies $z = z'$.

If $R(x, x') = \varrho(z, z')$, then $\varrho(z, z') \leq \varphi(\varrho(z, z')) < \varrho(z, z')$. This is a contradiction. Therefore $z = z'$ is a unique point of coincidence of h and g in X. If the pair

(h, g) is weakly compatible, then

$$hz = hgx = ghx = gz = r \text{ (say)}.$$

Thus, r is a point of coincidence of h and g in X. But z is a unique point of coincidence of h and g in X.

$$\Rightarrow hz = gz = z.$$

Therefore, z is a unique common fixed point of h and g. □

9.4 An Application

In this section, we will utilize Theorem 9.3.5 proved in Sect. 9.2 to show the existence of a unique common solution of integral equations. We can express the problem in an integral form as follows:

$$u(t) = \int_a^b h(t, u(s))d(s), t \in [a, b], \tag{9.12}$$

$$v(t) = \int_a^b g(t, v(s))d(s), t \in [a, b]. \tag{9.13}$$

Let $X = [0, +\infty]$, $E = \mathbb{R}$. Take $P = \{u(t) \in E : u(t) \geq 0\}$. Define $\varrho : X \times X \to \mathbb{R}$ as $\varrho(u(t), v(t)) = (\mid u(t) - v(t) \mid)$, $\forall u(t), v(t) \in X$, and $u(t) \neq v(t)$. Assume that (X, ϱ) is a complete cone b-metric space.

Now, we define the self-mappings h and g by

$$h(u(t)) = \begin{cases} \int_a^b K_1(t)d(s) \; if \; u(s) \in (0, a) \\ \int_a^b K_2(t, u(s))d(s) \; if \; u(s) \in [a, b], \end{cases}$$

$$g(v(t)) = \begin{cases} \int_a^b K_2(t)d(s) \; if \; v(s) \in (0, a) \\ \int_a^b K_3(t, v(s))d(s) \; if \; v(s) \in [a, b], \end{cases}$$

and $\varphi : P \to P$ by $\varphi(u(t)) = \alpha(u(t))$ for $\alpha \in [0, 1)$, where $K_1, K_2, K_3 : [a, b] \times [a, b] \to \mathbb{R}^+$. If $u(t)$ and $v(t)$ are the fixed points of $hu(t)$ and $gv(t)$ respectively, then these are solutions of integral equations (3.1) and (3.2) respectively. We will prove the existence of a unique common solution of integral Equations (3.1) and (3.2) by the use of Theorem 9.3.5.

Theorem 9.4.1 *Let $h, g : X \to X$ be two self-mappings of a complete cone b-metric space (X, ϱ). Suppose that (h, g) satisfies the property (E. A.) and $g(X)$ is a closed subspace of X and either $g(X)$ or $h(X)$ is a complete subspace of X. Suppose that* *there exists a continuous mapping* $\varphi : P \to P$ *satisfying* $\varphi(0) = 0$ *and* $\varphi(t) \succ 0$, *for* $t \in P \setminus \{0\}$ *such that*

$$\varrho(hu(t), hv(t)) \preceq \varphi(R(u(t), v(t)))\ \forall u(t), v(t) \in X, \tag{9.14}$$

where,

$$R(u(t), v(t)) \in \{\varrho(gu(t), gv(t)), \varrho(hu(t), gu(t)), \varrho(hv(t), gv(t)), \frac{1}{2}[\varrho(hu(t), gv(t)) + \varrho(hv(t), gu(t))]\}.$$

Assume that the following condition is satisfied:

(i)

$$|\int_a^b (K_2(t, u(s)) - K_2(t, v(s)))d(s)| \leq \alpha(|\int_a^b (K_3(t, u(s)) - K_3(t, v(s)))d(s)|). \tag{9.15}$$

for $\alpha \in [0, 1)$.

Then, the integral Equations (3.1) *and* (3.2) *have a unique common solution.*

Proof The map ϱ is defined as $\varrho : X \times X \to \mathbb{R}$ by

$$\varrho(u(t), v(t)) = (|\ u(t) - v(t)\ |), \forall u(t), v(t) \in X,$$

and

$$u(t) \neq v(t).$$

Now, using the definition of ϱ and the Condition (3.4), we consider the following cases:

Case (i) If $u(t), v(t) \in (0, a)$. In this case $\varrho(hu(t), hv(t)) = |\int_a^b K_1(t)d(s) - \int_a^b K_1(t)d(s)| = 0$. So, (3.3) holds.

Case (ii) If $u(t), v(t) \in [a, b]$. So, $\varrho(hu(t), hv(t)) = |\int_a^b (K_2(t, u(s)) - K_2(t, v(s)))d(s)| \leq \alpha(|\int_a^b (K_3(t, u(s)) - K_3(t, v(s)))d(s)|) = \alpha\varrho(gu(t), gv(t)) = \varphi(\varrho(gu(t), gv(t)))$ So, (3.3) holds.

Case (iii) If $u(s) = b$ and $v(s) = b$ then, $\varrho(hu(t), hv(t)) = |\int_a^b (K_2(t, b) - K_2(t, b))d(s)|$. So, (3.3) holds.

Hence, all conditions of the Theorem 9.3.5 are satisfied. So, the integral Equations (3.1) and (3.2) have a unique common solution.

□

Acknowledgements The authors are thankful to the learned reviewers for their valuable comments.

Conflict of interest The authors declare that they have no competing interests.

References

Aage, C. T., & Salunke, J. N. (2011). Fixed points of (Ψ, ϕ)-weak contractions in cone metric spaces. *Annals of Functional Analysis,2*(1), 59–71.

Aamri, M., & El Moutawakil, D. (2002). Some new common fixed point theorems under strict contractive conditions. *Journal of Mathematical Analysis and Applications,270*(1), 181–188.

Alsulami, H. H., Karapinar, E., & Rakocevic, V. (2017). Ciric type nonunique fixed point theorems on b-metric spaces. *Filomat,31*(11), 3147–3156.

Aydi, H., Bota, M. F., Karapinar, E., & Mitrovic, S. (2012). A fixed point theorem for set-valued quasi-contractions in b-metric spaces. *Fixed Point Theory Applications,1*, 1–8.

Babu, G. V. R., & Sailaja, P. D. (2012). Common fixed point theorems of Ćirić type weak contractions in cone metric spaces. *Thai Journal of Mathematics,10*(3), 517–533.

Banach, S. (1922). Sur les OperÌ'rations dans les ensembles abstraits et leur application aux eq́ations inteǵrales. *Fundamenta Mathematicae,3*, 133–181.

Ćirić, L., Hussain, N., & Cakic, N. (2010). Common fixed points for Ćirić type f-weak contraction with applications. *Math. (Debr.),76*(1–2), 31–49.

Huang, H., & Xu, S. (2013). Fixed point theorems of contractive mappings in cone b-metric spaces and applications. *Fixed Point Theory Application,1*, 1–10.

Huang, L. G., & Zhang, X. (2007). Cone metric spaces and fixed point theorems of contractive mappings. *Journal of Mathematical Analysis and Applications,332*(2), 1468–1476.

Huang, X., Zhu, C., & Wen, X. (2010). A common fixed point theorem in cone metric spaces. *International Journal of Mathematical Analysis,4*(13–16), 721–726.

Hussain, N., & Shah, M. H. (2011). KKM mappings in cone b-metric spaces. *Computers and Mathematics Applications,62*(4), 1677–1684.

Jungck, G. (1986). Compatible Mappings and Common Fixed Points. *International Journal of Mathematics and Mathematical Sciences,9*(4), 771–779.

Jungck, G., Radenovic, S., Radojevic, S., & Rakocevic, V. (2009). Common fixed point theorems for weakly compatible pairs on cone metric spaces. *Fixed Point Theory Application,1*, 1–13.

Jungck, G., & Rhoades, B. E. (1998). Fixed points for set valued functions without continuity. *Indian Journal of Pure and Applied Mathematics,29*, 227–238.

Karapinar, E., & Du, W. S. (2013). A note on b-cone metric and its related results: Generalizations or equivalence? *Fixed Point Theory Application,2013*, 210.

Karapinar, E., Mitrović, Z. D., Ozturk, A. , Radenovic, S. (2020). On a theorem of Ćirić in b-metric spaces. *Rendiconti del Circolo Matematico di Palermo,* II. Series. https://doi.org/10.1007/s12215-020-00491-9.

Kaneko, H., & Sessa, S. (1989). Fixed point theorems for compatible multi-valued and single-valued mappings. *International Journal of Mathematics and Mathematical Sciences,12*(2), 257–262.

Kumar, S., & Kapinga, E. (2023). Common fixed point theorems concerning f-contraction maps in complete cone b-metric space. *Caspian Journal of Mathematical Sciences (CJMS), 12*(2), 375–396. http://cjms.journals.umz.ac.ir.

Malhotra, S. K., Shukla, S., & Sen, R. (2012). Some coincidence and common fixed point theorems in cone metric spaces. *Bulletin of Mathematical Analysis and Applications,4*(2), 64–71.

Pauline, S., & Kumar, S. (2021). Common Fixed point theorems for T-Hardy-Rodgers contraction mappings in complete cone b-metric spaces with an application. *Topological Algebra Applications,9*, 105–117.

Rangamma, M., & Prudhvi, K. (2012). Common fixed points under contractive conditions for three maps in cone metric spaces. *Bulletin of Mathematical Analysis and Applications,4*(1), 174–180.

Singh, S. L., & Tomar, A. (2003). Weaker forms of commuting maps and existence of fixed points. *Journal of the Korean Society of Mathematical Education Series B, Pure and Applied Mathematics Journal, 10*(03), 145–161.

Sundaresan, K. (1971). Uniform convexity of Banach spaces $l(p_i)$. *Studia Mathematica,39*(3), 227–231.

Tomar, A., & Joshi, M. (2021). Results in strongly minihedral cone and scalar weighted cone metric spaces and applications. *Annales Mathematicae Silesianae,35*(2), 302–318. https://doi.org/10.2478/amsil-2021-0009

Tomar, A., Joshi, M., & Bhatt, V. (2021). Fixed point and a Cantilever Beam problem in a partial *b*-metric Space. *Acta Universitatis Sapientiae, Mathematica,13*(2), 506–518. https://doi.org/10.2478/ausm-2021-0032

Vetro, P. (2007). *Common fixed points in cone metric spaces* (pp. 464–468). LVI: Rendiconti Del Circolo Mathematico Di Palermo.

Wangwe, L., & Kumar, S. (2022). Common fixed point theorem for generalized F-Kannan Suzuki type mapping in TVS valued cone metric space with applications, *Journal of Mathematics, 2022*, Article ID 6504663, 17 pages. https://doi.org/10.1155/2022/6504663.

Chapter 10
Fixed Points of Closed Graph Operators on Partial Metric Spaces

G. Siva

Abstract The extendibility of the Banach iteration technique to some contraction mappings with variations in partial metric space domains to establish some fixed-point results has been discussed in this article. A map that satisfies the contraction condition has been proposed for an increasing sequence of subsets of a partial metric space, where one element of the sequence is mapped into the following member of the sequence. Additionally, some fixed-point theorems are established for contraction mappings of Banach, Kannan, Chatterjee, and Hardy–Roger types with closed graphs.

Keywords Closed graph · Metric function · Partial metric space

AMS Subject Classification Primary 54E40 · Secondary 54H25

10.1 Introduction

Kirk et al. (2003) proposed using a cycle of domains to derive various fixed-point theorems for metric spaces. Moorthy et al. (2008) considered an increasing sequence of subsets $\Theta_1 \subseteq \Theta_2 \subseteq \ldots$ of a metric space (Θ, d), and a map $G : \Theta \to \Theta$ satisfying a contraction condition such that $G(\Theta_i) \subseteq \Theta_{i+1}, \forall i$, and $\Theta = \bigcup_{j=1}^{\infty} \Theta_j$. There are some articles related to the cycle of domains; see Siva (2022a, b, 2023a, b).

The most active area of research is fixed-point theory, which has many uses in both pure and applied mathematics. Several writers have investigated fixed-point results in various generalized metric spaces. Matthews introduced the idea of partial metric spaces in Matthews (1992a), which is a generalization of traditional metric spaces in which every point's self-distance between points need not equal zero.

G. Siva (✉)
Department of Mathematics, Rajiv Gandhi National Institute of Youth Development, Sriperumbudur, India
e-mail: gsivamaths2012@gmail.com

© The Author(s), under exclusive license to Springer Nature Singapore Pte Ltd. 2026

A. Tomar and M. Jain (eds.), *Applications of Fixed-Point Theorem*, Industrial and Applied Mathematics, https://doi.org/10.1007/978-981-95-5160-6_10

He developed the notion of partial metrics in response to the drawbacks of metric functions in computer science. There are numerous uses for partial metric spaces in both computer science and mathematics. Matthews (1992b) discovered the partial metric contraction theorem after developing partial metric functions, making the partial metric function applicable to fixed-point theory. On fixed-point theory in partial metric spaces and its extensions, there are numerous articles; see Shobkolaei et al. (2011), Karapinar et al. (2013, 2018), Pant et al. (2017) and Tomar et al. (2017, 2021a, b, 2022a, b, 2023).

By using variations in partial metric space domains, we prove some more fixed-point theorems for different types of contraction mappings with closed graphs. Also, we generalize the main result of Moorthy et al. (2008).

10.2 Partial Metric Spaces

Let us give some preliminary and known results in this section. See Matthews (1992a, b) for further information.

Definition 10.2.1 (*Matthews* 1992b) Let $\Theta \neq \emptyset$ be a set. A partial metric is a mapping $p : \Theta \times \Theta \to \mathbb{R}^+ = [0, \infty)$ satisfying the following conditions.

(i) $k = l$ if and only if(or, iff) $p(k,k) = p(l,l) = p(k,l), \forall\, k, l \in \Theta$,
(ii) $p(k,k) \leq p(k,l), \forall\, k, l \in \Theta$,
(iii) $p(k,l) = p(l,k), \forall\, k, l \in \Theta$,
(iv) $p(k,l) \leq p(k,q) + p(q,l) - p(q,q), \forall\, k, l, q \in \Theta$.

The pair (Θ, p) is then referred to as a partial metric space (or, PMS).

It is clear that, if $p(k,l) = 0$, then $k = l$, because (i) and (ii). But if $k = l$, $p(k,l)$ may not be 0.

Remark 10.2.2 If p is a partial metric on Θ, then $d : \Theta \times \Theta \to [0, \infty)$ defined by $d(k,l) = 2p(k,l) - p(k,k) - p(l,l)$ is an usual metric on Θ.

Example 10.2.3 Let $\Theta = \mathbb{R}$ and $p(k,l) = e^{\max\{k,l\}}, \forall k, l \in \Theta$. Then Θ is a PMS.

Definition 10.2.4 (*Matthews* 1992a) Each partial metric p on Θ generates a T_0 topology τ_p on Θ which has as a base the family of open p-balls $\{B_p(k,\epsilon) : k \in \Theta; \epsilon > 0\}$, where $B_p(k,\epsilon) = \{l \in \Theta : p(k,l) < p(k,k) + \epsilon\}, \forall\, k \in \Theta$ and $\epsilon > 0$.

Lemma 10.2.5 (Matthews (1992a)) *Let (Θ, p) be a PMS, $\{k_n\}$ be a sequence in Θ and $k \in \Theta$. Then $k_n \to k (n \to \infty)$ iff $p(k_n, k) \to p(k,k) (n \to \infty)$. Moreover, if $p(k_n, k) \to p(k,k) (n \to \infty)$, then $p(k_n, k_n) \to p(k,k) (n \to \infty)$.*

Note that the limit of a convergent sequence is not unique in PMS, see Shobkolaei et al. (2011) for more details. So the next lemma shows that the limit is unique with some additional conditions.

Lemma 10.2.6 (Shobkolaei et al. (2011)) *Let* (Θ, p) *be a PMS, and* $\{k_n\}$ *be a sequence converges to two limits* k *and* l *in* Θ. *If* $\lim_{n\to\infty} p(k_n, k_n) = p(k, k) = p(l, l)$, *then* $k = l$

Lemma 10.2.7 (Matthews (1992a)) *Let* (Θ, p) *be a PMS and* $\{k_n\}$ *be a sequence in* Θ. *Then* $\{k_n\}$ *is a Cauchy sequence iff* $p(k_n, k_m)$ *exists and is finite whenever* $m, n \to \infty$.

Definition 10.2.8 (*Matthews* 1992a) A PMS (Θ, p) is said to be complete if every Cauchy sequence $\{k_n\}$ is converges to k in Θ such that $\lim_{n,m\to\infty} p(k_n, k_m) = p(k, k)$.

The next lemma shows that the partial metric function is continuous.

Lemma 10.2.9 (Shobkolaei et al. (2011)) *Let* (Θ, p) *be a PMS. Let* $\{k_n\}$ *and* $\{l_n\}$ *be two sequences in* Θ *such that* $k_n \to k$, $l_n \to l$ $(n \to \infty)$, $k, l \in \Theta$*(or,* $\lim_{n,\to\infty} p(k_n, k) = \lim_{n\to\infty} p(k_n, k_n) = p(k, k)$ *and* $\lim_{n,\to\infty} p(l_n, l) = \lim_{n\to\infty} p(l_n, l_n) = p(l, l)$*), then* $p(k_n, l_n) \to p(k, l)(n \to \infty)$.

Definition 10.2.10 Let $G : (\Theta, p) \to (\Theta, p)$ be a self-mapping on a PMS (Θ, p). If whenever $k_n \to k_0$ and $Gk_n \to l_0$ for some sequence $\{k_n\}$ in Θ and some k_0, l_0 in Θ, we have $l_0 = Gk_0$, then G is said to have a closed graph.

10.3 Main Results

Let's prove some fixed-point theorems in this section for various contractions on PMSs.

Theorem 10.3.1 *Let* (Θ, p) *be a complete PMS and* $G : \Theta \to \Theta$ *has a closed graph. Let* $\Theta_1 \subseteq \Theta_2 \subseteq ...$ *be subsets of* Θ *such that* $\Theta = \bigcup_{j=1}^{\infty} \Theta_j$, $G(\Theta_i) \subseteq \Theta_{i+1}$, $\forall i$, *and* $p(Gt, Gz) \leq \xi_i p(t, z)$, $\forall t, z \in \Theta_i$, $\forall i$, *where* $\xi_i \in (0, \infty)$ *are real positive constants such that* $\sum_{n=1}^{\infty} \xi_1\xi_2...\xi_n < \infty$. *Then, for any fixed* $t_1 \in \Theta$, $\{G^n t_1\}$ *converges to a fixed point. Moreover, if* $\xi_i \in (0, 1)$, $\forall i$, *then* G *has a unique fixed point(or, UFP) in* Θ.

Proof Fix $t_1 \in \Theta_1$, and define $t_{n+1} = Gt_n = G^n t_1$, $\forall\, n = 1, 2, 3, ...$ Then we have,

$$\begin{aligned} p(G^{n+1}t_1, G^n t_1) &\leq \xi_{n+1} p(G^n t_1, G^{n-1} t_1) \\ &\leq \xi_{n+1}\xi_n\xi_{n-1}...\xi_2 p(Gt_1, t_1). \end{aligned}$$

Further, for $1 \leq n < m$, we have

$$
\begin{aligned}
p(G^m t_1, G^n t_1) &\leq p(G^m t_1, G^{m-1} t_1) + p(G^{m-1} t_1, G^{m-2} t_1) + ... + p(G^{n+1} t_1, G^n t_1) \\
&\quad - \sum_{r=1}^{m-n-1} p(G^{m-r} t_1, G^{m-r} t_1) \\
&\leq p(G^m t_1, G^{m-1} t_1) + p(G^{m-1} t_1, G^{m-2} t_1) + ... + p(G^{n+1} t_1, G^n t_1) \\
&\leq \Big(\sum_{i=n}^{m-1} \xi_2 \xi_3 ... \xi_{i+1} \Big) p(G t_1, t_1).
\end{aligned}
$$

Therefore, $p(G^m t_1, G^n t_1) \to 0$ $(m, n \to \infty)$. Since Lemma 10.2.7, $\{G^m t_1\}_{m=1}^{\infty}$ is a Cauchy sequence in Θ. Let $\{G^m t_1\}_{m=1}^{\infty}$ converge to t^* in Θ, which is complete.

$$
p(t^*, t^*) = \lim_{m \to \infty} p(G^m t_1, t^*) = \lim_{m \to \infty} p(G^m t_1, G^m t_1) = 0. \tag{10.1}
$$

Note that $\{G^{m+1} t_1\}_{m=1}^{\infty}$ is also a Cauchy sequence and it converges to s^* in Θ.

$$
p(s^*, s^*) = \lim_{m \to \infty} p(G^m t_1, s^*) = \lim_{m \to \infty} p(G^m t_1, G^m t_1) = 0. \tag{10.2}
$$

Since (10.1) and (10.2), we have

$$
\lim_{n \to \infty} p(G^m t_1, G^m t_1) = p(t^*, t^*) = p(s^*, s^*). \tag{10.3}
$$

Therefore, $t^* = s^*$, because of Lemma 10.2.6. Also, the closed graph of G gives $G t^* = t^*$. Hence, we obtained a fixed point t^* of G.

These processes can be extended to the general case: $t_1 \in \Theta_n$, for some n.

Suppose, further that $\xi_i \in (0, 1)$, $\forall i$.

If $G t^* = t^*$, $G z^* = z^*$ in G, then let $t^*, z^* \in \Theta_n$, for some n, so we have

$$
p(t^*, z^*) = p(G t^*, G z^*) \leq \xi_n p(t^*, z^*).
$$

Then, $p(t^*, z^*) \leq (\xi_n)^m p(t^*, z^*), \forall m \in \mathbb{N}$. Since $(\xi_n)^m \to 0$ as $m \to \infty$, $p(t^*, z^*) = 0$. By $p(t^*, z^*) = p(z^*, z^*) = p(t^*, t^*) = 0$, we have $t^* = z^*$. Hence, G has a UFP. □

Theorem 2.1 of Moorthy et al. (2008) is thus generalized by the previous Theorem 10.3.1.

Example 10.3.2 Let $\Theta = \left[\frac{1}{4}, \infty\right)$. Let $p(t, z) = \max\{t, z\}, \forall t, z \in \Theta$. Then (Θ, p) is a complete PMS.

Let $\Theta_n = [\frac{1}{4}, n]$, and $\xi_n = \frac{n^2}{(n+1)^2} \in [\frac{1}{4}, 1)$, for $n = 1, 2, 3, ...$ Then $\sum_{n=1}^{\infty} \xi_1 \xi_2 ... \xi_n < \infty$.

Define $G : \Theta \to \Theta$ by $Gt = t^{\frac{1}{4}}$, if $t \in \Theta_n$, for $n \in \mathbb{N}$.
For $t, z \in \Theta_n$, we have

$$\begin{aligned} p(Gt, Gz) &= \max\left\{t^{\frac{1}{4}}, z^{\frac{1}{4}}\right\} \\ &= (p(t,z))^{\frac{1}{4}} \\ &\le p(t,z)^{\xi_n}, \ \forall n \in \mathbb{N}. \end{aligned}$$

Theorem 10.3.1's hypotheses are then fulfilled. Moreover, the UFP is 1.

Theorem 10.3.3 *Let (Θ, p) be a complete PMS and $G : \Theta \to \Theta$ has a closed graph. Let $\Theta_1 \subseteq \Theta_2 \subseteq ...$ be subsets of Θ such that $\Theta = \bigcup_{j=1}^{\infty} \Theta_j$, $G(\Theta_i) \subseteq \Theta_{i+1}$, $\forall i$, and $p(Gt, Gz) \le \xi_i(p(Gt, t) + p(Gz, z))$, $\forall t, z \in \Theta_i, \forall i$, where $\xi_i \in (0, 1)$ are real positive constants such that $\sum_{n=1}^{\infty} \vartheta_1\vartheta_2...\vartheta_n < \infty$, where $\vartheta_i = \frac{\xi_i}{1-\xi_i}$, $\forall i$. Then G has a UFP in Θ. Moreover, for any fixed $t_1 \in \Theta$, $\{G^n t_1\}$ converges to the UFP.*

Proof Fix $t_1 \in \Theta_1$, and define $t_{n+1} = Gt_n = G^n t_1$, $\forall\, n = 1, 2, ...$ Then we have

$$\begin{aligned} p(G^{n+1}t_1, G^n t_1) &\le \xi_{n+1}(p(G^{n+1}t_1, G^n t_1) + p(G^n t_1, G^{n-1}t_1)) \\ &= \xi_{n+1} p(G^{n+1}t_1, G^n t_1) + \xi_{n+1} p(G^n t_1, G^{n-1}t_1). \end{aligned}$$

Now, we get

$$\begin{aligned} p(G^{n+1}t_1, G^n t_1) &\le \frac{\xi_{n+1}}{1-\xi_{n+1}} p(G^n t_1, G^{n-1}t_1) \\ &= \vartheta_{n+1} p(G^n t_1, G^{n-1}t_1), \\ &\le \vartheta_{n+1}\vartheta_n\vartheta_{n-1}...\vartheta_2 p(Gt_1, t_1). \end{aligned}$$

Further, for $1 \le n < m$, we have

$$\begin{aligned} p(G^m t_1, G^n t_1) &\le p(G^m t_1, G^{m-1}t_1) + p(G^{m-1}t_1, G^{m-2}t_1) + ... + p(G^{n+1}t_1, G^n t_1) \\ &\quad - \sum_{r=1}^{m-n-1} p(G^{m-r}t_1, G^{m-r}t_1) \\ &\le p(G^m t_1, G^{m-1}t_1) + p(G^{m-1}t_1, G^{m-2}t_1) + ... + p(G^{n+1}t_1, G^n t_1) \\ &\le \Big(\sum_{i=n}^{m-1} \vartheta_2\vartheta_3...\vartheta_{i+1}\Big) p(Gt_1, t_1). \end{aligned}$$

Therefore, $p(G^m t_1, G^n t_1) \to 0$ $(m, n \to \infty)$. Since Lemma 10.2.7, $\{G^m t_1\}_{m=1}^{\infty}$ is a Cauchy sequence in Θ. Let $\{G^m t_1\}_{m=1}^{\infty}$ converge to t^* in Θ, which is complete.

$$p(t^*, t^*) = \lim_{m\to\infty} p(G^m t_1, t^*) = \lim_{m\to\infty} p(G^m t_1, G^m t_1) = 0. \tag{10.4}$$

Note that $\{G^{m+1} t_1\}_{m=1}^{\infty}$ is also a Cauchy sequence and it converges to s^* in Θ.

$$p(s^*, s^*) = \lim_{m\to\infty} p(G^m t_1, s^*) = \lim_{m\to\infty} p(G^m t_1, G^m t_1) = 0. \tag{10.5}$$

Since (10.4) and (10.5), we have

$$\lim_{n\to\infty} p(G^m t_1, G^m t_1) = p(t^*, t^*) = p(s^*, s^*) \tag{10.6}$$

Therefore, $t^* = s^*$, because of Lemma 10.2.6. Also, the closed graph of G gives $Gt^* = t^*$. Hence, we obtained a fixed point t^* of G.

These processes can be extended to the general case: $t_1 \in \Theta_n$, for some n.

If $Gt^* = t^*$, $Gz^* = z^*$ in G, then let $t^*, z^* \in \Theta_n$, for some n, so we have

$$p(t^*, z^*) = p(Gt^*, Gz^*) \leq \xi_n(p(Gt^*, t^*) + p(Gz^*, z^*)) = 0.$$

By $p(t^*, z^*) = p(z^*, z^*) = p(t^*, t^*) = 0$, we have $t^* = z^*$. Hence, G has a UFP. □

Theorem 10.3.4 *Let (Θ, p) be a complete PMS and $G : \Theta \to \Theta$ has a closed graph. Let $\Theta_1 \subseteq \Theta_2 \subseteq \ldots$ be subsets of Θ such that $\Theta = \bigcup_{j=1}^{\infty} \Theta_j$, $G(\Theta_i) \subseteq \Theta_{i+1}$, $\forall i$, and $p(Gt, Gz) \leq \xi_i(p(Gt, z) + p(Gz, t))$, $\forall t, z \in \Theta_i, \forall i$, where $\xi_i \in (0, \frac{1}{2})$ are real positive constants such that $\sum_{n=1}^{\infty} \vartheta_1\vartheta_2 \ldots \vartheta_n < \infty$, where $\vartheta_i = \frac{\xi_i}{1-\xi_i}$, $\forall i$. Then G has a UFP in Θ. Moreover, for any fixed $t_1 \in \Theta$, $\{G^n t_1\}$ converges to the UFP.*

Proof Fix $t_1 \in \Theta_1$, and define $t_{n+1} = Gt_n = G^n t_1, \forall\, n = 1, 2, 3, \ldots$. Then we have

$$p(G^{n+1} t_1, G^n t_1) \leq \xi_{n+1}(p(G^{n+1} t_1, G^{n-1} t_1) + p(G^n t_1, G^n t_1)).$$

Since $p(G^n t_1, G^n t_1) = 0$, we get

$$\begin{aligned} p(G^{n+1} t_1, G^n t_1) &\leq \xi_{n+1} p(G^{n+1} t_1, G^{n-1} t_1) \\ &\leq \xi_{n+1}(p(G^{n+1} t_1, G^n t_1) + p(G^n t_1, G^{n-1} t_1) - p(G^n t_1, G^n t_1)) \\ &\leq \xi_{n+1}(p(G^{n+1} t_1, G^n t_1) + p(G^n t_1, G^{n-1} t_1)) \\ &= \xi_{n+1} p(G^{n+1} t_1, G^n t_1) + \xi_{n+1} p(G^n t_1, G^{n-1} t_1). \end{aligned}$$

Now, we get

$$\begin{aligned} p(G^{n+1}t_1, G^n t_1) &\le \Big(\frac{\xi_{n+1}}{1-\xi_{n+1}}\Big) p(G^n t_1, G^{n-1}t_1) \\ &= \vartheta_{n+1} p(G^n t_1, G^{n-1}t_1), \\ &\le \vartheta_{n+1}\vartheta_n\vartheta_{n-1}...\vartheta_2 p(Gt_1, t_1). \end{aligned}$$

Further, for $1 \le n < m$, we have

$$\begin{aligned} p(G^m t_1, G^n t_1) &\le p(G^m t_1, G^{m-1}t_1) + p(G^{m-1}t_1, G^{m-2}t_1) + ... + p(G^{n+1}t_1, G^n t_1) \\ &\quad - \sum_{r=1}^{m-n-1} p(G^{m-r}t_1, G^{m-r}t_1) \\ &\le p(G^m t_1, G^{m-1}t_1) + p(G^{m-1}t_1, G^{m-2}t_1) + ... + p(G^{n+1}t_1, G^n t_1) \\ &\le \Big(\sum_{i=n}^{m-1} \vartheta_2\vartheta_3...\vartheta_{i+1}\Big) p(Gt_1, t_1). \end{aligned}$$

Therefore, $p(G^m t_1, G^n t_1) \to 0$ $(m, n \to \infty)$. Since Lemma 10.2.7, $\{G^m t_1\}_{m=1}^{\infty}$ is a Cauchy sequence in Θ. Let $\{G^m t_1\}_{m=1}^{\infty}$ converge to t^* in Θ, which is complete.

$$p(t^*, t^*) = \lim_{m\to\infty} p(G^m t_1, t^*) = \lim_{m\to\infty} p(G^m t_1, G^m t_1) = 0. \tag{10.7}$$

Note that $\{G^{m+1}t_1\}_{m=1}^{\infty}$ is also a Cauchy sequence and it converges to s^* in Θ.

$$p(s^*, s^*) = \lim_{m\to\infty} p(G^m t_1, s^*) = \lim_{m\to\infty} p(G^m t_1, G^m t_1) = 0. \tag{10.8}$$

Since (10.7) and (10.8), we have

$$\lim_{n\to\infty} p(G^m t_1, G^m t_1) = p(t^*, t^*) = p(s^*, s^*). \tag{10.9}$$

Therefore, $t^* = s^*$, because of Lemma 10.2.6. Also, the closed graph of G gives $Gt^* = t^*$. Hence, we obtained a fixed point t^* of G.

These processes can be extended to the general case: $t_1 \in \Theta_n$, for some n.

If $Gt^* = t^*$, $Gz^* = z^*$ in G, then let $t^*, z^* \in \Theta_n$, for some n, so we have

$$\begin{aligned} p(t^*, z^*) = p(Gt^*, Gz^*) &\le \xi_n(p(Gt^*, z^*) + p(Gz^*, t^*)) \\ &= 2\xi_n p(t^*, z^*). \end{aligned}$$

Then, $p(t^*, z^*) \le (2\xi_n)^m p(t^*, z^*)$, $\forall\ m \in \mathbb{N}$. Since $(2\xi_n)^m \to 0$ as $m \to \infty$, $p(t^*, z^*) = 0$. By $p(t^*, z^*) = p(z^*, z^*) = p(t^*, t^*) = 0$, we have $t^* = z^*$. Hence, G has a UFP. □

Theorem 10.3.5 *Let (Θ, P) be a complete PMS and $G : \Theta \to \Theta$ has a closed graph. Let $\Theta_1 \subseteq \Theta_2 \subseteq ...$ be subsets of Θ such that $\Theta = \bigcup_{j=1}^{\infty} \Theta_j$, $G(\Theta_i) \subseteq \Theta_{i+1}$, $\forall i$, and $p(Gt, Gz) \leq \xi_i p(t, z) + \mu_i p(z, Gt)$, $\forall t, z \in \Theta_i, \forall i$, where $\xi_i, \mu_i \in (0, 1)$ are real positive constants such that $\xi_i + \mu_i < 1$, $\forall i$, and $\sum_{n=1}^{\infty} \vartheta_1\vartheta_2...\vartheta_n < \infty$, where $\vartheta_i = \frac{\xi_i+\mu_i}{1-\mu_i}$, $\forall i$. Then G has a UFP in Θ. Moreover, for any fixed $t_1 \in \Theta$, $\{G^n t_1\}$ converges to the UFP.*

Proof Fix $t_1 \in \Theta_1$, and define $t_{n+1} = Gt_n = G^n t_1$, $\forall\, n = 1, 2, 3, ...$ Then we have

$$\begin{aligned}
p(G^{n+1}t_1, G^n t_1) &\leq \xi_{n+1} p(G^n t_1, G^{n-1}t_1) + \mu_{n+1} p(G^{n-1}t_1, G^{n+1}t_1) \\
&\leq \xi_{n+1} p(G^n t_1, G^{n-1}t_1) + \mu_{n+1} p(G^{n-1}t_1, G^n t_1) \\
&\quad + \mu_{n+1} p(G^n t_1, G^{n+1}t_1) - \mu_{n+1} p(G^n t_1, G^n t_1) \\
&\leq \xi_{n+1} p(G^n t_1, G^{n-1}t_1) + \mu_{n+1} p(G^{n-1}t_1, G^n t_1) \\
&\quad + \mu_{n+1} p(G^n t_1, G^{n+1}t_1) \\
&= (\xi_{n+1} + \mu_{n+1}) p(G^n t_1, G^{n-1}t_1) + \mu_{n+1} p(G^n t_1, G^{n+1}t_1).
\end{aligned}$$

Now, we get

$$\begin{aligned}
p(G^{n+1}t_1, G^n t_1) &\leq \left(\frac{\xi_{n+1} + \mu_{n+1}}{1 - \mu_{n+1}}\right) p(G^n t_1, G^{n-1}t_1) \\
&= \vartheta_{n+1} p(G^n t_1, G^{n-1}t_1), \\
&\leq \vartheta_{n+1}\vartheta_n\vartheta_{n-1}...\vartheta_2 p(Gt_1, t_1).
\end{aligned}$$

Further, for $1 \leq n < m$, we have

$$\begin{aligned}
p(G^m t_1, G^n t_1) &\leq p(G^m t_1, G^{m-1}t_1) + p(G^{m-1}t_1, G^{m-2}t_1) + ... + p(G^{n+1}t_1, G^n t_1) \\
&\quad - \sum_{r=1}^{m-n-1} p(G^{m-r}t_1, G^{m-r}t_1) \\
&\leq p(G^m t_1, G^{m-1}t_1) + p(G^{m-1}t_1, G^{m-2}t_1) + ... + p(G^{n+1}t_1, G^n t_1) \\
&\leq \left(\sum_{i=n}^{m-1} \vartheta_2\vartheta_3...\vartheta_{i+1}\right) p(Gt_1, t_1).
\end{aligned}$$

Therefore, $p(G^m t_1, G^n t_1) \to 0$ $(m, n \to \infty)$. Since Lemma 10.2.7, $\{G^m t_1\}_{m=1}^{\infty}$ is a Cauchy sequence in Θ. Let $\{G^m t_1\}_{m=1}^{\infty}$ converge to t^* in Θ, which is complete.

$$p(t^*, t^*) = \lim_{m\to\infty} p(G^m t_1, t^*) = \lim_{m\to\infty} p(G^m t_1, G^m t_1) = 0. \tag{10.10}$$

Note that $\{G^{m+1}t_1\}_{m=1}^{\infty}$ is also a Cauchy sequence and it converges to s^* in Θ.

$$p(s^*, s^*) = \lim_{m\to\infty} p(G^m t_1, s^*) = \lim_{m\to\infty} p(G^m t_1, G^m t_1) = 0. \tag{10.11}$$

Since (10.10) and (10.11), we have

$$\lim_{n\to\infty} p(G^m t_1, G^m t_1) = p(t^*, t^*) = p(s^*, s^*). \tag{10.12}$$

Therefore, $t^* = s^*$, because of Lemma 10.2.6. Also, the closed graph of G gives $Gt^* = t^*$. Hence, we obtained a fixed point t^* of G.

These processes can be extended to the general case: $t_1 \in \Theta_n$, for some n.

If $Gt^* = t^*$, $Gz^* = z^*$ in G, then let $t^*, z^* \in \Theta_n$, for some n, so we have

$$\begin{aligned} p(t^*, z^*) = p(Gt^*, Gz^*) &\le \xi_n p(t^*, z^*) + \mu_n p(z^*, Gt^*) \\ &\le (\xi_n + \mu_n) p(t^*, z^*). \end{aligned}$$

Then, $p(t^*, z^*) \le (\xi_n + \mu_n)^m p(t^*, z^*)$, $\forall\, m \in \mathbb{N}$. Since $(\xi_n + \mu_n)^m \to 0$ as $m \to \infty$, $p(t^*, z^*) = 0$. By $p(t^*, z^*) = p(z^*, z^*) = p(t^*, t^*) = 0$, we have $t^* = z^*$. Hence, G has a UFP. □

Theorem 10.3.6 *Let (Θ, p) be a complete PMS and $G : \Theta \to \Theta$ has a closed graph. Let $\Theta_1 \subseteq \Theta_2 \subseteq ...$ be subsets of Θ such that $\Theta = \bigcup_{j=1}^{\infty} \Theta_j$, $G(\Theta_i) \subseteq \Theta_{i+1}$, $\forall i$, and $p(Gt, Gz) \le \xi_i p(t, z) + \mu_i (p(Gt, t) + p(Gz, z)) + \nu_i (p(Gt, z) + p(Gz, t))$, $\forall t, z \in \Theta_i, \forall i$, where $\xi_i, \mu_i, \nu_i \in (0, \frac{1}{2})$ are real positive constants such that $\xi_i + 2\nu_i < 1$, $\forall i$, and $\sum_{n=1}^{\infty} \vartheta_1 \vartheta_2 ... \vartheta_n < \infty$, where $\vartheta_i = \frac{\xi_i + \mu_i + \nu_i}{1 - \mu_i - \nu_i}$, $\forall i$. Then G has a UFP in Θ. Moreover, for any fixed $t_1 \in \Theta$, $\{G^n t_1\}$ converges to the UFP.*

Proof Fix $t_1 \in \Theta_1$, and define $t_{n+1} = Gt_n = G^n t_1$, $\forall\, n = 1, 2, 3, ...$ Then we have

$$\begin{aligned} p(G^{n+1}t_1, G^n t_1) \le \xi_{n+1} p(G^n t_1, G^{n-1} t_1) + \mu_{n+1} (p(G^{n+1} t_1, G^n t_1) + p(G^n t_1, G^{n-1} t_1)) \\ + \nu_{n+1} (p(G^{n+1} t_1, G^{n-1} t_1) + p(G^n t_1, G^n t_1)). \end{aligned}$$

Since $p(G^n t_1, G^n t_1) = 0$, we get

$$\begin{aligned} p(G^{n+1}t_1, G^n t_1) &\le \xi_{n+1} p(G^n t_1, G^{n-1} t_1) + \mu_{n+1} (p(G^{n+1} t_1, G^n t_1) + p(G^n t_1, G^{n-1} t_1)) \\ &\quad + \nu_{n+1} (p(G^{n+1} t_1, G^n t_1) + p(G^n t_1, G^{n-1} t_1) - p(G^n t_1, G^n t_1)). \\ &\le \xi_{n+1} p(G^n t_1, G^{n-1} t_1) + \mu_{n+1} (p(G^{n+1} t_1, G^n t_1) + p(G^n t_1, G^{n-1} t_1)) \\ &\quad + \nu_{n+1} (p(G^{n+1} t_1, G^n t_1) + p(G^n t_1, G^{n-1} t_1)). \end{aligned}$$

Now, we get

$$\begin{aligned} p(G^{n+1}t_1, G^n t_1) &\leq \Big(\frac{\xi_{n+1} + \mu_{n+1} + \nu_{n+1}}{1 - \mu_{n+1} - \nu_{n+1}}\Big) p(G^n t_1, G^{n-1} t_1) \\ &= \vartheta_{n+1} p(G^n t_1, G^{n-1} t_1), \\ &\leq \vartheta_{n+1}\vartheta_n\vartheta_{n-1}...\vartheta_2 p(Gt_1, t_1). \end{aligned}$$

Further, for $1 \leq n < m$, we have

$$\begin{aligned} p(G^m t_1, G^n t_1) &\leq p(G^m t_1, G^{m-1} t_1) + p(G^{m-1} t_1, G^{m-2} t_1) + ... + p(G^{n+1} t_1, G^n t_1) \\ &\quad - \sum_{r=1}^{m-n-1} p(G^{m-r} t_1, G^{m-r} t_1) \\ &\leq p(G^m t_1, G^{m-1} t_1) + p(G^{m-1} t_1, G^{m-2} t_1) + ... + p(G^{n+1} t_1, G^n t_1) \\ &\leq \Big(\sum_{i=n}^{m-1} \vartheta_2\vartheta_3...\vartheta_{i+1}\Big) p(Gt_1, t_1). \end{aligned}$$

Therefore, $p(G^m t_1, G^n t_1) \to 0$ $(m, n \to \infty)$. Since Lemma 10.2.7, $\{G^m t_1\}_{m=1}^{\infty}$ is a Cauchy sequence in Θ. Let $\{G^m t_1\}_{m=1}^{\infty}$ converge to t^* in Θ, which is complete.

$$p(t^*, t^*) = \lim_{m\to\infty} p(G^m t_1, t^*) = \lim_{m\to\infty} p(G^m t_1, G^m t_1) = 0. \tag{10.13}$$

Note that $\{G^{m+1} t_1\}_{m=1}^{\infty}$ is also a Cauchy sequence and it converges to s^* in Θ.

$$p(s^*, s^*) = \lim_{m\to\infty} p(G^m t_1, s^*) = \lim_{m\to\infty} p(G^m t_1, G^m t_1) = 0. \tag{10.14}$$

Since (10.13) and (10.14), we have

$$\lim_{n\to\infty} p(G^m t_1, G^m t_1) = p(t^*, t^*) = p(s^*, s^*). \tag{10.15}$$

Therefore, $t^* = s^*$, because of Lemma 10.2.6. Also, the closed graph of G gives $Gt^* = t^*$. Hence, we obtained a fixed point t^* of G.

These processes can be extended to the general case: $t_1 \in \Theta_n$, for some n.

If $Gt^* = t^*$, $Gz^* = z^*$ in G, then let $t^*, z^* \in \Theta_n$, for some n, so we have

$$\begin{aligned} p(t^*, z^*) &= p(Gt^*, Gz^*) \\ &\leq \xi_n p(t^*, z^*) + \mu_n(p(Gt^*, t^*) + p(Gz^*, z^*)) + \nu_n(p(Gt^*, z^*) + p(Gz^*, t^*)) \\ &\leq (\xi_n + 2\nu_n) p(t^*, z^*). \end{aligned}$$

Then, $p(t^*, z^*) \leq (\xi_n + 2\nu_n)^m p(t^*, z^*)$, $\forall\, m \in \mathbb{N}$. Since $(\xi_n + 2\nu_n)^m \to 0$ as $m \to \infty$, $p(t^*, z^*) = 0$. By $p(t^*, z^*) = p(z^*, z^*) = p(t^*, t^*) = 0$, we have $t^* = z^*$. Hence, G has a UFP. □

Theorem 10.3.7 *Let (Θ, p) be a complete PMS and $G : \Theta \to \Theta$ has a closed graph. Let $\xi_i \in (0, 1)$, $\forall i$, such that for $1 \leq n \leq m$, $\frac{\xi_1\xi_2...\xi_m}{1-\xi_n} \to 0$ as $n \to \infty$, and let $\Theta_1 \subseteq \Theta_2 \subseteq ...$ be subsets of Θ such that $G(\Theta_i) \subseteq \Theta_{i+1}$, $\forall i$, and $p(Gt, Gz) \leq \xi_i p(t, z)$, $\forall t \in \Theta_i$, $\forall z \in \Theta$, $\forall i$. Suppose $t_1 \in \bigcup_{j=1}^{\infty} \Theta_j$. Then $\{G^n t_1\}$ converges to the fixed point of G in Θ. If $\Theta = \bigcup_{j=1}^{\infty} \Theta_j$, then G has a UFP in Θ.*

Proof Fix $t_1 \in \Theta_1$, and define $t_{n+1} = Gt_n = G^n t_1$, $\forall\, n \in \mathbb{N}$. Then for each n, we have

$$\begin{aligned} p(G^{n+1}t_1, G^n t_1) &\leq \xi_n p(G^n t_1, G^{n-1} t_1) \\ &\leq \xi_n \xi_{n-1} ... \xi_1 p(Gt_1, t_1). \end{aligned}$$

Therefore, $p(t_{n+1}, t_n) \leq \xi_{n-1}\xi_{n-2}...\xi_1 p(Gt_1, t_1)$. Further, for $1 \leq n < m$, we have

$$\begin{aligned} p(t_n, t_m) &\leq p(t_n, t_{n+1}) + p(t_{n+1}, t_{m+1}) + p(t_m, t_{m+1}) \\ &\quad - \sum_{r=1}^{m-n-1} p(t_{m-r}, t_{m-r}) \\ &\leq p(t_n, t_{n+1}) + p(t_{n+1}, t_{m+1}) + p(t_m, t_{m+1}) \\ &\leq \xi_1\xi_2...\xi_{n-1} p(Gt_1, t_1) + \xi_{n+1} p(t_n, t_m) + \xi_1\xi_2...\xi_{m-1} p(Gt_1, t_1) \end{aligned}$$

so that,

$$p(t_n, t_m) \leq \frac{(\xi_1\xi_2...\xi_{n-1}) + (\xi_1\xi_2...\xi_{m-1})}{(1 - \xi_{n+1})} p(Gt_1, t_1).$$

Therefore, $p(t_n, t_m) \to 0$ $(n, m \to \infty)$. Since Lemma 10.2.7, $\{t_n\}$ is a Cauchy sequence. By Θ is complete, $\{t_n\} \to t^*$, for some t^* in Θ.

$$p(t^*, t^*) = \lim_{n\to\infty} p(t_n, t^*) = \lim_{n\to\infty} p(t_n, t_n) = 0. \tag{10.16}$$

Also $\{Gt_n\} = t_{n+1} \to s^*$.

$$p(s^*, s^*) = \lim_{m\to\infty} p(t_{n+1}, s^*) = \lim_{m\to\infty} p(t_{n+1}, t_{n+1}) = 0. \tag{10.17}$$

Since (10.16) and (10.17), we have

$$\lim_{n\to\infty} p(t_n, t_n) = p(t^*, t^*) = p(s^*, s^*). \tag{10.18}$$

Therefore, $t^* = s^*$, because of Lemma 10.2.6. Also, the closed graph of G gives $Gt^* = t^*$. Hence, we obtained a fixed point t^* of G.

These processes can be extended to the general case: $t_1 \in \Theta_n$, for some n.

Suppose now $\Theta = \bigcup_{j=1}^{\infty} \Theta_j$. If $Gt^* = t^*$, $Gz^* = z^*$ in G, then let $t^*, z^* \in \Theta_n$, for some n, so we have

$$p(t^*, z^*) = p(Gt^*, Gz^*) \leq \xi_n p(t^*, z^*) \\ \leq (\xi_n)^m p(t^*, z^*), \ \forall\, m > 1.$$

Therefore, $p(t^*, z^*) = 0$, because $(\xi_n)^m \to 0$ as $m \to \infty$. By $p(t^*, z^*) = p(t^*, t^*) = p(z^*, z^*) = 0$, we have $t^* = z^*$. Hence, G has a UFP, when $\Theta = \bigcup_{j=1}^{\infty} \Theta_j$. □

Theorem 10.3.8 *Let (Θ, p) be a complete PMS and $G : \Theta \to \Theta$ has a closed graph. Suppose $p(t, z) \leq \alpha, \forall t, z \in \Theta$ and for some $\alpha \in [0, \infty)$. Let $\xi_i \in (0, 1), \forall i$, be such that $\xi_1\xi_2...\xi_n \to 0$ as $n \to \infty$. Suppose $\Theta_1 \subseteq \Theta_2 \subseteq ...$ be subsets of Θ such that $G(\Theta_i) \subseteq \Theta_{i+1}, \forall i$, and $p(Gt, Gz) \leq \xi_i p(t, z), \forall t \in \Theta_i, \forall z \in \bigcup_{j=1}^{\infty} \Theta_j, \forall i$. Let $t_1 \in \bigcup_{j=1}^{\infty} \Theta_j$. Then the sequence $\{G^n t_1\}$ converges to a unique fixed point of G in Θ. If $\Theta = \bigcup_{j=1}^{\infty} \Theta_j$, then G has a UFP in Θ.*

Proof Fix $t_1, z_1 \in \Theta_1$. Define $t_{n+1} = Gt_n = G^n t_1$, and $z_{n+1} = Gz_n = G^n z_1, \forall\, n \in \mathbb{N}$. For $m < n$, we have

$$p(t_n, z_m) = p(Gt_{n-1}, Gz_{m-1}) \leq \xi_{m-1} p(t_{n-1}, z_{m-1}) \\ \leq \xi_{m-1}\xi_{m-2}...\xi_2\xi_1 p(t_{n-m+1}, z_1) \\ \leq \xi_{m-1}\xi_{m-2}...\xi_2\xi_1 \alpha,$$

because $p(t, z) \leq \alpha, \forall t, z \in \Theta$. Hence, $p(t_n, z_m) \to 0$ as $m, n \to \infty$. Also $p(t_n, t_m) \to 0$, and $p(z_n, z_m) \to 0$ as $m, n \to \infty$. So, $\{t_n\}$ and $\{z_n\}$ are Cauchy sequences in Θ, because of Lemma 10.2.7. By (Θ, d) is complete, $\{t_n\}$ and $\{z_n\}$ converges to a unique point t^* in Θ, because of Lemma 10.2.9. Since $\{t_n\} \to t^*$,

$$p(t^*, t^*) = \lim_{n\to\infty} p(t_n, t^*) = \lim_{n\to\infty} p(t_n, t_n) = 0. \tag{10.19}$$

Also $\{Gt_n\} = t_{n+1} \to s^*$.

$$p(s^*, s^*) = \lim_{m\to\infty} p(t_{n+1}, s^*) = \lim_{m\to\infty} p(t_{n+1}, t_{n+1}) = 0. \tag{10.20}$$

Since (10.19) and (10.20), we have

$$\lim_{n\to\infty} p(t_n, t_n) = p(t^*, t^*) = p(s^*, s^*). \tag{10.21}$$

Therefore, $t^* = s^*$, because of Lemma 10.2.6. Also, the closed graph of G gives $Gt^* = t^*$. Hence, we obtained a fixed point t^* of G.

These processes can be extended to the general case: $t_1, z_1 \in \Theta_n$, for some n.

Suppose now $\Theta = \bigcup_{j=1}^{\infty} \Theta_j$. If $Gt^* = t^*$, $Gz^* = z^*$ in G, then let $t^*, z^* \in \Theta_n$, for some n, so we have

$$p(t^*, z^*) = p(Gt^*, Gz^*) \le \xi_n p(t^*, z^*) \le (\xi_n)^m p(t^*, z^*), \ \forall\, m > 1.$$

So, $p(t^*, z^*) = 0$, because $(\xi_n)^m \to 0$ as $m \to \infty$. By $p(t^*, z^*) = p(z^*, z^*) = p(t^*, t^*) = 0$, we have $t^* = z^*$. Therefore, G has a UFP, when $\Theta = \bigcup_{j=1}^{\infty} \Theta_j$. □

10.4 Conclusions

Equations, including differential equations, integral equations, and algebraic equations, are typically solved using fixed-point iteration techniques. So, in relation to fixed point iteration methods, we must find all feasible fixed-point results.

References

Karapinar, E., Shatanawi, W., & Tas K. (2013). Fixed point theorem on partial metric spaces involving rational expressions. *Miskolc Mathematical Notes, 14*, 135–142.

Karapinar, E., Agarwal, R. P., & Aydi, H. (2018). Interpolative Reich-Rus-Ciric type contractions on partial metric spaces. *Mathematics, 6*, 256.

Kirk, W. A., Srinivasan, P. S., & Veeramani, P. (2003). Fixed points for mappings satisfying cyclical conditions. *Fixed Point Theory, 4*, 78–89.

Matthews, S. G. (1992b). Partial metric topology. In S. Andima (Ed.), *Papers from the 8th Summer Conference at Queens College, New York, June 18–20, 1992*. New York. *New York Academy of Sciences, 728*, 183–197.

Matthews, S. G. (1992). *Partial metric topology* (p. 212). Research Report: University of Warwick.

Moorthy, C. G., & Raj, P. X. (2008). Contraction mappings with variations in domain. *Journal of Analysis, 16*, 53–58.

Pant, R., Shukla, R., Nashine, H. H., & Panicker, R. (2017). Some new fixed point theorems in partial metric spaces with application. *Journal of Function Spaces,* 1072750

Shobkolaei, N., Vaezpour, S. M., & Sedghi, S. (2011). A common fixed point theorem on ordered partial metric spaces. *Journal of Basic and Applied Scientific Research, 12*, 3433–3439.

Siva, G. (2022a). Fixed point theorems of contraction mappings with variations in cone metric space domains. *Asian-European Journal of Mathematics,* 1–14.

Siva, G. (2023). Fixed points of closed graph operators on N-cone metric spaces. *Indian Journal of Mathematics, 65*, 53–71.

Siva, G. (2023). Fixed points of multiplicative closed graph operators on b-multiplicative metric spaces. *Malaya Journal of Matematik, 12*, 21–30.

Siva, G. (2022). Fixed points of contraction mappings with variations in S-metric space domains. *Mathematics Student, 91*, 173–185.

Tomar, A., & Joshi, M. (2022). On existence of fixed points and applications to a boundary value problem and a Matrix equation in C^*- Algebra Valued Partial Metric Spaces. *Acta Universitatis Sapientiae, Mathematica, 14*(2), 341–355. https://doi.org/10.2478/ausm-2022-0023

Tomar, A., Joshi, M., & Bhatt, V. (2021). Fixed point and a Cantilever beam problem in a partial b-metric space. *Acta Universitatis Sapientiae, Mathematica, 13*(2), 506–518. https://doi.org/10.2478/ausm-2021-0032

Tomar, A., Joshi, M., & Padaliya, S. K. (2022b). Fixed point to fixed circle and activation function in partial metric space. *Journal of Applied Analysis, 28*(1), 57–66. Article number: 000010151520212057. https://doi.org/10.1515/jaa-2021-2057

Tomar, A., Kumar. D., Sharma, R., & Joshi, M. (2023). Results via partial-b metric and solution of a pair of elliptic boundary value problem. *Sahand Communications in Mathematical Analysis, 20*(04), 205–223. https://doi.org/10.22130/scma.2023.563638.1187

Tomar, A., Beloul, S., Sharma, R., & Upadhyay, S. (2017). Common fixed point theorems via generalized condition (B) in quasi-partial metric space and applications. *Demonstratio Mathematica, 50*, 278–298.

Tomar, A., Joshi, M., & Deep, A. (2021). Fixed points and its applications in C^*- algebra valued partial metric space. *TWMS Journal of Applied and Engineering Mathematics, 11*(2), 329–340.

Chapter 11
Some Common Fixed Point Theorems for Hardy–Rogers-Type Contraction in Complete Cone *b*-Metric Spaces

Santosh Kumar and Emmanuel Kapinga

Abstract This chapter explores some common fixed point theorems for Hardy–Rogers-type contractions in complete cone b-metric spaces. We demonstrate the existence and uniqueness of fixed points in these spaces, extending the known results from metric spaces to cone b-metric spaces. Additionally, the findings are supported by a relevant application.

Keywords Cone metric spaces · Cone b-metric spaces · Hardy–Rogers type of contraction mappings · Common fixed point

AMS Subject Classification 47H10 · 54H25 · 46T99

11.1 Introduction

Cone metric spaces, which became a generalization of metric spaces, were first presented by Huang and Zhang (2007). Also, they established a few fixed point theorems for contractive mappings that improved on previous findings about fixed points in metric space. Cone b-metric spaces were first introduced by Hussain and Shah (2011) as a generalization of cone metric spaces and b-metric spaces. Certain topological properties were established in a cone b-metric space and some recent findings regarding KKM mappings in such a framework were enhanced. Abbas et al. (2012) investigated the requirements for the existence of a fixed point of mapping sat-

S. Kumar (✉)
Department of Mathematics, School of Physical Sciences, North Eastern Hill University, Shillong, Meghalaya, India
e-mail: drsengar2002@gmail.com

E. Kapinga
Department of Mathematics, College of Natural and Applied Sciences, University of Dar es Salaam, Dar es Salaam, Tanzania
e-mail: emanuel.kapinga10@gmail.com

© The Author(s), under exclusive license to Springer Nature Singapore Pte Ltd. 2026

A. Tomar and M. Jain (eds.), *Applications of Fixed-Point Theorem*, Industrial and Applied Mathematics, https://doi.org/10.1007/978-981-95-5160-6_11

isfying T-Hardy–Rogers-type contractions in the context of partially ordered partial metric spaces. In the same space, they also demonstrated some fixed point theorems for a T-Hardy–Rogers contraction. They used their findings to study periodic point issues for such mappings. A few fixed point theorems for T-Hardy–Rodgers contraction mappings in complete cone b-metric spaces have recently been established by Shashi and Kumar (2021) without the assumption of normalcy requirements.

The first generalization of Hardy–Rogers-type contractions in metric-like spaces was introduced by Patil et al. (2020). Furthermore, he used this technique to investigate the existence and uniqueness of solutions for the classical boundary value problems and generalized fractional boundary value problems while taking into account the key characteristics of the associated Green's functions. Several authors have recently modified the fixed point results in various ways. Some of them are useful to readers Pragati et al. (2022, 2021), Kumar and Sholastica (2021), Lucas and Kumar (2022), Kumar et al. (2020), Tomar et al. (2021), Tomar and Joshi (2021), Tomar and Sharma (2019), Kumar (2021). The main result of the paper is the establishment of modified conditions for Hardy–Rogers' fixed point theorem, along with the introduction of some advanced applications.

11.2 Preliminaries

To establish the results, the following definitions, lemmas, and theorems are frequently employed:

Huang and Zhang (2007) introduced the concept of a cone in Banach space. Using the concept of cone, they introduced the definition of cone metric space.

Definition 11.2.1 (Huang and Zhang (2007)) Let Θ be a nonempty set, $\mathfrak{E}$ be a real Banach space and assume that the mapping $\eth : \Theta \times \Theta \to \mathfrak{E}$ satisfies the following conditions:

(d1) $0 \leq \eth(\theta, \mu)$ and $\eth(\theta, \mu) = 0$ if and only if $\theta = \mu$ for all θ and μ in Θ,
(d2) $\eth(\theta, \mu) = \eth(\mu, \theta)$ for all θ and μ in Θ,
(d3) $\eth(\theta, \mu) \leq \eth(\theta, \omega) + \eth(\omega, \mu)$ for all θ, μ and ω in Θ.

Then $\eth$ is called cone metric and $(\Theta, \eth)$ is called a cone metric space.

Definition 11.2.2 (Huang and Zhang (2007)) A cone $\mathcal{P}$ is said to be normal if there exists $\mathcal{M} > 0$ such that $0 \leq \theta \leq \mu$ implies $\|\theta\| \leq \mathcal{M}\|\mu\|, \forall \theta, \mu \in \mathcal{P}$. The least value of $\mathcal{M}$ is called the normal constant.

Definition 11.2.3 (Bakhtin (1989)) Let Θ be a non empty set and $s \geq 1$ be a given real number. A function $\eth : \Theta \times \Theta \to [0, \infty)$ is called b-metric if it satisfies the following properties for each $\theta, \mu, \omega \in \Theta$:

(b1): $\eth(\theta, \mu) = 0 \Leftrightarrow \theta = \mu$,
(b2): $\eth(\theta, \mu) = \eth(\mu, \theta)$,

(b3): $\eth(\theta, \omega) \leq s[\eth(\theta, \mu) + \eth(\mu, \omega)]$.

The pair $(\Theta, \eth)$ is called a b-metric space.

Cone b-metric space was first developed by Hussain and Shah (2011) as a generalization of cone b-metric space. The following is the definition:

Definition 11.2.4 (Hussain and Shah (2011)) Let Θ be a nonempty set and $s \geq 1$ be a given real number. A mapping $\eth : \Theta \times \Theta \to \mathfrak{E}$ is said to be cone b-metric if and only if for all $\theta, \mu, \omega \in \Theta$, the following conditions are satisfied:

(i) $0 \leq \eth(\theta, \mu)$ with $\theta \neq \mu$ and $\eth(\theta, \mu) = 0$ if and only if $\theta = \mu$,
(ii) $\eth(\theta, \mu) = \eth(\mu, \theta)$,
(iii) $\eth(\theta, \mu) \leq s[\eth(\theta, \omega) + \eth(\omega, \mu)]$.

The pair $(\Theta, \eth)$ is called a cone b-metric space.

Cone b-metric space is the generalization of cone metric space since any cone metric space is also a cone b-metric space, however, the reverse may not always be true.

Huang and Zhang (2007) established the idea of a Cauchy sequence in cone metric space, and their definition was as follows:

Definition 11.2.5 (Huang and Zhang (2007)) The sequence $\{\theta_n\}$ in a cone metric space $(\Theta, \eth)$ is said to be Cauchy if for every $c \in \mathfrak{E}$ with $0 \ll c$ there is a natural number $\mathbb{N}$ such that $\eth(\theta_m, \theta_n) \ll c$ for all $n, m \geq \mathbb{N}$.

Also, Huang and Zhang (2007) introduced the definition of a cone metric space to be complete as follows:

Definition 11.2.6 (Huang and Zhang (2007)) The cone metric space $(\Theta, \eth)$ is said to be complete if every Cauchy sequence in Θ is convergent.

The following theorems are due to Huang and Zhang (2007) in a cone metric space:

Theorem 11.2.7 (Huang and Zhang (2007)) *Let $(\Theta, \eth)$ be a complete cone metric space, $\mathcal{P}$ be a normal cone with normal constant K. Suppose the mapping $\mathcal{T} : \Theta \to \Theta$ satisfies the contractive condition*

$$\eth(\mathcal{T}\theta, \mathcal{T}\mu) \preceq k\eth(\theta, \mu), \forall \theta, \mu \in \Theta,$$

where $k \in [0, 1)$ is a constant. Then $\mathcal{T}$ has a unique fixed point in Θ. Moreover, for any $\theta \in \Theta$, iterative sequence $\{\mathcal{T}^n\theta\}$ converges to the fixed point.

Theorem 11.2.8 (Huang and Zhang (2007)) *Let $(\Theta, \eth)$ be a complete cone metric space, $\mathcal{P}$ a normal cone with normal constant K. Suppose the mapping $\mathcal{T} : \Theta \to \Theta$ satisfies the contractive condition*

$$\eth(\mathcal{T}\theta, \mathcal{T}\mu) \preceq k\eth(\mathcal{T}\theta, \theta) + \eth(\mathcal{T}\mu, \mu), \forall \theta, \mu \in \Theta,$$

where $k \in [0, 1)$ is a constant. Then $\mathcal{T}$ has a unique fixed point in Θ. Moreover, for any $\theta \in \Theta$, iterative sequence $\{\mathcal{T}^n\theta\}$ converges to the fixed point.

Huang and Wen (2010) presented the following lemma, which stipulates the requirements for a sequence to be a Cauchy sequence in cone metric space as follows:

Lemma 11.2.9 (Huang and Wen (2010)) *Let $(\Theta, \eth)$ be a cone metric space, $\mathcal{P}$ be a normal cone with a normal constant K. Let $\{\theta_n\}$ be a sequence in Θ. Then $\{\theta_n\}$ is a Cauchy sequence if and only if $\eth(\theta_n, \theta_m) \to 0$ for $n, m \to \infty$.*

The following theorem is due to Patil et al. (2020) on a common fixed point for Hardy–Rogers type of contraction in complete metric spaces.

Theorem 11.2.10 (Patil et al. (2020)) *Let $\mathcal{L}$ be complete metric space and let $\mathcal{F} : \mathcal{L} \longrightarrow \mathbb{R}$ be the continuous self-mapping on $\mathcal{L}$. Suppose $\mathcal{F}$ satisfies the condition*

$$\eth(\mathcal{F}\theta, \mathcal{F}\mu) < \ell_1\eth(\theta, \mathcal{F}\theta) + \ell_2\eth(\mu, \mathcal{F}\mu) + \ell_3\eth(\theta, \mathcal{F}\mu) + \ell_4\eth(\mu, \mathcal{F}\theta) + \ell_5\eth(\theta, \mu),$$

for all $\theta, \mu \in \mathcal{L}, \theta \neq \mu$ and for some ℓ_1, ℓ_2,ℓ_3, ℓ_4 $\ell_5 \in (0, 1)$ such that

$$\sum_{r=1}^{5} \ell_r = \varsigma < 1.$$

Then $\mathcal{F}$ has a unique fixed point.

In this paper, we extend the common fixed point theorems by Patil et al. (2020) to complete cone b-metric space. Examples and applications help to illustrate the main conclusions.

11.3 Main Results

In this part, we present an extension of Theorem 11.2.10 due to Patil et al. (2020) to complete cone b-metric spaces to obtain the common fixed point for Hardy–Rogers type of contraction in complete cone b-metric space.

Theorem 11.3.1 *Let $(\Theta, \eth)$ be a complete cone b-metric space and let $\mathcal{F} : \Theta \times \Theta \to \mathfrak{E}$ be the continuous self-mapping on Θ with $s \geq 1$ and $s\ell < 1$. Suppose $\mathcal{F}$ satisfying the condition:*

$$\eth(\mathcal{F}\theta, \mathcal{F}\mu) < \ell_1\eth(\theta, \mathcal{F}\theta) + \ell_2\eth(\mu, \mathcal{F}\mu) + \ell_3\eth(\theta, \mathcal{F}\mu) + \ell_4\eth(\mu, \mathcal{F}\theta) + \ell_5\eth(\theta, \mu),$$

for all $\theta, \mu \in \Theta, \theta \neq \mu$ *and for some* $\ell_1, \ell_2, \ell_3, \ell_4$ *and* $\ell_5 \in (0, 1)$ *is such that*

$$\sum_{r=1}^{5} \ell_r = \varsigma < 1.$$

Then $\mathcal{F}$ *has a unique fixed point.*

Proof Suppose that $\theta_0 \in \Theta$ is an arbitrary point in Θ and the sequence of iteration for $\mathcal{F}$ at θ_0 such that,

$$\mathcal{F}(\theta_{n-1}) = \theta_n. \tag{11.1}$$

Let $\theta_{n-1} \neq \theta_n$ for all $n \in \mathbb{N}$.
Then,

$$\begin{aligned}\eth(\theta_{n-1}, \theta_n) &= \eth(\mathcal{F}(\theta_{n-2}), \mathcal{F}(\theta_{n-1}))\\ &< \ell_1\eth(\theta_{n-2}, \mathcal{F}(\theta_{n-2})) + \ell_2\eth(\theta_{n-1}, \mathcal{F}(\theta_{n-1})) + \ell_3\eth(\theta_{n-2}, \mathcal{F}(\theta_{n-1}))\\ &\quad + \ell_4\eth(\theta_{n-1}, \mathcal{F}(\theta_{n-2})) + \ell_5\eth(\theta_{n-2}, \theta_{n-1}).\end{aligned}$$

Using (11.1), we get

$$\begin{aligned}\eth(\theta_{n-1}, \theta_n) \leq{}& \ell_1\eth(\theta_{n-2}, \theta_{n-1}) + \ell_2\eth(\theta_{n-1}, \theta_n) + \ell_3\eth(\theta_{n-2}, \theta_n) + \ell_4\eth(\theta_{n-1}, \theta_{n-1})\\ &+ \ell_5\eth(\theta_{n-2}, \theta_{n-1}).\end{aligned}$$

Applying the triangular inequality for some

$$\theta_{n-2} \leq \theta_{n-1} \leq \theta_n,$$

we obtain

$$\begin{aligned}\eth(\theta_{n-1}, \theta_n) &\leq \ell_1\eth(\theta_{n-2}, \theta_{n-1}) + \ell_2\eth(\theta_{n-1}, \theta_n) + s(\ell_3\eth(\theta_{n-2}, \theta_{n-1}) +\\ &\quad \ell_3\eth(\theta_{n-1}, \theta_n)) + \ell_5\eth(\theta_{n-2}, \theta_n),\\ &\leq \ell_1\eth(\theta_{n-2}, \theta_{n-1}) + \ell_2\eth(\theta_{n-1}, \theta_n) + s\ell_3\eth(\theta_{n-2}, \theta_{n-1}) +\\ &\quad s\ell_3\eth(\theta_n, \theta_{n-1}) + \ell_5\eth(\theta_{n-2}, \theta_{n-1}).\end{aligned}$$

Collecting like terms of the above inequality and applying (ii) of Definition (11.2.4), we get

$$\eth(\theta_{n-1},\theta_n) - \ell_2\eth(\theta_{n-1},\theta_n) - s\ell_3\eth(\theta_{n-1},\theta_n) \le \ell_1\eth(\theta_{n-2},\theta_{n-1}) + s\ell_3\eth(\theta_{n-2},\theta_{n-1}) + \ell_5\eth(\theta_{n-2},\theta_{n-1}),$$

$$(1-\ell_2-s\ell_3)\eth(\theta_{n-1},\theta_n) \le (\varsigma-\ell_2-\ell_4)\eth(\theta_{n-2},\theta_{n-1}),$$

$$eth(\theta_{n-1},\theta_n) \le \Big(\frac{\varsigma-\ell_2-\ell_4}{1-\ell_2-s\ell_3}\Big)\eth(\theta_{n-2},\theta_{n-1}.$$

Let $K = (\frac{\varsigma-\ell_2-\ell_4}{1-\ell_2-s\ell_3})$. Then $\eth(\theta_{n-1},\theta_n) \le K\eth(\theta_{n-2},\theta_{n-1})$.

Repeating the inequality above, we obtain

$$\eth(\theta_{n-1},\theta_n) \le K^n\eth(\theta_{n-2},\theta_{n-1}).$$

Then we have $\eth(\theta_{n-1},\theta_n) \to 0 \quad$ as $n \to \infty$.

Claiming that, $\{\theta_n\}$ is a Cauchy sequence in Θ and provided that Θ is a complete cone b-metric space, then each Cauchy sequence converges in Θ. Therefore there exists $\theta_1 \in \Theta$ such that $\theta_n \to \theta_1$. Also since $\mathcal{F}$ is a continuous mapping, we have

$$\mathcal{F}(\lim_{n\longrightarrow\infty}\theta_n) = \mathcal{F}(\theta_1),$$

$$\lim_{n\longrightarrow\infty}\theta_n = \theta_1.$$

Hence θ_1 is a fixed point of $\mathcal{F}$ in Θ.

We show that θ_1 is a unique fixed point in Θ. Suppose that there exist two fixed points θ_1 and θ_2 in Θ such that $\theta_1 \ne \theta_2$, $\mathcal{F}(\theta_1) = \theta_1$ and $\mathcal{F}(\theta_2) = \theta_2$.

Then,

$$\eth(\theta_1,\theta_2) < \ell_1\eth(\theta_1,\mathcal{F}\theta_1) + \ell_2\eth(\theta_2,\mathcal{F}\theta_2) + \ell_3\eth(\theta_1,\mathcal{F}\theta_2) + \ell_4\eth(\theta_2,\mathcal{F}\theta_2) + \ell_5\eth(\theta_1,\theta_2),$$
$$< (\varsigma-\ell_1-\ell_2)\eth(\theta_1,\theta_2),$$

which proves that $\theta_1 = \theta_2$.

Hence θ_1 is a unique fixed point of $\mathcal{F}$ in Θ. □

Now, we will extend the theorem above for a pair of maps to prove the existence of common fixed points.

Theorem 11.3.2 *Let $(\Theta,\eth)$ be a complete cone b-metric space and let $\mathcal{F},\mathcal{G}$: $\Theta \longrightarrow \Theta$ be the continuous self-mappings on Θ with $s \ge 1$ and $s\ell < 1$. Suppose $\mathcal{F},\mathcal{G}$ satisfying the condition;*

$$\eth(\mathcal{F}\theta,\mathcal{G}\mu) < \ell_1\eth(\theta,\mathcal{F}\theta) + \ell_2\eth(\mu,\mathcal{G}\mu) + \ell_3\eth(\theta,\mathcal{G}\mu) + \ell_4\eth(\mu,\mathcal{F}\theta) + \ell_5\eth(\theta,\mu),$$

for all $\theta,\mu \in \Theta, \theta \ne \mu$ and for some $\ell_1, \ell_2,\ell_3, \ell_4\,\ell_5 \in (0,1)$ is such that $\sum_{r=1}^{5}\ell_r < 1$. Then $\mathcal{F}$ and $\mathcal{G}$ have a unique fixed point.

Proof Let θ_0 be an arbitrary element in Θ. By defining a sequence θ_n in such a way that,

$$\theta_{n+1} = \mathcal{G}\theta_n,$$

for all $n = 0, 1, 2, 3, ...$

Considering,

$$\begin{aligned}
\eth(\mathcal{F}\theta_n, \mathcal{F}\theta_{n+1}) &= \eth(\mathcal{F}\mathcal{G}\theta_{n-1}, \mathcal{F}\mathcal{G}\theta_n),\\
&< \ell_1\eth(\mathcal{G}\theta_{n-1}, \mathcal{F}\mathcal{G}\theta_{n-1}) + \ell_2\eth(\mathcal{G}\theta_n, \mathcal{F}\mathcal{G}\theta_n) + \ell_3\eth(\mathcal{G}\theta_{n-1}, \mathcal{F}\mathcal{G}\theta_n)\\
&+ \ell_4\eth(\mathcal{G}\theta_n, \mathcal{F}\mathcal{G}\theta_{n-1}) + \ell_5\eth(\mathcal{G}\theta_{n-1}, \mathcal{G}\theta_n),\\
&\le \ell_1\eth(\theta_n, \mathcal{F}\theta_n) + \ell_2\eth(\theta_{n+1}, \mathcal{F}\theta_{n+1}) + \ell_3\eth(\theta_n, \mathcal{F}\theta_{n+1})\\
&+ \ell_4\eth(\theta_{n+1}, \mathcal{F}\theta_n) + \ell_5\eth(\theta_n, \theta_{n+1}).
\end{aligned}$$

By the use of (ii) and (iii) of Definition (11.2.4), we have

$$\begin{aligned}
\eth(\mathcal{F}\theta_{n+1}, \mathcal{F}\theta_n) &\le \ell_1\eth(\theta_n, \mathcal{F}\theta_n) + \ell_2\eth(\theta_{n+1}, \mathcal{F}\theta_{n+1}) + s\{\ell_3\eth(\theta_n, \mathcal{F}\theta_{n+1})\\
&\quad + \ell_3\eth(\mathcal{F}\theta_n, \mathcal{F}\theta_{n+1})\} + \ell_4\eth(\theta_{n+1}, \mathcal{F}\theta_n) + \ell_5\eth(\theta_n, \theta_{n+1}),
\end{aligned}$$

$$\begin{aligned}
&\le \ell_1\eth(\theta_n, \mathcal{F}\theta_n) + s\{\ell_2\eth(\theta_n, \mathcal{F}\theta_n) + \ell_2\eth(\mathcal{F}\theta_n, \mathcal{F}\theta_{n+1})\}\\
&\quad + s\{\ell_3\eth(\theta_n, \mathcal{F}\theta_n) + \ell_3\eth(\mathcal{F}\theta_n, \mathcal{F}\theta_{n+1})\} + \ell_4\eth(\theta_{n+1}, \mathcal{F}\theta_n) + \ell_5\eth(\theta_n, \mathcal{F}\theta_n)\\
&\quad \eth(\mathcal{F}\theta_{n+1}, \mathcal{F}\theta_n) - s\ell_2\eth(\mathcal{F}\theta_n, \mathcal{F}\theta_{n+1}) - s\ell_3\eth(\mathcal{F}\theta_n, \mathcal{F}\theta_{n+1}),\\
&\le \ell_1\eth(\theta_n, \mathcal{F}\theta_n) + s\ell_2\eth(\theta_n, \mathcal{F}\theta_n) + s\ell_3\eth(\theta_n, \mathcal{F}\theta_n) + \ell_5\eth(\theta_n, \mathcal{F}\theta_n)\\
&\quad (1 - s\ell_2 - s\ell_3)\eth(\mathcal{F}\theta_n, \mathcal{F}\theta_{n+1}),\\
&\le (\ell_1 + s\ell_2 + s\ell_3 + \ell_5)\eth(\theta_n, \mathcal{F}\theta_n),\\
\eth(\mathcal{F}\theta_n, \mathcal{F}\theta_{n+1}) &\le \left(\frac{\ell_1 + s\ell_2 + s\ell_3 + \ell_5}{1 - s\ell_2 - s\ell_3}\right)eth(\theta_n, \mathcal{F}\theta_n).
\end{aligned}$$

Let $K = \left(\frac{\ell_1+s\ell_2+s\ell_3+\ell_5}{1-s\ell_2-s\ell_3}\right)$.

Hence,

$$\begin{aligned}
\eth(\mathcal{F}\theta_n, \mathcal{F}\theta_{n+1}) &\le K\eth(\theta_n, \mathcal{F}\theta_n),\\
&\le K\eth(\theta_n, \theta_{n+1}),\\
&\le K^n\eth(\theta_0, \theta_1).
\end{aligned}$$

Thus we have $\{\theta_n\}$ is a Cauchy sequence in Θ. We can also show that $\{\mu_n\}$ is a Cauchy sequence in Θ by applying the same technique as above. Since $(\Theta, \eth)$ is a complete cone b-metric space, there exists a common fixed point in Θ.

Suppose,

$$\theta_0 = \lim_{n\longrightarrow\infty} \theta_n, \quad \mu_0 = \lim_{n\longrightarrow\infty} \mu_n,$$

then

$$\eth(\theta_n, \theta_0) \to 0 \quad \text{as} \quad n \to \infty \quad \text{and} \quad \eth(\mu_n, \mu_0) \to 0 \quad \text{as} \quad n \to \infty.$$

Since $\mathcal{F}$ and $\mathcal{G}$ are continuous mappings, we have

$$\begin{aligned} \eth(\mathcal{F}\theta_n, \mathcal{F}\theta_0) &\longrightarrow 0 \quad as \quad n \longrightarrow \infty, \\ \eth(\mathcal{G}\theta_n, \mathcal{G}\theta_0) &\longrightarrow 0 \quad as \quad n \longrightarrow \infty. \end{aligned}$$

Hence,

$$\begin{aligned} \eth(\theta_0, \mathcal{F}\theta_0) &= \eth(\mathcal{F}^{-}(\mathcal{F}\theta_0), \mathcal{F}\theta_0), \\ &< \ell_1\eth(\mathcal{F}^{-}(\mathcal{F}\theta_0), \mathcal{F}\theta_0) + \ell_2\eth(\theta_0, \mathcal{F}\theta_0) + \ell_3\eth(\mathcal{F}\theta_0, \mathcal{F}\theta_0) \\ &\quad + \ell_4\eth(\theta_0, (\mathcal{F}^{-}(\mathcal{F}\theta_0)) + \ell_5\eth(\mathcal{F}\theta_0, \theta_0), \\ &\leq \ell_1\eth(\theta_0, \mathcal{F}\theta_0) + \ell_2\eth(\theta_0, \mathcal{F}\theta_0) + \ell_5\eth(\theta_0, \mathcal{F}\theta_0), \\ &\leq \quad (\ell_1 + \ell_2 + \ell_5)\eth(\theta_0, \mathcal{F}\theta_0), \end{aligned}$$

which gives $\mathcal{F}(\theta_0) = \theta_0$.

By using the same procedure as applied above, we have

$$\begin{aligned} \eth(\theta_0, \mathcal{G}\theta_0) &= \eth(\mathcal{G}^{-}(\mathcal{G}\theta_0), \mathcal{G}\theta_0), \\ &< \ell_1\eth(\mathcal{G}^{-}(\mathcal{G}\theta_0), \mathcal{G}\theta_0) + \ell_2\eth(\theta_0, \mathcal{G}\theta_0) + \ell_3\eth(\mathcal{G}\theta_0, \mathcal{G}\theta_0), \\ &\quad + \ell_4\eth(\theta_0, (\mathcal{G}^{-}(\mathcal{G}\theta_0)) + \ell_5\eth(\mathcal{G}\theta_0, \theta_0), \\ &\leq \ell_1\eth(\theta_0, \mathcal{G}\theta_0) + \ell_2\eth(\theta_0, \mathcal{G}\theta_0) + \ell_5\eth(\theta_0, \mathcal{G}\theta_0), \\ &\leq \quad (\ell_1 + \ell_2 + \ell_5)\eth(\theta_0, \mathcal{G}\theta_0), \end{aligned}$$

which gives $\mathcal{G}(\theta_0) = \theta_0$.

Now we will show that θ_0 is a common fixed point of $\mathcal{F}$ and $\mathcal{G}$ in Θ.

$$\begin{aligned} \eth(\theta_0, \mu_0) &< \ell_1\eth(\theta_0, \mathcal{F}\theta_0) + \ell_2\eth(\mu_0, \mathcal{G}\mu_0) + \ell_3\eth(\theta_0, \mathcal{G}\mu_0) \\ &\quad + \ell_4\eth(\mu_0, \mathcal{F}\theta_0) + \ell_5\eth(\theta_0, \mu_0), \\ &\leq \quad \ell_3\eth(\theta_0, \mu_0) + \ell_4\eth(\theta_0, \mu_0) + \ell_5\eth(\theta_0, \mu_0), \\ &\leq \quad (\ell_3 + \ell_4 + \ell_5)\eth(\theta_0, \mu_0), \end{aligned}$$

which shows that $\theta_0 = \mu_0$ and hence θ_0 is a common fixed point of $\mathcal{F}$ and $\mathcal{G}$ in Θ.

To show that θ_0 is a unique fixed point in Θ, let θ_0 be another fixed point in Θ such that $\mathcal{F}(\omega_0) = \omega_0$ and $\mathcal{G}(\omega_0) = \omega_0$.

Hence,

$$\begin{aligned}\eth(\theta_0,\omega_0) &= \eth(\mathcal{F}(\theta_0),\mathcal{F}(\omega_0)),\\ &< \ell_1\eth(\theta_0,\mathcal{F}\theta_0)+\ell_2\eth(\omega_0,\mathcal{G}\omega_0)+\ell_3\eth(\theta_0,\mathcal{G}\omega_0)+\ell_4\eth(\omega_0,\mathcal{F}\theta_0)+\ell_5\eth(\theta_0,\omega_0),\\ &\le (\ell_3+\ell_4+\ell_5)\eth(\theta_0,\omega_0).\end{aligned}$$

As observed above, we obtain that $\theta_0 = \omega_0$.

Thus, θ_0 is a unique fixed point of $\mathcal{F}$ and $\mathcal{G}$ in a complete cone b-metric space. □

Imdad et al. (2002) introduced the fixed point results for a family of mappings in their work. Employing the concept given by them, we are considering the family of continuous self-mappings in a complete cone b-metric space to prove that the unique fixed point exists for such maps.

Theorem 11.3.3 *Let $\{\mathcal{S}_1,\mathcal{S}_2,...\mathcal{S}_m\}$ and $\{\mathcal{T}_1,\mathcal{T}_2,...\mathcal{T}_n\}$ be two families of self-mappings of a cone b-metric space $(\Theta,\eth)$ with $\mathcal{S}=\mathcal{S}_1\mathcal{S}_2...\mathcal{S}_m$ and $\mathcal{T}=\mathcal{T}_1\mathcal{T}_2....\mathcal{T}_n$ satisfying the following conditions:*

(i) $\mathcal{S}(\Theta)\subset\mathcal{T}(\Theta)$.
(ii) $(\mathcal{S},\mathcal{T})$ *have a point of coincidence.*
(iii) *Also let $\alpha\ge 1$ and $\alpha\ell<1$ and suppose that $\mathcal{S}$ and $\mathcal{T}$ satisfy the condition,*

$$\eth(\mathcal{S}\theta,\mathcal{T}\mu)<\ell_1\eth(\theta,\mathcal{S}\theta)+\ell_2\eth(\mu,\mathcal{T}\mu)+\ell_3\eth(\theta,\mathcal{T}\mu)+\ell_4\eth(\mu,\mathcal{S}\theta)+\ell_5\eth(\theta,\mu),$$

for all $\theta,\mu\in\Theta,\theta\neq\mu$ and $\ell_1,\ell_2,\ell_3,\ell_4,\ell_5\in(0,1)$ such that $\sum_{r=1}^{5}\ell_r<1$.

Moreover, if $\mathcal{S}_i\mathcal{S}_j=\mathcal{S}_j\mathcal{S}_i$, $\mathcal{T}_k\mathcal{T}_l=\mathcal{T}_l\mathcal{T}_k$ and $\mathcal{S}_i\mathcal{T}_k=\mathcal{T}_k\mathcal{S}_i$ for all $i,j\in I_1=\{1,2,...,m\}$, $k,l\in I_2=\{1,2,...n\}$. Then (for all $i\in I_1$, $k\in I_2$) $\mathcal{S}_i$ and $\mathcal{T}_k$ have a common fixed point.

Proof Let θ_0 be an arbitrary point in Θ. Suppose that one of $\mathcal{S}(\Theta)$ and $\mathcal{T}(\Theta)$ is a complete subspace of a complete cone b-metric space. Also, suppose that the pair $(\mathcal{S},\mathcal{T})$ is coincidentally commuting. Then by assuming that $\mathcal{S}(\Theta)\subset\mathcal{T}(\Theta)$, we can define a sequence

$$\{\mathcal{S}\theta_0,\mathcal{T}\theta_1,\mathcal{S}\theta_2,\mathcal{T}\theta_3,...\mathcal{S}\theta_{2n},\mathcal{T}\theta_{2n+1},...\}. \tag{11.2}$$

Considering the steps used in Imdad et al. (2002), we can show that (11.2) is a Cauchy sequence in Θ, hence $\mathcal{T}(\Theta)$ has a limit u. Let $\theta\in\mathcal{T}^{-1}u$ then $\mathcal{T}\theta=u$. Since $\mathcal{S}(\Theta)\subset\mathcal{T}(\Theta)$, then $\mathcal{S}\theta=u$. We need to show that $\mathcal{T}\theta=u$. Assume that $\eth(u,\mathcal{T}\theta)>0$. Setting $\theta=\theta_{2n}$ and $\mu=\theta$ in (11.3.3), we get

$$\begin{aligned}\eth(\mathcal{S}\theta_{2n},\mathcal{T}\theta) &< \ell_1\eth(\theta_{2n},\mathcal{S}\theta_{2n})+\ell_2\eth(\theta,\mathcal{T}\theta)+\ell_3\eth(\theta_{2n},\mathcal{S}\theta)+\ell_4\eth(\theta,\mathcal{S}\theta_{2n})\\ &+ \ell_5\eth(\theta_{2n},\theta),\end{aligned}$$

which on letting $n \to \infty$ reduce to

$$\begin{aligned}\eth(u, \mathcal{T}\theta) &< \ell_1\eth(u, \mathcal{S}u) + \ell_2\eth(\theta, \mathcal{T}\theta) + \ell_3\eth(u, \mathcal{S}\theta) + \ell_4\eth(\theta, u) + \ell_5\eth(u, \theta), \\ &< \ell_1\eth(u, \mathcal{S}u) + \alpha\ell_2\{\eth(u, \theta) + \eth(u, \mathcal{T}\theta)\} + \alpha\ell_3\{\eth(u, \theta) + \eth(\theta, \mathcal{S}\theta)\} \\ &+ \ \ell_4\eth(u, \theta) + \ell_5\eth(u, \theta).\end{aligned}$$

Collecting like terms gives

$$\begin{aligned}(1 - \alpha\ell_2)\eth(u, \mathcal{T}\theta) &< \ell_1\eth(u, \mathcal{S}u) + (\alpha\ell_2 + \alpha\ell_3 + \ell_4 + \ell_5)\eth(u, \theta) \\ &+ \ \alpha\ell_3\eth(\theta, \mathcal{S}\theta), \\ \eth(u, \mathcal{T}\theta) &< \frac{\ell_1\eth(u, \mathcal{S}u) + (\alpha\ell_2 + \alpha\ell_3 + \ell_4 + \ell_5)\eth(u, \theta) + \alpha\ell_3\eth(\theta, \mathcal{S}\theta)}{1 - \alpha\ell_2},\end{aligned}$$

which is a contradiction. Therefore $\mathcal{T}\theta = u$ and similarly $\mathcal{S}\theta = u$, hence evidently u is a common fixed point of $\mathcal{S}$ and $\mathcal{T}$.

Now,

$$\begin{aligned}\mathcal{S}(\mathcal{S}_i\theta) &= (\mathcal{S}_1\mathcal{S}_2 ... \mathcal{S}_m)\mathcal{S}_i)\theta = (\mathcal{S}_1\mathcal{S}_2 ... \mathcal{S}_{m-1})((\mathcal{S}_m\mathcal{S}_i)\theta), \\ &= (\mathcal{S}_1 ... \mathcal{S}_{m-1})(\mathcal{S}_i\mathcal{S}_m\theta) = (\mathcal{S}_1 ... \mathcal{S}_{m-2})(\mathcal{S}_{m-1}\mathcal{S}_i(\mathcal{S}_m\theta)), \\ &= (\mathcal{S}_1 ... \mathcal{S}_{m-2})(\mathcal{S}_i\mathcal{S}_{m-1}(\mathcal{S}_m\theta)) = (\mathcal{S}_1 ... \mathcal{S}_{m-3})\mathcal{S}_i((\mathcal{S}_{m-2})(\mathcal{S}_m\theta)) = ..., \\ &= (\mathcal{S}_i\mathcal{S}_1\mathcal{S}_2 ... \mathcal{S}_m)\theta = \mathcal{S}_i(\mathcal{S}\theta), \\ &= \mathcal{S}_i\theta.\end{aligned}$$

Similarly, one can show that
$\mathcal{S}(\mathcal{T}_k\theta) = \mathcal{T}_k(\mathcal{S}\theta) = \mathcal{T}_k\theta, \mathcal{T}(\mathcal{S}_i)\theta = \mathcal{S}_i(\mathcal{T}\theta) = \mathcal{S}_i\theta$.
Now appealing to the uniqueness of a common fixed point of $\mathcal{S}$ and $\mathcal{T}$, one can get (for all i, k) $\theta = \mathcal{S}_i\theta = \mathcal{T}_k\theta$ which shows that θ is a common fixed point of $\mathcal{S}$ and $\mathcal{T}$. □

From Theorem 2.1 above, we can reduce the right-hand side items of the condition to get the following corollary to prove the existence of a unique fixed point.

Corollary 11.3.4 *Let* $(\Theta, \eth)$ *be a complete cone b-metric space and* $\mathcal{F} : \Theta \longrightarrow \Theta$ *is a continuous self-mapping on* Θ *with* $s \geq 1$ *and* $s\ell < 1$. *Suppose* $\mathcal{F}$ *satisfying the following condition:*

$$\eth(\mathcal{F}\theta, \mathcal{F}\mu) < \ell_1\eth(\theta, \mathcal{F}\theta) + \ell_2\eth(\mu, \mathcal{F}\mu) + \ell_3\eth(\theta, \mu),$$

for all $\theta, \mu \in \Theta, \theta \neq \mu$ *and for some* ℓ_1, ℓ_2 *and* $\ell_3 \in (0, 1)$ *such that* $\sum_{r=1}^{3} \ell_r = \varsigma < 1$ *then* $\mathcal{F}$ *has a unique fixed point in* Θ.

Proof Let $\theta_0 \in \Theta$ be an arbitrary point and suppose that a sequence of iteration for $\mathcal{F}$ at θ_0 as

$$\mathcal{F}(\theta_{n-1}) = \theta_n. \tag{11.3}$$

Let $\theta_{n-1} \neq \theta_n$ for all $n \in \mathbb{N}$.
Then,

$$\begin{aligned}\eth(\theta_{n-1}, \theta_n) &= \eth(\mathcal{F}(\theta_{n-2}), \mathcal{F}(\theta_{n-1})),\\ &< \ell_1\eth(\theta_{n-2}, \mathcal{F}(\theta_{n-2})) + \ell_2\eth(\theta_{n-1}, \mathcal{F}(\theta_{n-1})) + \ell_3\eth(\theta_{n-2}, \theta_{n-1}).\end{aligned}$$

Using (11.3), we have

$$\eth(\theta_{n-1}, \theta_n) \leq \ell_1\eth(\theta_{n-2}, \theta_{n-1}) + \ell_2\eth(\theta_{n-1}, \theta_n) + \ell_3\eth(\theta_{n-2}, \theta_{n-1}). \tag{11.4}$$

Collecting like terms in (11.4) above, we get

$$\begin{aligned}\eth(\theta_{n-1}, \theta_n) - \ell_2\eth(\theta_{n-1}, \theta_n) &\leq \ell_1\eth(\theta_{n-2}, \theta_{n-1}) + \ell_3\eth(\theta_{n-2}, \theta_{n-1}),\\ (1 - \ell_2)\eth(\theta_{n-1}, \theta_n) &\leq (\ell_1 + \ell_3)\eth(\theta_{n-2}, \theta_{n-1}),\\ \eth(\theta_{n-1}, \theta_n) &\leq \frac{\ell_1 + \ell_3}{1 - \ell_2}\eth(\theta_{n-2}, \theta_{n-1}).\end{aligned}$$

Let $k = (\frac{\ell_1+\ell_3}{1-\ell_2})$.
Then,

$$\begin{aligned}\eth(\theta_{n-1}, \theta_n) &\leq k\eth(\theta_{n-2}, \theta_{n-1}),\\ \eth(\theta_{n-1}, \theta_n) &\leq k^n\eth(\theta_{n-2}, \theta_{n-1}).\end{aligned}$$

Hence as $n \to \infty$, we have

$$\eth(\theta_{n-1}, \theta_n) \to 0.$$

Thus, on claiming that $\{\theta_n\} \in \Theta$ is a Cauchy sequence and since Θ is a complete space, then there exists $\theta \in \Theta$ such that $\theta_n \to \theta$ and from a continuous mapping $\mathcal{F}$, we have

$$\begin{aligned}\mathcal{F}(\lim_{n\to\infty} \theta_n) &= \mathcal{F}(\theta),\\ \lim_{n\to\infty} \theta_n &= \theta.\end{aligned}$$

Thus, θ is a fixed point of $\mathcal{F}$ in Θ.
To show that θ is a unique fixed point in Θ, let two distinct points θ and μ in Θ such that $\theta \neq \mu$, $\mathcal{F}(\theta) = \theta$ and $\mathcal{F}(\mu) = \mu$.

Then,

$$\begin{aligned}\eth(\theta, \mu) &< \ell_1 \eth(\theta, \mathcal{F}\theta) + \ell_2 \eth(\mu, \mathcal{F}\mu) + \ell_3 \eth(\theta, \mu),\\ &< \ell_3 \eth(\theta, \mu),\\ &< (\varsigma - \ell_1 - \ell_2)\eth(\theta, \mu),\end{aligned}$$

which shows that $\theta = \mu$. Therefore θ is a unique fixed point of $\mathcal{F}$ in a complete cone b-metric space. □

Also from Theorem 11.3.2 above, we can reduce the right-hand side items of the condition to get the following corollary to prove the existence of a unique fixed point.

Corollary 11.3.5 *Let* $(\Theta, \eth)$ *be a complete cone b-metric space and let* $\mathcal{F}, \mathcal{G} : \Theta \times \Theta \to \mathfrak{E}$ *be two continuous self-mappings on* Θ *with* $s \geq 1$ *and* $s\ell < 1$*. Suppose* $\mathcal{F}, \mathcal{G}$ *satisfying the condition,*

$$\eth(\mathcal{F}\theta, \mathcal{G}\mu) \leq \ell_1 \eth(\theta, \mathcal{F}\theta) + \ell_2 \eth(\mu, \mathcal{G}\mu) + \ell_3 \eth(\theta, \mu),$$

for all $\theta, \mu \in \Theta, \theta \neq \mu$ *and* $\ell_1, \ell_2, \ell_3 \in (0, 1)$ *is such that* $\sum_{r=1}^{3} \ell_r = \varsigma < 1$ *then* $\mathcal{F}$ *and* $\mathcal{G}$ *have a unique fixed point in* Θ.

Proof Let θ_0 be an arbitrary point and define a sequence $\{\theta_n\}$ as

$$\theta_{n+1} = \mathcal{G}\theta_n, \tag{11.5}$$

for all 0,1.....
Then,

$$\begin{aligned}\eth(\mathcal{F}\theta_n, \mathcal{F}\theta_{n+1}) &= \eth(\mathcal{F}\mathcal{G}\theta_{n-1}, \mathcal{F}\mathcal{G}\theta_n),\\ &< \ell_1 \eth(\mathcal{G}\theta_{n-1}, \mathcal{F}\mathcal{G}\theta_{n-1}) + \ell_2 \eth(\mathcal{G}\theta_n, \mathcal{F}\mathcal{G}\theta_n) + \ell_3 \eth(\mathcal{G}\theta_{n-1}, \mathcal{G}\theta_n),\\ &< \ell_1 \eth(\theta_n, \mathcal{F}\theta_n) + \ell_2 \eth(\theta_{n+1}, \mathcal{F}\theta_{n+1}) + \ell_3 \eth(\theta_n, \theta_{n+1}),\\ &< \ell_1 \eth(\theta_n, \mathcal{F}\theta_n) + \ell_2 \eth(\mathcal{F}\theta_n, \mathcal{F}\theta_{n+1}) + \ell_3 \eth(\theta_n, \mathcal{F}\theta_n).\end{aligned}$$

Collecting like terms, we get

$$\begin{aligned}\eth(\mathcal{F}\theta_n, \mathcal{F}\theta_{n+1}) - \ell_2 \eth(\mathcal{F}\theta_n, \mathcal{F}\theta_{n+1}) &< (\ell_1 + \ell_3)\eth(\theta_n, \mathcal{F}\theta_n),\\ (1 - \ell_2)\eth(\mathcal{F}\theta_n, \mathcal{F}\theta_{n+1}) &< (\ell_1 + \ell_3)\eth(\theta_n, \mathcal{F}\theta_n),\\ \eth(\mathcal{F}\theta_n, \mathcal{F}\theta_{n+1}) &< \frac{\ell_1 + \ell_3}{1 - \ell_2}\eth(\theta_n, \mathcal{F}\theta_n),\\ \eth(\mathcal{F}\theta_n, \mathcal{F}\theta_{n+1}) &< \frac{\ell_1 + \ell_3}{1 - \ell_2}\eth(\theta_n, \theta_{n+1}).\end{aligned}$$

Assuming that,

$$k = \frac{\ell_1 + \ell_3}{1 - \ell_2}.$$

Then,

$$\eth(\mathcal{F}\theta_n, \mathcal{F}\theta_{n+1}) < k\eth(\theta_n, \theta_{n+1}).$$

Repeating this, we get

$$\eth(\mathcal{F}\theta_n, \mathcal{F}\theta_{n+1}) < k^n\eth(\theta_n, \theta_{n+1})$$

as $n \to \infty$ we observe that,

$$\eth(\mathcal{F}\theta_n, \mathcal{F}\theta_{n+1}) \to 0,$$

which suggests that the sequence $\{\theta_n\}$ is a Cauchy sequence in a cone b-metric space, and since the space is complete, then there exists $\theta \in \Theta$ such that $\theta_n \to \theta$ and since $\mathcal{F}$ and $\mathcal{G}$ are continuous mappings, we have

$$\mathcal{F}(\lim_{n\to\infty} \theta_n) = \mathcal{F}(\theta) \quad and \quad \mathcal{G}(\lim_{n\to\infty} \theta_n) = \mathcal{G}(\theta).$$

Hence,

$$\lim_{n\to\infty} \theta_n = \theta \quad and \quad \lim_{n\to\infty} \theta_n = \theta.$$

Therefore, θ is a fixed point of $\mathcal{F}$ and $\mathcal{G}$ in Θ.

Now, we will show that θ is a unique fixed point of $\mathcal{F}$ and $\mathcal{G}$ in a complete cone b-metric space.

Suppose that there exists two distinct points θ and μ such that $\mathcal{F}(\theta) = \theta$ and $\mathcal{F}(\mu) = \mu$ and also $\mathcal{G}(\theta) = \theta$ and $\mathcal{G}(\mu) = \mu$.
Then,

$$\begin{aligned}\eth(\theta, \mu) &< \ell_1\eth(\theta, \mathcal{F}\theta) + \ell_2\eth(\mu, \mathcal{G}\mu) + \ell_3\eth(\theta, \mu),\\ \eth(\theta, \mu) &< \ell_3\eth(\theta, \mu).\end{aligned}$$

Hence $\theta = \mu$, which proves that θ is a unique fixed point of $\mathcal{F}$ and $\mathcal{G}$ in a complete cone b-metric space. □

11.4 Applications

In this section, we use the obtained results to show the existence and uniqueness of common fixed points for mapping with the exception of continuity conditions.

Theorem 11.4.1 *Let* $(\Theta, \eth)$ *be a complete cone b-metric space and let* $\mathcal{F}_{k1}$,$\mathcal{F}_{k2}$ *be two self-mappings on* Θ *with* $s > 1$ *and* $s\ell < 1$ *satisfying the following:*

$$\eth(\mathcal{F}_{k1}(\theta), \mathcal{F}_{k2}(\mu)) < \ell_1\eth(\theta, \mathcal{F}_{k1}(\theta)) + \ell_2\eth(\mu, \mathcal{F}_{k2}(\mu)) + \ell_3\eth(\theta, \mathcal{F}_{k2}(\mu))$$
$$+ \ell_4\eth(\mu, \mathcal{F}_{k1}(\theta)) + \ell_5\eth(\theta, \mu),$$

for all $\theta, \mu \in \Theta$ *with* $\theta \neq \mu$ *and* $\sum_{r=1}^{5} \ell_r = \varsigma < 1$. *Suppose that* $f_{k1} f_{k2} = f_{k2} f_{k1}$ *is continuous then* f_{k1} *and* f_{k2} *have a unique common fixed point in* Θ.

Proof Let $\theta_n = \mathcal{F}_{k1}(\theta_{n-1})$, $\theta_{n+1} = \mathcal{F}_{k2}(\theta_n)$ and
$\mathcal{F}_{k1}(\theta_{n-1}) \neq \mathcal{F}_{k2}(\theta_{n-1}), \theta_n \neq \theta_{n-1}$ for all $n \in N$.
Hence,

$$\eth(\theta_{2n+1}, \theta_{2n}) = \eth(\mathcal{F}_{k1}(\theta_{2n}), \mathcal{F}_{k2}(\theta_{2n-1})),$$
$$< \ell_1\eth(\theta_{2n}, \theta_{2n+1}) + \ell_2\eth(\theta_{2n-1}, \theta_{2n}) + \ell_3\eth(\theta_{2n}, \theta_{2n})$$
$$+ \ell_4\eth(\theta_{2n-1}, \theta_{2n+1}) + \ell_5\eth(\theta_{2n}, \theta_{2n-1}).$$

Applying the definition of triangle inequality on above, we have

$$\eth(\theta_{2n+1}, \theta_{2n}) < \ell_1\eth(\theta_{2n}, \theta_{2n+1}) + \ell_2\eth(\theta_{2n-1}, \theta_{2n})$$
$$+ s\ell_4[\eth(\theta_{2n-1}, \theta_{2n}) + \eth(\theta_{2n}, \theta_{2n+1})] + \ell_5\eth(\theta_{2n}, \theta_{2n-1}).$$

Then collecting like terms and applying the symmetric property, we get

$$\eth(\theta_{2n+1}, \theta_{2n}) - \ell_1\eth(\theta_{2n+1}, \theta_{2n}) - s\ell_4\eth(\theta_{2n+1}, \theta_{2n}) < \ell_2\eth(\theta_{2n}, \theta_{2n-1})$$
$$+ s\ell_4\eth(\theta_{2n}, \theta_{2n-1}) + \ell_5\eth(\theta_{2n}, \theta_{2n-1}),$$
$$(1 - \ell_1 - s\ell_4)\eth(\theta_{2n+1}, \theta_{2n}) < (\ell_2 + s\ell_4 + \ell_5)\eth(\theta_{2n}, \theta_{2n-1}),$$
$$\eth(\theta_{2n+1}, \theta_{2n}) \leq \Big(\frac{\ell_2 + s\ell_4 + s\ell_5}{1 - s\ell_1 - s\ell_4}\Big)(\eth(\theta_{2n}, \theta_{2n-1})).$$

Hence from the above, we obtain the following:

$$\eth(\theta_{2n+1}, \theta_{2n}) \leq \left(\frac{\ell_2 + s\ell_4 + s\ell_5}{1 - s\ell_1 - s\ell_4}\right)^{2n} \eth(\theta_1, \theta_0).$$

Therefore,

$$\mathcal{F}_{k1}\mathcal{F}_{k2}(\theta) = \mathcal{F}_{k2}\mathcal{F}_{k1}(\theta) = \mathcal{F}_{k1}\mathcal{F}_{k2}(\lim_{n\to\infty} \theta_{n_k}) = \lim_{n\to\infty} \theta_{n_{k+1}} = \theta.$$

Let θ be a fixed point of $\mathcal{F}_{k1}\mathcal{F}_{k2} \in \Theta$ such that $\mathcal{F}_{k1}\mathcal{F}_{k2}(\theta) = \theta$.

We show that θ is a fixed point of $\mathcal{F}_{k1}$ and $\mathcal{F}_{k2}$ in Θ.

Hence $\mathcal{F}_{k1}(\theta) = \theta$ and $\mathcal{F}_{k2}(\theta) = \theta$.

Let,

$$\mathcal{F}_{k1}(\theta) \neq \theta \quad and \quad \mathcal{F}_{k2}(\theta) \neq \theta.$$

Then,

$$\begin{aligned}\eth(\theta, \mathcal{F}_{k1}(\theta)) &= \eth(\mathcal{F}_{k1} f_{k2}(\theta), \mathcal{F}_{k1}(\theta)),\\ &< \ell_1\eth(\mathcal{F}_{k1}(\theta), \mathcal{F}_{k2}\mathcal{F}_{k1}(\theta)) + \ell_2\eth(\theta, \mathcal{F}_{k1}(\theta)) + \ell_3\eth(\mathcal{F}_{k1}(\theta), \mathcal{F}_{k1}(\theta))\\ &+ \ell_4\eth(\theta, \mathcal{F}_{k1}\mathcal{F}_{k1}(\theta)) + \ell_5\eth(\mathcal{F}_{k1}(\theta), \theta).\end{aligned}$$

From the above, we obtain θ is a fixed point of $\mathcal{F}_{k1}(\theta)$ in Θ.

By applying the same technique, we get $\mathcal{F}_{k2}(\theta) = \theta$. Thus $\mathcal{F}_{k1}$ and $\mathcal{F}_{k2}$ have a fixed point in Θ. Now we show that a fixed point is unique.

Let $\mu \in \Theta$ such that $\mu \neq \theta$ is another fixed point of $\mathcal{F}_{k1}$ and $\mathcal{F}_{k2}$ in Θ such that,

$$\begin{aligned}\eth(\theta, \mu) &= \eth(\mathcal{F}_{k1}(\theta), \mathcal{F}_{k2}(\mu)),\\ &< \ell_1\eth(\theta, \mathcal{F}_{k1}(\theta)) + \ell_2\eth(\mu, \mathcal{F}_{k1}(\mu)) + \ell_3\eth(\theta, \mathcal{F}_{k2}(\mu)) + \ell_4\eth(\mu, \mathcal{F}_{k1}(\theta)) + \ell_5\eth(\theta, \mu).\end{aligned}$$

Therefore $\theta = \mu$ proves that θ is a unique fixed point. □

From the above theorem, we can also prove the following theorem to show the existence of uniqueness of a fixed point in a complete cone b-metric space.

Theorem 11.4.2 *Let* $(\Theta, \eth)$ *be a complete cone b-metric space and let* $\mathcal{F}_{k1}\mathcal{F}_{k2}$ *be two self-mappings on* Θ *with* $s > 1$ *and* $s\ell < 1$ *satisfying the following:*

$$\eth(\mathcal{F}_{k1}(\theta), \mathcal{F}_{k2}(\mu)) < \ell_1\eth(\theta, \mathcal{F}_{k1}(\theta)) + \ell_2\eth(\mu, \mathcal{F}_{k1}(\mu)) + \ell_3\eth(\theta, \mu),$$

for all $\theta, \mu \in \Theta$ *with* $\theta \neq \mu$ *and* $\sum_{r=1}^{3} \ell_r = \varsigma < 1$. *Suppose that* $\mathcal{F}_{k1}\mathcal{F}_{k2} = \mathcal{F}_{k1}\mathcal{F}_{k2}$ *is continuous, then* $\mathcal{F}_{k1}$ *and* $\mathcal{F}_{k2}$ *have a common fixed point in* Θ.

Proof Let $\theta_n = \mathcal{F}_{k1}(\theta_{n-1}), \theta_{n+1} = \mathcal{F}_{k2}(\theta_n)$ and $\mathcal{F}_{k1}(\theta_{n-1}) \neq \mathcal{F}_{k2}(\theta_{n-1}), \theta_n \neq \theta_{n-1}$ for all $n \in \mathbb{N}$.
Then,

$$\begin{aligned}\eth(\theta_{2n+1}, \theta_{2n}) &= \eth(\mathcal{F}_{k1}\theta_{2n}, \mathcal{F}_{k2}\theta_{2n-1}),\\ &< \ell_1\eth(\theta_{2n}, \mathcal{F}_{k1}\theta_{2n}) + \ell_2\eth(\theta_{2n-1}, \mathcal{F}_{k2}\theta_{2n-1}) + \ell_3\eth(\theta_{2n}, \theta_{2n-1}).\end{aligned}$$

Applying the definition above, we get

$$\eth(\theta_{2n+1}, \theta_{2n}) < \ell_1\eth(\theta_{2n}, \theta_{2n+1}) + \ell_2\eth(\theta_{2n-1}, \theta_{2n}) + \ell_3\eth(\theta_{2n}, \theta_{2n-1}).$$

Collecting like terms gives

$$\begin{aligned}\eth(\theta_{2n+1}, \theta_{2n}) - \ell_1\eth(\theta_{2n}, \theta_{2n+1}) &< \ell_2\eth(\theta_{2n-1}, \theta_{2n}) + \ell_3\eth(\theta_{2n-1}, \theta_{2n}),\\ \eth(\theta_{2n+1}, \theta_{2n}) - \ell_1\eth(\theta_{2n}, \theta_{2n+1}) &< (\ell_2 + \ell_3)\eth(\theta_{2n-1}, \theta_{2n}),\\ (1 - \ell_1)\eth(\theta_{2n}, \theta_{2n+1}) &< (\ell_2 + \ell_3)\eth(\theta_{2n-1}, \theta_{2n}),\\ \eth(\theta_{2n}, \theta_{2n+1}) &\leq \left(\frac{\ell_2 + \ell_3}{1 - \ell_1}\right)\eth(\theta_{2n-1}, \theta_{2n}).\end{aligned}$$

Repeating the above inequality, we get

$$\eth(\theta_{2n}, \theta_{2n+1}) \leq \Big(\frac{\ell_2 + \ell_3}{1 - \ell_1}\Big)\eth(\theta_{2n-1}, \theta_{2n})^{2n}\eth(\theta_{2n-1}, \theta_{2n}).$$

Thus we have

$$\mathcal{F}_{k1}\mathcal{F}_{k2}(\theta) = \mathcal{F}_{k2}\mathcal{F}_{k1}(\theta) = \mathcal{F}_{k1}\mathcal{F}_{k2}(\lim_{n\to\infty} \theta_{n_k}) = \lim_{n\to\infty} \theta_{n_{k+1}}.$$

Suppose that θ is a fixed point of $\mathcal{F}_{k1}\mathcal{F}_{k2} \in \Theta$ such that $\mathcal{F}_{k1}\mathcal{F}_{k2}(\theta) = \theta$.
We show that θ is a fixed point of $\mathcal{F}_{k1}$ and $\mathcal{F}_{k2}$ in Θ so that $\mathcal{F}_{k1}(\theta) = \theta$ and $\mathcal{F}_{k2}(\theta) = \theta$.
Therefore let,

$$\mathcal{F}_{k1}(\theta) \neq \theta \quad and \quad \mathcal{F}_{k2}(\theta) \neq \theta.$$

Hence,

$$\begin{aligned}\eth(\theta, \mathcal{F}_{k1}\theta) &= \eth(\mathcal{F}_{k2}\mathcal{F}_{k1}\theta, \mathcal{F}_{k1}\theta),\\ &< \ell_1\eth(\mathcal{F}_{k1}\theta, \mathcal{F}_{k2}\mathcal{F}_{k1}\theta) + \ell_2\eth(\theta, \mathcal{F}_{k1}\theta) + \ell_3\eth(\mathcal{F}_{k1}\theta, \mathcal{F}_{k1}\theta),\end{aligned}$$

which proves that,

$$\mathcal{F}_{k1}(\theta) = \theta.$$

Also considering the following:

$$\begin{aligned}\eth(\theta, \mathcal{F}_{k2}\theta) &= \eth(\mathcal{F}_{k1}\mathcal{F}_{k2}\theta, \mathcal{F}_{k2}\theta),\\ &< \ell_1\eth(\mathcal{F}_{k2}\theta, \mathcal{F}_{k1}\mathcal{F}_{k2}\theta) + \ell_2\eth(\theta, \mathcal{F}_{k2}\theta) + \ell_3\eth(\mathcal{F}_{k2}\theta, \mathcal{F}_{k2}\theta),\end{aligned}$$

then it proves that,

$$\mathcal{F}_{k2}(\theta) = \theta.$$

Therefore we observe that $\mathcal{F}_{k1}(\theta) = \theta$ and $\mathcal{F}_{k2}(\theta) = \theta$. So both $\mathcal{F}_{k1}$ and $\mathcal{F}_{k2}$ have a common fixed point.

We show that a common fixed point is unique in a complete cone b-metric space. Let $\mu \in \Theta$ such that $\mu \neq \theta$ is also a fixed point of $\mathcal{F}_{k1}$ and $\mathcal{F}_{k2}$ such that,

$$\begin{aligned}\eth(\theta, \mu) &< \eth(\mathcal{F}_{k1}\theta, \mathcal{F}_{k2}\mu),\\ &< \ell_1\eth(\theta, \mathcal{F}_{k1}\theta) + \ell_2\eth(\mu, \mathcal{F}_{k2}\mu) + \ell_3\eth(\theta, \mu).\end{aligned}$$

Therefore, this proves that $\theta = \mu$ so θ is a unique fixed point in Θ. □

Acknowledgements The authors are thankful to the learned reviewers for their valuable comments.

Conflict of Interest The authors declare that they have no competing interests

References

Abbas, M., Aydi, H., Radenović, S. (2012). Fixed point of T-Hardy-Rogers contractive mappings in partially ordered partial metric spaces. *International Journal of Mathematics*. https://doi.org/10.1155/2012/313675.

Abbas, M., & Jungck, G. (2008). Common fixed point results for noncommuting mappings without continuity in cone metric spaces. *Journal of Mathematical Analysis and Applications,341*(1), 416–420.

Bakhtin, I. A. (1989). The contraction mapping principle in almost metric spaces. *Functional Analysis,30*, 26–37.

Banach, S. (1922). Sur les OperÃŒÂ'ations dans les ensembles abstraits et leur application aux éqations intégrales. *Fundamenta Mathematicae,3*, 133–181.

Cosentino, V., & Vetro, P. (2014). Fixed point result for F-contractive mappings of hardy-rogers-type. *Filomat,28*, 715–722.

Dubey, A. K., Shukla, R., & Dubey, R. P. (2013). Common fixed point theorems for generalized contractive mappings on cone metric spaces. *International Journal of Applied Mathematical Research,2*(2), 269.

Gautam, P., Kumar, S., Verma, S., & Gulati, S. (2022). On some w-Interpolative contractions of Suzuki type mappings in Quasi-partial b-metric space. *Journal of Function Spaces, 2022*, Article Id. 9158199, 12 Pages.

Gautam, P., Kumar, S., Verma, S., & Gupta, G. (2021). Non-unique Fixed Point results via Kannan F-contraction on quasi-partial b-metric space. *Journal of Function Spaces,1*, 1–10. https://doi.org/10.1155/2021/2163108

Goswami, N., Haokip, N., & Mishra, V. N. (2019). F-contractive type mappings in b-metric spaces and some related fixed point results. *Fixed Point Theory Application,1*, 13.

Huang, L. G., & Zhang, X. (2019). Cone metric spaces and fixed point theorems of contractive mappings. *Journal of Mathematical Analysis and Applications,332*(2), 1468–1476.

Huang, H., & Xu, S. (2013). Fixed point theorems of contractive mappings in cone b-metric spaces and applications. *Fixed Point Theory Applications,1*, 112.

Huang, L. G., & Zhang, X. (2007). Cone metric spaces and fixed point theorems of contractive mappings. *Journal of Mathematical Analysis and Applications,332*(2), 1468–1476.

Huang, X., Zhu, C., & Wen, X. (2010). A common fixed point theorem in cone metric spaces. *International Journal of Analysis and Applications,4*(13–16), 721–726.

Hussain, N., & Shah, M. H. (2011). KKM mappings in cone b-metric spaces. *Computers and Mathematics Applications,62*, 1677–1684.

Imdad, M., Kumar, S., & Khan, M. S. (2002). Remarks on some fixed point theorems satisfying implicit relations. *Radovi Matematicki,11*(1), 135–143.

Kumar, S. (2021). Fixed points and continuity for a pair of contractive maps in metric spaces with application to nonlinear volterra-integral equations. *Journal of Function Spaces, 2021*, Article ID 9982217, 13 pages. https://doi.org/10.1155/2021/9982217.

Kumar, D., Tomar, A., & Sharma, R. (2020). Well-posedness and data dependence of strict fixed point for δ-Hardy Roger type contraction and applications. *Applied Mathematics E-Notes,20*, 46–54.

Kumar, S., & Sholastica, L. (2021). On some fixed point theorems for multivalued F-contractions in partial metric spaces. *Demonstrating Mathematical,54*, 151–161.

Patil, J., Hadan, B., Abdo, M. S., Bachhav, A., & Chaudari, A. (2020). A fixed point theorem for Hardy Roger type on generalized fractional differential equations. *Advances in the Theory of Nonlinear Analysis and its Applications,4*(4), 407–420.

Pauline, S., & Kumar, S. (2021). Common fixed point theorems for T-Hardy-Rodgers contraction mappings in complete cone b-metric spaces with an application. *Topological Algebra and its Applications,9*, 105–117.

Popa, V. (1999). Some fixed point theorems for compatible mappings satisfying 11 an implicit relation. *Demonstrations in Mathematics,32*(1), 157–163.

Rezapour, S., & Hamlbarani, R. (2008). Some notes on the paper Cone metric spaces and fixed point theorems of contractive mappings. *Journal of Mathematical Analysis and Applications,345*(2), 719–724.

Sharma, N. (2017). Fixed point theorem in cone b–metric spaces using contractive mappings. *Global Journal of Pure and Applied Mathematics,13*(7), 2997–3004.

Tomar, A., & Joshi, M. (2021). Results in strongly minihedral cone and scalar weighted cone metric spaces and applications. *Annales Mathematicae Silesianae,35*(2), 302–318. https://doi.org/10.2478/amsil-2021-0009

Tomar, A., Joshi, M., & Bhatt, V. (2021). Fixed point and a cantilever beam problem in a partial b-metric space. *Acta Universitatis Sapientiae Mathematica, 13*(2), 506–518. https://doi.org/10.2478/ausm-2021-0032.

Tomar, A., & Sharma, R. (2019). Almost alpha-Hardy-Rogers-F-contractions and applications. *Armenian Journal of Mathematics, 11*(11), 1–19. https://doi.org/10.52737/18291163-2019.11.11-1-19.

Wangwe, L., & Kumar, S. (2022). Fixed point results for interpolative ψ-Hardy-Rogers type contraction mappings in quasi-partial b-metric space with some applications. *Journal of Analysis*, 18 Pages. https://doi.org/10.1007/s41478-022-00456-4 .

Chapter 12
Generalized Weak Integral Contractive Condition with an Application

Dinesh Rani Bansal, Kavita, Sanjay Kumar, and Manish Jain

Abstract In this chapter, we establish the existence of a common fixed point for pairs of sub-compatible and reciprocal continuous mappings (or compatible and subsequential continuous mappings) that satisfy a generalized (ψ, ϕ)-weak integral contractive condition involving cubic terms of distance functions. Additionally, a sufficient condition is discussed for the existence of a solution to certain functional equations arising in dynamic programming.

Keywords Generalized weak integral contractive condition · Compatible mappings · Sub-compatible mappings · Reciprocal continuous · Subsequential continuous · Functional equations

AMS Subject Classification 47H10 · 54H25 · 90C39

12.1 Introduction and Preliminaries

The basic tool of "fixed point theory" (**FPT**), which is known as the "Banach contraction principle" (**BCP**) Banach (1922), guarantees the existence of a unique fixed point (**FP**) for every contraction mapping defined on a complete metric space (**CMS**). It behaves as a mathematical tool to solve various practical problems arising during the mathematical formulation of these problems. Over a hundred years, researchers have made efforts to extend, generalize, and improve the **BCP** in various directions.

D. R. Bansal
Department of Mathematics, Vaish College, Rohtak, Haryana, India

Kavita (✉) · S. Kumar
Department of Mathematics, Deenbandhu Chhotu Ram University of Science and Technology, Murthal (Sonepat), Haryana, India
e-mail: kvtlather@gmail.com

M. Jain
Department of Mathematics, Ahir College, Rewari, Haryana, India

© The Author(s), under exclusive license to Springer Nature Singapore Pte Ltd. 2026
A. Tomar and M. Jain (eds.), *Applications of Fixed-Point Theorem*, Industrial and Applied Mathematics, https://doi.org/10.1007/978-981-95-5160-6_12

Jungck (1976) was the first to prove a common fixed point (**CFP**) theorem for commuting mappings. This theorem opened the door for researchers to generalize the **BCP** for pair(s) of mappings satisfying a set of minimal commutative conditions involving compatible (**Comp**) mappings or their variants, weak commutative mappings or their variants, and more.

All over this chapter, we consider the ensuing mathematical notations:
$\mathbb{R} \equiv$ the set of all real numbers.
$\mathbb{R}^+ \equiv [0, \infty)$.
$(\mathcal{M}, \rho) \equiv$ a metric space (**MS**)
$(\gamma, \sigma) \equiv$ a pair of self-mappings defined on an **MS** $\mathcal{M}$.
$(\alpha, \gamma) \equiv$ a pair of self-mappings defined on an **MS** $\mathcal{M}$.
$(\beta, \sigma) \equiv$ a pair of self-mappings defined on an **MS** $\mathcal{M}$.
$\mathcal{X} \equiv \{$ sequence $\{\mathfrak{z}_j\} \subset \mathcal{M}$ such that (s.t.) $\gamma\mathfrak{z}_j \to r$ and $\sigma\mathfrak{z}_j \to r, r \in \mathcal{M}$ as $j \to \infty\}$
$\mathcal{Y} \equiv \{$ sequence $\{\mathfrak{z}_j\} \subset \mathcal{M}$ s.t. $\alpha\mathfrak{z}_j \to r$ and $\gamma\mathfrak{z}_j \to r, r \in \mathcal{M}$ as $j \to \infty$ $\}$
$\mathcal{Z} \equiv \{$ sequence $\{\mathrm{u}_j\} \subset \mathcal{M}$ s.t. $\beta\mathrm{u}_j$ and $\sigma\mathrm{u}_j \to r \in \mathcal{M}$ as $j \to \infty\}$
$\mathfrak{C}(\gamma, \sigma) \equiv \{a \in \mathcal{M}$ s.t. $\gamma a = \sigma a\}$
$\Omega \equiv \{ \phi : \mathbb{R}^+ \to \mathbb{R}^+$ s.t. ϕ is a continuous function with $\phi(0) = 0$ and $\phi(\kappa) > 0$, for each $\kappa > 0$ $\}$
$\mathbb{L} \equiv \{\varsigma : \mathbb{R}^+ \to \mathbb{R}^+$ s.t. ς is "Lebesgue integrable" and "summable on each compact subset of $\mathbb{R}^+$" with $\int_0^{\epsilon} \varsigma(\kappa)\, d\kappa > 0$, for each $\epsilon > 0\}$.

First, we recall some basic definitions as follows:

Definition 12.1.1 (Jungck (1986)) The pair (γ, σ) is known as **Comp** if $\rho(\gamma\sigma\mathfrak{z}_j, \sigma\gamma\mathfrak{z}_j) \to 0$ as $j \to \infty$, whenever the sequence $\{\mathfrak{z}_j\} \in \mathcal{X}$.

Definition 12.1.2 (Jungck (1986)) The pair (γ, σ) is known as non-compatible (**non-Comp**) if $\exists$ a sequence $\{\mathfrak{z}_j\} \in \mathcal{X}$ s.t. $\rho(\gamma\sigma\mathfrak{z}_j, \sigma\gamma\mathfrak{z}_j)$ either tends to a non-zero point or does not exist as $j \to \infty$.

Jungck (1996) presented the notion of weakly compatible (**WC**) as follows:

Definition 12.1.3 (Jungck (1996)) The pair (γ, σ) is known as **WC** if $\gamma\sigma\mathfrak{z} = \sigma\gamma\mathfrak{z}$, whenever $\mathfrak{z} \in \mathfrak{C}(\gamma, \sigma)$.

For a detailed discussion on weaker forms of commutativity, refer to Singh and Tomar (2003).

Aamri and Moutawakil (2002) presented a generalization of **non-Comp** mappings in the form of the property $(E.A.)$.

Definition 12.1.4 (Aamri and Moutawakil (2002)) The pair (γ, σ) is said to satisfy property $(E.A.)$ if $\mathcal{X} \neq \emptyset$.

Liu et al. (2005) presented the idea of common property $(E.A.)$ as follows:

Definition 12.1.5 (Liu et al. (2005)) The pairs (α, γ) and (β, σ) are said to satisfy common property $(E.A.)$ if $\exists$ two sequences $\{\mathfrak{z}_j\}$ and $\{\mathfrak{u}_j\}$ in $\mathcal{M}$ s.t.

$$\alpha\mathfrak{z}_j \to r, \gamma\mathfrak{z}_j \to r, \beta\mathfrak{u}_j \to r \text{ and } \sigma\mathfrak{u}_j \to r, for some r \in \mathcal{M} \text{ as } j \to \infty.$$

Al-Thagafi and Shahzad (2008) gave the notion of occasionally weakly compatible (***OWC***) mappings to generalize the notion of **WC** mappings. These mappings do not necessarily commute at all coincidence points.

Definition 12.1.6 (Al-Thagafi and Shahzad (2008)) The pair (γ, σ) is called **OWC** if $\exists\, \mathfrak{z} \in \mathfrak{C}(\gamma, \sigma)$ with $\gamma\sigma\mathfrak{z} = \sigma\gamma\mathfrak{z}$.

Bouhadjera and Godet-Thobie (2011) gave the concepts of sub-compatible (**Sub-Comp**) and subsequential continuous (**SSC**) mappings.

Definition 12.1.7 (Bouhadjera and Godet-Thobie (2011)) The pair (γ, σ) is called **Sub-Comp** if $\exists$ a sequence $\{\mathfrak{z}_j\} \in \mathcal{X}$ s.t.

$$\lim_{j\to\infty}\rho(\gamma\sigma\mathfrak{z}_j, \sigma\gamma\mathfrak{z}_j) = 0.$$

Definition 12.1.8 (Bouhadjera and Godet-Thobie (2011)) A pair (γ, σ) is called **SSC** if $\exists$ a sequence $\{\mathfrak{z}_j\} \in \mathcal{X}$ satisfying $\lim_{j\to\infty} \gamma\sigma\mathfrak{z}_j = \gamma r$ and $\lim_{j\to\infty} \sigma\gamma\mathfrak{z}_j = \sigma r$, $r \in \mathcal{M}$.

Remark 12.1.9 **Comp** mappings enjoying the property $(E.A.)$ are **Sub-Comp**, but the converse may not hold, see (Rouzkard et al., 2012, Example 2.3).

Remark 12.1.10 A pair of ***OWC*** mappings is **Sub-Comp**, but reverse implication need not hold, see (Bouhadjera & Godet-Thobie, 2011, Example 1.2).

Definition 12.1.11 (Pant (1999)) A pair (γ, σ) is called reciprocal continuous (**RC**) if $\lim_{j\to\infty} \gamma\sigma\mathfrak{z}_j = \gamma r$ and $\lim_{j\to\infty} \sigma\gamma\mathfrak{z}_j = \sigma r$, whenever sequence $\{\mathfrak{z}_j\} \in \mathcal{X}$, for some $r \in \mathcal{M}$.

Remark 12.1.12 If the pair (γ, σ) is continuous then it is **RC** also, but the converse may not be true, see (Kim & Bisht, 2014, Example 2.2).

Remark 12.1.13 The notion of **SSC** and **RC** are independent of each other, see (Kim & Bisht, Kim et al. (2014), Examples 2.3–2.4).

Remark 12.1.14 For an in-depth exploration of continuity variants, refer to Tomar and Karapinar (2016).

Khan et al. (1984) presented the notion of altering the distance/control function. An altering distance function is an increasing continuous function $\phi : \mathbb{R}^+ \to \mathbb{R}^+$ which vanishes only at zero. Researchers tried to generalize the **BCP** by using a control function. In this direction, Boyd and Wong (1969) gave "ϕ contraction" as follows:

γ is said to be ϕ contraction mapping if $\rho(\gamma\mathfrak{z}, \gamma\mathfrak{u}) \leq \phi(\rho(\mathfrak{z}, \mathfrak{u})), \forall\, \mathfrak{z}, \mathfrak{u} \in \mathcal{M}$, where $\phi : \mathbb{R}^+ \to \mathbb{R}^+$ is an upper semi-continuous function from right s.t. $0 \leq \phi(\kappa) < \kappa, \forall\, \kappa > 0$.

Alber and Guerre-Delabriere (1997) gave the idea of "$\phi-$weak contraction" in "Hilbert spaces" and Rhoades (2001) proved some results of $\phi-$weak contraction in **CMS**.

A self-mapping γ of a **CMS** $\mathcal{M}$ is said to be a $\phi-$ weak contraction if for each $\mathfrak{z}, \mathfrak{u} \in \mathcal{M}$, $\exists$ a function $\phi \in \Omega$ with

$$\rho(\gamma\mathfrak{z}, \gamma\mathfrak{u}) \leq \rho(\mathfrak{z}, \mathfrak{u}) - \phi(\rho(\mathfrak{z}, \mathfrak{u})). \tag{12.1}$$

Murthy and Prasad (2013) obtained an **FP** theorem for a new type of contraction: "weak contraction involving cubic terms of distance functions."

Theorem 12.1.15 (Murthy and Prasad (2013)) *Let γ be a self-mapping of a* ***CMS*** *$\mathcal{M}$ satisfying*

$$\begin{aligned}[1 + \mathcal{L}\rho(\mathfrak{z}, \mathfrak{u})]\rho^2(\gamma\mathfrak{z}, \gamma\mathfrak{u}) \leq \mathcal{L}\max\Big\{&\frac{1}{2}[\rho^2(\mathfrak{z}, \gamma\mathfrak{z})\rho(\mathfrak{u}, \gamma\mathfrak{u}) + \rho(\mathfrak{z}, \gamma\mathfrak{z})\rho^2(\mathfrak{u}, \gamma\mathfrak{u})],\\ &\rho(\mathfrak{z}, \gamma\mathfrak{z})\rho(\mathfrak{z}, \gamma\mathfrak{u})\rho(\mathfrak{u}, \gamma\mathfrak{z}), \rho(\mathfrak{z}, \gamma\mathfrak{u})\rho(\mathfrak{u}, \gamma\mathfrak{z})\rho(\mathfrak{u}, \gamma\mathfrak{u})\Big\} + m(\mathfrak{z}, \mathfrak{u}) - \phi(m(\mathfrak{z}, \mathfrak{u})),\end{aligned} \tag{12.2}$$

where

$$\begin{aligned}\mathfrak{m}(\mathfrak{z}, \mathfrak{u}) = \max\Big\{&\rho^2(\mathfrak{z}, \mathfrak{u}), \rho(\mathfrak{z}, \gamma\mathfrak{z})\rho(\mathfrak{u}, \gamma\mathfrak{u}), \rho(\mathfrak{z}, \gamma\mathfrak{u})\rho(\mathfrak{u}, \gamma\mathfrak{z}),\\ &\frac{1}{2}[\rho(\mathfrak{z}, \gamma\mathfrak{z})\rho(\mathfrak{z}, \gamma\mathfrak{u}) + \rho(\mathfrak{u}, \gamma\mathfrak{z})\rho(\mathfrak{u}, \gamma\mathfrak{u})]\Big\},\end{aligned} \tag{12.3}$$

$\mathcal{L} \in \mathbb{R}^+$ and $\phi \in \Omega$. Then γ has a unique ***FP*** *in $\mathcal{M}$.*

Kavita and Kumar (2022) generalized the weak contraction (12.2) formulated in Murthy and Prasad (2013) by introducing a "generalized $(\psi, \phi)-$weak contraction involving cubic terms of distance functions."

In 2002, an integral form of the **BCP** was given by Branciari (2002).

Theorem 12.1.16 (Branciari (2002)) *Let $(\mathcal{M}, \rho)$ be a* ***CMS****, $\mathfrak{c} \in (0, 1)$ and $\gamma : \mathcal{M} \to \mathcal{M}$ be a mapping s.t. for each $\mathfrak{z}, \mathfrak{u} \in \mathcal{M}$*

$$\int_0^{\rho(\gamma\mathfrak{z}, \gamma\mathfrak{u})} \varsigma(\kappa)d\kappa < \mathfrak{c} \int_0^{\rho(\mathfrak{z}, \mathfrak{u})} \varsigma(\kappa)d\kappa,$$

where $\varsigma \in \mathbb{L}$. Then γ has a unique ***FP*** *$r \in \mathcal{M}$ and $\lim_{j\to\infty} \gamma^j\mathfrak{z} = r$, for each $\mathfrak{z} \in \mathcal{M}$.*

In recent years, researchers have been fascinated by the study of **CFPs** for pairs of mappings satisfying contractive conditions of integral type, leading to the proof of several **FP** theorems for more general integral contractive mappings (see, e.g., Altun et al., 2007; Bari & Vetro, 2008; Chauhan et al., 2014; Djoudi & Aliouche, 2007; Samet & Vetro, 2012; Vijayaraju et al., 2005).

This chapter aims to establish the existence of the unique **CFP** for pairs of **Sub-Comp** and **RC** mappings (or **Comp** and **SSC** mappings) satisfying a generalized (ψ, ϕ)-weak contractive condition of integral type, involving cubic terms of distance functions, which improve upon Theorem 12.1.15 and the results of Branciari (2002).

12.2 Main Result

Let Ξ denote the collection of all the functions $\psi : [0, \infty)^4 \to [0, \infty)$ satisfying

(ψ_1) ψ is non-decreasing and upper semi-continuous in each coordinate,
(ψ_2) $\Delta(\kappa) = \max\{\psi(\kappa, \kappa, 0, 0), \psi(0, 0, 0, \kappa), \psi(0, 0, \kappa, 0), \psi(\kappa, \kappa, \kappa, \kappa)\} \leq \kappa$, for each $\kappa > 0$.

Now, we proceed to the main result.

Theorem 12.2.1 *Let α, β, γ, and σ be self-mappings of an* **MS** $(\mathcal{M}, \rho)$ *s.t.* $\forall\, \mathfrak{z}, \mathfrak{u} \in \mathcal{M}$, $\exists$ *a function* $\phi \in \Omega$, *a function* $\psi \in \Xi$ *and a real number* $\mathcal{L} \geq 0$ *satisfying*

$$\begin{aligned}\Big[1+\mathcal{L}\int_0^{\rho(\alpha\mathfrak{z},\beta\mathfrak{u})}\varsigma(\kappa)d\kappa\Big]\Big(\int_0^{\rho(\gamma\mathfrak{z},\sigma\mathfrak{u})}\varsigma(\kappa)d\kappa\Big)^2 \leq \mathcal{L}\psi\Big(\Big(\int_0^{\rho(\alpha\mathfrak{z},\gamma\mathfrak{z})}\varsigma(\kappa)d\kappa\Big)^2\cdot\int_0^{\rho(\beta\mathfrak{u},\sigma\mathfrak{u})}\varsigma(\kappa)d\kappa,\\ \int_0^{\rho(\alpha\mathfrak{z},\gamma\mathfrak{z})}\varsigma(\kappa)d\kappa\cdot\Big(\int_0^{\rho(\beta\mathfrak{u},\sigma\mathfrak{u})}\varsigma(\kappa)d\kappa\Big)^2,\\ \int_0^{\rho(\alpha\mathfrak{z},\gamma\mathfrak{z})}\varsigma(\kappa)d\kappa\cdot\int_0^{\rho(\alpha\mathfrak{z},\sigma\mathfrak{u})}\varsigma(\kappa)d\kappa\cdot\int_0^{\rho(\beta\mathfrak{u},\gamma\mathfrak{z})}\varsigma(\kappa)d\kappa,\\ \int_0^{\rho(\alpha\mathfrak{z},\sigma\mathfrak{u})}\varsigma(\kappa)d\kappa\cdot\int_0^{\rho(\beta\mathfrak{u},\gamma\mathfrak{z})}\varsigma(\kappa)d\kappa\cdot\int_0^{\rho(\beta\mathfrak{u},\sigma\mathfrak{u})}\varsigma(\kappa)d\kappa\Big)\\ +\mathfrak{m}(\alpha\mathfrak{z},\beta\mathfrak{u})-\phi(\mathfrak{m}(\alpha\mathfrak{z},\beta\mathfrak{u})),\end{aligned} \tag{12.4}$$

where

$$\begin{aligned}\mathfrak{m}(\alpha\mathfrak{z},\beta\mathfrak{u}) = \max\Big\{\Big(\int_0^{\rho(\alpha\mathfrak{z},\beta\mathfrak{u})}\varsigma(\kappa)d\kappa\Big)^2, \int_0^{\rho(\alpha\mathfrak{z},\gamma\mathfrak{z})}\varsigma(\kappa)d\kappa\cdot\int_0^{\rho(\beta\mathfrak{u},\sigma\mathfrak{u})}\varsigma(\kappa)d\kappa,\\ \int_0^{\rho(\alpha\mathfrak{z},\sigma\mathfrak{u})}\varsigma(\kappa)d\kappa\cdot\int_0^{\rho(\beta\mathfrak{u},\gamma\mathfrak{z})}\varsigma(\kappa)d\kappa,\\ \frac{1}{2}\Big[\int_0^{\rho(\alpha\mathfrak{z},\gamma\mathfrak{z})}\varsigma(\kappa)d\kappa\cdot\int_0^{\rho(\alpha\mathfrak{z},\sigma\mathfrak{u})}\varsigma(\kappa)d\kappa\\ +\int_0^{\rho(\beta\mathfrak{u},\gamma\mathfrak{z})}\varsigma(\kappa)d\kappa\cdot\int_0^{\rho(\beta\mathfrak{u},\sigma\mathfrak{u})}\varsigma(\kappa)d\kappa\Big]\Big\},\end{aligned} \tag{12.5}$$

*and $\varsigma \in \mathbb{L}$. Suppose (α, γ) and (β, σ) are either **Comp** and **SSC** or **Sub-Comp** and **RC**. Then α, β, γ, and σ have a unique **CFP** in $\mathcal{M}$.*

Proof Case (i) Consider (α, γ) is **SSC** and **Comp**, then $\exists$ a sequence $\{\mathfrak{z}_j\} \in \mathcal{Y}$ s.t. $\rho(\alpha\varkappa, \gamma\varkappa) = \lim_{j\to\infty} \rho(\alpha\gamma\mathfrak{z}_j, \gamma\alpha\mathfrak{z}_j) = 0$, which gives that $\alpha\varkappa = \gamma\varkappa$, i.e., $\varkappa \in \mathfrak{C}(\alpha, \gamma)$. Also, (β, σ) is **SSC** and **Comp**, therefore, $\exists$ a sequence $\{\mathfrak{u}_j\} \in \mathcal{Z}$ s.t. $\rho(\beta\vartheta, \sigma\vartheta) = \lim_{j\to\infty} \rho(\beta\mathcal{T}\mathfrak{u}_j, \sigma\beta\mathfrak{u}_j) = 0$, which gives that $\beta\vartheta = \sigma\vartheta$, i.e., $\vartheta \in \mathfrak{C}(\beta, \sigma)$.

Next, claim that $\varkappa = \vartheta$. Taking $\mathfrak{z} = \mathfrak{z}_j$ and $\mathfrak{u} = \mathfrak{u}_j$ in inequalities (12.4) and (12.5) and letting $j \to \infty$, we have

$$\begin{aligned}\Big[1+\mathcal{L}\int_0^{\rho(\varkappa,\vartheta)} \varsigma(\kappa)d\kappa\Big]\Big(\int_0^{\rho(\varkappa,\vartheta)} \varsigma(\kappa)d\kappa\Big)^2 \le \mathcal{L}\psi\Big(&\Big(\int_0^{\rho(\varkappa,\varkappa)} \varsigma(\kappa)d\kappa\Big)^2 \cdot \int_0^{\rho(\vartheta,\vartheta)} \varsigma(\kappa)d\kappa,\\ &\int_0^{\rho(\varkappa,\varkappa)} \varsigma(\kappa)d\kappa \cdot \Big(\int_0^{\rho(\vartheta,\vartheta)} \varsigma(\kappa)d\kappa\Big)^2,\\ &\int_0^{\rho(\varkappa,\varkappa)} \varsigma(\kappa)d\kappa \cdot \int_0^{\rho(\varkappa,\vartheta)} \varsigma(\kappa)d\kappa \cdot \int_0^{\rho(\vartheta,\varkappa)} \varsigma(\kappa)d\kappa,\\ &\int_0^{\rho(\varkappa,\vartheta)} \varsigma(\kappa)d\kappa \cdot \int_0^{\rho(\vartheta,\varkappa)} \varsigma(\kappa)d\kappa \cdot \int_0^{\rho(\vartheta,\vartheta)} \varsigma(\kappa)d\kappa\Big)\\ &+ \mathfrak{m}(\varkappa, \vartheta) - \phi(\mathfrak{m}(\varkappa, \vartheta)),\end{aligned}$$

where

$$\begin{aligned}\mathfrak{m}(\varkappa, \vartheta) = \max\Big\{&\Big(\int_0^{\rho(\varkappa,\vartheta)} \varsigma(\kappa)d\kappa\Big)^2, \int_0^{\rho(\varkappa,\varkappa)} \varsigma(\kappa)d\kappa \cdot \int_0^{\rho(\vartheta,\vartheta)} \varsigma(\kappa)d\kappa,\\ &\int_0^{\rho(\varkappa,\vartheta)} \varsigma(\kappa)d\kappa \cdot \int_0^{\rho(\vartheta,\varkappa)} \varsigma(\kappa)d\kappa, \frac{1}{2}\Big[\int_0^{\rho(\varkappa,\varkappa)} \varsigma(\kappa)d\kappa \cdot \int_0^{\rho(\varkappa,\vartheta)} \varsigma(\kappa)d\kappa\\ &+ \int_0^{\rho(\vartheta,\varkappa)} \varsigma(\kappa)d\kappa \cdot \int_0^{\rho(\vartheta,\vartheta)} \varsigma(\kappa)d\kappa\Big]\Big\} = \Big(\int_0^{\rho(\varkappa,\vartheta)} \varsigma(\kappa)d\kappa\Big)^2.\end{aligned}$$

After simplification, we get

$$\mathcal{L}\Big(\int_0^{\rho(\varkappa,\vartheta)} \varsigma(\kappa)d\kappa\Big)^3 + \phi\Big(\Big(\int_0^{\rho(\varkappa,\vartheta)} \varsigma(\kappa)d\kappa\Big)^2\Big) \le 0,$$

which holds only for $\rho(\varkappa, \vartheta) = 0$, i.e., $\varkappa = \vartheta$. Next, suppose $\gamma\varkappa \neq \varkappa$. Taking $\mathfrak{z} = \varkappa$ and $\mathfrak{u} = \mathfrak{u}_j$ in inequalities (12.4) and (12.5) and letting $j \to \infty$, we have

$$\Big[1+\mathcal{L}\int_0^{\rho(\gamma\varkappa,\varkappa)}\varsigma(\kappa)d\kappa\Big]\Big(\int_0^{\rho(\gamma\varkappa,\varkappa)}\varsigma(\kappa)d\kappa\Big)^2 \le \mathcal{L}\psi\Big(\Big(\int_0^{\rho(\gamma\varkappa,\gamma\varkappa)}\varsigma(\kappa)d\kappa\Big)^2\cdot\int_0^{\rho(\varkappa,\varkappa)}\varsigma(\kappa)d\kappa,$$
$$\int_0^{\rho(\gamma\varkappa,\gamma\varkappa)}\varsigma(\kappa)d\kappa\cdot\Big(\int_0^{\rho(\varkappa,\varkappa)}\varsigma(\kappa)d\kappa\Big)^2,$$
$$\int_0^{\rho(\gamma\varkappa,\gamma\varkappa)}\varsigma(\kappa)d\kappa\cdot\int_0^{\rho(\gamma\varkappa,\varkappa)}\varsigma(\kappa)d\kappa\cdot\int_0^{\rho(\varkappa,\gamma\varkappa)}\varsigma(\kappa)d\kappa,$$
$$\int_0^{\rho(\gamma\varkappa,\varkappa)}\varsigma(\kappa)d\kappa\cdot\int_0^{\rho(\varkappa,\gamma\varkappa)}\varsigma(\kappa)d\kappa\cdot\int_0^{\rho(\varkappa,\varkappa)}\varsigma(\kappa)d\kappa\Big)$$
$$+\,\mathfrak{m}(\gamma\varkappa,\varkappa)-\phi(\mathfrak{m}(\gamma\varkappa,\varkappa)),$$

where

$$\mathfrak{m}(\gamma\varkappa,\varkappa)=\max\Big\{\Big(\int_0^{\rho(\gamma\varkappa,\varkappa)}\varsigma(\kappa)d\kappa\Big)^2,\int_0^{\rho(\gamma\varkappa,\gamma\varkappa)}\varsigma(\kappa)d\kappa\cdot\int_0^{\rho(\varkappa,\varkappa)}\varsigma(\kappa)d\kappa,$$
$$\int_0^{\rho(\gamma\varkappa,\varkappa)}\varsigma(\kappa)d\kappa\cdot\int_0^{\rho(\varkappa,\gamma\varkappa)}\varsigma(\kappa)d\kappa,$$
$$\frac{1}{2}\Big[\int_0^{\rho(\gamma\varkappa,\gamma\varkappa)}\varsigma(\kappa)d\kappa\cdot\int_0^{\rho(\gamma\varkappa,\varkappa)}\varsigma(\kappa)d\kappa+\int_0^{\rho(\varkappa,\gamma\varkappa)}\varsigma(\kappa)d\kappa\cdot\int_0^{\rho(\varkappa,\varkappa)}\varsigma(\kappa)d\kappa\Big]\Big\}$$
$$=\Big(\int_0^{\rho(\gamma\varkappa,\varkappa)}\varsigma(\kappa)d\kappa\Big)^2.$$

On solving, we conclude that

$$\mathcal{L}\Big(\int_0^{\rho(\gamma\varkappa,\varkappa)}\varsigma(\kappa)d\kappa\Big)^3+\phi\Big(\Big(\int_0^{\rho(\gamma\varkappa,\varkappa)}\varsigma(\kappa)d\kappa\Big)^2\Big)\le 0,$$

which holds only if $\rho(\gamma\varkappa,\varkappa)=0$, i.e., $\gamma\varkappa=\varkappa$, a contradiction. Hence, $\gamma\varkappa=\varkappa$. Next, taking $\mathfrak{z}=\mathfrak{u}=\varkappa$ in inequalities (12.4) and (12.5) and solving, we get

$$\mathcal{L}\Big(\int_0^{\rho(\varkappa,\beta\varkappa)}\varsigma(\kappa)d\kappa\Big)^3+\phi\Big(\Big(\int_0^{\rho(\varkappa,\beta\varkappa)}\varsigma(\kappa)d\kappa\Big)^2\Big)\le 0,$$

which is true only for $\rho(\varkappa,\beta\varkappa)=0$, i.e., $\beta\varkappa=\varkappa$. Hence, $\alpha\varkappa=\gamma\varkappa=\beta\varkappa=\sigma\varkappa=\varkappa$, i.e., $\varkappa$ is a **CFP** of α, γ, β and σ. The uniqueness follows easily.

Case (ii) Suppose (α,γ) is **Sub-Comp**, then $\exists$ a sequence $\{\mathfrak{z}_j\}\in\mathcal{Y}$ s.t. $\lim_{j\to\infty}\alpha\mathfrak{z}_j=\lim_{j\to\infty}\gamma\mathfrak{z}_j=w$, for some $w\in\mathcal{M}$ and $\lim_{j\to\infty}\rho(\alpha\gamma\mathfrak{z}_j,\gamma\alpha\mathfrak{z}_j)=0$. Also, reciprocal continuity of $(\alpha,\gamma)\implies \mathcal{A}w=\gamma w$, i.e., $w\in\mathfrak{C}(\alpha,\gamma)$.

Also, (β,σ) is **Sub-Comp**, $\exists$ a sequence $\{\mathfrak{u}_j\}\in\mathcal{Z}$ s.t. $\lim_{j\to\infty}\beta\mathfrak{u}_j=\lim_{j\to\infty}\sigma\mathfrak{u}_j=\varkappa$, for some $\varkappa\in\mathcal{M}$ and $\lim_{j\to\infty}\rho(\beta\sigma\mathfrak{u}_j,\sigma\beta\mathfrak{u}_j)=0$. Reciprocal continuity of $(\beta,\sigma)\implies\beta\varkappa=\sigma\varkappa$, i.e., $\varkappa\in\mathfrak{C}(\beta,\sigma)$. The rest of the proof follows from case (i). □

Setting $\varsigma(\kappa) = 1$ in Theorem 12.2.1, we have the following result.

Corollary 12.2.2 *Let α, β, γ, and σ be self-mappings of an **MS** $(\mathcal{M}, \rho)$ s.t. $\forall$ $\mathfrak{z}, \mathfrak{u} \in \mathcal{M}$, $\exists$ a function $\phi \in \Omega$, a function $\psi \in \Xi$ and a real number $\mathcal{L} \geq 0$ s.t.*

$$\begin{aligned}\rho^2(\gamma\mathfrak{z}, \sigma\mathfrak{u}) \leq \mathcal{L}\,\psi\Big(&\rho^2(\alpha\mathfrak{z}, \gamma\mathfrak{z})\rho(\beta\mathfrak{u}, \sigma\mathfrak{u}), \rho(\alpha\mathfrak{z}, \gamma\mathfrak{z})\rho^2(\beta\mathfrak{u}, \sigma\mathfrak{u}),\\ &\rho(\alpha\mathfrak{z}, \gamma\mathfrak{z})\rho(\alpha\mathfrak{z}, \sigma\mathfrak{u})\rho(\beta\mathfrak{u}, \gamma\mathfrak{z}),\\ &\rho(\alpha\mathfrak{z}, \sigma\mathfrak{u})\rho(\beta\mathfrak{u}, \gamma\mathfrak{z})\rho(\beta\mathfrak{u}, \sigma\mathfrak{u})\Big)\\ &+ \mathfrak{m}(\alpha\mathfrak{z}, \beta\mathfrak{u}) - \phi(\mathfrak{m}(\alpha\mathfrak{z}, \beta\mathfrak{u})),\end{aligned} \tag{12.6}$$

where

$$\begin{aligned}\mathfrak{m}(\alpha\mathfrak{z}, \beta\mathfrak{u}) = \max\Big\{&\rho^2(\alpha\mathfrak{z}, \beta\mathfrak{u}), \rho(\alpha\mathfrak{z}, \gamma\mathfrak{z})\rho(\beta\mathfrak{u}, \sigma\mathfrak{u}), \rho(\alpha\mathfrak{z}, \sigma\mathfrak{u})\rho(\beta\mathfrak{u}, \gamma\mathfrak{z}),\\ &\frac{1}{2}[\rho(\alpha\mathfrak{z}, \gamma\mathfrak{z})\rho(\alpha\mathfrak{z}, \sigma\mathfrak{u}) + \rho(\beta\mathfrak{u}, \gamma\mathfrak{z})\rho(\beta\mathfrak{u}, \sigma\mathfrak{u})]\Big\}.\end{aligned} \tag{12.7}$$

*Suppose (α, γ) and (β, σ) are either **Comp** and **SSC** or **Sub-Comp** and **RC**. Then α, β, γ, and σ have a unique **CFP** in $\mathcal{M}$.*

On substituting $\alpha = \beta$ and $\gamma = \sigma$ in Theorem 12.2.1 and Corollary 12.2.2, one can deduce the following results:

Corollary 12.2.3 *Let α and γ be self-mappings of an **MS** $(\mathcal{M}, \rho)$ s.t. $\forall$ $\mathfrak{z}, \mathfrak{u} \in \mathcal{M}$, $\exists$ a function $\phi \in \Omega$, a function $\psi \in \Xi$ and a real number $\mathcal{L} \geq 0$ s.t.*

$$\begin{aligned}\Big[1+\mathcal{L}\int_0^{\rho(\alpha\mathfrak{z},\alpha\mathfrak{u})} \varsigma(\kappa)d\kappa\Big]\Big(\int_0^{\rho(\gamma\mathfrak{z},\gamma\mathfrak{u})} \varsigma(\kappa)d\kappa\Big)^2 \leq \mathcal{L}\psi\Big(&\Big(\int_0^{\rho(\alpha\mathfrak{z},\gamma\mathfrak{z})} \varsigma(\kappa)d\kappa\Big)^2 \cdot \int_0^{\rho(\alpha\mathfrak{u},\gamma\mathfrak{u})} \varsigma(\kappa)d\kappa,\\ &\int_0^{\rho(\alpha\mathfrak{z},\gamma\mathfrak{z})} \varsigma(\kappa)d\kappa \cdot \Big(\int_0^{\rho(\alpha\mathfrak{u},\gamma\mathfrak{u})} \varsigma(\kappa)d\kappa\Big)^2,\\ &\int_0^{\rho(\alpha\mathfrak{z},\gamma\mathfrak{z})} \varsigma(\kappa)d\kappa \cdot \int_0^{\rho(\alpha\mathfrak{z},\gamma\mathfrak{u})} \varsigma(\kappa)d\kappa \cdot \int_0^{\rho(\alpha\mathfrak{u},\gamma\mathfrak{z})} \varsigma(\kappa)d\kappa,\\ &\int_0^{\rho(\alpha\mathfrak{z},\gamma\mathfrak{u})} \varsigma(\kappa)d\kappa \cdot \int_0^{\rho(\alpha\mathfrak{u},\gamma\mathfrak{z})} \varsigma(\kappa)d\kappa \cdot \int_0^{\rho(\alpha\mathfrak{u},\gamma\mathfrak{u})} \varsigma(\kappa)d\kappa\Big)\\ &+ \mathfrak{m}(\alpha\mathfrak{z}, \alpha\mathfrak{u}) - \phi(\mathfrak{m}(\alpha\mathfrak{z}, \alpha\mathfrak{u})),\end{aligned} \tag{12.8}$$

where

$$\mathfrak{m}(\alpha\mathfrak{z}, \alpha\mathfrak{u}) = \max\Big\{\Big(\int_0^{\rho(\alpha\mathfrak{z},\alpha\mathfrak{u})} \varsigma(\kappa)d\kappa\Big)^2, \int_0^{\rho(\alpha\mathfrak{z},\gamma\mathfrak{z})} \varsigma(\kappa)d\kappa \cdot \int_0^{\rho(\alpha\mathfrak{u},\gamma\mathfrak{u})} \varsigma(\kappa)d\kappa,$$
$$\int_0^{\rho(\alpha\mathfrak{z},\gamma\mathfrak{u})} \varsigma(\kappa)d\kappa \cdot \int_0^{\rho(\alpha\mathfrak{u},\gamma\mathfrak{z})} \varsigma(\kappa)d\kappa,$$
$$\frac{1}{2}[\int_0^{\rho(\alpha\mathfrak{z},\gamma\mathfrak{z})} \varsigma(\kappa)d\kappa \cdot \int_0^{\rho(\alpha\mathfrak{z},\gamma\mathfrak{u})} \varsigma(\kappa)d\kappa$$
$$+ \int_0^{\rho(\alpha\mathfrak{u},\gamma\mathfrak{z})} \varsigma(\kappa)d\kappa \cdot \int_0^{\rho(\alpha\mathfrak{u},\gamma\mathfrak{u})} \varsigma(\kappa)d\kappa]\Big\}, \tag{12.9}$$

and $\varsigma \in \mathbb{L}$. Then α and γ have a unique ***CFP*** *in $\mathcal{M}$, if (α, γ) is either* ***Comp*** *and subsequential or* ***Sub-Comp*** *continuous and* ***RC****.*

Corollary 12.2.4 *Let α and γ be self-mappings of an* ***MS*** *$(\mathcal{M}, \rho)$ s.t. $\forall\ \mathfrak{z}, \mathfrak{u} \in \mathcal{M}$, $\exists$ a function $\phi \in \Omega$, a function $\psi \in \Xi$ and a real number $\mathcal{L} \geq 0$ s.t.*

$$[1 + \mathcal{L}\,\rho(\alpha\mathfrak{z}, \alpha\mathfrak{u})]\rho^2(\gamma\mathfrak{z}, \gamma\mathfrak{u}) \leq \mathfrak{u}\mathcal{L}\,\psi\Big(\rho^2(\alpha\mathfrak{z}, \gamma\mathfrak{z})\rho(\alpha\mathfrak{u}, \gamma\mathfrak{u}), \rho(\alpha\mathfrak{z}, \gamma\mathfrak{z})\rho^2(\alpha\mathfrak{u}, \gamma\mathfrak{u}),$$
$$\rho(\alpha\mathfrak{z}, \gamma\mathfrak{z})\rho(\alpha\mathfrak{z}, \gamma\mathfrak{u})\rho(\alpha\mathfrak{u}, \gamma\mathfrak{z}), \rho(\alpha\mathfrak{z}, \gamma\mathfrak{u})\rho(\alpha\mathfrak{u}, \gamma\mathfrak{z})\rho(\alpha\mathfrak{u}, \gamma\mathfrak{u})\Big)$$
$$+ \mathfrak{m}(\alpha\mathfrak{z}, \alpha\mathfrak{u}) - \phi(\mathfrak{m}(\alpha\mathfrak{z}, \alpha\mathfrak{u})),$$

where

$$\mathfrak{m}(\alpha\mathfrak{z}, \alpha\mathfrak{u}) = \max\Big\{\rho^2(\alpha\mathfrak{z}, \alpha\mathfrak{u}), \rho(\alpha\mathfrak{z}, \gamma\mathfrak{z})\rho(\alpha\mathfrak{u}, \gamma\mathfrak{u}), \rho(\alpha\mathfrak{z}, \gamma\mathfrak{u})\rho(\alpha\mathfrak{u}, \gamma\mathfrak{z}),$$
$$\frac{1}{2}[\rho(\alpha\mathfrak{z}, \gamma\mathfrak{z})\rho(\alpha\mathfrak{z}, \gamma\mathfrak{u}) + \rho(\alpha\mathfrak{u}, \gamma\mathfrak{z})\rho(\alpha\mathfrak{u}, \gamma\mathfrak{u})]\Big\}.$$

Then α and γ have a unique ***CFP*** *in $\mathcal{M}$, if (α, γ) is either* ***Comp*** *and* ***SSC*** *or* ***Sub-Comp*** *and* ***RC****.*

Example 12.2.5 Let $(\mathcal{M}, \rho)$ be a usual **MS**, where $\mathcal{M} = [0, 10)$. Define $\alpha, \beta, \gamma, \sigma : \mathcal{M} \to \mathcal{M}$ by $\alpha\mathfrak{z} = 0, \gamma\mathfrak{z} = 0, \mathfrak{z} = 0, \alpha\mathfrak{z} = 8, \gamma\mathfrak{z} = 4, \mathfrak{z} \in (0, 2.5], \alpha\mathfrak{z} = \mathfrak{z} - 2.5, \gamma\mathfrak{z} = 0, \mathfrak{z} \in (2.5, 10)$, $\beta\mathfrak{z} = 0, \sigma\mathfrak{z} = 0, \mathfrak{z} = 0, \beta\mathfrak{z} = 4, \sigma\mathfrak{z} = 2, \mathfrak{z} \in (0, 10]$. Also, define $\phi : \mathbb{R}^+ \to \mathbb{R}^+$ by $\phi(\kappa) = \frac{3\kappa}{2}$, for $\kappa \geq 0$ and $\psi : \mathbb{R}^{+4} \to \mathbb{R}^+$ by $\psi(\mathfrak{s}_1, \mathfrak{s}_2, \mathfrak{s}_3, \mathfrak{s}_4) = \max\{\mathfrak{s}_1, \mathfrak{s}_2, \mathfrak{s}_3, \mathfrak{s}_4\}$, $\mathfrak{s}_i \geq 0, i = 1, 2, 3, 4$. Consider a sequence $\{\mathfrak{z}_j\} = \{2.5 + \frac{1}{j}\}$ in $\mathcal{M}$. Then $\alpha\mathfrak{z}_j \to 0$ and $\gamma\mathfrak{z}_j \to 0$ as $j \to \infty$, but $\lim_{j\to\infty} \rho(\alpha\gamma\mathfrak{z}_j, \gamma\alpha\mathfrak{z}_j) = 0$. Take another sequence $\{\mathfrak{z}_j\} = \{0\}$, then (α, γ) and (β, σ) are **Sub-Comp** and **RC**. Also, all the conditions of Corollary 12.2.2 hold and 0 is the unique **CFP** of α, β, γ and σ.

12.3 Application

Let $\mathtt{U}$ and $\mathtt{D}$ denote "Banach spaces," $\mathtt{W} \subset \mathtt{U}$, $\mathtt{P} \subset \mathtt{D}$ are "state space and decision space," respectively. Let $C(\mathtt{W}) = \{\lambda : \mathtt{W} \to \mathbb{R}, \lambda is continuous\}$. Let $\rho(\lambda, \mu) = \sup\{|\lambda(\nu) - \mu(\nu)| : \nu \in \mathtt{W}\}$ for any $\lambda, \mu \in C(\mathtt{W})$. Obviously, $(C(\mathtt{W}), \rho)$ is a **CMS**.

A basic form of functional equation (**FE**) was given by Bellman and Lee (1978) as follows:

$$\mathcal{N}(\nu) = \underset{\upsilon}{opt}\ \mathcal{E}(\nu, \omega, \mathcal{N}(\delta(\nu, \omega))),$$

where $\nu \in \mathtt{W}$, $\omega \in \mathtt{P}$, δ is the transformation process, the opt denotes max or min and $\mathcal{N}(\nu)$ is the optimized value starting with ν.

Consider the following **FE**s arising in dynamic programming (DP) (see Baskaran and Subrahmanayam (1986), Bellman and Lee (1978)):

$$\mathcal{H}_i(\nu) = \sup_{\omega \in \mathtt{P}} \mathcal{F}_i(\nu, \omega, \mathcal{H}_i(\delta(\nu, \omega))), \nu \in \mathtt{W} \tag{12.10}$$

$$\mathcal{N}_i(\nu) = \sup_{\omega \in \mathtt{P}} \mathcal{E}_i(\nu, \omega, \mathcal{N}_i(\delta(\nu, \omega))), \nu \in \mathtt{W}, \tag{12.11}$$

where $\delta : \mathtt{W} \times \mathtt{P} \to \mathtt{W}$ and $\mathcal{J}_i, \mathcal{E}_i : \mathtt{W} \times \mathtt{P} \times \mathbb{R} \to \mathbb{R}$, $i = 1, 2$.

Theorem 12.3.1 *Let* $\mathcal{J}_i, \mathcal{E}_i : \mathtt{W} \times \mathtt{P} \times \mathbb{R} \to \mathbb{R} (i = 1, 2)$ *be bounded. Define the mappings* $\mathcal{A}_i, \mathcal{R}_i : C(\mathtt{W}) \to C(\mathtt{W})$*, as follows:*

$$\begin{aligned} \mathcal{A}_i\lambda(\nu) &= \sup_{\omega \in \mathtt{P}} \mathcal{J}_i(\nu, \omega, \lambda(\delta(\nu, \omega))), \\ \mathcal{R}_i\mu(\nu) &= \sup_{\omega \in \mathtt{P}} \mathcal{E}_i(\nu, \omega, \mu(\delta(\nu, \omega))), \end{aligned} \tag{12.12}$$

$\forall\, \nu \in \mathtt{W}$, $\lambda, \mu \in C(\mathtt{W})$, $i = 1, 2$. *Also, the following conditions hold:*

(a) $\forall\, \nu, t \in \mathtt{W}, \omega \in \mathtt{P}, \lambda, \mu \in C(\mathtt{W})$,

$$\begin{aligned} \left|\mathcal{J}_1(\nu, \omega, \lambda(t)) - \mathcal{J}_2(\nu, \omega, \mu(t))\right|^2 \le \mathtt{Q}^{-1}\Big(&\mathcal{L}\,\psi\big(\rho^2(\mathcal{R}_1\lambda, \mathcal{A}_1\lambda)\rho(\mathcal{R}_2\mu, \mathcal{A}_2\mu), \\ &\rho(\mathcal{R}_1\lambda, \mathcal{A}_1\lambda)\rho^2(\mathcal{R}_2\mu, \mathcal{A}_2\mu), \\ &\rho(\mathcal{R}_1\lambda, \mathcal{A}_1\lambda)\rho(\mathcal{R}_1\lambda, \mathcal{A}_2\mu)\rho(\mathcal{R}_2\mu, \mathcal{A}_1\lambda), \\ &\rho(\mathcal{R}_1\lambda, \mathcal{A}_2\mu)\rho(\mathcal{R}_2\mu, \mathcal{A}_1\lambda)\rho(\mathcal{R}_2\mu, \mathcal{A}_2\mu)\big) + \\ &\mathfrak{m}(\mathcal{R}_1\lambda, \mathcal{R}_2\mu) - \phi(\mathfrak{m}(\mathcal{R}_1\lambda, \mathcal{R}_2\mu))\Big), \end{aligned}$$

where

$$\mathrm{m}(\mathcal{R}_1\lambda, \mathcal{R}_2\mu) = \max\Big\{\rho^2(\mathcal{R}_1\lambda, \mathcal{R}_2\mu), \rho(\mathcal{R}_1\lambda, \mathcal{A}_1\lambda)\rho(\mathcal{R}_2\mu, \mathcal{A}_2\mu), \rho(\mathcal{R}_1\lambda, \mathcal{A}_2\mu)\rho(\mathcal{R}_2\mu, \mathcal{A}_1\lambda)$$
$$12[\rho(\mathcal{R}_1\lambda, \mathcal{A}_1\lambda)\rho(\mathcal{R}_1\lambda, \mathcal{A}_2\mu) + \rho(\mathcal{R}_2\mu, \mathcal{A}_1\lambda)\rho(\mathcal{R}_2\mu, \mathcal{A}_2\mu)]\Big\},$$

$\mathtt{Q} = [1 + \mathcal{L}\rho(\mathcal{R}_1\lambda, \mathcal{R}_2\mu)]$, $\phi \in \Omega$, $\psi \in \Xi$, $\mathcal{L}$ *is a positive real number;*

(b) $\exists$ *a sequence* $\{\lambda_j\}$ *of* $C(\mathtt{W})$ *s.t.* $\lim_{j\to\infty} \mathcal{R}_1\lambda_j = \lim_{j\to\infty} \mathcal{A}_1\lambda_j = \lambda$, *for some* $\lambda \in C(\mathtt{W})$ *which satisfies*

$$\lim_{j\to\infty} \sup_{\nu\in\mathtt{W}} |\mathcal{R}_1\mathcal{A}_1\lambda_j(\nu) - \mathcal{A}_1\mathcal{R}_1\lambda_j(\nu)| = 0;$$

(c) $\exists$ *a sequence* $\{\mu_j\}$ *of* $C(\mathtt{W})$ *s.t.* $\lim_{j\to\infty} \mathcal{R}_2\mu_j = \lim_{j\to\infty} \mathcal{A}_2\mu_j = \mu$, *for some* $\mu \in C(\mathtt{W})$ *which implies that*

$$\lim_{j\to\infty} \sup_{\nu\in\mathtt{W}} |\mathcal{R}_2\mathcal{A}_2\mu_j(\nu) - \mathcal{A}_2\mathcal{R}_2\mu_j(\nu)| = 0.$$

Then the system of **FE***s (12.10) and (12.11) has a unique common solution in* $C(\mathtt{W})$.

Proof As $\mathcal{J}_i, \mathcal{E}_i (i = 1, 2)$ are continuous and the mappings defined by (12.12) are continuous and so **RC**. By conditions b and c, $(\mathcal{A}_1, \mathcal{R}_1)$ and $(\mathcal{A}_2, \mathcal{R}_2)$ are **Sub-Comp**.
For $\eta > 0$, $\nu \in \mathtt{W}$ and $\mu_1, \mu_2 \in C(\mathtt{W})$, $\exists\, \omega_1, \omega_2 \in \mathtt{P}$ s.t.

$$\mathcal{A}_i\mu_i(\nu) < \mathcal{J}_i(\nu, \omega_i, \mu_i(\nu_i)) + \eta, \tag{12.13}$$

where $\nu_i = \delta(\nu, \omega_i)$, $i = 1, 2$. Also, we have

$$\mathcal{A}_1\mu_1(\nu) \geq \mathcal{J}_1(\nu, \omega_2, \mu_1(\nu_2)), \tag{12.14}$$
$$\mathcal{A}_2\mu_2(\nu) \geq \mathcal{J}_2(\nu, \omega_1, \mu_2(\nu_1)). \tag{12.15}$$

From inequalities (12.13), (12.15) and condition a, we have

$$\begin{aligned}(\mathcal{A}_1\mu_1(\nu) - \mathcal{A}_2\mu_2(\nu))^2 &< (\mathcal{J}_1(\nu, \omega_1, \mu_1(\nu_1)) - \mathcal{J}_2(\nu, \omega_1, \mu_2(\nu_1)) + \eta)^2 \\ &= (\mathcal{J}_1(\nu, \omega_1, \mu_1(\nu_1)) - \mathcal{J}_2(\nu, \omega_1, \mu_2(\nu_1)))^2 + \xi, \\ &\leq \mathcal{M}^{-1}\Big(\mathcal{L}\psi\big(\rho^2(\mathcal{R}_1\mu_1, \mathcal{A}_1\mu_1)\rho(\mathcal{R}_2\mu_2, \mathcal{A}_2\mu_2), \\ &\rho(\mathcal{R}_1\mu_1, \mathcal{A}_1\mu_1)\rho^2(\mathcal{R}_2\mu_2, \mathcal{A}_2\mu_2), \\ &\rho(\mathcal{R}_1\mu_1, \mathcal{A}_1\mu_1)\rho(\mathcal{R}_1\mu_1, \mathcal{A}_2\mu_2)\rho(\mathcal{R}_2\mu_2, \mathcal{A}_1\mu_1), \\ &\rho(\mathcal{R}_1\mu_1, \mathcal{A}_2\mu_2)\rho(\mathcal{R}_2\mu_2, \mathcal{A}_1\mu_1)\rho(\mathcal{R}_2\mu_2, \mathcal{A}_2\mu_2)\big) + \\ &\mathrm{m}(\mathcal{R}_1\mu_1, \mathcal{R}_2\mu_2) - \phi(\mathrm{m}(\mathcal{R}_1\mu_1, \mathcal{R}_2\mu_2))\Big) + \xi,\end{aligned} \tag{12.16}$$

where $\xi = \eta^2 + 2\eta(\mathcal{J}_1 - \mathcal{J}2)$.

From inequalities (12.13), (12.14) and condition a, we have

$$\begin{aligned}(\mathcal{A}_1\mu_1(\nu)-\mathcal{A}_2\mu_2(\nu))^2 &> (\mathcal{J}_1(\nu,\omega_2,\mu_1(\nu_2))-\mathcal{J}_2(\nu,\omega_2,\mu_2(\nu_2))-\eta)^2\\ &= (\mathcal{J}_1(\nu,\omega_1,\mu_1(\nu_1))-\mathcal{J}_2(\nu,\omega_1,\mu_2(\nu_1)))^2+\xi_1,\\ &\geq -\mathtt{Q}^{-1}\Big(\mathcal{L}\psi(\rho^2(\mathcal{R}_1\mu_1,\mathcal{A}_1\mu_1)\rho(\mathcal{R}_2\mu_2,\mathcal{A}_2\mu_2),\\ &\rho(\mathcal{R}_1\mu_1,\mathcal{A}_1\mu_1)\rho^2(\mathcal{R}_2\mu_2,\mathcal{A}_2\mu_2),\\ &\rho(\mathcal{R}_1\mu_1,\mathcal{A}_1\mu_1)\rho(\mathcal{R}_1\mu_1,\mathcal{A}_2\mu_2)\rho(\mathcal{R}_2\mu_2,\mathcal{A}_1\mu_1),\\ &\rho(\mathcal{R}_1\mu_1,\mathcal{A}_2\mu_2)\rho(\mathcal{R}_2\mu_2,\mathcal{A}_1\mu_1)\rho(\mathcal{R}_2\mu_2,\mathcal{A}_2\mu_2))+\\ &\mathtt{m}(\mathcal{R}_1\mu_1,\mathcal{R}_2\mu_2)-\phi(\mathtt{m}(\mathcal{R}_1\mu_1,\mathcal{R}_2\mu_2))\Big)-\xi,\end{aligned} \tag{12.17}$$

where $\xi_1=\eta^2-2\eta(\mathcal{J}_1-\mathcal{J}2)<\xi$.

From inequalities (12.16) and (12.17), we obtain

$$\begin{aligned}|\mathcal{A}_1\mu_1(\nu)-\mathcal{A}_2\mu_2(\nu)|^2 &\leq \mathtt{Q}^{-1}\Big(\mathcal{L}\psi(\rho^2(\mathcal{R}_1\mu_1,\mathcal{A}_1\mu_1)\rho(\mathcal{R}_2\mu_2,\mathcal{A}_2\mu_2),\\ &\rho(\mathcal{R}_1\mu_1,\mathcal{A}_1\mu_1)\rho^2(\mathcal{R}_2\mu_2,\mathcal{A}_2\mu_2),\\ &\rho(\mathcal{R}_1\mu_1,\mathcal{A}_1\mu_1)\rho(\mathcal{R}_1\mu_1,\mathcal{A}_2\mu_2)\rho(\mathcal{R}_2\mu_2,\mathcal{A}_1\mu_1),\\ &\rho(\mathcal{R}_1\mu_1,\mathcal{A}_2\mu_2)\rho(\mathcal{R}_2\mu_2,\mathcal{A}_1\mu_1)\rho(\mathcal{R}_2\mu_2,\mathcal{A}_2\mu_2))+\\ &\mathtt{m}(\mathcal{R}_1\mu_1,\mathcal{R}_2\mu_2)-\phi(\mathtt{m}(\mathcal{R}_1\mu_1,\mathcal{R}_2\mu_2))\Big)+\xi.\end{aligned} \tag{12.18}$$

As $\eta>0$ is arbitrary, so ξ is negligible and inequality (12.18) is true for all $\nu\in\mathtt{W}$, taking supremum, we get

$$\begin{aligned}[1+\mathcal{L}\rho(\mathcal{R}_1\mu_1,\mathcal{R}_2\mu_2)]\rho^2(\mathcal{A}_1\mu_1,\mathcal{A}_2\mu_2) &\leq \mathcal{L}\psi(\rho^2(\mathcal{R}_1\mu_1,\mathcal{A}_1\mu_1)\rho(\mathcal{R}_2\mu_2,\mathcal{A}_2\mu_2),\\ &\rho(\mathcal{R}_1\mu_1,\mathcal{A}_1\mu_1)\rho^2(\mathcal{R}_2\mu_2,\mathcal{A}_2\mu_2),\\ &\rho(\mathcal{R}_1\mu_1,\mathcal{A}_1\mu_1)\rho(\mathcal{R}_1\mu_1,\mathcal{A}_2\mu_2)\rho(\mathcal{R}_2\mu_2,\mathcal{A}_1\mu_1),\\ &\rho(\mathcal{R}_1\mu_1,\mathcal{A}_2\mu_2)\rho(\mathcal{R}_2\mu_2,\mathcal{A}_1\mu_1)\rho(\mathcal{R}_2\mu_2,\mathcal{A}_2\mu_2))+\\ &\mathtt{m}(\mathcal{R}_1\mu_1,\mathcal{R}_2\mu_2)-\phi(\mathtt{m}(\mathcal{R}_1\mu_1,\mathcal{R}_2\mu_2)).\end{aligned}$$

All the conditions of Corollary 12.2.2 are satisfied. So, $\mathcal{A}_1$, $\mathcal{A}_2$, $\mathcal{R}_1$ and $\mathcal{R}_2$ have a unique **CFP** $\mu^*\in C(\mathtt{W})$, i.e., $\mu^*(\nu)$ is a unique common solution of the system of **FE**s (12.10) and (12.11). □

Remark 12.3.2 Corollary 12.2.1 is also applicable if we replace conditions b and c in Theorem 12.3.1 with the following conditions, respectively:

(d) For any sequence $\{\mu_j\}$ of $C(\mathtt{W})$ s.t. $\lim_{j\to\infty} \mathcal{R}_1\mu_j = \lim_{j\to\infty} \mathcal{A}_1\mu_j = \mu^*$, for some $\mu^* \in C(\mathtt{W})$ then

$$\lim_{j\to\infty} \sup_{\nu\in\mathtt{W}} |\mathcal{R}_1\mathcal{A}_1\mu_j(\nu) - \mathcal{A}_1\mathcal{R}_1\mu_j(\nu)| = 0.$$

(e) For any sequence $\{\lambda_j\}$ of $C(\mathtt{W})$ s.t. $\lim_{j\to\infty} \mathcal{R}_2\lambda_j = \lim_{j\to\infty} \mathcal{A}_2\lambda_j = \lambda^*$, for some $\lambda^* \in C(\mathtt{W})$ then

$$\lim_{j\to\infty} \sup_{\nu\in\mathtt{W}} |\mathcal{R}_2\mathcal{A}_2\lambda_j(\nu) - \mathcal{A}_2\mathcal{R}_2\lambda_j(\nu)| = 0.$$

The rest of the conditions are as it is in Theorem 12.3.1.

12.4 Conclusion

In this chapter, the existence of a unique **CFP** for pairs of **Sub-Comp** and **RC** mappings (or **Comp** and **SSC** mappings) satisfying a generalized (ψ, ϕ)-weak contractive condition of integral type involving cubic terms of distance functions has been established, improving Theorem 12.1.15 and the results of Branciari (2002) in several aspects. Theorem 12.3.1 illustrates the application of the obtained result in finding a common solution to the system of **FE**s arising in DP.

Acknowledgements The authors are very thankful to the referees for their valuable suggestions.

References

Aamri, M., & Moutawakil, D. E. I. (2002). Some new common fixed point theorems under strict contractive conditions. *Journal of Mathematical Analysis and Applications,270*, 181–188.

Alber, Y. I., & Guerre-Delabriere, S. (1997). Principle of weakly contractive maps in Hilbert spaces. *New Results in Operator Theory Advances and Applications,98*, 7–22.

Al-Thagafi, M. A., & Shahzad, N. (2008). Generalized $I-$nonexpansive self maps and invariant approximations. *ACT Mathematical Sin,24*, 867–876.

Altun, I., Turkoglu, D., & Rhoades, B. E. (2007). Fixed points of weakly compatible maps satisfying a general contractive condition of integral type. *Fixed Point Theory Application*, Art. id 017301, 9 pages.

Banach, S. (1922). Surles opérations dans les ensembles abstraites et leur applications aux équations intégrales. *Fundamenta Mathematicae,3*, 133–181.

Bari, C. D., & Vetro, C. (2008). Common fixed points theorems for weakly compatible maps satisfying a general contractive condition. *International Journal of Mathematics and Mathematical Sciences* 2008: Art. id 891375, 8 pages.

Baskaran, R., & Subrahmanayam, P. V. (1986). A note on the solution of a class of functional equations. *Applied Analysis,22*, 235–241.

Bellman, R., & Lee, B. S. (1978). Functional equations arising in dynamic programming. *Aequationes Mathematicae,17*, 1–18.

Bouhadjera, H., & Godet-Thobie, C. (2011). Common fixed point theorems for pairs of subcompatible maps. arxiv:0906.3159v2/ [math.FA].

Boyd, D. W., & Wong, J. S. W. (1969). On nonlinear contractions. *Proceedings of the American Mathematical Society,20*(2), 458–464.

Branciari, A. (2002). A fixed point theorem for mappings satisfying a general contractive condition of integral type. *International Journal of Mathematics and Mathematical Sciences,29*(9), 531–536.

Chauhan, S., Shatanawi, W., Radenovć, S., & Abu-Irwaq, I. (2014). Variants of sub-sequentially continuous mappings and integral-type fixed point results. *Rendiconti del Circolo Matematico di Palermo,63*, 53–72. https://doi.org/10.1007/s12215-013-0141-7

Djoudi, A., & Aliouche, A. (2007). Common fixed point theorems of Gregus type for weakly compatible mappings satisfying contractive conditions of integral. *(Journal of Mathematical Analysis and Applications,329*(1), 31–45.

Jungck, G. (1976). Commuting mappings and fixed point. *American Mathematical Monthly,83*, 261–263.

Jungck, G. (1986). Compatible mappings and common fixed points. *International Journal of Mathematics and Mathematical Sciences,9*, 771–779.

Jungck, G. (1996). Common fixed points for noncontinuous non-self maps on non-metric spaces. *Far East Journal of Mathematical Sciences,4*(2), 199–215.

Kang, S. M., Cho, Y. J., & Jungck, G. (1990). Common fixed points of compatible mappings. *International Journal of Mathematics and Mathematical Sciences,13*(1), 61–66.

Kavita, & Kumar, S. (2022). Fixed points for Intimate mappings. *Journal of Mathematics and Computer Science, 12*, Art. id 48, 21 pages.

Kim, J. K., & Bisht, R. K. (2014). A critical remark on subsequential continuity. *Nonlinear Functional Analysis and Applications,19*(1), 127–130.

Khan, M. S., Swalek, M., & Sessa, S. (1984). Fixed point theorems by altering distances between two points. *Bulletin of the Australian Mathematical Society,30*, 1–9.

Liu, Y., Wu, J., & Li, Z. (2005). Common fixed point of single valued and multivalued maps. *International Journal of Mathematics and Mathematical Sciences,19*, 3045–3055.

Murthy, P. P., & Prasad, K. N. V. V. V. (2013). Weak contraction condition involving cubic terms of $d(x, y)$ under the fixed point consideration. *Journal of Mathematics and the Arts ID,967045*, 5 pages.

Pant, R. P. (1999). A common fixed point theorem under a new condition. *Indian Journal of Pure and Applied Mathematics,30*(2), 147–152.

Rhoades, B. E. (2001). Some theorems on weakly contractive maps. *Nonlinear Analysis,47*(4), 2683–2693.

Rouzkard, F., Imdad, M., & Nashine, H. K. (2012). New common fixed point theorems in convex metric spaces. *Bulletin of the Belgian Mathematical Society,19*(2), 311–328.

Samet, B., & Vetro, C. (2012). An integral version of Ćirić's fixed point theorem. *Mediterranean Journal of Mathematics,9*, 225–238.

Singh, S. L., & Tomar, A. (2003). Weaker forms of commuting maps and existence of fixed points. *Journal of the Korean Society of Mathematical Education Series B, Pure and Applied Mathematics,10*(03), 145–161.

Tomar, A., & Karapinar, E. (2016). On variants of continuity and existence of fixed point via Meir-Keeler contractions in MC-spaces. *Journal of Advanced Mathematical Studies,9*(2), 348–359.

Vijayaraju, P., Rhoades, B. E., & Mohanraj, R. (2005). A fixed point theorem for a pair of maps satisfying a general contractive condition of integral type. *International Journal of Mathematical Sciences,15*, 2359–2364.

Chapter 13
Advancements in the Banach Contraction Principle Within Metric Spaces

Sanjay Kumar, Kavita, Manish Jain, and Rajesh Kumar

Abstract This chapter aims to study the development of the Banach contraction principle in metric spaces and discusses results for compatible-type mappings that satisfy a contraction condition involving cubic terms of distance functions. Some applications are also presented.

Keywords Fixed point · Compatible mappings · Weakly compatible mappings · $R-$weakly commuting mappings · Property $(E.A)$

AMS Subject Classification 47H10 · 54H25

13.1 Introduction

One of the useful tools of non-linear analysis is fixed point theory (**FPT**), which deals with the existence of solutions and their uniqueness. A self-mapping γ defined on a non-empty set Ξ is said to have a fixed point (**FP**) if $\gamma\kappa = \kappa$ for some $\kappa \in \Xi$. **FPT** is mainly concerned with metric **FPT** and topological **FPT**. Brouwer (1912) proved an **FP** theorem, which states that every continuous self-mapping on a unit ball $\mathbb{R}^n$ has a unique **FP**. The Brouwer's theorem falls under the category of topological **FPT**. In 1972, Kenneth Joseph Arrow, along with John Hicks and in 1994, John Nash won the Nobel Prize in economics by using Brouwer's **FP** Theorem. We restrict our discussion to metric **FPT**.

S. Kumar · Kavita (✉)
Department of Mathematics, Deenbandhu Chhotu Ram University of Science and Technology, Murthal (Sonepat), Haryana, India
e-mail: kvtlather@gmail.com

M. Jain
Department of Mathematics, Ahir College, Rewari, India

R. Kumar
Department of Mathematics, Institute of Higher Learning, BPS Mahila Vishwavidyalaya, Khanpur Kalan (Sonepat), Haryana, India

© The Author(s), under exclusive license to Springer Nature Singapore Pte Ltd. 2026

A. Tomar and M. Jain (eds.), *Applications of Fixed-Point Theorem*, Industrial and Applied Mathematics, https://doi.org/10.1007/978-981-95-5160-6_13

The well-known result in metric **FPT** is the "Banach contraction principle" (**BCP**) Banach (1922), which is based on an iteration process and provides a sufficient condition for the existence and uniqueness of solutions to various mathematical models, such as differential equations, fractional differential equations, partial differential equations, variational inequalities, control theory, dynamic programming (**DP**), and many more phenomena arising in multiple fields across the universe.

Picard (1890) showed the existence of a solution of a non-linear equation.

Let $\gamma : [\mathfrak{a}, \mathfrak{b}] \to \mathbb{R}$ be a continuous and differentiable function. If $\exists$ a positive number $\mathcal{L} < 1$ such that(s.t.)

$$|\gamma'(\kappa)| < \mathcal{L}, \ \forall \kappa \in (\mathfrak{a}, \mathfrak{b}), \tag{13.1}$$

then the sequence $\{\kappa_i\}$ in $(\mathfrak{a}, \mathfrak{b})$ defined by

$$\kappa_{i+1} = \gamma \kappa_i, \ \forall i \geq 0 \tag{13.2}$$

converges to a solution of the equation defined by $\gamma\kappa = \kappa$.

The iterative sequence defined by (13.2) is known as the "Picard iteration sequence".

BCP in the setting of metric space (**MS**) can be stated as follows:

Let (Ξ, ρ) be a complete metric space (**CMS**) and $\gamma : \Xi \to \Xi$ be a contraction mapping, i.e., a mapping satisfying

$$\rho(\gamma\kappa, \gamma\vartheta) \leq \mathcal{L}\rho(\kappa, \vartheta), \ \forall \kappa, \vartheta \in \Xi, \tag{13.3}$$

where $0 \leq \mathcal{L} < 1$. Then γ has a unique **FP**.

It is interesting to note that no discernible similarity exists between the Picard theorem and the Banach **FP** theorem. In the Picard Theorem, the image of the ith-iteration (say) is bounded by a fixed positive number $\mathcal{L} \in (0, 1)$. However, in the case of the **BCP**, the distance between successive image points is less than the distance between the corresponding points, scaled by a factor $\mathcal{L}$, where $\mathcal{L}$ lies in $[0, 1)$. Though it is noteworthy that any mapping satisfying the Banach contraction (**BC**) condition is a Picard operator and is continuous, a Picard operator is not necessarily continuous in general.

Remark 13.1.1 *γ is said to be*

(i) contractive mapping if inequality ($\leq$) in (13.3) is replaced by strict inequality ($<$) and take $\mathcal{L} = 1$.
(ii) non-expansive mapping if (13.3) holds for $\mathcal{L} = 1$.
(iii) Lipschitzian mapping if (13.3) holds for fixed $\mathcal{L} < \infty$.

Remark 13.1.2 *The following implications hold:*

"Contraction ⇒ contractive ⇒ non-expansive ⇒ Lipschitzian." But the converse implication may not be true. For this, one can go through the following examples:

(i) Let (Ξ, ρ) *be a usual* ***MS*** *with* $\Xi = \mathbb{R}$*. Define* $\gamma : \Xi \to \Xi$ *by* $\gamma\kappa = \kappa + 1$, $\forall \kappa \in \Xi$*. Here,* γ *is an expansive mapping without having an* ***FP****.*

(ii) Let (Ξ, ρ) *be a Euclidean* ***MS****, where* $\Xi = \mathbb{R} \times \mathbb{R}$*. Define* $\gamma : \Xi \to \Xi$ *by* $\gamma(\kappa, \vartheta) = (\kappa, 0)$, $\forall \kappa, \vartheta \in \mathbb{R}$*. Then,* γ *has a non-expansive property with an* ***FP*** $(\kappa, 0)$.

(iii) Let $\Xi = (0, \infty)$ *be a* ***CMS*** *with absolute value. Consider the mapping* $\gamma : \Xi \to \Xi$ *by* $\gamma\kappa = \kappa + \frac{1}{\kappa}$*. Then,* γ *is a contractive mapping while* γ *is not a contraction.*

(iv) Let $\Xi = \mathbb{R}$ *with a usual metric* ρ*. Define* $\gamma : \Xi \to \Xi$ *by* $\gamma\kappa = \kappa$, $\forall \kappa \in \Xi$*. Here,* γ *is a non-expansive mapping, but neither contractive nor contraction.*

(v) Let (Ξ, ρ) *be a usual* ***MS*** *with* $\Xi = \mathbb{R}$*. Define* $\gamma : \Xi \to \Xi$ *by* $\gamma\kappa = 2\kappa$, $\forall \kappa \in \Xi$*. Here,* γ *is the Lipschitzian mapping, but neither of non-expansive, contractive, and contraction.*

Remark 13.1.3 *If* γ *is a contraction, then* γ^n *is also a contraction. But not conversely.*

Let $\Xi = [0, 4]$ *with the usual metric* ρ*. Define* $\gamma : \Xi \to \Xi$ *by*

$$\gamma\kappa = \begin{cases} 0, & \kappa \in [0, 2]; \\ 2, & \kappa \in (2, 4]. \end{cases}$$

Here, $\gamma^2\kappa = 0$, $\forall \kappa \in [0, 2]$ *and* γ^2 *is a contraction. But* γ *is not continuous, so* γ *is not a contraction.*

Remark 13.1.4 *Some mappings are not contractions but have a unique* ***FP****. For instance* $\gamma\kappa = 1 - \kappa$ *for* $\kappa \in \mathbb{R}$*. Here* γ *is not a contraction but has a unique* ***FP*** $\kappa = \frac{1}{2}$*, but* $\gamma^2\kappa = \kappa$, $\forall \kappa \in \mathbb{R}$ *is enriched with* ***FPs****.*

13.2 Various Types of Contraction/Contractive Mappings

Rhoades (1977) presented a beautiful survey on several contractions defined on **MS**. A brief description is given below.

Let (Ξ, ρ) be a compact **MS** and γ be a self-mapping on Ξ satisfying

$$\rho(\gamma\kappa, \gamma\vartheta) < \rho(\kappa, \vartheta)\,, \forall \kappa, \vartheta \in \Xi \text{ with } \kappa \neq \vartheta. \tag{13.4}$$

The inequality (13.4) is known as an "Edelstein contraction" (**EC**). Edelstein (1962) proved an **FP** theorem using inequality (13.4): "A self mapping γ is defined on a compact **MS** Ξ satisfying (13.4) has a unique **FP** in Ξ".

In (13.4), if compactness is replaced by completeness, then γ may or may not have an **FP**.

Let (Ξ, ρ) be the usual **MS** where $\Xi = [1, \infty)$. Define $\gamma : \Xi \to \Xi$ by $\gamma\kappa = \kappa + \frac{1}{\kappa}$, $\forall\, \kappa \in \Xi$.

$$\rho(\gamma\kappa, \gamma\vartheta) = |\gamma\kappa - \gamma\vartheta| = \left|(\kappa + \frac{1}{\kappa}) - (\vartheta + \frac{1}{\vartheta})\right| = \left|(\kappa - \vartheta)(1 - \frac{1}{\kappa\vartheta})\right| < |\kappa - \vartheta| = \rho(\kappa, \vartheta),$$

$\forall\, \kappa, \vartheta \in \Xi$ with $\kappa \neq \vartheta$.

Hence, γ satisfies condition (13.4). But γ has no **FP**, since $\kappa + \frac{1}{\kappa} \neq \kappa$.

We note that every **BC** is an **EC** but not conversely:

Example 13.2.1 *Let (Ξ, ρ) be the usual **MS** where $\Xi = [0, 1]$. Clearly, Ξ is a **CMS**. Define a mapping $\gamma : \Xi \to \Xi$ by $\gamma\kappa = \frac{\kappa}{\kappa+1}$, $\forall\, \kappa \in \Xi$.*

$$\rho(\gamma\kappa, \gamma\vartheta) = \left|\frac{\kappa}{\kappa + 1} - \frac{\vartheta}{\vartheta + 1}\right| = \left|\frac{\kappa - \vartheta}{(\kappa + 1)(\vartheta + 1)}\right| < |\kappa - \vartheta| = \rho(\kappa, \vartheta).$$

*Here, γ satisfies **EC** (13.4) and has an **FP** at $\kappa = 0$. Also, γ is continuous but not **BC**.*

From **BC** (13.3), one can conclude that γ is continuous. In fact, it is uniformly continuous, since on taking $\rho(\kappa, \vartheta) < \frac{\epsilon}{\mathcal{L}}$, $\forall\, \kappa, \vartheta \in \Xi$, we have $\rho(\gamma\kappa, \gamma\vartheta) \leq \epsilon$. But the reverse implication may not hold, see Example 13.2.1.

Now a natural question arises: "Does there exist any contraction that does not force the mapping to be continuous at the **FP**?". Kannan (1968) provided a positive answer to this question, stating that $\exists\, \mathcal{L} \in (0, \frac{1}{2})$ s.t.

$$\rho(\gamma\kappa, \gamma\vartheta) \leq \mathcal{L}[\rho(\kappa, \gamma\kappa) + \rho(\vartheta, \gamma\vartheta)],\ \forall\, \kappa, \vartheta \in \Xi. \tag{13.5}$$

The contraction (13.5) is called the "Kannan contraction" (**KC**). Using the contraction (13.5) Kannan proved an **FP** theorem which states, "Let (Ξ, ρ) be a **CMS** and $\gamma : \Xi \to \Xi$ be a Kannan contraction. Then γ has a unique **FP**."

The following illustration shows that "A mapping may be continuous at the **FP** but not continuous on the whole domain".

Let (Ξ, ρ) be the usual **MS** where $\Xi = [0, 4]$. Define a mapping $\gamma : \Xi \to \Xi$ by

$$\gamma\kappa = \begin{cases} \frac{\kappa}{8}, & \kappa \text{ is rational;} \\ 0, & \kappa \text{ is irrational.} \end{cases}$$

We note that $\rho(\gamma\kappa, \gamma\vartheta) \leq \frac{1}{8}[\rho(\kappa, \gamma\kappa) + \rho(\vartheta, \gamma\vartheta)]$, $\forall\, \kappa, \vartheta \in \Xi$ and $\kappa = 0$ is a unique **FP** of γ. But γ is discontinuous at all the points except $\kappa = 0$.

Remark 13.2.2 *The conditions (13.3) and (13.5) are independent:*

Let (Ξ, ρ) *be the usual* **MS** *with* $\Xi = [0, 1]$*. Define a mapping* $\gamma : \Xi \to \Xi$ *by*

$$\gamma\kappa = \begin{cases} \frac{\kappa}{4}, & \kappa \in [0, \frac{1}{2}); \\ \frac{\kappa}{5}, & \kappa \in [\frac{1}{2}, 1]. \end{cases}$$

Here, γ *is discontinuous at* $\kappa = \frac{1}{2}$*, so condition (13.3) is not satisfied, but condition (13.5) is satisfied for* $\mathcal{L} = \frac{4}{9}$*.*

Further, define the mapping γ *by* $\gamma\kappa = \frac{\kappa}{3}$*,* $\forall\, \kappa \in \Xi$*. Clearly, condition (13.3) is satisfied but condition (13.5) is not satisfied for* $\kappa = \frac{1}{3}, \vartheta = 0$*.*

It is observed that **KC** characterizes metric completeness, while **BC** does not (see Subrahmanyam (1975)).

Meir et al. (1969) introduced a contraction and called it the Meir-Keeler contraction (**MKC**): "for any $\epsilon > 0$, $\exists\, \theta > 0$ s.t.

$$\epsilon \leq \rho(\kappa, \vartheta) < \epsilon + \theta \implies \rho(\gamma\kappa, \gamma\vartheta) < \epsilon. \tag{13.6}$$

MKC implies **EC** (13.4)."

Kannan's theorem has various generalizations. We are stating one of the generalizations given by Reich (1971) as follows:

"Let γ be a self-mapping defined on a **CMS** (Ξ, ρ) satisfying

$$\rho(\gamma\kappa, \gamma\vartheta) \leq a\, \rho(\kappa, \gamma\kappa) + b\, \rho(\vartheta, \gamma\vartheta) + c\, \rho(\kappa, \vartheta),\ \forall\, \kappa, \vartheta \in \Xi, \tag{13.7}$$

where a, b, c are non-negative numbers with $a + b + c < 1$. Then γ has a unique **FP** $\vartheta \in \Xi$ and for any $\kappa \in \Xi$, the sequence of iterates $\{\gamma^n\kappa\}$ converges to ϑ."

Here, we note that (13.7) is stronger than **BC** and **KC**:

Let (Ξ, ρ) be the usual **MS** with $\Xi = [0, 1]]$ and a mapping $\gamma : \Xi \to \Xi$ is defined by

$$\gamma\kappa = \begin{cases} \frac{\kappa}{3}, & \kappa \in [0, 1); \\ \frac{1}{6}, & \kappa = 1. \end{cases}$$

Clearly, γ does not satisfy the **BC** (13.3) as γ is not continuous at $\kappa = 1$. Further, **KC** (13.5) is also not satisfied for $\kappa = 0, \vartheta = \frac{1}{3}$. But it satisfies the condition (13.7) for $a = \frac{1}{6}, b = \frac{1}{9}, c = \frac{1}{3}$.

Also, Ćirić (1971) introduced the contractive condition as follows:

"$\exists$ non-negative numbers $a, b, c, e \in [0, \infty)$ with $a + b + c + 2e < 1$ s.t.

$$\rho(\gamma\kappa, \gamma\vartheta) \leq a\rho(\kappa, \vartheta) + b\rho(\kappa, \gamma\kappa) + c\rho(\vartheta, \gamma\vartheta) + e[\rho(\kappa, \gamma\vartheta) + \rho(\vartheta, \gamma\kappa)],\ \forall\, \kappa, \vartheta \in \Xi." \tag{13.8}$$

One can easily deduce conditions (13.3), (13.5) from (13.7) and (13.8).

Chatterjea (1972) gave a contractive condition analogous to **KC** (13.5) in the following form:
"$\exists$ a point $c \in (0, \frac{1}{2})$ s.t.

$$\rho(\gamma\kappa, \gamma\vartheta) \leq c[\rho(\kappa, \gamma\kappa) + \rho(\vartheta, \gamma\vartheta)], \ \forall\, \kappa, \vartheta \in \Xi." \tag{13.9}$$

Zamfirescu (1972) obtained an **FP** theorem by combining contractive conditions (13.3), (13.5) and (13.9) in the following form:

"Let (Ξ, ρ) be a **CMS** and $\gamma : \Xi \to \Xi$ be a mapping for which $\exists$ real numbers a, b and c with $0 \leq a < 1$, $0 < b, c < \frac{1}{2}$ s.t. for each pair $\kappa, \vartheta \in \Xi$, at least one of the following is true

(i) $\rho(\gamma\kappa, \gamma\vartheta) \leq a\rho(\kappa, \vartheta)$;
(ii) $\rho(\gamma\kappa, \gamma\vartheta) \leq b[\rho(\kappa, \gamma\kappa) + \rho(\vartheta, \gamma\vartheta)]$;
(iii) $\rho(\gamma\kappa, \gamma\vartheta) \leq c[\rho(\kappa, \gamma\vartheta) + \rho(\vartheta, \gamma\kappa)]$."

Hardy et al. (1973) introduced a contractive condition as follows:
"$\exists$ numbers $a, b, c, e, h \in [0, \infty)$ with $a + b + c + e + h < 1$ s.t.

$$\rho(\gamma\kappa, \gamma\vartheta) \leq a\rho(\kappa, \gamma\kappa) + b\,\rho(\vartheta, \gamma\vartheta) + c\,\rho(\kappa, \gamma\vartheta) + e\,\rho(\vartheta, \gamma\kappa) + h\,\rho(\kappa, \vartheta), \tag{13.10}$$

$\forall\, \kappa, \vartheta \in \Xi$."

Berinde (1922) introduced the contractive condition of the following form:
"$\exists\, 0 \leq \mathfrak{h} < 1$ and $\mathcal{L} \geq 0$ s.t.

$$\rho(\gamma\kappa, \gamma\vartheta) \leq \mathfrak{h}\,\rho(\kappa, \vartheta) + \mathcal{L}\,\rho(\vartheta, \gamma\kappa)." \tag{13.11}$$

Many authors have replaced the constant $\mathcal{L}$ with an appropriate real-valued function to generalize and extend the **BCP**, marking the beginning of a new era with the use of functions known as comparison functions. Some authors have considered these functions to be upper semi-continuous, while others have treated them as lower semi-continuous.

Semi-continuity is weaker than continuity and is a useful concept for studying the maxima and minima of discontinuous functions. It generalizes one-sided continuity, particularly from left to right.

A real function $\phi : A \to \mathbb{R}$, $A \subset \mathbb{R}$ is said to be lower semi-continuous at κ_0 if $\forall\, \epsilon > 0$, $\exists\, \theta > 0$ s.t.

$$|\kappa - \kappa_0| < \theta \implies \phi(\kappa) > \phi(\kappa_0) - \epsilon, \ \forall\, \kappa \in A$$

and ϕ is upper semi-continuous at κ_0 if $\forall\, \epsilon > 0$, $\exists\, \theta > 0$ s.t.

$$|\kappa - \kappa_0| < \theta \implies \phi(\kappa) < \phi(\kappa_0) + \epsilon, \ \forall\, \kappa \in A.$$

A function is continuous at κ_0 if it is both upper and lower semi-continuous at κ_0.

In the case of topological spaces, these notions can be materialized as: Let A be a topological space and $\phi : A \to \mathbb{R}$ is lower semi-continuous (upper semi-continuous) at u_0 if for each $\epsilon > 0$, $\exists$ a neighborhood U of κ_0 s.t. $\kappa \in U$ implies $\phi(\kappa) > \phi(\kappa_0) - \epsilon(\phi(\kappa) < \phi(\kappa_0) + \epsilon)$.
ϕ is lower semi-continuous if $\forall\, a \in \mathbb{R}$, the set $\{\kappa \in \Xi : \phi(\kappa) > a\}$ is open in Ξ.
ϕ is upper semi-continuous if $\forall\, a \in \mathbb{R}$, the set $\{\kappa \in \Xi : \phi(\kappa) < a\}$ is open in Ξ.
The floor functions $\phi(\kappa) =$ greatest integer less than or equal to a given real number κ is every upper semi-continuous.

The ceiling function $\phi(\kappa) =$ least integer greater than or equal to κ is lower semi-continuous.

Example 13.2.3 *(i) Define $\phi : \mathbb{R} \to \mathbb{R}$ as*

$$\phi(\kappa) = \begin{cases} \kappa^2, & \kappa \neq 0 \\ -1, & \kappa = 0, \end{cases}$$

then ϕ is the lower semi-continuous function.

Define $\phi : \mathbb{R} \to \mathbb{R}$ as

$$\phi(\kappa) = \begin{cases} \kappa^2, & \kappa \neq 0 \\ 1, & \kappa = 0, \end{cases}$$

then ϕ is the upper semi-continuous function.
(ii) Let $\phi, \psi : \mathbb{R} \to \mathbb{R}$ be defined by $\phi\kappa \neq \psi\kappa$, $\forall\, \kappa \in \mathbb{R}$

$$\delta(\kappa) = \begin{cases} \phi(\kappa), & \kappa \text{ is rational} \\ \psi(\kappa), & \kappa \text{ is irrational}, \end{cases}$$

δ is the upper semi-continuous at all rational points and the lower semi-continuous at all irrational points while neither left nor right is continuous.

Remark 13.2.4 *For a continuous function ϕ, increasing its value at a certain point κ_0 to $\phi(\kappa_0) + c$ for some $c > 0$ results in the function being upper semi-continuous. Conversely, decreasing its value to $\phi(\kappa_0) - c$ makes the function lower semi-continuous.*

Now, we discuss some of the generalizations of the **BCP** by using auxiliary functions.

Rakotch (1962) generalized the **BCP** as follows: "$\forall\, \kappa, \vartheta \in \Xi$, $\rho(\gamma\kappa, \gamma\vartheta) \leq \phi(\rho(\kappa, \vartheta))\rho(\kappa, \vartheta)$, where $\phi : (0, \infty) \to [0, 1)$ is a decreasing function."

Boyd et al. (1969) introduced ϕ contraction condition: "$\exists$ an upper semi-continuous function from the right $\phi : [0, \infty) \to [0, \infty)$ with $0 \leq \phi(t) < t$, $\forall\, t > 0$ s.t. $\rho(\gamma\kappa, \gamma\vartheta) \leq \phi(\rho(\kappa, \vartheta))$, $\forall\, \kappa, \vartheta \in \Xi$."

Ćirić (1974) introduced the contractive condition of the form: "$\exists\, 0 \leq \mathfrak{h} < 1$ s.t.

$$\rho(\gamma\kappa, \gamma\vartheta) \leq \mathfrak{h} \max\{\rho(\kappa, \gamma\kappa), \rho(\vartheta, \gamma\vartheta), \rho(\kappa, \gamma\vartheta), \rho(\vartheta, \gamma\kappa), \rho(\kappa, \vartheta)\}, \quad (13.12)$$

$\forall\, \kappa, \vartheta \in \Xi$."

Carisiti (1976) introduced the following contraction condition: "$\exists$ a lower semi-continuous function $\phi : \Xi \to [0, \infty)$ s.t.

$$\rho(\kappa, \gamma\kappa) \leq \phi(\kappa) - \phi(\gamma\kappa)." \quad (13.13)$$

Alber et al. (1997) introduced the concept of $\phi-$weak contraction in Hilbert spaces. Analyzing this notion, Rhoades (2001) extended and proved this result in the **CMS** in 2001: "$\exists$ a function $\phi : [0, \infty) \to [0, \infty)$ with $\phi(t) > 0$, $\forall\, t > 0$ and $\phi(0) = 0$ s.t. for each $\kappa, \vartheta \in \Xi$ s.t.

$$\rho(\gamma\kappa, \gamma\vartheta) \leq \rho(\kappa, \vartheta) - \phi(\rho(\kappa, \vartheta))."$$

The weak contraction mapping was further generalized by Dutta et al. (2008) for $(\psi - \phi)$-contractions and thereby, proved an important **FP** theorem which states, "A self-mapping γ on an **MS** (Ξ, ρ) has an **FP** if it satisfies the inequality

$$\psi(\rho(\gamma\kappa, \gamma\vartheta)) \leq \psi(\rho(\kappa, \vartheta)) - \phi(\rho(\kappa, \vartheta)),\ \forall\, \kappa, \vartheta \in \Xi,$$

where $\phi, \psi : [0, \infty) \to [0, \infty)$ are both continuous and non-decreasing functions with $\psi(t) = \phi(t) = 0$ if and only if $t = 0$."

Dorić (2009) proved an **FP** theorem for generalized $(\psi - \phi)-$ contractions which states, "If a self mapping γ of an **MS** (Ξ, ρ) satisfies the following condition:

$$\psi(\rho(\gamma\kappa, \gamma\vartheta)) \leq \psi(m(\kappa, \vartheta)) - \phi(m(\kappa, \vartheta)),\ \ \forall\, \kappa, \vartheta \in \Xi,$$

where $m(\kappa, \vartheta) = \max\{\rho(\kappa, \vartheta), \rho(\kappa, \gamma\kappa), \rho(\vartheta, \gamma\vartheta), \frac{\rho(\kappa,\gamma\vartheta)+\rho(\vartheta,\gamma\kappa)}{2}\}$, $\psi : [0, \infty] \to [0, \infty]$ s.t. $\psi(t) < t$, $\forall\, t > 0$ and $\psi(t) = 0$ if and only if $t = 0$ and $\phi : [0, \infty] \to [0, \infty]$ s.t. ϕ is a lower semi-continuous function vanishing only at $t = 0$, then γ has a unique **FP**."

13.3 Various Types of Minimal Commutative Mappings

Several interesting and elegant common fixed point (**CFP**) theorems for a pair (or family) of mappings in **MS** have been obtained by various authors. The central concern in **FPT** is to find a minimal set of sufficient conditions that ensure the existence of **FP/CFP** for contractive type mappings. Generally, **CFP** theorems for contractive type mappings involve the following steps:

(a) a commutative type condition,
(b) containment of range spaces,
(c) continuity of one of the mappings,
(d) a contractive/contraction condition.

Goebel (1968) obtained a coincidence theorem for a pair of mappings in the following form.

Theorem 13.3.1 *Let (Ξ, ρ) be an **MS** and $\mathcal{B}$ be an arbitrary set. Let $\beta, \gamma : \mathcal{B} \to \Xi$ be two mappings s.t. $\beta(\mathcal{B}) \subseteq \gamma(\mathcal{B})$, where $\gamma(\mathcal{B})$ is a complete subspace of Ξ and*

$$\rho(\beta\kappa, \gamma\vartheta) \leq \mathfrak{h}\rho(\gamma\kappa, \gamma\vartheta), \tag{13.14}$$

$\forall\, \kappa, \vartheta \in \mathcal{B}$, $\mathfrak{h} \in [0, 1)$. *Then $\exists$ a point $w \in \mathcal{B}$ s.t. $\beta w = \gamma w$.*

Definition 13.3.2 *(Jungck (1976)) Let (Ξ, ρ) be an **MS** and $\beta, \gamma : \Xi \to \Xi$ be two mappings. β and γ are said to be commuting mappings on Ξ if $\forall\, \kappa \in \Xi$, $\beta\gamma\kappa = \gamma\beta\kappa$.*

Jungck (1976) obtained an **FP** theorem for a pair of commuting mappings in the following form.

Theorem 13.3.3 (Jungck (1976) *Let (Ξ, ρ) be a **CMS** and $\beta, \gamma : \Xi \to \Xi$ be two commuting mappings s.t. $\beta(\Xi) \subseteq \gamma(\Xi)$. For all $\kappa, \vartheta \in \Xi$, $\mathfrak{h} \in [0, 1)$, mappings β and γ satisfy the condition (13.14). If γ is continuous, then β and γ have a unique **CFP**.*

Sessa (1982) introduced the notion of weakly commuting (**WC**) mappings and researchers started utilizing weak conditions of commutativity to improve **CFP** theorems.

Definition 13.3.4 *Two self-mappings β and γ of an **MS** (Ξ, ρ) are said to be **WC** mappings if $\rho(\beta\gamma\kappa, \gamma\beta\kappa) \leq \rho(\beta\kappa, \gamma\kappa)$, $\forall\, \kappa \in \Xi$.*

Compatible Mappings and its Variants

Jungck (1986) introduced the concept of compatible (**Comp**) mappings, which was "a turning point" in the "**FP** arena". It opened doors for researchers in the metric **FPT**, leading to the introduction of various forms of **Comp** mappings, often referred to as minimal commuting mappings. For details on weaker forms of commuting mappings, one may refer to Singh et al. (2003).

Definition 13.3.5 *Let (Ξ, ρ) be an **MS**. A pair (β, γ) of self-mappings defined on Ξ is said to be*

(i) ***Comp*** *(Jungck, 1986) if $\lim_{i\to\infty} \rho(\beta\gamma\kappa_i, \gamma\beta\kappa_i) = 0$,*
(ii) ***Comp*** *of type (A) (Jungck et al., 1993) if $\lim_{i\to\infty} \rho(\beta\beta\kappa_i, \gamma\beta\kappa_i) = 0$ and $\lim_{i\to\infty} \rho(\gamma\gamma\kappa_i, \beta\gamma\kappa_i) = 0$,*

(iii) ***Comp*** *of type* (B) *(Pathak & Khan, 1995) if*

$$\lim_{i\to\infty}\rho(\beta\gamma\kappa_i,\gamma\gamma\kappa_i)\le\frac{1}{2}[\lim_{i\to\infty}\rho(\beta\gamma\kappa_i,\beta z)+\lim_{i\to\infty}\rho(\beta z,\beta\beta\kappa_i)],$$

and

$$\lim_{i\to\infty}\rho(\gamma\beta\kappa_i,\beta\beta\kappa_i)\le\frac{1}{2}[\lim_{i\to\infty}\rho(\gamma\beta\kappa_i,\gamma z)+\lim_{i\to\infty}\rho(\gamma z,\gamma\gamma\kappa_i)],$$

(iv) ***Comp*** *of type* (C) *(Pathak et al., 1998) if*

$$\lim_{i\to\infty}\rho(\beta\gamma\kappa_i,\gamma\gamma\kappa_i)\le\frac{1}{3}[\lim_{i\to\infty}\rho(\beta\gamma\kappa_i,\beta z)+\lim_{i\to\infty}\rho(\beta z,\beta\beta\kappa_i)+\lim_{i\to\infty}\rho(\beta z,\gamma\gamma\kappa_i)]$$

and

$$\lim_{i\to\infty}\rho(\gamma\beta\kappa_i,\beta\beta\kappa_i)\le\frac{1}{3}[\lim_{i\to\infty}\rho(\gamma\beta\kappa_i,\gamma z)+\lim_{i\to\infty}\rho(\gamma z,\gamma\gamma\kappa_i)+\lim_{i\to\infty}\rho(\gamma z,\beta\beta\kappa_i)],$$

(v) ***Comp*** *of type* (P) *(Pathak & Khan, 1995) if* $\lim_{i\to\infty}\rho(\beta\beta\kappa_i,\gamma\gamma\kappa_i)=0$,

(vi) ***Comp*** *of type* (R) *(Rohen & Singh, 2008) if* $\lim_{i\to\infty}\rho(\beta\gamma\kappa_i,\gamma\beta\kappa_i)=0$ *and* $\lim_{i\to\infty}\rho(\beta\beta\kappa_i,\gamma\gamma\kappa_i)=0$,

(vii) ***Comp*** *of type* (K) *(Jha et al., 2014) if* $\lim_{i\to\infty}\beta\beta\kappa_i=\gamma z$ *and* $\lim_{i\to\infty}\gamma\gamma\kappa_i=\beta z$,

(viii) ***Comp*** *of type* (E) *(Singh & Singh, 2007) if* $\lim_{i\to\infty}\beta\beta\kappa_i=\lim_{i\to\infty}\beta\gamma\kappa_i=\gamma z$ *and* $\lim_{i\to\infty}\gamma\gamma\kappa_i=\lim_{i\to\infty}\gamma\beta\kappa_i=\beta z$,

whenever $\exists$ *a sequence* $\{\kappa_i\}$ *in* Ξ *s.t.* $\lim_{i\to\infty}\beta\kappa_i=\lim_{i\to\infty}\gamma\kappa_i=z$, *for* $z\in\Xi$.

(xi) *sub-compatible* (**Sub-Comp**) *(Bouhadjera & Godet-Thobie, 2011) if* $\exists$ *a sequence* $\{\kappa_i\}\in\Xi$ *s.t.* $\lim_{i\to\infty}\rho(\gamma\sigma\kappa_i,\sigma\gamma\kappa_i)=0$.

Shahu et al. (2001) introduced the concept of intimate mappings generalizing the concept of **Comp** of type (A) mappings as follows.

Definition 13.3.6 *Let* (Ξ,ρ) *be an* ***MS****. A pair* (β,γ) *of self-mappings defined on* Ξ *is said to be*

(a) γ*-intimate if*

$$\theta\rho(\gamma\beta\kappa_i,\gamma\kappa_i)\le\theta\rho(\beta\beta\kappa_i,\beta\kappa_i),$$

(a) β-intimate if

$$\theta\rho(\beta\gamma\kappa_i, \beta\kappa_i) \leq \theta\rho(\gamma\gamma\kappa_i, \gamma\kappa_i),$$

where $\theta = \liminf$ *or* $\limsup$ *and* $\{\kappa_i\}$ *is a sequence in* Ξ *s.t.* $\lim_i \beta\kappa_i = \lim_i \gamma\kappa_i = z$, *for some* $z \in \Xi$.

Now, we have a brief description of the limit inferior and the limit superior:
(i) If $< a_n >$ is a bounded sequence.
(a) The least limit point of $< a_n >$ is called limit inferior.
(b) The greatest limit point of $< a_n >$ is called limit superior.
(ii) If $< a_n >$ is an unbounded sequence.
(a) If $< a_n >$ is not bounded above then

$$\limsup a_n = \infty.$$

(b) If $< a_n >$ is not bounded below then

$$\liminf a_n = -\infty.$$

(iii) If $< a_n >$ may or may not be a bounded sequence.
(a) Let $< a_n >$ be a sequence of real numbers then

$$\liminf a_n = \sup_n \inf\{a_n, a_{n+1}, ...\} = \sup_n A_n, \text{ where}$$

$$A_1 = \inf\{a_1, a_2, ...\} = \alpha_1,$$

$$A_2 = \inf\{a_2, a_3, ...\} = \alpha_2,$$

$$\vdots$$

$$A_n = \inf\{a_n, a_{n+1}, ...\} = \alpha_n,$$

$$\liminf a_n = \sup\{\alpha_1, \alpha_2, ..., \alpha_n\}.$$

(b)

$$\limsup a_n = \inf_n \sup\{a_n, a_{n+1}, ...\} = \inf_n A_n, \text{ where}$$

$$B_1 = \sup\{a_1, a_2, ...\} = \beta_1,$$

$$B_2 = \sup\{a_2, a_3, ...\} = \beta_2,$$

$$\vdots$$

$$B_n = \sup\{a_n, a_{n+1}, ...\} = \beta_n,$$

$$\limsup a_n = \inf\{\alpha_1, \alpha_2, ..., \alpha_n\}.$$

Example 13.3.7 *(i) Consider the sequence* $< a_n >= \frac{1}{n}$.
Range set $\mathsf{S} = \{1, \frac{1}{2}, \frac{1}{3}, ...\}$.
S *is bounded as 0 is the lower bound and 1 is the upper bound.*
Set of limit points= $\{0\}$.
$\limsup a_n = \liminf a_n = 0$.
(ii) Consider the sequence $< a_n >=< n >$.
Range set $\mathsf{S} = \{1, 2, 3, ...\}$. S *is not bounded above.*
Then $\limsup a_n = \infty$.
Now $\sup\inf a_n = \sup_n A_n$, *where*

$$A_n = \inf\{a_n, a_{n+1}, a_{n+2}, ...\}$$

$$A_1 = \inf\{1, 2, 3, ...\} = 1,$$

$$A_2 = \inf\{2, 3, 4, ...\} = 2,$$

$$\vdots$$

$$A_n = \inf\{n, n+1, n+2, ...\} = n.$$

$$\sup\{1, 2, 3, ...\} = \infty.$$

$$\liminf a_n = \infty.$$

(iii) Consider the sequence $a_n =< \frac{(-1)^n}{n^2}$.
Range set $\mathsf{S} = \{-1, \frac{1}{2^2}, \frac{-1}{3^2}, \frac{1}{4^2}, ...\}$.
Set of limit points =$\{0\}$.
$\limsup a_n = \liminf a_n = 0$.
(iv) Let $a_n =< n(1 + (-1)^n))$.
Range set S= $\{0, 4, 0, 8, 0, 12, ...\}$.

Since it is not bounded above, so $\limsup a_n = \infty$.
$A_1 = \inf\{0, 4, 0, 8, ...\} = 0$.
$A_2 = \inf\{4, 0, 8, ...\} = 0$.
...
Hence, $\liminf a_n = 0$.

Definition 13.3.8 *Let* (Ξ, ρ) *be an* ***MS***. *A pair* (β, γ) *of self-mappings defined on* Ξ *is said to be*

(i) *non-compatible (non-**Comp**) (Jungck, 1986) if* $\exists$ *a sequence* $\{\kappa_i\}$ *in* Ξ *s.t.* $\lim_{i\to\infty} \beta\kappa_i = \lim_{i\to\infty} \gamma\kappa_i$, *but* $\lim_{i\to\infty} \rho(\beta\gamma\kappa_i, \gamma\beta\kappa_i)$ *is either non-zero or does not exist,*
(ii) *weakly compatible (weakly **Comp**) (Jungck, 1996) if the pair commutes on the set of coincidence points, i.e.,* $\beta\gamma\kappa = \gamma\beta\kappa$, *whenever* $\beta\kappa = \gamma\kappa$ *for* $\kappa \in \Xi$,
(iii) *conditionally commuting (Pant & Pant, 2010) if the pair commutes on a non-empty subset of the set of coincidence points whenever the set of coincidences is non-empty,*
(iv) *conditionally compatible (conditionally **Comp**) (Pant & Bisht, 2012) if and only if there exists a sequence* $\{\kappa_i\}$ *in* Ξ *s.t.* $\lim_{i\to\infty} \beta\kappa_i = \lim_{i\to\infty} \gamma\kappa_i$ *and*

$$\lim_{i\to\infty} \rho(\beta\gamma\kappa_n, \gamma\beta\kappa_i) = 0,$$

whenever the set of sequences $\{\vartheta_i\} \in \Xi$ *satisfying* $\lim_{i\to\infty} \beta\vartheta_i = \lim_{i\to\infty} \gamma\vartheta_i$ *is a non-empty set,*
(v) *faintly compatible (faintly **Comp**) (Bisht & Shahzad, 2013) if and only if the pair* (β, γ) *is conditionally **Comp** and* β *and* γ *commute on a non-empty subset of coincidence points whenever the set of coincidences is non-empty.*

One can observe that weak commutative and **Comp** mappings differ in one respect: weak commutativity is essentially a point property, whereas **Comp** uses the "machinery of sequences".

Now, we highlight the relationship among various types of **Comp** mappings which are useful for the main result.

Proposition 13.3.9 (Jungck et al. (1993)) *Suppose that self-mappings* β *and* γ *of an* ***MS*** (Ξ, ρ) *are* ***Comp*** *mappings of type* (A). *If one of the mappings is continuous, then* β *and* γ *are* ***Comp***.

Proposition 13.3.10 (Jungck (1986)) *Let self-mappings* β *and* γ *of an* ***MS*** (Ξ, ρ) *are* ***Comp***. *If*

(i) $\beta z = \gamma z$, *then* $\beta\gamma z = \beta\beta z = \gamma\gamma z = \gamma\beta z$, *for some* $z \in \Xi$.
(ii) $\lim_{i\to\infty} \beta\kappa_i = \lim_{i\to\infty} \gamma\kappa_i = z$ *for some* $z \in \Xi$, *then*

(a) $\lim_{i\to\infty} \gamma\beta\kappa_i = \beta z$, *if* β *is continuous at* z,
(b) $\lim_{i\to\infty} \beta\gamma\kappa_i = \gamma z$, *if* γ *is continuous at* z,
(c) $\beta\gamma z = \gamma\beta z$ *and* $\beta z = \gamma z$, *if* β *and* γ *are continuous at* z.

Proposition 13.3.11 (Pathak et al. (1995)) *Every pair of* ***Comp*** *mappings of type* (A) *is* ***Comp*** *of type* (B).

Proposition 13.3.12 (Pathak et al. (1995), Pathak et al. (1995), Pathak et al. (1998)) *Suppose that self-mappings* β *and* γ *of an* ***MS*** (Ξ, ρ) *are* ***Comp*** *mappings of type* (B) *or type* (C) *or type* (P) *on* Ξ. *If for some* z *in* Ξ,

(i) $\beta z = \gamma z$, *then* $\beta\gamma z = \beta\beta z = \gamma\gamma z = \gamma\beta z$.
(ii) $\lim_{i\to\infty} \beta\kappa_i = \lim_{i\to\infty} \gamma\kappa_i = z$, *then*

(a) $\lim_{i\to\infty} \gamma\gamma\kappa_i = \beta z$, *if* β *is continuous at* z;
(b) $\lim_{i\to\infty} \beta\beta\kappa_i = \gamma z$, *if* γ *is continuous at* z;
(c) $\beta\gamma z = \gamma\beta z$ *and* $\beta z = \gamma z$, *if* β *and* γ *are continuous at* z.

R–Weak Commuting Mappings and Their Variants

Pant (1994) introduced the notion of R-weakly commuting (**RWC**) mappings in an **MS**. Pathak et al. (1997) extended the **RWC** mappings to the **RWC** mappings of type (A_β) and **RWC** mappings of type (A_γ). Furthermore, Kumar et al. (2009) introduced the notion of **RWC** of type (P). The main application of **RWC** mappings of type (A_β) or type (A_γ) is to study **CFP**s for non-**Comp** mappings.

Definition 13.3.13 *Let* (Ξ, ρ) *be an* ***MS***. *A pair* (β, γ) *of self-mappings defined on* Ξ *is said to be*

(i) ***RWC*** *(Pant, 1994) if* $\exists$ *some real number* $R > 0$ *s.t.*

$$\rho(\beta\gamma\kappa, \gamma\beta\kappa) \leq R\rho(\beta\kappa, \gamma\kappa)\,, \forall\, \kappa \in \Xi.$$

(ii) *pointwise* ***RWC*** *(Pant, 1994) if for a given* $\kappa \in \Xi$, $\exists$ *a real number* $R > 0$ *s.t.*

$$\rho(\beta\gamma\kappa, \gamma\beta\kappa) \leq R\,\rho(\beta\kappa, \gamma\kappa).$$

(iii) ***RWC*** *of type* (A_β) *(Pathak et al., 1997) if* $\exists$ *some real number* $R > 0$ *s.t.*

$$\rho(\beta\gamma\kappa, \gamma\gamma\kappa) \leq R\,\rho(\beta\kappa, \gamma\kappa)\,, \forall\, \kappa \in \Xi.$$

(iv) ***RWC*** *of type* (A_γ) *(Pathak et al., 1997) if* $\exists$ *some real number* $R > 0$ *s.t.*

$$\rho(\gamma\beta\kappa, \beta\beta\kappa) \leq R\,\rho(\beta\kappa, \gamma\kappa)\,, \forall\,\kappa \in \Xi.$$

(v) ***RWC*** *mappings of type* (P) *(Kumar & Garg, 2009) if* $\exists$ *a real number* $R > 0$ *s.t.*

$$\rho(\beta\beta\kappa, \gamma\gamma\kappa) \leq R\,\rho(\beta\kappa, \gamma\kappa)\,, \forall\,\kappa \in \Xi.$$

Remark 13.3.14 *The notions of* ***RWC*** *and* ***RWC*** *of type* (A_β) *or type* (A_γ) *are independent of each other (see, Pathak, 1997). However, at a coincidence point,* $R-$*weak commuting and all analogous notions of* $R-$*weak commuting are equivalent to each other and imply commutativity, (see, Agarwal et al., 2014). Both* ***Comp*** *and non-****Comp*** *mappings imply* ***RWC*** *of type* (A_β) *or* (A_γ)*,(see, Pant & Bisht, 2013, 2012).*

Example 13.3.15 *Consider the usual* ***MS*** (Ξ, ρ) *with* $\Xi = [-1, 1]$*. Define* $\beta, \gamma : \Xi \to \Xi$ *by* $\beta\kappa = \kappa$ *and* $\gamma\kappa = |\kappa| - 1$*. Then*

$$\rho(\beta\kappa, \gamma\kappa) = 1,\ \rho(\beta\gamma\kappa, \gamma\beta\kappa) = 2(1 - |\kappa|),\ \rho(\beta\gamma\kappa, \gamma\gamma\kappa) = 1,$$

$$\rho(\beta\beta\kappa, \gamma\gamma\kappa) = 2\,|\kappa|,\ \rho(\gamma\beta\kappa, \beta\beta\kappa) = 1,$$

$\forall\,\kappa \in \Xi.$

It can be concluded that

(i) The pair (β, γ) is not **WC**.
(ii) for $R = 2$ the pair (β, γ) is **RWC**, **RWC** of type (P), **RWC** of type (A_γ) and **RWC** of type (A_β).
(iii) for $R = \frac{3}{2}$, the pair (β, γ) is **RWC** of type (A_β) but not **RWC** of type (P) and **RWC**.

Example 13.3.16 *Let* (Ξ, ρ) *be the usual* ***MS*** *with* $\Xi = [0, 1]$ *and* β, γ *be two self-mapping of* Ξ *defined by* $\beta\kappa = \kappa$ *and* $\gamma\kappa = \kappa^2$*. Then,*

$$\rho(\beta\kappa, \gamma\kappa) = |\kappa(\kappa - 1)|,\ \rho(\beta\gamma\kappa, \gamma\beta\kappa) = 0,\ \rho(\beta\gamma\kappa, \gamma\gamma\kappa) = |\kappa^2(\kappa - 1)(\kappa + 1)|,$$

$$\rho(\beta\beta\kappa, \gamma\gamma\kappa) = |\kappa(\kappa - 1)(\kappa^2 + \kappa + 1)|,\ \rho(\gamma\beta\kappa, \beta\beta\kappa) = |\kappa(\kappa - 1)|,$$

$\forall\,\kappa \in \Xi.$

We infer that

(i) The pair (β, γ) is **RWC** $R > 0$.
(ii) for $R = 3$, the pair (β, γ) is **RWC** of type (P), **RWC** of type (A_γ) and **RWC** of type (A_β).
(iii) for $R = 2$, the pair (β, γ) is **RWC** of type (A_β) and **RWC** of type (A_γ) but not **RWC** of type (P) (for this take $\kappa = \frac{3}{4}$).

Remark 13.3.17 *(Bouhadjera (2005)) Notions of* ***Comp*** *mappings and their variants (type(A), type(B), type (C), type(P)) are independent of each other. However, all types of* ***Comp*** *mappings become equivalent when the continuity of all mappings is considered.*

Properties and Various Forms of Continuous Mappings

A natural question now arises: "How can **FP** theorems be extended to the setting of non-complete **MS** and without the continuity of the involved mappings over the entire space, say, Ξ". A partial answer can be given using "property $(E.A)$".

Aamri et al. (2002) generalized the notion of non-**Comp** mappings in the **MS** by defining "the property $(E.A)$". A major benefit of this property is that it ensures the convergence of desired sequences without completeness. It was pointed out in Aamri et al. (2002) that "property $(E.A)$" buys containment of ranges without any continuity requirements besides minimizing the commutativity conditions of the mappings to the commutativity at their points of coincidence. Moreover, it allows replacing the completeness requirement of the space with a more natural condition: the closedness of the range.

Definition 13.3.18 *A pair of self-mappings β and γ on an* ***MS*** *(Ξ, ρ) is said to*

(i) satisfy property (E.A) (Aamri & Moutawakil, 2002) if $\exists$ a sequence $\{\kappa_i\} \in \Xi$ s.t.

$$\lim_{i\to\infty}\beta\kappa_i = \lim_{i\to\infty}\gamma\kappa_i = t, \text{ for some } t \in \Xi.$$

(ii) satisfy the common limit in the range of γ property, denoted by (CLR_γ) (Sintunavarat & Kumam, 2012), if $\exists$ a sequence $\{\kappa_i\} \in \Xi$ s.t.

$$\lim_{n\to\infty} \beta\kappa_i = \lim_{n\to\infty} \gamma\kappa_i = \gamma t, \text{ for some } t \in \Xi.$$

Definition 13.3.19 *Two pairs (α, β) and (σ, γ) of self-mappings defined on a metric space (Ξ, ρ) satisfy*

(i) common property $(E.A)$ (Liu et al., 2005) if there exist two sequences $\{\kappa_i\}$ and $\{\vartheta_i\}$ in Ξ with

$$\lim_{i\to\infty} \alpha\kappa_i = \lim_{i\to\infty} \beta\kappa_i = \lim_{i\to\infty} \sigma\vartheta_i = \lim_{i\to\infty} \gamma\vartheta_i = r, \text{ for some } r \text{ in } \Xi.$$

(ii) the (CLR) *property w.r.t. mappings* α *and* σ, *i.e.,* $(CLR_{\alpha\sigma})$ *(Imdad et al., 2012) if there exist two sequences* $\{\kappa_i\}$ *and* $\{\vartheta_i\}$ *in* Ξ *s.t.*

$$\lim_{i\to\infty} \alpha\kappa_i = \lim_{i\to\infty} \beta\kappa_i = \lim_{i\to\infty} \sigma\vartheta_i = \lim_{i\to\infty} \gamma\vartheta_i = r,$$

for some $r \in \alpha(\Xi) \cap \sigma(\Xi)$.

A pair (β, γ) that satisfies "the property $(E.A)$" along with the closedness of the subspace $\gamma(\Xi)$ always satisfies "the (CLR_γ) property", see (Sintunavarat et al. (2012), Examples 2.16–2.17). However, "the property (CLR)" does not impose any conditions on the closedness of the space or subspace.

Example 13.3.20 *Let* $\Xi = \mathbb{R}^+ \cup \{0\}$ *and* ρ *be the usual metric on* Ξ. *Define* $\beta, \gamma : \Xi \to \Xi$ *by* $\beta\kappa = 0$, *if* $0 < \kappa \leq 1$ *and* $\beta\kappa = 1$, *if* $\kappa > 1$ *or* $\kappa = 0$ *and* $\gamma\kappa = [\kappa]$, *the greatest integer that is greater than or equal to* κ, $\forall\, \kappa \in \Xi$. *Consider a sequence* $\kappa_i = 1 + \frac{1}{i}$, $i \geq 2$. *Then we have* $\lim_{i\to\infty} \beta\kappa_i = \lim_{i\to\infty} \gamma\kappa_i = 1$. *Similarly for the sequence* $\vartheta_i = 1 - \frac{1}{i}$, $i \geq 2$, *we have* $\lim_{i\to\infty} \beta\vartheta_i = \lim_{i\to\infty} \gamma\vartheta_i = 0$.*Thus* β *and* γ *satisfy property* $(E.A)$.
However β *and* γ *are not weakly* ***Comp*** *as* β *and* γ *do not commute at each coincidence points* $\kappa_1 \in (0, 1)$ *and* $\kappa_2 \in (1, 2)$. *Moreover,* β *and* γ *commute at* $\kappa = 0, 1, 2, \ldots$ *but none of these points are coincidence points of* β *and* γ.

Pant (1999) introduced the concept of reciprocal continuity and initiated the study of **FP** for discontinuous mappings.

Definition 13.3.21 *The pair* (β, γ) *is said to be*

*(i) reciprocally continuous (**RC**) (Pant, 1999) if* $\lim_{i\to\infty} \beta\gamma\kappa_i = \beta z$ *and* $\lim_{i\to\infty} \gamma\beta\kappa_i = \gamma z$, *for any sequence* $\{\kappa_i\} \in \Xi$ *s.t.* $\lim_{i\to\infty} \beta\kappa_i = \lim_{i\to\infty} \gamma\kappa_i = z$,

*(ii) subsequential continuous (**SSC**) (Bouhadjera & Godet-Thobie, 2011), if* $\exists$ *a sequence* $\{\kappa_i\} \in \Xi$ *satisfying* $\lim_{i\to\infty} \beta\kappa_i = \lim_{i\to\infty} \gamma\kappa_i = z$, $z \in \Xi$ *s.t.* $\lim_{i\to\infty} \beta\gamma\kappa_i = \beta z$ *and* $\lim_{i\to\infty} \gamma\beta\kappa_i = \gamma z$, $z \in \Xi$,

(iii) sequentially continuous of type (A_β) *(**SC**(**A**$_\beta$)) (Gopal et al., 2012) if and only if* $\exists$ *a sequence* $\{\kappa_i\} \in \Xi$ *satisfying* $\lim_{i\to\infty} \beta\kappa_i = \lim_{i\to\infty} \gamma\kappa_i = z$ *s.t.* $\lim_{i\to\infty} \beta\gamma\kappa_i = \beta z$ *and* $\lim_{i\to\infty} \gamma\gamma\kappa_i = \gamma z$, $z \in \Xi$,

(iv) sequentially continuous of type (A_γ) *(**SC**(**A**$_\gamma$)) (Gopal et al., 2012) if and only if* $\exists$ *a sequence* $\{\kappa_i\} \in \Xi$ *satisfying* $\lim_{i\to\infty} \beta\kappa_i = \lim_{i\to\infty} \gamma\kappa_i = z$ *s.t.* $\lim_{i\to\infty} \gamma\beta\kappa_i = \gamma z$ *and* $\lim_{i\to\infty} \beta\beta\kappa_i = \beta z$, $z \in \Xi$,

(v) sequentially continuous of type (P) *(**SC**(**P**)) (Chauhan et al., 2014) if and only if* $\exists$ *a sequence* $\{\kappa_i\} \in \Xi$ *satisfying* $\lim_{i\to\infty} \beta\kappa_i = \lim_{i\to\infty} \gamma\kappa_i = z$ *s.t.* $\lim_{i\to\infty} \beta\beta\kappa_i = \beta z$ *and* $\lim_{i\to\infty} \gamma\gamma\kappa_i = \gamma z$, $z \in \Xi$.

Remark 13.3.22 *If the pair* (β, γ) *is continuous then it is* ***RC*** *and* ***SSC*** *also, but the converse is not true in general, see (Kim et al. (2004), Example 2.2).*

Remark 13.3.23 *Notions of* ***SSC*** *and* ***RC*** *mappings are independent of each other, see (Kim et al. (2004), Examples 2.3–2.4). Also, it may be noted that the concept of* ***SSC*** *and* $\mathbf{SC}(\mathbf{A}_\beta)$ *and* $\mathbf{SC}(\mathbf{A}_\gamma)$ *mappings are independent of each other, see (Gopal et al. (2012), Example 2.1–2.2).*

Remark 13.3.24 *For details on variants of continuity, one may refer to Tomar et al. (2016).*

13.4 Implicit Relations

Several metrical **FP** theorems can be unified naturally by using an implicit contraction-type condition in place of the usual explicit contractive conditions. Popa (1997), Popa (1999) initiated this direction of research, which has produced consistent literature on **FP**, **CFP** and coincidence point theorems for both single-valued and multivalued mappings in various abstract spaces.

For proving results, Popa (1997), Popa (1999) considered $\mathbb{H}$ to be the set of all continuous functions $\mathcal{H} : \mathbb{R}^{+6} \to \mathbb{R}$ satisfying the following conditions:

($\mathcal{H}_1$) $\mathcal{H}$ is non-increasing in 5th and 6th variables,
($\mathcal{H}_2$) $\exists\, h \in (0, 1)$ s.t. for $\kappa, \vartheta \geq 0$ with

$$\mathcal{H}(\kappa, \vartheta, \vartheta, \kappa, \kappa + \vartheta, 0) \leq 0 \text{ or } \mathcal{H}(\kappa, \vartheta, \kappa, \vartheta, 0, \kappa + \vartheta) \leq 0 \text{ implies } \kappa \leq h\vartheta,$$

($\mathcal{H}_3$) $\mathcal{H}(\kappa, \kappa, 0, 0, \kappa, \kappa) > 0,\ \forall\, \kappa > 0.$

Ali et al. (2008) defined a new implicit function to enhance the domain of applicability, which includes several known contraction conditions such as Ćirić quasi-contraction, generalized contraction, ϕ-type contraction, and rational inequality. In the process, many known results are enriched and improved.

Ali et al. (2008) defined $\mathbb{H}$ as the family of lower semi-continuous functions $\mathcal{H} : \mathbb{R}_+^6 \to \mathbb{R}$ satisfying the following conditions:

($\mathcal{H}_1$): $\mathcal{H}(\kappa, 0, \kappa, 0, 0, \kappa) > 0,\ \forall\, \kappa > 0,$
($\mathcal{H}_2$): $\mathcal{H}(\kappa, 0, 0, \kappa, \kappa, 0) > 0,\ \forall\, \kappa > 0,$
($\mathcal{H}_3$): $\mathcal{H}(\kappa, \kappa, 0, 0, \kappa, \kappa) > 0,\ \forall\, \kappa > 0.$

Example 13.4.1 *Define* $\mathcal{H} : \mathbb{R}_+^6 \to \mathbb{R}$ *as*

$$\mathcal{H}(\kappa_1, \kappa_2, \cdots, \kappa_6) = \kappa_1 - \varkappa \max\{\kappa_2, \kappa_3, \kappa_4, \kappa_5, \kappa_6\},$$

where $\varkappa \in [0, 1)$.

($\mathcal{H}_1$): $\mathcal{H}(\kappa, 0, \kappa, 0, 0, \kappa) = \kappa(1 - \varkappa) > 0,\ \forall\, \kappa > 0,$
($\mathcal{H}_2$): $\mathcal{H}(\kappa, 0, 0, \kappa, \kappa, 0) = \kappa(1 - \varkappa) > 0,\ \forall\, \kappa > 0,$
($\mathcal{H}_3$): $\mathcal{H}(\kappa, \kappa, 0, 0, \kappa, \kappa) = \kappa(1 - \varkappa) > 0,\ \forall\, \kappa > 0.$

Example 13.4.2 *Define* $\mathcal{H} : \mathbb{R}_+^6 \to \mathbb{R}$ *as*

$$\mathcal{H}(\kappa_1, \kappa_2, \ldots, \kappa_6) = \kappa_1 - \varkappa \max\{\kappa_2, \kappa_3, \kappa_4, \kappa_5, \kappa_6\},$$

where $\varkappa \in [0, 1)$.

($\mathcal{H}_1$): $\mathcal{H}(\kappa, 0, \kappa, 0, 0, \kappa) = \kappa(1 - \varkappa) > 0, \ \forall \kappa > 0,$
($\mathcal{H}_2$): $\mathcal{H}(\kappa, 0, 0, \kappa, \kappa, 0) = \kappa > 0, \ \forall \kappa > 0,$
($\mathcal{H}_3$): $\mathcal{H}(\kappa, \kappa, 0, 0, \kappa, \kappa) = \kappa(1 - \varkappa) > 0, \ \forall \kappa > 0.$

Example 13.4.3 *Define* $\mathcal{H}(\kappa_1, \kappa_2, \cdots, \kappa_6) : \mathbb{R}_+^6 \to \mathbb{R}$ *as*

$$\mathcal{H}(\kappa_1, \kappa_2, \ldots, \kappa_6) = \kappa_1 - \varkappa[\max\{\kappa_2^2, \kappa_3\kappa_4, \kappa_5\kappa_6, \kappa_3\kappa_5, \kappa_4\kappa_6\}]^{\frac{1}{2}},$$

where $\varkappa \in [0, 1)$.

Using implicit relation, Ali et al. (2008) proved the following theorem:

Theorem 13.4.4 *Let* (Ξ, ρ) *be an* ***MS*** *and* α, β, γ *and* σ *be four self-mappings on* Ξ *satisfying the following conditions:*
(a) $\mathcal{H} \in \mathbb{F}$ *and* $\forall \kappa, \vartheta \in \Xi$

$$\mathcal{H}(\rho(\alpha\kappa, \sigma\vartheta), \rho(\beta\kappa, \gamma\vartheta), \rho(\alpha\kappa, \beta\kappa), \rho(\sigma\vartheta, \gamma\vartheta), \rho(\beta\kappa, \sigma\vartheta), \rho(\gamma\vartheta, \alpha\kappa) \leq 0, \tag{13.15}$$

(b) pairs (α, β) *and* (σ, γ) *enjoy the common property (E.A),*
(c) $\beta(\Xi)$ *and* $\gamma(\Xi)$ *are closed subsets of* Ξ.
Then the pair (α, β) *as well as* (σ, γ) *have a coincidence point. Moreover,* α, σ, β *and* γ *have a unique* ***CFP*** *provided both the pairs* (α, β) *and* (σ, γ) *are weakly* ***Comp***.

The conclusions of Theorem 13.4.4 remain true if inequality (13.15) is replaced by one of the following contraction conditions: $\forall \kappa, \vartheta \in \Xi$,
(a_1)

$$\rho(\alpha\kappa, \sigma\vartheta) \leq \varkappa \max\{\rho(\beta\kappa, \gamma\vartheta), \rho(\alpha\kappa, \beta\kappa), \rho(\sigma\vartheta, \gamma\vartheta), \rho(\beta\kappa, \sigma\vartheta), \rho(\gamma\vartheta, \alpha\kappa)\}, \text{ where } \varkappa \in [0, 1).$$

(a_2)

$$\rho(\alpha\kappa, \sigma\vartheta) \leq \varkappa \max\{\rho(\beta\kappa, \gamma\vartheta), \rho(\alpha\kappa, \beta\kappa), \rho(\alpha\kappa, \beta\kappa)\rho(\beta\kappa, \sigma\vartheta), \rho(\sigma\vartheta, \gamma\vartheta)\rho(\gamma\vartheta, \alpha\kappa)\},$$
where $\varkappa \in [0, 1)$.

(a_3)

$$\rho(\alpha\kappa, \sigma\vartheta) \leq \varkappa[\max\{\rho^2(\beta\kappa, \gamma\vartheta), \rho(\alpha\kappa, \beta\kappa)\rho(\sigma\vartheta, \gamma\vartheta), \rho(\beta\kappa, \sigma\vartheta)\rho(\gamma\vartheta, \alpha\kappa), \rho(\alpha\kappa, \beta\kappa)\rho(\beta\kappa, \sigma\vartheta), \rho(\sigma\vartheta, \gamma\vartheta)\rho(\gamma\vartheta, \alpha\kappa)\}]^{\frac{1}{2}}, \text{ where } \varkappa \in [0, 1).$$

(a_4)

$$\rho(\alpha\kappa, \sigma\vartheta) \leq a[b \max\{\rho(\beta\kappa, \gamma\vartheta), \rho(\alpha\kappa, \beta\kappa), \rho(\sigma\vartheta, \gamma\vartheta), \rho(\beta\kappa, \sigma\vartheta), \rho(\gamma\vartheta, \alpha\kappa)\} (1-b)(\max\{\rho^2(\beta\kappa, \gamma\vartheta), \rho(\alpha\kappa, \beta\kappa)\rho(\sigma\vartheta, \gamma\vartheta), \rho(\beta\kappa, \sigma\vartheta)\rho(\gamma\vartheta, f\kappa), \rho(\alpha\kappa, \beta\kappa)\rho(\gamma\vartheta, \alpha\kappa), \rho(\sigma\vartheta, \gamma\vartheta)\rho(\beta\kappa, \sigma\vartheta)\})^{\frac{1}{2}}], \text{ where } a \in [0, 1) \text{ and } b \geq 0.$$

(a_5)

$$\begin{aligned}\rho^2(\alpha\kappa, \sigma\vartheta) \leq & a \max\{\rho^2(\beta\kappa, \gamma\vartheta), \rho^2(\alpha\kappa, \beta\kappa), \rho^2(\sigma\vartheta, \gamma\vartheta)\} \\ & + b \max\{\rho(\alpha\kappa, \beta\kappa)\rho(\beta\kappa, \sigma\vartheta), \rho(\sigma\vartheta, \gamma\vartheta)\rho(\gamma\vartheta, \alpha\kappa)\} \\ & + c\, \rho(\beta\kappa, \sigma\vartheta)\rho(\gamma\vartheta, \alpha\kappa),\end{aligned}$$

where $a, b, c \geq 0$ and $a + c < 1$.

(a_6)

$$\begin{aligned}(1 + a\rho(\beta\kappa, \gamma\vartheta))\rho(\alpha\kappa, \sigma\vartheta) \leq & a \max\{\rho(\alpha\kappa, \beta\kappa)\rho(\sigma\vartheta, \gamma\vartheta), \rho(\beta\kappa, \sigma\vartheta)\rho(\gamma\vartheta, \alpha\kappa)\} \\ & + b \max\{\rho(\beta\kappa, \gamma\vartheta), \rho(\alpha\kappa, \beta\kappa), \rho(\sigma\vartheta, \gamma\vartheta), \rho(\beta\kappa, \sigma\vartheta), \rho(\gamma\vartheta, \alpha\kappa)\},\end{aligned}$$

where $a \geq 0$ and $b \in [0, 1)$.

(a_7)

$$\begin{aligned}\rho(\alpha\kappa, \sigma\vartheta) \leq & a\rho(\beta\kappa, \gamma\vartheta) + b \max\{\rho(\alpha\kappa, \beta\kappa), \rho(\sigma\vartheta, \gamma\vartheta)\} \\ & + c \max\{\rho(\alpha\kappa, \beta\kappa) + \rho(\sigma\vartheta, \gamma\vartheta), \rho(\beta\kappa, \sigma\vartheta) + \rho(\gamma\vartheta, \alpha\kappa)\},\end{aligned}$$

where $a, b, c \geq 0$ and $a + b + 2c < 1$.

(a_8)

$$\rho(\alpha\kappa, \sigma\vartheta) \leq \phi(\max\{\rho(\beta\kappa, \gamma\vartheta), \rho(\alpha\kappa, \beta\kappa), \rho(\sigma\vartheta, \gamma\vartheta), \rho(\beta\kappa, \sigma\vartheta), \rho(\gamma\vartheta, \alpha\kappa)\}),$$

where $\phi : \mathbb{R}^+ \to \mathbb{R}$ is an upper semi-continuous function s.t. $\phi(\varsigma) < \varsigma \, \forall \varsigma > 0$ and $\phi(0) = 0$.

(a_9)

$$\rho(\alpha\kappa, \sigma\vartheta) \leq \phi(\rho(\beta\kappa, \gamma\vartheta), \rho(\alpha\kappa, \beta\kappa), \rho(\sigma\vartheta, \gamma\vartheta), \rho(\beta\kappa, \sigma\vartheta), \rho(\gamma\vartheta, \alpha\kappa)),$$

where $\phi : \mathbb{R}^5_+ \to \mathbb{R}$ is an upper semi-continuous function s.t.

$$\max\{\phi(0, \varsigma, 0, 0, \varsigma), \phi(0, 0, \varsigma, \varsigma, 0), \phi(\varsigma, 0, 0, \varsigma, \varsigma)\} < \varsigma, \text{ for each } \varsigma > 0.$$

(a_{10})

$$\rho^2(\alpha\kappa,\sigma\vartheta) \leq \phi(\rho^2(\beta\kappa,\gamma\vartheta),\rho(\alpha\kappa,\beta\kappa)\rho(\sigma\vartheta,\gamma\vartheta),\rho(\beta\kappa,\sigma\vartheta)\rho(\gamma\vartheta,\alpha\kappa),$$
$$\rho(\alpha\kappa,\beta\kappa)\rho(\gamma\vartheta,\alpha\kappa),\rho(\sigma\vartheta,\gamma\vartheta)\rho(\beta\kappa,\sigma\vartheta)),$$

where $\phi:\mathbb{R}_+^5 \to \mathbb{R}$ is an upper semi-continuous function s.t.

$$\max\{\phi(0,0,0,\varsigma,0),\phi(0,0,0,0,\varsigma),\phi(\varsigma,0,\varsigma,0,0)\} < \varsigma, \text{ for each } \varsigma > 0.$$

(a_{11})

$$\rho^3(\alpha\kappa,\sigma\vartheta) \leq a\,\rho^2(\alpha\kappa,\sigma\vartheta)\rho(\beta\kappa,\gamma\vartheta) + b\,\rho(\alpha\kappa,\sigma\vartheta)\rho(\alpha\kappa,\beta\kappa)\rho(\sigma\vartheta,\gamma\vartheta)$$
$$+ c\,\rho^2(\beta\kappa,\sigma\vartheta)\rho(\gamma\vartheta,\alpha\kappa) + d\,\rho(\beta\kappa,\sigma\vartheta)\rho^2(\gamma\vartheta,\alpha\kappa),$$

where $a, b, c, d \geq 0$ and $a + c + d < 1$.

Remark 13.4.5 *The contraction conditions* (a_1) *to* (a_{11}) *are well-known and generalize many relevant results of metric* ***FPT***.

13.5 Applications of BCP

Application to Initial Value Problem

First, we discuss the applications of **FPT** to the existence of unique solutions for certain first-order initial value problems (**IVP**) (Saxe, 2009). In particular, consider

$$\mathfrak{u}'(\mathfrak{t}) = \mathrm{F}(\mathfrak{t},\mathfrak{u}(\mathfrak{t})),\ \mathfrak{u}(0) = \mathfrak{u}_0, \tag{13.16}$$

where $\mathrm{F} : J \times \mathbb{R}^n \Rightarrow \mathbb{R}^n$ and $J = [a, b]$.

Here F takes values in $\mathbb{R}^n$, so (13.16) is a system of first-order equations. First, suppose that F is continuous. Clearly, $\mathfrak{u} \in C^1(J)$ is a solution of (13.16) if and only if $u \in C(J)$ is a solution of

$$\mathfrak{u}(\mathfrak{t}) = \mathfrak{u}_0 + \int_0^{\mathfrak{t}} \mathrm{F}(\mathfrak{s},\mathfrak{u}(\mathfrak{s}))d\mathfrak{s}, \tag{13.17}$$

where $C^1(J)$ is the Banach space(**BS**) of functions $\mathfrak{u}$ with continuous first-order derivative on J assembled with the norm $|\mathfrak{u}|_1 = \max\{\sup \mathfrak{t} \in J|\mathfrak{u}(\mathfrak{t})|, \sup \mathfrak{t} \in J|\mathfrak{u}'(\mathfrak{t})|\}$ and $C(J)$ is the **BS** of continuous functions $\mathfrak{u}$ on J assembled with the norm $|\mathfrak{u}|_0 = \sup \mathfrak{t} \in J|\mathfrak{u}(\mathfrak{t})|$.

Define an operator $\gamma : C(J) \to C(J)$ by

$$\gamma \mathfrak{u}(\mathfrak{t}) = \mathfrak{u}_0 + \int_0^{\mathfrak{t}} \mathrm{F}(\mathfrak{s}, \mathfrak{u}(\mathfrak{s}))d\mathfrak{s}. \tag{13.18}$$

Evidently, $\mathfrak{u}$ is a solution of (13.16) if and only if $\mathfrak{u}$ is an **FP** of the γ defined by (13.18).

Application to Linear Equations

Although there are various other methods to solve such a system, such as the well-known "Gauss elimination method", an indirect approach, namely the iteration method, may be more useful.

To apply the **BCP**, a **CMS** and a contraction mapping defined on it are required.

Consider the set $\mathfrak{X}$ of all ordered $n-$tuples of real numbers s.t.

$$\mathfrak{u} = (\mathfrak{u}_1, \mathfrak{u}_2, \ldots, \mathfrak{u}_n),\ \mathfrak{v} = (\mathfrak{v}_1, \mathfrak{v}_2, \ldots, \mathfrak{v}_n),\ \mathfrak{z} = (\mathfrak{z}_1, \mathfrak{z}_2, \ldots, \mathfrak{z}_n).$$

Define a metric ρ on $\mathfrak{X}$ as follows:

$$\rho(\mathfrak{u}, \mathfrak{v}) = \max_i |\mathfrak{u}_i - \mathfrak{v}_i| \tag{13.19}$$

Clearly, $(\mathfrak{X}, \rho)$ is a **CMS**. Define a mapping $\gamma : \mathfrak{X} \to \mathfrak{X}$ by

$$\mathfrak{v} = \gamma \mathfrak{u} = C\mathfrak{u} + b, \tag{13.20}$$

where $C = (C_{jk})$ is a fixed real $n \times n$ matrix and $b \in \mathfrak{X}$ is a fixed vector.

In writing component form, we have

$$\mathfrak{v}_j = \Sigma_{k=1}^{n} C_{jk}\mathfrak{u}_{ki} + B_j \ (j = 1, 2, \ldots, n). \tag{13.21}$$

where $b = (B_j)$.

Clearly, finding a solution to (13.21) is equivalent to finding an **FP** of γ, and to find a unique **FP** of γ, the **BCP** is applied.

Remark 13.5.1 *Several authors have applied **FP** results to find sufficient conditions for the existence of common solutions to certain functional equations (**FE**s) arising in **DP** (Pathak, 2005; Pathak et al., 1995). Some authors have used **FP** methods for the stability of **FE**s. Baker (1991) was the first author to use a variant of the **BCP** to discuss the "Hyers-Ulam stability" of an **FE** in a single variable; however, many authors followed Radu's approach (Radu et al., 2003). The "Hyers-Ulam stability" of differential equations, see (Bojor, 2012; Jung, 2010) and integral equations, see (Castro & Ramos, 2009; Jung, 2007) can also be proved by the **FP** method.*

13.6 Main Results

Let Ψ be a collection of all continuous functions $\psi : [0, \infty)^4 \to [0, \infty)$ satisfying the following conditions:

(ψ_1) ψ is non-decreasing in each coordinate variable with $\psi(0, 0, 0, 0) = 0$,
(ψ_2) $\Delta(\varsigma) = \max\{\psi(\varsigma, \varsigma, 0, 0), \psi(0, 0, 0, \varsigma), \psi(0, 0, \varsigma, 0), \psi(\varsigma, \varsigma, \varsigma, \varsigma)\} < \varsigma$, for each $\varsigma > 0$.

Let (Ξ, ρ) be an **MS** and α, β, γ, and σ be four self-mappings on Ξ satisfying the following conditions:

(C_1) $\beta(\Xi) \subset \sigma(\Xi)$ and $\gamma(\Xi) \subset \alpha(\Xi)$,
(C_2) $\forall \kappa, \vartheta \in \Xi$, $\exists$ a function $\psi \in \Psi$ s.t.

$$\rho^3(\beta\kappa, \gamma\vartheta) \leq \psi\Big(\rho^2(\alpha\kappa, \beta\kappa)\rho(\sigma\vartheta, \gamma\vartheta), \rho(\alpha\kappa, \beta\kappa)\rho^2(\sigma\vartheta, \gamma\vartheta),$$
$$\rho(\alpha\kappa, \beta\kappa)\rho(\alpha\kappa, \gamma\vartheta)\rho(\sigma\vartheta, \beta\kappa), \rho(\alpha\kappa, \gamma\vartheta)\rho(\sigma\vartheta, \beta\kappa)\rho(\sigma\vartheta, \gamma\vartheta)\Big),$$

(C_3) one of α, β, γ, and σ is continuous.

Now, we state some **CFP** theorems for **Comp** mappings and its variants.

Theorem 13.6.1 *Let (Ξ, ρ) be a **CMS**. Let α, β, γ, and σ be four self-mappings on Ξ satisfying the conditions (C_1)-(C_3). Then α, β, γ, and σ have a unique **CFP** in Ξ. If (α, β), and (σ, γ) are pairs of one of the following:*

*(a) **Comp** mappings*
*(b) **Comp** mappings of type (A)*
*(c) **Comp** mappings of type (B)*
*(d) **Comp** mappings of type (C)*
*(e) **Comp** mappings of type (P).*

Proof The proof follows on similar lines of the paper of Kavita et al. (2023). □

Now, we state **FP** theorems for weakly **Comp** mappings and property $(E.A)$ and (CLR) property.

Theorem 13.6.2 *Let (Ξ, ρ) be an **MS**. Let α, β, γ, and σ be four self-mappings on Ξ satisfying the conditions (C_1),(C_2). If one of the subspaces $\alpha(\Xi)$ or $\sigma(\Xi)$ or $\beta(\Xi)$ or $\gamma(\Xi)$ is complete. Then α, β, γ, and σ have a unique **CFP** in Ξ provided that the pairs (α, β) and (σ, γ) are weakly **Comp**.*

Theorem 13.6.3 *Let (Ξ, ρ) be an **MS**. Let α, β, γ, and σ be four self-mappings on Ξ satisfying the conditions (C_1),(C_2). If one of the subspaces $\alpha(\Xi)$ or $\sigma(\Xi)$ or $\beta(\Xi)$ or $\gamma(\Xi)$ is closed. Then α, β, γ, and σ have a unique **CFP** in Ξ provided that the pairs (α, β) and (σ, γ) are weakly **Comp** satisfying common property $(E.A.)$.*

Theorem 13.6.4 *Let α, β, γ, and σ be self mappings of an* ***MS*** (Ξ, ρ) *satisfying the assertion* (C_2). *If* (α, β) *and* (σ, γ) *enjoy the property* $(CLR_{\alpha\sigma})$ *and are weakly* ***Comp****, then α, β, γ, and σ have a unique* ***CFP*** *in* Ξ.

Remark 13.6.5 *Proof of Theorems 13.6.2–13.6.4 follow on the similar lines of the papers of Kavita et al. (2023) and Murthy et al. (2022).*

Let $\mathtt{Y}, \mathtt{D}$ denote **BS**s, $\mathtt{W} \subset \mathtt{Y}$, $\mathtt{P} \subset \mathtt{D}$ are "state space and decision space" respectively. Let $\mathbb{B}(\mathtt{W}) = \{\lambda : \mathtt{W} \to \mathbb{R}, \lambda is bounded\}$, where $\mathbb{R}$ is the set of all real numbers.

The basic form of the **FE** given by Bellman et al. (1978) is as follows:

$$\mathcal{N}(\mu) = \operatorname*{opt}_{v} \mathcal{T}(\mu, v, \mathcal{N}(\delta(\mu, v))),$$

where $\mu \in \mathtt{W}$, $v \in \mathtt{P}$, δ is the transformation process, $\mathcal{N}(\mu)$ is the optimal return with initial state μ and the opt denotes max or min.

Now, we find the sufficient conditions for the existence of a common solution to the following **FE**s, which arises in **DP**:

$$\mathcal{M}_i(\mu) = \sup_{\omega \in \mathtt{P}} \mathcal{S}_i(\mu, \omega, \mathcal{M}_i(\delta(\mu, \omega))), \mu \in \mathtt{W} \tag{13.22}$$

$$\mathcal{N}_i(\mu) = \sup_{\omega \in \mathtt{P}} \mathcal{T}_i(\mu, \omega, \mathcal{N}_i(\delta(\mu, \omega))), \mu \in \mathtt{W}, \tag{13.23}$$

where $\delta : \mathtt{W} \times \mathtt{P} \to \mathtt{W}$ and $\mathcal{S}_i, \mathcal{T}_i : \mathtt{W} \times \mathtt{P} \times \mathbb{R} \to \mathbb{R}$, $i = 1, 2$.

Theorem 13.6.6 *Suppose that the following conditions hold:*

(D_1) $\mathcal{S}_1, \mathcal{S}_2, \mathcal{T}_1$ *and* $\mathcal{T}_2$ *are bounded;*
(D_2) $\forall\, \mu, t \in \mathtt{W}, \omega \in \mathtt{P}, \lambda, \nu \in \mathbb{B}(\mathtt{W})$, $\exists$ *a function* $\psi \in \Psi$ *s.t.*

$$\begin{aligned}\big|\mathcal{S}_1(\mu, \omega, \lambda(t)) - \mathcal{S}_2(\mu, \omega, \nu(t))\big|^3 \le \psi\big(&\rho^2(\mathcal{R}_1\lambda, \mathcal{A}_1\lambda)\rho(\mathcal{R}_2\nu, \mathcal{A}_2\nu),\\ &\rho(\mathcal{R}_1\lambda, \mathcal{A}_1\lambda)\rho^2(\mathcal{R}_2\nu, \mathcal{A}_2\nu),\\ &\rho(\mathcal{R}_1\lambda, \mathcal{A}_1\lambda)\rho(\mathcal{R}_1\lambda, \mathcal{A}_2\nu)\rho(\mathcal{R}_2\nu, \mathcal{A}_1\lambda)\\ &\rho(\mathcal{R}_1\lambda, \mathcal{A}_2\nu)\rho(\mathcal{R}_2\nu, \mathcal{A}_1\lambda)\rho(\mathcal{R}_2\nu, \mathcal{A}_2\nu)\big),\end{aligned}$$

where the mappings $\mathcal{A}_i, \mathcal{R}_i$, $(i = 1, 2)$ *are defined as follows,* $\forall\, \mu \in \mathtt{W}$, $\lambda, \nu \in \mathbb{B}(\mathtt{W})$

$$\begin{aligned}\mathcal{A}_i\lambda(\mu) &= \sup_{\omega \in \mathtt{P}} \mathcal{S}_i(\mu, \omega, \lambda(\delta(\mu, \omega))),\\ \mathcal{R}_i\nu(\mu) &= \sup_{\omega \in \mathtt{P}} \mathcal{T}_i(\mu, \omega, \nu(\delta(\mu, \omega)));\end{aligned} \tag{13.24}$$

(D_3) *either* $\exists\, \mathcal{A}_i \in \{\mathcal{A}_1, \mathcal{A}_2\}$ *s.t. for any sequence* $\{\nu_j\} \subset \mathbb{B}(\mathtt{W})$ *and* $\nu \in \mathbb{B}(\mathtt{W})$

$$\lim_{j\to\infty} \sup_{\mu\in\mathtt{W}} |\nu_j(\mu) - \nu(\mu)| = 0 implies that \lim_{j\to\infty} \sup_{\mu\in\mathtt{W}} |\mathcal{A}_i\nu_j(\mu) - \mathcal{A}_i\nu(\mu)| = 0$$

or $\exists\, \mathcal{R}_i \in \{\mathcal{R}_1, \mathcal{R}_2\}$ *s.t. for any sequence* $\{\nu_j\} \subset \mathbb{B}(\mathtt{W})$ *and* $\nu \in \mathbb{B}(\mathtt{W})$

$$\lim_{j\to\infty} \sup_{\mu\in\mathtt{W}} |\nu_j(\mu) - \nu(\mu)| = 0, implies that \lim_{j\to\infty} \sup_{\mu\in\mathtt{W}} |\mathcal{R}_i\nu_j(\mu) - \mathcal{R}_i\nu(\mu)| = 0;$$

(D_4) *for any* $\lambda \in \mathbb{B}(\mathtt{W})$, $\exists\, \nu_1, \nu_2 \in \mathbb{B}(\mathtt{W})$ *s.t.*

$$\mathcal{A}_1\lambda(\mu) = \mathcal{R}_2\nu_1(\mu),\ \mathcal{A}_2\lambda(\mu) = \mathcal{R}_1\nu_2(\mu),\ \mu \in \mathtt{W};$$

(D_5) *for any* $i \in \{1, 2\}$,

$$\lim_{j\to\infty} \sup_{\mu\in\mathtt{W}} |\mathcal{R}_i\mathcal{A}_i\nu_j(\mu) - \mathcal{A}_i\mathcal{R}_i\nu_j(\mu)| = 0,$$

whenever $\{\nu_j\}$ *is a sequence of* $\mathbb{B}(\mathtt{W})$ *s.t.*

$$\lim_{j\to\infty} \mathcal{R}_i\nu_j = \lim_{j\to\infty} \mathcal{A}_i\nu_j = \lambda,\ \lambda \in \mathbb{B}(\mathtt{W}).$$

Then the system of ***FE****s (13.22) and (13.23) has a unique common solution in* $\mathbb{B}(\mathtt{W})$.

Proof Let $\rho(\lambda, \nu) = \sup\{|\lambda(\mu) - \nu(\mu)| : \mu \in \mathtt{W}\}$ for any $\lambda, \nu \in \mathbb{B}(\mathtt{W})$. Obviously, $(\mathbb{B}(\mathtt{W}), \rho)$ is a **CMS**. By conditions $(D_1) - (D_5)$, $\mathcal{A}_i, \mathcal{R}_i$ are self-mappings of $\mathbb{B}(\mathtt{W})$. One of $\mathcal{A}_i, \mathcal{R}_i$ is continuous for $i \in \{1, 2\}$, $\mathcal{A}_1(\mathbb{B}(\mathtt{W})) \subset \mathcal{R}_2(\mathbb{B}(\mathtt{W}))$ and $\mathcal{A}_2(\mathbb{B}(\mathtt{W})) \subset \mathcal{R}_1(\mathbb{B}(\mathtt{W}))$ and the pairs of mappings $(\mathcal{A}_1, \mathcal{R}_1)$ and $(\mathcal{A}_2, \mathcal{R}_2)$ are **Comp**.

For $\eta > 0$, $\mu \in \mathtt{W}$ and $\nu_1, \nu_2 \in \mathbb{B}(\mathtt{W})$, $\exists\, \omega_1, \omega_2 \in \mathtt{P}$ s.t.

$$\mathcal{A}_i\nu_i(\mu) < \mathcal{S}_i(\mu, \omega_i, \nu_i(\mu_i)) + \eta, \tag{13.25}$$

where $\mu_i = \delta(\mu, \omega_i)$, $i = 1, 2$. Also, we have

$$\mathcal{A}_1\nu_1(\mu) \geq \mathcal{S}_1(\mu, \omega_2, \nu_1(\mu_2)), \tag{13.26}$$
$$\mathcal{A}_2\nu_2(\mu) \geq \mathcal{S}_2(\mu, \omega_1, \nu_2(\mu_1)). \tag{13.27}$$

From inequalities (13.25),(13.27) and (D_2), we have

$$
\begin{aligned}
(\mathcal{A}_1\nu_1(\mu)-\mathcal{A}_2\nu_2(\mu))^3 <&(\mathcal{S}_1(\mu,\omega_1,\nu_1(\mu_1))-\mathcal{S}_2(\mu,\omega_1,\nu_2(\mu_1))+\eta)^3\\
&=(\mathcal{S}_1(\mu,\omega_1,\nu_1(\mu_1))-\mathcal{S}_2(\mu,\omega_1,\nu_2(\mu_1)))^3+\xi\\
&\leq \psi\big(\rho^2(\mathcal{R}_1\nu_1,\mathcal{A}_1\nu_1)\rho(\mathcal{R}_2\nu_2,\mathcal{A}_2\nu_2),\\
&\rho(\mathcal{R}_1\nu_1,\mathcal{A}_1\nu_1)\rho^2(\mathcal{R}_2\nu_2,\mathcal{A}_2\nu_2),\\
&\rho(\mathcal{R}_1\nu_1,\mathcal{A}_1\nu_1)\rho(\mathcal{R}_1\nu_1,\mathcal{A}_2\nu_2)\rho(\mathcal{R}_2\nu_2,\mathcal{A}_1\nu_1)\\
&\rho(\mathcal{R}_1\nu_1,\mathcal{A}_2\nu_2)\rho(\mathcal{R}_2\nu_2,\mathcal{A}_1\nu_1)\rho(\mathcal{R}_2\nu_2,\mathcal{A}_2\nu_2)\big)+\xi,
\end{aligned}
\tag{13.28}
$$

where $\xi=\eta^3+3(\mathcal{S}_1-\mathcal{S}_2)(\mathcal{S}_1-\mathcal{S}_2+\eta)\eta$.

From inequalities (13.25), (13.26), and (D_2), we have

$$
\begin{aligned}
(\mathcal{A}_1\nu_1(\mu)-\mathcal{A}_2\nu_2(\mu))^3 >&(\mathcal{S}_1(\mu,\omega_2,\nu_1(\mu_2))-\mathcal{S}_2(\mu,\omega_2,\nu_2(\mu_2))-\eta)^3\\
&=(\mathcal{S}_1(\mu,\omega_2,\nu_1(\mu_2))-\mathcal{S}_2(\mu,\omega_2,\nu_2(\mu_2))^3-\xi_1\\
&>(\mathcal{S}_1(\mu,\omega_2,\nu_1(\mu_2))-\mathcal{S}_2(\mu,\omega_2,\nu_2(\mu_2))^3-\xi\\
&\geq -\psi\big(\rho^2(\mathcal{R}_1\nu_1,\mathcal{A}_1\nu_1)\rho(\mathcal{R}_2\nu_2,\mathcal{A}_2\nu_2),\\
&\rho(\mathcal{R}_1\nu_1,\mathcal{A}_1\nu_1)\rho^2(\mathcal{R}_2\nu_2,\mathcal{A}_2\nu_2),\\
&\rho(\mathcal{R}_1\nu_1,\mathcal{A}_1\nu_1)\rho(\mathcal{R}_1\nu_1,\mathcal{A}_2\nu_2)\rho(\mathcal{R}_2\nu_2,\mathcal{A}_1\nu_1)\\
&\rho(\mathcal{R}_1\nu_1,\mathcal{A}_2\nu_2)\rho(\mathcal{R}_2\nu_2,\mathcal{A}_1\nu_1)\rho(\mathcal{R}_2\nu_2,\mathcal{A}_2\nu_2)\big)-\xi,
\end{aligned}
\tag{13.29}
$$

where $\xi_1=\eta^3+3(\mathcal{S}_1-\mathcal{S}_2)(\mathcal{S}_1-\mathcal{S}_2-\eta)\eta<\xi$.
From inequalities (13.28) and (13.29), we obtain

$$
\begin{aligned}
|\mathcal{A}_1\nu_1(\mu)-\mathcal{A}_2\nu_2(\mu)|^3 \leq&\psi\big(\rho^2(\mathcal{R}_1\nu_1,\mathcal{A}_1\nu_1)\rho(\mathcal{R}_2\nu_2,\mathcal{A}_2\nu_2),\\
&\rho(\mathcal{R}_1\nu_1,\mathcal{A}_1\nu_1)\rho^2(\mathcal{R}_2\nu_2,\mathcal{A}_2\nu_2),\\
&\rho(\mathcal{R}_1\nu_1,\mathcal{A}_1\nu_1)\rho(\mathcal{R}_1\nu_1,\mathcal{A}_2\nu_2)\rho(\mathcal{R}_2\nu_2,\mathcal{A}_1\nu_1)\\
&\rho(\mathcal{R}_1\nu_1,\mathcal{A}_2\nu_2)\rho(\mathcal{R}_2\nu_2,\mathcal{A}_1\nu_1)\rho(\mathcal{R}_2\nu_2,\mathcal{A}_2\nu_2)\big)+\xi,
\end{aligned}
\tag{13.30}
$$

As $\eta>0$ is arbitrary and inequality (13.30) is true $\forall\,\mu\in\mathtt{W}$, taking supremum, we get

$$
\begin{aligned}
\rho^3(\mathcal{A}_1\nu_1,\mathcal{A}_2\nu_2)\leq&\psi\big(\rho^2(\mathcal{R}_1\nu_1,\mathcal{A}_1\nu_1)\rho(\mathcal{R}_2\nu_2,\mathcal{A}_2\nu_2),\\
&\rho(\mathcal{R}_1\nu_1,\mathcal{A}_1\nu_1)\rho^2(\mathcal{R}_2\nu_2,\mathcal{A}_2\nu_2),\\
&\rho(\mathcal{R}_1\nu_1,\mathcal{A}_1\nu_1)\rho(\mathcal{R}_1\nu_1,\mathcal{A}_2\nu_2)\rho(\mathcal{R}_2\nu_2,\mathcal{A}_1\nu_1)\\
&\rho(\mathcal{R}_1\nu_1,\mathcal{A}_2\nu_2)\rho(\mathcal{R}_2\nu_2,\mathcal{A}_1\nu_1)\rho(\mathcal{R}_2\nu_2,\mathcal{A}_2\nu_2)\big).
\end{aligned}
\tag{13.31}
$$

Therefore, Theorem 13.6.1 (a) applies, where $\mathcal{A}_1,\mathcal{A}_2,\mathcal{R}_1,\mathcal{R}_2$ correspond to the mappings $\beta,\gamma,\alpha,\sigma$ respectively. So, $\mathcal{A}_1$, $\mathcal{A}_2$, $\mathcal{R}_1$ and $\mathcal{R}_2$ have a unique **CFP** $\nu^*\in\mathbb{B}(\mathtt{W})$, i.e., $\nu^*(\mu)$ is a unique common solution of the system of **FE**s (13.22) and (13.23). □

13.7 Conclusion

A brief study of various contractions and the development of variants of **Comp** and **RWC** mappings has been discussed, which are useful in generalizing the **BCP**. Additionally, the derivation of some well-known contractions from the implicit relations has been discussed, along with their applications.

Acknowledgements The authors are thankful to the reviewers for their suggestions to improve this chapter.

References

Aamri, M., & Moutawakil, D. E. I. (2002). Some new common fixed point theorems under strict contractive conditions. *Journal of Mathematical Analysis and Applications, 270*(1), 181–188.

Agarwal, R. P., Bisht, R. K. & Shahzad, N. (2014). A comparison of various noncommuting conditions in metric fixed point theory and their applications. *Fixed Point Theory Applications* 2014: Art. id 38.

Alber, Y. I., & Guerre-Delabriere, S. (1997). Principle of weakly contractive maps in Hilbert spaces. *Nonlinear Analysis, 98*, 7–22.

Ali, J., & Imdad, M. (2008). An implicit function implies several contraction conditions. *Sarajevo Journal of Mathematics, 4*(17), 269–285.

Baker, J. A. (1991). The stability of certain functional equations. *Proceedings of the American Mathematical Society, 112*(3), 729–732.

Banach, S. (1922). Sur les operations dans les ensembles abstraits et leur applications aux equations integrales. *Foundations of Mathematics, 3*, 133–181.

Bellman, R., & Lee, B. S. (1978). Functional equations arising in dynamic programming. *Aequationes Mathematicae, 17*, 1–18.

Berinde, V. (1922). Approximating fixed points of weak contractions using the Picard iteration. *Nonlinear Analysis Forum, 9*, 45–53.

Bisht, R. K., & Shahzad, N. (2013). Fainlty compatible mappings and Common fixed points. *Fixed Point Theory Applications*. https://doi.org/10.1186/1687-1812-2013-156

Bojor, F. (2012). Note on the stability of first order linear differential equations. *Opuscula Mathematica, 32*(1), 67–74.

Bouhadjera, H., & Godet-Thobie, C. (2011). Common fixed point theorems for pairs of subcompatible maps. arXiv:0906.3159v2 [math.FA], [new version].

Bouhadjera, H. (2005). Common fixed point theorems for compatible mappings of type (C). *Sarajevo Journal of Mathematics, 1*(4), 261–270.

Boyd, D. W., & Wong, J. S. W. (1969). On non-linear contractions. *Proceedings of the American Mathematical Society, 20*, 458–464.

Brouwer, L. (1912). Uber abbildungen von mannigfaltigkeiten. *Mathematische Annalen, 71*, 97–115.

Caristi, J. (1976). Fixed point theorems for mappings satisfying inwardness conditions. *Transactions of the American Mathematical Society, 215*, 241–251.

Castro, L. P., & Ramos, A. (2009). Hyers–Ulam-Rassias stability for a class of nonlinear Volterra integral equations. *Banach Journal of Mathematical Analysis, 3*(1), 36–43.

Chatterjea, S. K. (1972). Fixed point theorems. C.R. *Academy of Bulgare Sciences, 25*, 727–730.

Chauhan, S., Shatanawi, W., Radenovć, S., & Abu-Irwaq, I. (2014). Variants of sub-sequentially continuous mappings and integral-type fixed point results. *Rendiconti del Circolo Matematico di Palermo, 63*, 53–72.

Dorić, D. (2009). Common fixed point for generalized (ψ, ϕ)-weak contractions. *Applied Mathematics Letters, 22*, 1896–1900.

Dutta, P. N., & Chaudhury, B. S. (2008). A generalization of contraction principle in metric spaces. *Fixed Point Theory and Applications*, 2008: Art id 406368.

Edelstein, M. (1962). On fixed and periodic points under contraction mappings. *Journal of the London Mathematical Society, 37*, 74–79.

Goebel, K. (1968). A coincidence theorem. *Bulletin de l'Académie Polonaise des Sciences, Série des Sciences Mathématiques, Astronomiques et Physiques, 16*, 733–735.

Gopal, D., Imdad, M., & Abbas, M. (2012). Metrical common fixed point theorems without completeness and closedness. *Fixed Point Theory Application*, 2012: Art. id 13.

Hary, G. E., & Rogers, T. D. ((1973)"A). generalization of a fixed point theorem of Reich.". *Canadian Mathematical Bulletin,16*, 201–206.

Imdad, M., Pant, B. D., & Chauhan, S. (2012). Fixed point theorems in Menger spaces using the (CLR_{ST}) property and applications. *Journal of Nonlinear Analysis and Optimization, 3*(2), 225–237.

Jha, K., Popa, V., & Manandhar, K. B. (2014). Common fixed point theorem for compatible of type (K) in metric space. *International Journal of Mathematical Sciences and Applications, 8*, 383–391.

Jung, S. M. (2007). A fixed point approach to the stability of a Volterra integral equation. *Fixed Point Theory Application*, 2007: Art. id 57064.

Jung, S. M. (2010). A fixed point approach to the stability of differential equations $y' = F(x, y)$. *Bulletin of the Malaysian Mathematical Sciences Society, 33*(1), 47–56.

Jungck, G. (1976). Commuting mappings and fixed points. *The American Mathematical Monthly, 83*, 261–263.

Jungck, G. (1986). Compatible mappings and common fixed points. *International Journal of Mathematics and Mathematical Sciences, 9*, 771–779.

Jungck, G. (1996). Common fixed points for non continuous non-self maps on non-metric spaces. *Far East Journal of Mathematical Sciences, 4*(2), 199–215.

Jungck, G., Murthy, P. P., & Cho, Y. J. (1993). Compatible mappings of type (A) and common fixed points. *Mathematics of Japanese, 38*, 381–390.

Kannan, R. (1968). Some results on fixed points. *Bulletin of the Calcutta Mathematical Society, 62*, 71–76.

Kavita, & Kumar, S. (2023). Fixed point for weakly compatible mappings along with common limit range property. *Bulletin of the Calcutta Mathematical Society,115*(3), 363–374.

Kavita, K. S., Kumar, R., & Nikita. (2023). Compatible mappings and its variants satisfying generalized (ψ, ϕ)-weak contraction. *Ratio Mathematica,48*, 368–399.

Kim, J. K., & Bisht, R. K. (2004). A critical remark on subsequential continuity. *Nonlinear Functional Analysis and Applications, 19*(1), 127–130.

Kumar, S., & Garg, S. K. (2009). Expansion mappings theorems in metric spaces. *International Journal of Contemporary Mathematical Sciences, 4*(36), 1749–1758.

Liu, Y., Wu, J., & Li, Z. (2005). Common fixed point of single valued and multivalued maps. *International Journal of Mathematics and Mathematical Sciences, 19*, 3045–3055.

Lj, B., & Ćirić. (1971). Generalized contractions and fixed-point theorems. *Publications de l'Institut Mathématique (Beograd),12*(26), 19–26.

Lj, B., & Ćirić. (1974). A generalization of Banach's contraction principle. *Proceedings of the American Mathematical Society,45*, 267–273.

Meir, A., & Keeler, K. (1969). A theorem on contraction mappings. *Journal of Mathematical Analysis and Applications, 28*, 326–329.

Murthy, P. P., Kumar, R., Kumar, S., Sahu, O., Mitrovic, Z. D., & George, R. (2022). New results for weakly compatible (WC) and R-weakly commuting (RWC) mappings with an application in dynamic programming. *Fractal Fractional, 6*(12), 13 pages.

Pant, V., & Bisht, R. K. (2013). A new continuity condition and fixed point theorems with applications. *Rev: Revista de la Real Academia de Ciencias Exactas Fisicas y Naturales Serie A-Matematicas*. https://doi.org/10.1007/s13398-013-0132-8.

Pant, V., & Pant, R.P. (2010). Common fixed points of conditionally commuting maps. *Fixed Point Theory, 1*, 113–118.

Pant, R. P. (1994). Common fixed points of non-commuting mappings. *Journal of Mathematical Analysis and Applications, 188*(2), 436–440.

Pant, R. P. (1999). A common fixed point theorem under a new condition. *Indian Journal of Pure and Applied Mathematics, 30*(2), 147–152.

Pant, R. P., & Bisht, R. K. (2012). Occasionally weakly compatible mappings and fixed points. *Bulletin of the Belgian Mathematical Society, 19*, 655–661.

Pant, R. P., & Bisht, R. K. (2012). Common fixed point theorems under a new continuity condition. *Annali dell'Università di Ferrara, 58*, 127–141.

Pathak, H. K. (2005). Applications of fixed point technique in solving certain dynamic programming and variational inequalities. *Nonlinear Analysis*. https://doi.org/10.1016/j.na.2005.03.033

Pathak, H. K., & Khan, M. S. (1995). Compatible mappings of type (B) and common fixed point theorems of Gregus type. *Czechoslovak Mathematical Journal, 45*, 685–698.

Pathak, H. K., Cho, Y. J., Kang, S. M., & Lee, B. S. (1995). Fixed point theorems for compatible mappings of type (P) and applications to dynamic programming. *Matematiche, 50*, 15–33.

Pathak, H. K., Cho, Y. J., & Kang, S. M. (1997). Remarks on R-weakly commuting maps and common fixed point theorems. *Bulletin of the Korean Mathematical Society, 34*, 247–257.

Pathak, H. K., Cho, Y. J., Kang, S. M., & Madharia, B. (1998). Compatible mappings of type (C) and common fixed point theorems of Gregus type. *Demonstratio Mathematica, 31*, 499–518.

Picard, E. (1890). Memoire sur la theorie des equations aux derivees partielles et la methode des approximations successives. *Journal de Mathématiques Pures et Appliquées, 6*, 145–210.

Popa, V. (1997). Fixed point theorems for implicit contractive mappings. *Stud Cerc St Ser Mat Univ Bacău., 7*, 127–133.

Popa, V. (1999). Some fixed point theorems for compatible mappings satisfying an implicit relation. *Demonstratio Mathematica, 32*, 157–163.

Radu, V. (2003). The fixed point alternative and the stability of functional equations. *Fixed Point Theory, 4*(1), 91–96.

Rakotch, E. (1962). A note on contraction mappings. *Proceedings of the American Mathematical Society, 13*, 459–462.

Reich, E. (1971). Some remarks concerning contraction mappings. *Canadian Mathematical Bulletin, 14*, 121–124.

Rhoades, B. E. (1977). A comparison of various definitions of contractive mappings. *Transactions of the American Mathematical Society, 226*, 257–290.

Rhoades, B. E. (2001). Some theorems on weakly contractive maps. *Nonlinear Analysis, 47*(4), 2683–2693.

Rohen, Y., & Singh, M. R. (2008). Common fixed point of compatible mappings of type (R) in complete metric spaces. *International Journal of Mathematical Sciences Engineering Applications, 2*(4), 295–303.

Saxe, K. (2009). *Beginning Functional Analysis* (pp. 160–164). New York, Inc.: Springer.

Sessa, S. (1982). On weak commutativity conditions of mappings in fixed point consideration. *Publ. Inst. Math. Belg., 32*, 149–153.

Shahu, D. R., Dhagat, V. B., & Srivastava, M. (2001). Fixed points with intimate mappings (I). *Bulletin of the Calcutta Mathematical Society, 93*, 107–114.

Singh, S. L., & Tomar, A. (2003). Weaker forms of commuting maps and existence of fixed points. *Journal of the Korean Society of Mathematical Education Series B, Pure and Applied Mathematics, 10*(03), 145–161.

Singh, M. R., & Singh, Y. M. (2007). Compatible mappings of type (E) and common fixed point theorem of Meir-Keeler type. *International Journal of Mathematical Sciences and Engineering Applications, 1*, 299–315.

Sintunavarat, W., & Kumam, P. (2012). Common fixed points for $R-$weakly commuting mappings in fuzzy metric spaces. *Annali dell'Università di Ferrara. Sezione VII. Scienze Matematiche, 58*(2), 389–406.

Subrahmanyam, V. (1975). Completeness and fixed points. *Monatshefte für Mathematik, 80*, 325–330.

Tomar, A., & Karapinar, E. (2016). On variants of continuity and existence of fixed point via Meir-Keeler contractions in MC-spaces. *Journal of Advanced Mathematical Studies, 9*(2), 348–359.

Zamfirescu, T. (1972). Fixed point theorems in metric spaces. *Archiv der Mathematik (Basel), 23*, 292–298.

Chapter 14
Admissible Mappings with Implicit Relations in Metric Spaces

Kavita, Sanjay Kumar, Manish Jain, and Sushma Devi

Abstract In this chapter, we introduce an implicit type contractive condition, and by using this condition, we discuss the existence and uniqueness of a common fixed point for a pair of admissible mappings. The work presented here unifies and extends several existing results. Further, as an application of our main result, a sufficient condition for the existence of a solution to a nonlinear fractional differential equation is also incorporated here.

Keywords Weakly compatible mappings · Implicit relations · Admissible mappings · Fractional differential equation

AMS Subject Classification 47H10 · 54H25 · 26A33

14.1 Introduction and Preliminaries

Fixed point (**FP**) theory is an active area of research with various applications in diverse fields such as optimization, variational inequalities, fractional calculus, etc. Samet et al. (2012) introduced the concept of "α-admissible (α-**Adm**) mappings" and "$\alpha - \psi$-contractive mappings" to generalize and improve various existing results in the literature, including the Banach contraction (**BC**) (Banach, 1922). Furthermore, Karapinar et al. (2013) refined this concept, and Sahi et al. (2015) extended it to pairs of mappings.

Kavita (✉) · S. Kumar
Department of Mathematics, Deenbandhu Chhotu Ram University of Science and Technology, Murthal (Sonepat), Haryana, India
e-mail: kvtlather@gmail.com

M. Jain
Department of Mathematics, Ahir College, Rewari, India

S. Devi
Kanya Mahavidyalaya, Kharkhoda (Sonepat), Haryana, India

© The Author(s), under exclusive license to Springer Nature Singapore Pte Ltd. 2026

A. Tomar and M. Jain (eds.), *Applications of Fixed-Point Theorem*, Industrial and Applied Mathematics, https://doi.org/10.1007/978-981-95-5160-6_14

Definition 14.1.1 (Samet et al. (2012)) Let $\alpha : \Xi \times \Xi \to [0, \infty)$ be a function. A mapping $\Omega : \Xi \to \Xi$ is called α-**Adm** if $\forall\ \kappa, \vartheta \in \Xi,\ \alpha(\kappa, \vartheta) \geq 1$ gives $\alpha(\Omega\kappa, \Omega\vartheta) \geq 1$.

Definition 14.1.2 (Karapinar et al. (2013)) A self mapping $\Omega : \Xi \to \Xi$ is called triangular α-**Adm** if

(i) Ω is α-**Adm**,
(ii) for $\kappa, \vartheta, r \in \Xi, \alpha(\kappa, r) \geq 1, \alpha(r, \vartheta) \geq 1$ gives $\alpha(\kappa, \vartheta) \geq 1$.

Definition 14.1.3 (Sahi et al. (2015)) Let $\Omega, \Theta : \Xi \to \Xi$ and $\alpha : \Xi \times \Xi \to [0, \infty)$. Ω is called α-**Adm** with respect to Θ if $\forall\ \kappa, \vartheta \in \Xi, \alpha(\Theta\kappa, \Theta\vartheta) \geq 1 \implies \alpha(\Omega\kappa, \Omega\vartheta) \geq 1$.

Definition 14.1.4 (Jungck (1996)) Let $(\Xi, \mathtt{m})$ be a metric space (**MS**). A pair (Ω, Θ) of self mappings defined on Ξ is said to be weakly compatible (**WC**) if for some $r \in \Xi, \Omega r = \Theta r \implies \Omega\Theta r = \Theta\Omega r$.

For a comprehensive analysis of weaker forms of commutativity, see the work of Singh and Tomar (2003).

Popa (1999) used implicit functions (**IF**s) to prove an **FP** theorem in the setting of a complete metric space (**CMS**) rather than contraction conditions. These **IF**s are useful due to their unifying power to cover many contraction conditions, which is clear from the examples presented in Popa (1999).

Meanwhile, fractional calculus and fractional differential equations (**FDE**s)/ fractional integral equations (**FIE**s) fascinated researchers, and many significant results demonstrating the applicability of **FP** theory in the existence and uniqueness of solutions for **FDE**s/**FIE**s have already been presented in Alsaedi et al. (2016), Agarwal et al. (2012), Aydi et al. (2014), Baleanu et al. (2013), Jain et al. (2021), Karapinar and Samet (2012), Karapinar et al. (2013, 2019) and the references cited therein.

14.2 Theorems with Several Consequences

In this section, a new modified version of **IF**s and examples, as presented in Aliouche and Djoudi (2007), Popa and Mocanu (2009), Jain et al. (2021), are discussed.

Let $\mathbb{H}$ be a collection of all continuous functions $H : [0, \infty)^6 \to (-\infty, \infty)$ satisfying the following conditions:

(H_1) $H(\mu_1, \mu_2, \ldots, \mu_6)$ is non-increasing in variables $\mu_2, \mu_3, \ldots, \mu_6$;
(H_2) $\forall \varkappa, \upsilon \geq 0, \eta \geq 1, \exists 0 < \lambda < 1$ such that(s.t.) $H(\eta\varkappa, \upsilon, \upsilon, \varkappa, \varkappa + \upsilon, 0) \leq 0$ or $H(\eta\varkappa, \upsilon, 0, \varkappa + \upsilon, \varkappa, \upsilon) \leq 0 \implies \eta\varkappa \leq \lambda\, \upsilon$;
(H_3) for $\varkappa > 0$ and $\eta \geq 1$, $H(\eta\varkappa, 0, \varkappa, 0, 0, \varkappa) > 0$ or $H(\eta\varkappa, 0, 0, \varkappa, 0, \varkappa) > 0$.

Example 14.2.1 Let $H(\mu_1, \mu_2, \mu_3, \mu_4, \mu_5, \mu_6) = \mu_1 - \sigma(\max\{\mu_2, \frac{\mu_3+\mu_4}{2}, \frac{\mu_5+\mu_6}{2}\})$, where $\sigma : [0, \infty) \to [0, \infty)$ is non-decreasing function with $\sigma(0) = 0$ and $\sigma(\mu) < \mu$, for $\mu > 0$. Here, (H_1) is obvious.
(H_2): $H(\eta\varkappa, \upsilon, \upsilon, \varkappa, \varkappa + \upsilon, 0) = \eta\varkappa - \sigma(\max\{\upsilon, \frac{\upsilon+\varkappa}{2}, \frac{\varkappa+\upsilon}{2}\}) \leq 0$. If $\upsilon \leq \varkappa$, then $\upsilon \leq \frac{\varkappa+\upsilon}{2} \leq \varkappa$, so $\eta\varkappa < \sigma(\varkappa) < \varkappa$, a contradiction. Thus, $\varkappa < \upsilon$ and $\eta\varkappa \leq \sigma(\upsilon) = \frac{\upsilon}{2}$, i.e., $\eta\varkappa \leq \lambda\upsilon$, where $\lambda \in (0, 1)$.
(H_3) : $H(\eta\varkappa, 0, \varkappa, 0, 0, \varkappa) = \eta\varkappa - \sigma(\max\{0, \frac{\varkappa}{2}, \frac{\varkappa}{2}\}) = \eta\varkappa - \sigma(\varkappa) > 0$, for $\varkappa > 0$.

Example 14.2.2 Let $H(\mu_1, \mu_2, \mu_3, \mu_4, \mu_5, \mu_6) = \mu_1 - \lambda \max\{\mu_2, \frac{\mu_3+\mu_4}{2}, \frac{\mu_5+\mu_6}{2}\}$, $\lambda \in (0, 1)$. Similarly as Example 14.2.1.

Example 14.2.3 Let $H(\mu_1, \mu_2, \mu_3, \mu_4, \mu_5, \mu_6) = \mu_1 - \mathfrak{a}\,\mu_2 - \mathfrak{b}\,[\mu_3 + \mu_4] - \mathfrak{c}\,[\mu_5 + \mu_6]$, where $\mathfrak{a}, \mathfrak{b}, \mathfrak{c} \geq 0$ with $\mathfrak{a} + 2\,\mathfrak{b} + 2\,\mathfrak{c} < 1$.
(H_1) is obvious.
(H_2) : $H(\eta\varkappa, \upsilon, \upsilon, \varkappa, \varkappa + \upsilon, 0) = \eta\varkappa - \mathfrak{a}\,\upsilon - \mathfrak{b}\,[\upsilon + \varkappa] - \mathfrak{c}\,[\varkappa + \upsilon] \leq 0$. Then $(\eta - \mathfrak{c})\varkappa \leq (\mathfrak{a} + \mathfrak{b} + \mathfrak{c})\upsilon$, i.e., $\eta\varkappa \leq \lambda\,\upsilon$, where $\mathfrak{c} = 0$, $\lambda = \mathfrak{a} + \mathfrak{b} \in (0, 1)$, $\eta \geq 1$.
(H_3): $H(\eta\varkappa, 0, \varkappa, 0, 0, \varkappa) = \eta\varkappa - (\mathfrak{b} + \mathfrak{c})\varkappa > 0$.

Example 14.2.4 Let $H(\mu_1, \mu_2, \mu_3, \mu_4, \mu_5, \mu_6) = \mu_1 - \sigma(\mu_2, \mu_3, \mu_4, \mu_5, \mu_6)$, where
$\sigma : [0, \infty)^5 \to [0, \infty)$ is an upper semi continuous function s.t.
$\max\{\sigma(0, \mu, 0, 0, \mu), \sigma(0, 0, \mu, \mu, 0), \sigma(\mu, 0, \mu, \mu, \mu)\} < \mu$, for each $\mu > 0$.

Let $\mathbb{N}$, $\mathbb{W}$, $\mathbb{R}$, $\mathbb{R}^+$, $\mathrm{C}(\Omega, \Theta)$, and $\mathrm{Fix}(\Omega, \Theta)$ denote the sets of natural numbers, whole numbers, real numbers, positive real numbers, coincidence points of the mappings Ω and Θ, and the common fixed points (**CFP**) of Ω and Θ, respectively. Let I represent the identity mapping.

First, we establish the coincidence theorem for a pair of "α-**Adm** mappings."

Theorem 14.2.5 *Let* (Ξ, m) *be a* ***CMS*** *and* $\alpha : \Xi \times \Xi \to [0, \infty)$ *be a function. Let* $\Omega, \Theta : \Xi \to \Xi$ *be two mappings s.t.* $\forall\, \kappa, \vartheta \in \Xi$, $\exists\, H \in \mathbb{H}$ *with*

$$H\big(\alpha(\Theta\kappa, \Theta\vartheta)\mathrm{m}(\Omega\kappa, \Omega\vartheta), \mathrm{m}(\Theta\kappa, \Theta\vartheta), \mathrm{m}(\Omega\kappa, \Theta\kappa), \mathrm{m}(\Omega\vartheta, \Theta\vartheta), \mathrm{m}(\Theta\kappa, \Omega\vartheta), \mathrm{m}(\Theta\vartheta, \Omega\kappa)\big) \leq 0. \tag{14.1}$$

Also, Suppose

(i) $\Omega(\Xi) \subset \Theta(\Xi)$, $\Theta(\Xi)$ *is a closed subspace;*
(ii) Ω *is an* α-***Adm*** *mapping with respect to* Θ*;*
(iii) $\exists\, \kappa_0 \in \Xi$ *s.t.* $\alpha(\Theta\kappa_0, \Omega\kappa_0) \geq 1$*;*
(iv) *if* $\{\Theta\kappa_i\}$ *is a sequence in* Ξ *with* $\alpha(\Theta\kappa_i, \Theta\kappa_{i+1}) \geq 1$, $\forall\, i$ *and* $\lim_{i\to\infty} \Theta\kappa_i = \Theta r \in \Theta(\Xi)$, *then* $\exists$ *a subsequence* $\{\Theta\kappa_{i(k)}\}$ *of* $\{\Theta\kappa_i\}$ *with* $\alpha(\Theta\kappa_{i(k)}, \Theta r) \geq 1$, $\forall\, k$.

Then $\mathrm{C}(\Omega, \Theta) \neq \emptyset$.

Proof Let $\kappa_0 \in \Xi$ be arbitrary s.t. $\alpha(\Theta\kappa_0, \Omega\kappa_0) \geq 1$. As $\Omega(\Xi) \subset \Theta(\Xi)$, so, $\Omega\kappa_0 = \Theta\kappa_1$ for $\kappa_1 \in \Xi$. For κ_1, $\exists\, \kappa_2 \in \Xi$ s.t. $\Omega\kappa_1 = \Theta\kappa_2$. Inductively, $\exists\, \kappa_{i+1} \in \Xi$ s.t.

$$\upsilon_i = \Omega\kappa_i = \Theta\kappa_{i+1}, i \in \mathbb{W}. \tag{14.2}$$

As Ω is an α-**Adm** mapping with respect to Θ and $\alpha(\Theta\kappa_0, \Theta\kappa_1) = \alpha(\Theta\kappa_0, \Omega\kappa_0) \geq 1$, so, $\alpha(\Omega\kappa_0, \Omega\kappa_1) \geq 1$, i.e., $\alpha(\Theta\kappa_1, \Theta\kappa_2) \geq 1$. Inductively, we get

$$\alpha(\Theta\kappa_i, \Theta\kappa_{i+1}) \geq 1 \forall\, i \in \mathbb{W}. \tag{14.3}$$

If $\Omega\kappa_i = \Omega\kappa_{i+1}$ for some i, then from equation (14.2), $\Omega\kappa_{i+1} = \Theta\kappa_{i+1}$, i.e., $\kappa_{i+1} \in C(\Omega, \Theta)$. Suppose $\Omega\kappa_i \neq \Omega\kappa_{i+1}$, i.e., $\mathrm{m}(\Omega\kappa_i, \Omega\kappa_{i+1}) > 0 \;\forall\; i$. From inequalities (14.1) and (14.3), we have

$$H\big(\alpha(\Theta\kappa_i, \Theta\kappa_{i+1})\mathrm{m}(\Omega\kappa_i, \Omega\kappa_{i+1}), \mathrm{m}(\Theta\kappa_i, \Theta\kappa_{i+1}), \mathrm{m}(\Omega\kappa_i, \Theta\kappa_i), \mathrm{m}(\Omega\kappa_{i+1}, \Theta\kappa_{i+1}), \mathrm{m}(\Theta\kappa_i, \Omega\kappa_{i+1}), \mathrm{m}(\Theta\kappa_{i+1}, \Omega\kappa_i)\big) \leq 0.$$

Using equation (14.2), we have

$$H\big(\alpha(\vartheta_{i-1}, \vartheta_i)\mathrm{m}(\vartheta_i, \vartheta_{i+1}), \mathrm{m}(\vartheta_{i-1}, \vartheta_i), \mathrm{m}(\vartheta_i, \vartheta_{i-1}), \mathrm{m}(\vartheta_{i+1}, \vartheta_i), \mathrm{m}(\vartheta_{i+1}, \vartheta_{i-1}), \mathrm{m}(\vartheta_i, \vartheta_i)\big) \leq 0.$$

Using triangular inequality, we have

$$\mathrm{m}(\vartheta_{i+1}, \vartheta_{i-1}) \leq \mathrm{m}(\vartheta_{i+1}, \vartheta_i) + \mathrm{m}(\vartheta_i, \vartheta_{i-1}). \tag{14.4}$$

From inequality (14.4) and property (H_1),

$$H\big(\alpha(\vartheta_{i-1}, \vartheta_i)\mathrm{m}(\vartheta_i, \vartheta_{i+1}), \mathrm{m}(\vartheta_{i-1}, \vartheta_i), \mathrm{m}(\vartheta_i, \vartheta_{i-1}), \mathrm{m}(\vartheta_{i+1}, \vartheta_i), \mathrm{m}(\vartheta_{i+1}, \vartheta_i) + \mathrm{m}(\vartheta_i, \vartheta_{i-1}), 0\big) \leq 0.$$

By property (H_2), $\exists\, \lambda \in (0, 1)$ s.t.

$$\mathrm{m}(\vartheta_i, \vartheta_{i+1}) \leq \alpha(\vartheta_{i-1}, \vartheta_i)\mathrm{m}(\vartheta_i, \vartheta_{i+1}) \leq \lambda\, \mathrm{m}(\vartheta_{i-1}, \vartheta_i).$$

Inductively, we get $\mathrm{m}(\vartheta_i, \vartheta_{i+1}) \leq \lambda^i\, \mathrm{m}(\vartheta_0, \vartheta_1), \forall\, i \geq 1$.
For $j > 0$,

$$\begin{aligned}\mathrm{m}(\vartheta_i, \vartheta_{i+j}) &\leq \mathrm{m}(\vartheta_i, \vartheta_{i+1}) + \mathrm{m}(\vartheta_{i+1}, \vartheta_{i+2}) + \ldots + \mathrm{m}(\vartheta_{i+j-1}, \vartheta_{i+j}) \\ &\leq \lambda^i \mathrm{m}(\vartheta_0, \vartheta_1) + \lambda^{i+1}\mathrm{m}(\vartheta_0, \vartheta_1) + \ldots + \lambda^{i+j-1} \mathrm{m}(\vartheta_0, \vartheta_1) \\ &= \mathrm{m}(\vartheta_0, \vartheta_1) \sum_{n=i}^{i+j-1} \lambda^n \leq \mathrm{m}(\vartheta_0, \vartheta_1) \sum_{n=i}^{\infty} \lambda^n \\ &= \frac{\lambda^i}{1-\lambda}\mathrm{m}(\vartheta_0, \vartheta_1).\end{aligned}$$

As $i \to \infty, \mathrm{m}(\vartheta_i, \vartheta_{i+j}) \to 0$. Hence, the sequence $\{\vartheta_i\}$ is a Cauchy sequence in Ξ. As Ξ is a **CMS**, $\Theta(\Xi)$ is closed, so $\Theta(\Xi)$ is complete and $\vartheta_i = \Omega\kappa_i = \Theta\kappa_{i+1} \in \Theta(\Xi)$, for each i, $\exists\, r \in \Xi$ s.t.

$$\lim_{i\to\infty} \Theta\kappa_i = \Theta r. \tag{14.5}$$

From inequality (14.3), Eq. (14.5) and assertion (iv), we have

$$\alpha(\Theta\kappa_{i(k)}, \Theta r) \geq 1, \ \forall\, k. \tag{14.6}$$

Suppose $\mathrm{m}(\Omega r, \Theta r) > 0$. Taking $\vartheta = \kappa_{i(k)}, \kappa = r$ in (14.1), we get

$$H\big(\alpha(\Theta r, \Theta\kappa_{i(k)})\mathrm{m}(\Omega r, \Omega\kappa_{i(k)}), \mathrm{m}(\Theta r, \Theta\kappa_{i(k)}), \mathrm{m}(\Omega r, \Theta r), \mathrm{m}(\Omega\kappa_{i(k)}, \Theta\kappa_{i(k)}), \mathrm{m}(\Theta r, \Omega\kappa_{i(k)}), \mathrm{m}(\Theta\kappa_{i(k)}, \Omega r)\big) \leq 0.$$

Proceeding with $k \to \infty$, we have

$$H\big(\lim_{k\to\infty} \alpha(\Theta r, \Theta\kappa_{i(k)})\mathrm{m}(\Omega r, \Theta r), 0, \mathrm{m}(\Omega r, \Theta r), 0, 0, \mathrm{m}(\Theta r, \Omega r)\big) \leq 0, \tag{14.7}$$

a contradiction to the property (H_3). Therefore, $\mathrm{m}(\Omega r, \Theta r) = 0$, i.e., $\Omega r = \Theta r$. Hence, $r \in \mathrm{C}(\Omega, \Theta)$, i.e., $\mathrm{C}(\Omega, \Theta) \neq \emptyset$. □

Next, we investigate the following **CFP** theorem:

Theorem 14.2.6 *In addition to all the conditions of Theorem 14.2.5, suppose that* $\forall\, \varkappa, \upsilon \in \mathrm{C}(\Omega, \Theta)$, $\exists\, r \in \Xi$ *s.t.* $\alpha(\Theta\varkappa, \Theta r) \geq 1$ *and* $\alpha(\Theta\upsilon, \Theta r) \geq 1$ *and* (Ω, Θ) *is* ***WC****. Then, Fix*(Ω, Θ) *has a unique element of* Ξ.

Proof From Theorem 14.2.5, $\mathrm{C}(\Omega, \Theta) \neq \emptyset$.
(a) Suppose $\varkappa, \upsilon \in \mathrm{C}(\Omega, \Theta)$. We claim that $\Theta\varkappa = \Theta\upsilon$. By assertion, $\exists\, r_0 \in \Xi$ s.t.

$$\alpha(\Theta\varkappa, \Theta r_0) \geq 1 \text{ and } \alpha(\Theta\upsilon, \Theta r_0) \geq 1. \tag{14.8}$$

As $\Omega(\Xi) \subset \Theta(\Xi)$, $\exists\, r_1 \in \Xi$ s.t. $\Omega r_0 = \Theta r_1$. For this r_1, $\exists\, r_2 \in \Xi$ s.t. $\Omega r_1 = \Theta r_2$. Inductively, define a sequence $\{r_i\} \in \Xi$, by

$$\Omega r_i = \Theta r_{i+1}, \ \forall\, i \geq 0. \tag{14.9}$$

As Ω is an $\alpha-$ **Adm** mapping with respect to Θ, so inequality (14.8) gives that

$$\alpha(\Omega\varkappa, \Omega r_0) \geq 1 \text{ and } \alpha(\Omega\upsilon, \Omega r_0) \geq 1 \tag{14.10}$$

From equation (14.9) and inequality (14.10), we have

$$\alpha(\Theta\varkappa, \Theta r_1) = \alpha(\Omega\varkappa, \Omega r_0) \geq 1 \text{ and } \alpha(\Theta\upsilon, \Theta r_1) = \alpha(\Omega\upsilon, \Omega r_0) \geq 1$$

Using induction, we have

$$\alpha(\Theta\varkappa, \Theta r_i) \geq 1 \text{ and } \alpha(\Theta\upsilon, \Theta r_i) \geq 1. \tag{14.11}$$

From inequality (14.1), equation (14.9) and inequality (14.11), we have

$$H\big(\alpha(\Theta\varkappa, \Theta r_i)\mathrm{m}(\Theta\varkappa, \Theta r_{i+1}), \mathrm{m}(\Theta\varkappa, \Theta r_i), 0, \mathrm{m}(\Theta r_{i+1}, \Theta r_i), \mathrm{m}(\Theta\varkappa, \Theta r_{i+1}), \mathrm{m}(\Theta r_i, \Theta\varkappa)\big) \leq 0. \tag{14.12}$$

Using triangular inequality and property of (H_1), equality (14.12) gives

$$H\big(\alpha(\Theta\varkappa, \Theta r_i)\mathrm{m}(\Theta\varkappa, \Theta r_{i+1}), \mathrm{m}(\Theta\varkappa, \Theta r_i), 0, \mathrm{m}(\Theta r_{i+1}, \Theta\varkappa) + \mathrm{m}(\Theta\varkappa, \Theta r_i), \mathrm{m}(\Theta\varkappa, \Theta r_{i+1}), \mathrm{m}(\Theta r_i, \Theta\varkappa)\big) \leq 0.$$

By property (H_2), $\exists\, \lambda \in (0, 1)$ s.t.

$$\mathrm{m}(\Theta\varkappa, \Theta r_{i+1}) \leq \alpha(\Theta\varkappa, \Theta r_i)\mathrm{m}(\Theta\varkappa, \Theta r_{i+1}) \leq \lambda\, \mathrm{m}(\Theta\varkappa, \Theta r_i).$$

But $\mathrm{m}(\Theta\varkappa, \Theta r_{i+1}) \leq \lambda\, \mathrm{m}(\Theta\varkappa, \Theta r_i) \leq \ldots \leq \lambda^i\, \mathrm{m}(\Theta\varkappa, \Theta r_0)\ \forall\, i \geq 0$.

Letting $i \to \infty$, we have $\lim_{i\to\infty} \mathrm{m}(\Theta\varkappa, \Theta r_i) = 0$. Similarly, $\lim_{i\to\infty} \mathrm{m}(\Theta\upsilon, \Theta r_i) = 0$.

By uniqueness of limit, we have

$$\Theta\varkappa = \Theta\upsilon. \tag{14.13}$$

(b) Now, $\varkappa \in \mathrm{C}(\Omega, \Theta)$ and (Ω, Θ) is **WC**, so, we have

$$\Theta\Theta\varkappa = \Theta\Omega\varkappa = \Omega\Theta\varkappa. \tag{14.14}$$

Let $\Theta\varkappa = s$, then equation (14.14) $\implies \Theta s = \Omega s$, i.e., $s \in \mathrm{C}(\Omega, \Theta)$. From (a), we have $\Theta s = \Theta\varkappa$. Hence, $\Omega s = \Theta s = \Theta\varkappa = s$, i.e., $s \in \mathrm{Fix}(\Omega, \Theta)$.

For uniqueness: Suppose $t \in \mathrm{Fix}(\Omega, \Theta)$. Obviously, $t \in \mathrm{C}(\Omega, \Theta)$. Following (a), we have $\Theta s = \Theta t$ which gives $s = t$. Hence, $\mathrm{Fix}(\Omega, \Theta) = \{s\}$. □

Taking $\Theta = I$ in Theorems 14.2.5 and 14.2.6, the following **FP** theorem is obtained:

Theorem 14.2.7 *Let* (Ξ, m) *be a* ***CMS*** *and* $\alpha : \Xi \times \Xi \to [0, \infty)$ *be a function. Let* $\Omega : \Xi \to \Xi$ *be an* α*-**Adm** mapping s.t.*$\forall\, \kappa, \vartheta \in \Xi$, $\exists\, H \in \mathbb{H}$ *with*

$$H\big(\alpha(\kappa, \vartheta)\mathrm{m}(\Omega\kappa, \Omega\vartheta), \mathrm{m}(\kappa, \vartheta), \mathrm{m}(\Omega\kappa, \kappa), \mathrm{m}(\Omega\vartheta, \vartheta), \mathrm{m}(\kappa, \Omega\vartheta), \mathrm{m}(\vartheta, \Omega\kappa)\big) \leq 0. \tag{14.15}$$

Also, the following assertion holds:

(i) $\exists\, \kappa_0 \in \Xi$ *s.t.* $\alpha(\kappa_0, \Omega\kappa_0) \geq 1$,
(ii) *if* $\{\kappa_i\} \subset \Xi$ *is a sequence with* $\alpha(\kappa_i, \kappa_{i+1}) \geq 1, \forall\, i$ *and* $\lim_{i\to\infty} \kappa_i = r \in \Xi$, *then* $\exists$ *a subsequence* $\{\kappa_{i(k)}\} \subset \{\kappa_i\}$ *with* $\alpha(\kappa_{i(k)}, r) \geq 1$, *for all* k,
(iii) $\forall\, \kappa, \vartheta \in \Xi$, $\exists\, r \in \Xi$ *s.t.* $\alpha(\kappa, r) \geq 1$ *and* $\alpha(\vartheta, r) \geq 1$.

Then, the set of ***FP*** *of* Ω, *that is, Fix*(Ω) *is a singleton set.*

14.3 Consequences

Some Well-Known Fixed Point Theorems

Corollary 14.3.1 *The outcomes of Theorem 14.2.5 and Theorem 14.2.6 still hold if the condition (14.1) is replaced by any one of the following:*

(i) $\alpha(\Theta\kappa, \Theta\vartheta)\mathrm{m}(\Omega\kappa, \Omega\vartheta) \leq \sigma(\max\{\mathrm{m}(\Theta\kappa, \Theta\vartheta), \mathrm{m}(\Omega\kappa, \Theta\kappa), \mathrm{m}(\Omega\vartheta, \Theta\vartheta), \mathrm{m}(\Theta\kappa, \Omega\vartheta), \mathrm{m}(\Theta\vartheta, \Omega\kappa)\})$, *where* $\sigma : [0, \infty) \to [0, \infty)$ *is non-decreasing function with* $\sigma(0) = 0$ *and* $\sigma(\mu) < \mu$, *for* $\mu > 0$;
(ii) $\alpha(\Theta\kappa, \Theta\vartheta)\mathrm{m}(\Omega\kappa, \Omega\vartheta) \leq \lambda \max\{\mathrm{m}(\Theta\kappa, \Theta\vartheta), \mathrm{m}(\Omega\kappa, \Theta\kappa), \mathrm{m}(\Omega\vartheta, \Theta\vartheta), \mathrm{m}(\Theta\kappa, \Omega\vartheta), \mathrm{m}(\Theta\vartheta, \Omega\kappa)\}$, *where* $\lambda \in (0, 1)$;
(iii) $\alpha(\Theta\kappa, \Theta\vartheta)\mathrm{m}(\Omega\kappa, \Omega\vartheta) \leq \mathfrak{a}\,\mathrm{m}(\Theta\kappa, \Theta\vartheta) + \mathfrak{b}\,[\mathrm{m}(\Omega\kappa, \Theta\kappa) + \mathrm{m}(\Omega\vartheta, \Theta\vartheta)] + \mathfrak{c}\,[\mathrm{m}(\Theta\kappa, \Omega\vartheta) + \mathrm{m}(\Theta\vartheta, \Omega\kappa)]$, *where* $\mathfrak{a}, \mathfrak{b}, \mathfrak{c} \geq 0$ *with* $0 < \mathfrak{a} + 2\mathfrak{b} + 2\mathfrak{c} < 1$;
(iv) $\alpha(\Theta\kappa, \Theta\vartheta)\mathrm{m}(\Omega\kappa, \Omega\vartheta) \leq \sigma(\mathrm{m}(\Theta\kappa, \Theta\vartheta), \mathrm{m}(\Omega\kappa, \Theta\kappa), \mathrm{m}(\Omega\vartheta, \Theta\vartheta), \mathrm{m}(\Theta\kappa, \Omega\vartheta), \mathrm{m}(\Theta\vartheta, \Omega\kappa))$, *where* $\sigma : [0, \infty)^5 \to [0, \infty)$ *is an upper semi continuous function with* $\max\{\sigma(0, \mu, 0, 0, \mu), \sigma(0, 0, \mu, \mu, 0), \sigma(\mu, 0, \mu, \mu, \mu)\} < \mu$, *for each* $\mu > 0$.

Proof The proof follows from Examples 14.2.1–14.2.4, Theorems 14.2.5 and 14.2.6. □

Remarks:

1. From Corollary 14.3.1(i), we get the result of Sahi et al. (2015).
2. Taking $\alpha(\kappa, \vartheta) = 1, \forall\, \kappa, \vartheta \in \Xi$ and $\Theta = I$ in Corollary 14.3.1(ii)-(iv), we get the results of Ćirić (1971), Hardy and Rogers (1973) and Chugh and Kumar (2001), respectively.
3. Taking $\Theta = I$ Corollary 14.3.1(i) gives the result of Karapinar and Samet (2012).

Fixed Point and Metric Spaces Equipped With Partial Drder

In the last two decades, the study of **FP** theorems of contractive mappings in **MS** equipped with partial order (P.O.) has developed rapidly. Turinici (1986) presented the first result in this direction, which was applied to matrix equations by Ran and Reurings (2004). Later, various authors established the existence of **FP** for contractive mappings in P.O. **MS**; see, for example, Bhaskar and Lakshmikantham (2006), Jain et al. (2013), Nieto and Lopez (2005), Samet (2010). In this section, we apply Theorems 14.2.5 to 14.2.7 to derive various **FP** results for a mapping equipped with a P.O. First, we list some basic definitions as follows:

Definition 14.3.2 (Karapinar and Samet (2012)) Let $(\Xi, \preccurlyeq)$ be a P.O. set and Ω be a self mapping on Ξ. Ω is said to be non-decreasing with respect to $\preccurlyeq$ if $\Omega\kappa \preccurlyeq \Omega\vartheta$, for $\kappa, \vartheta \in \Xi$ with $\kappa \preccurlyeq \vartheta$.

Definition 14.3.3 (Karapinar and Samet (2012)) Let $(\Xi, \preccurlyeq)$ be a P.O. set, the sequence $\{\kappa_i\} \subset \Xi$ is said to be non-decreasing with respect to $\preccurlyeq$ if $\kappa_i \preccurlyeq \kappa_{i+1}$, $\forall\, i$.

Definition 14.3.4 (Karapinar and Samet (2012)) Let $(\Xi, \mathtt{m})$ be an **MS** equipped with a P.O. relation $\preccurlyeq$. Ξ is said to be regular if for every non-decreasing sequence $\{\kappa_i\} \subset \Xi$ s.t. $\lim_{i\to\infty} \kappa_i = r \in \Xi$, $\exists$ a subsequence $\{\kappa_{i(k)}\} \subset \{\kappa_i\}$ with $\kappa_{i(k)} \preccurlyeq r, \forall\, k$.

Definition 14.3.5 (Ćirić et al. (2008)) Let $(\Xi, \preccurlyeq)$ be a P.O. set with metric $\mathtt{m}$ and $\Omega, \Theta : \Xi \to \Xi$ be two mappings. Ω is called $\Theta-$non-decreasing mapping if for $\kappa, \vartheta \in \Xi$

$$\Theta\kappa \preccurlyeq \Theta\vartheta \quad \Longrightarrow \quad \Omega\kappa \preccurlyeq \Omega\vartheta. \tag{14.16}$$

Definition 14.3.6 (Sahi et al. (2015)) Let $(\Xi, \mathtt{m})$ be an **MS** equipped with a P.O. relation $\preccurlyeq$ and Ω be a self mapping defined on Ξ. Then Ξ is said to be Ω-regular if for every non-decreasing sequence $\{\Omega\kappa_i\} \subset \Xi$ with $\lim_{i\to\infty} \Omega\kappa_i = \Omega r \in \Xi$, $\exists$ a subsequence $\{\Omega\kappa_{i(k)}\} \subset \{\Omega\kappa_i\}$ s.t. $\Omega\kappa_{i(k)} \preccurlyeq \Omega r, \forall\, k$.

Theorem 14.3.7 *Let $(\Xi, \preccurlyeq, \mathtt{m})$ be a **CMS** equipped with P.O. relation $\preccurlyeq$. Let $\Omega, \Theta : \Xi \to \Xi$ be two mappings s.t. $\Omega(\Xi) \subset \Theta(\Xi)$ and $\Theta(\Xi)$ is closed. Suppose Ω be $\Theta-$non-decreasing mapping with respect to $\preccurlyeq$ s.t.$\forall\, \kappa, \vartheta \in \Xi$ with $\Theta\kappa \preccurlyeq \Theta\vartheta$, $\exists$ a function $H \in \mathbb{H}$ satisfying*

$$H\big(\mathtt{m}(\Omega\kappa, \Omega\vartheta), \mathtt{m}(\Theta\kappa, \Theta\vartheta), \mathtt{m}(\Omega\kappa, \Theta\kappa), \mathtt{m}(\Omega\vartheta, \Theta\vartheta), \mathtt{m}(\Theta\kappa, \Omega\vartheta), \mathtt{m}(\Theta\vartheta, \Omega\kappa)\big) \leq 0. \tag{14.17}$$

*Also, if $(\Xi, \preccurlyeq, \mathtt{m})$ is $\Theta-$regular and $\exists\, \kappa_0 \in \Xi$ s.t. $\Theta\kappa_0 \preccurlyeq \Omega\kappa_0$. Then $\mathrm{C}(\Omega, \Theta) \neq \emptyset$. Moreover, if $\forall\, \kappa, \vartheta \in \mathrm{C}(\Omega, \Theta)$, $\exists\, r \in \Xi$ s.t. $\Theta\kappa \preccurlyeq \Theta r$ and $\Theta\vartheta \preccurlyeq \Theta r$ and (Ω, Θ) is **WC**, then Fix(Ω, Θ) has a unique element of Ξ.*

Proof Define a function $\alpha : \Xi \times \Xi \to [0, \infty)$ as follows

$$\alpha(\kappa, \vartheta) = \begin{cases} 1, & \Theta\kappa \preccurlyeq \Theta\vartheta \text{ or } \Theta\vartheta \preccurlyeq \Theta\kappa; \\ 0, & \text{otherwise} . \end{cases}$$

Obviously, (Ω, Θ) satisfy (14.1). Let $\kappa_0 \in \Xi$ be s.t. $\Theta\kappa_0 \preccurlyeq \Omega\kappa_0$. From definition of α, $\alpha(\Theta\kappa_0, \Omega\kappa_0) \geq 1$. As Ω is a $\Theta-$non-decreasing mapping with respect to $\preccurlyeq$, so, $\forall\kappa, \vartheta \in \Xi$,

$$\alpha(\Theta\kappa, \Theta\vartheta) \geq 1 \implies \Theta\kappa \preccurlyeq \Theta\vartheta \text{ or } \Theta\vartheta \preccurlyeq \Theta\kappa \implies \Omega\kappa \preccurlyeq \Omega\vartheta \text{ or } \Omega\vartheta \preccurlyeq \Omega\kappa.$$

By definition of α, $\alpha(\Omega\kappa, \Omega\vartheta) \geq 1$. Hence, Ω is an α-**Adm** mapping with respect to Θ.

Let $\{\Theta\kappa_i\} \subset \Xi$ be a sequence with $\alpha(\Theta\kappa_i, \Theta\kappa_{i+1}) \geq 1$, $\forall\ i$ and $\lim_{i\to\infty} \Theta\kappa_i = \Theta r \in \Xi$. As $(\Xi, \preccurlyeq, \mathrm{m})$ is $\Theta-$regular, $\exists$ a subsequence $\{\Theta\kappa_{i(k)}\} \subset \{\Theta\kappa_i\}$ satisfying $\Theta\kappa_{i(k)} \preccurlyeq \Theta r, \forall k$. From definition of α, $\alpha(\Theta\kappa_{i(k)}, \Theta r) \geq 1$. Applying Theorem 14.2.5 $\mathrm{C}(\Omega,\Theta) \neq \emptyset$.

Further, let $\kappa, \vartheta \in \mathrm{C}(\Omega, \Theta)$, then by given condition $\exists\ r \in \Xi$ s.t. $\Theta\kappa \preccurlyeq \Theta r$ and $\Theta\vartheta \preccurlyeq \Theta r$. By definition of α, we have $\alpha(\Theta\kappa, \Theta r) \geq 1$ and $\alpha(\Theta\vartheta, \Theta r) \geq 1$. Following Theorem 14.2.6, the existence and uniqueness of a **CFP** of mappings Ω and Θ is established. □

Cyclic Contractive Mappings and Fixed Point

Kirk et al. (2003) introduced the notion of "cyclic mappings and cyclic contractions" to generalize the **BC** as follows:

Definition 14.3.8 (Kirk et al. (2003)) Let (Ξ, m) be a **CMS** and $X, Y (\neq \emptyset) \subset \Xi$. A mapping $\Omega : X \cup Y \to X \cup Y$ is called

(i) cyclic if $\Omega(X) \subseteq Y$ and $\Omega(Y) \subseteq X$.
(ii) cyclic contraction if $\exists\, 0 < \mathfrak{c} < 1$ s.t. $\mathrm{m}(\Omega\kappa, \Omega\vartheta) \leq \mathfrak{c}\,\mathrm{m}(\kappa, \vartheta), \forall\, \kappa \in X$ and $\vartheta \in Y$.

After the result of Kirk et al. (2003) various **FP** results have appeared in the context of cyclic contractive mappings, for example (Agarwal et al. (2012); Karapinar (2011); Pacurar and Rus (2010); Petric (2010)).

Theorem 14.3.9 *Let (Ξ, m) be a **CMS** and $B_1, B_2(\neq \emptyset) \subset \Xi$ be s.t. $V = B_1 \cup B_2$. Let $\Omega, \Theta : V \to V$ be two mappings s.t.*

(i) $\Omega(B_1) \subseteq \Theta(B_2)$,$\Omega(B_2) \subseteq \Theta(B_1)$;
(ii) Θ is one-one and $\Theta(B_1), \Theta(B_2)$ are closed,
(iii) $\forall\ (\kappa, \vartheta) \in B_1 \times B_2$, there exists a function $H \in \mathbb{H}$ s.t.

$$H\big(\mathrm{m}(\Omega\kappa, \Omega\vartheta), \mathrm{m}(\Theta\kappa, \Theta\vartheta), \mathrm{m}(\Omega\kappa, \Theta\kappa), \mathrm{m}(\Omega\vartheta, \Theta\vartheta), \mathrm{m}(\Theta\kappa, \Omega\vartheta), \mathrm{m}(\Theta\vartheta, \Omega\kappa)\big) \leq 0.$$

*Then $\mathrm{C}(\Omega, \Theta)$ has atleast one point $r \in B_1 \cap B_2$. Moreover, if (Ω, Θ) is **WC**, then $Fix(\Omega, \Theta)$ has a unique element of $B_1 \cap B_2$.*

Proof As Θ is one-to-one, so condition (iii) is equivalent to

$$H\big(\mathrm{m}(\Omega\kappa, \Omega\vartheta), \mathrm{m}(\Theta\kappa, \Theta\vartheta), \mathrm{m}(\Omega\kappa, \Theta\kappa), \mathrm{m}(\Omega\vartheta, \Theta\vartheta), \mathrm{m}(\Omega\kappa, \Theta\vartheta)\mathrm{m}(\Omega\vartheta, \Theta\kappa)\big) \leq 0, \quad (14.18)$$

$\forall\ (\Theta\kappa, \Theta\vartheta) \in \Theta(B_1) \times \Theta(B_2)$. As (Ξ, m) is a **CMS** and B_1, B_2 are closed subsets of Ξ, so $V = B_1 \cup B_2$ is complete with respect to metric m.

Suppose $\alpha : V \times V \to [0, \infty)$ is a function defined by

$$\alpha(\kappa, \vartheta) = \begin{cases} 1, & (\kappa, \vartheta) \in (\Theta(B_1) \times \Theta(B_2)) \cup (\Theta(B_2) \times \Theta(B_1)); \\ 0, & \text{otherwise}. \end{cases}$$

From condition (14.18) and definition of α, (Ω, Θ) satisfies

$$H\big(\mathrm{m}(\Omega\kappa, \Omega\vartheta), \mathrm{m}(\Theta\kappa, \Theta\vartheta), \mathrm{m}(\Omega\kappa, \Theta\kappa), \mathrm{m}(\Omega\vartheta, \Theta\vartheta), \mathrm{m}(\Omega\kappa, \Theta\vartheta), \mathrm{m}(\Omega\vartheta, \Theta\kappa)\big) \leq 0,$$

$\forall\ \Theta\kappa \in \Theta(B_1), \Theta\vartheta \in \Theta(B_2)$ with $\alpha(\Theta\kappa, \Theta\vartheta) \geq 1$. From condition (i), $\Omega(V) \subseteq \Theta(V)$. Moreover, $\Theta(V)$ is closed.

Next, we claim that Ω is α-**Adm** with respect to Θ. Let $(\Theta\kappa, \Theta\vartheta) \in V \times V$ s.t. $\alpha(\Theta\kappa, \Theta\vartheta) \geq 1$, i.e., $(\Theta\kappa, \Theta\vartheta) \in (\Theta(B_1) \times \Theta(B_2)) \cup (\Theta(B_2) \times \Theta(B_1))$. Θ is one-to-one implies that $(\kappa, \vartheta) \in (B_1 \times B_2) \cup (B_2 \times B_1)$. Also, $\Omega(B_1) \subseteq \Theta(B_2)$ and $\Omega(B_2) \subseteq \Theta(B_1)$ gives that $(\Omega\kappa, \Omega\vartheta) \in (\Theta(B_1) \times \Theta(B_2)) \cup (\Theta(B_2) \times \Theta(B_1))$. Definition of $\alpha \implies \alpha(\Omega\kappa, \Omega\vartheta) \geq 1$. Hence, Ω is α-**Adm** with respect to Θ.

Suppose $\{\Theta\kappa_i\}$ be a sequence in Ξ s.t. $\alpha(\Theta\kappa_i, \Theta\kappa_{i+1}) \geq 1$, $\forall\ i$ and $\lim_{i\to\infty} \Theta\kappa_i = \Theta r \in \Theta(\Xi)$. Definition of $\alpha \implies (\Theta\kappa_i, \Theta\kappa_{i+1}) \in (\Theta(B_1) \times \Theta(B_2)) \cup (\Theta(B_2) \times \Theta(B_1))$.

As $(\Theta(B_1) \times \Theta(B_2)) \cup (\Theta(B_2) \times \Theta(B_1))$ is a closed set with respect to Euclidean metric, so $(\Theta r, \Theta r) \in (\Theta(B_1) \times \Theta(B_2)) \cup (\Theta(B_2) \times \Theta(B_1))$, which gives that $\Theta r \in \Theta(B_1) \cap \Theta(B_2)$. Hence, the definition of $\alpha \implies \alpha(\Theta\kappa_i, \Theta r) \geq 1$, $\forall\ i$.

Let $x \in B_1$ be arbitrary. From condition (i), $\Omega x \in \Theta(B_2)$. Also, $\Theta x \in \Theta(B_1)$. Therefore, $(\Theta x, \Omega x) \in \Theta(B_1) \times \Theta(B_2)$. Definition of $\alpha \implies \alpha(\Theta x, \Omega x) \geq 1$. Applying Theorem 14.2.5, Ω and Θ have a coincidence point $r \in V = B_1 \cup B_2$. Suppose $r \in B_1$. From assertion (ii), $\Omega r \in \Theta(B_2)$, but $\Omega r = \Theta r \in \Theta(B_1)$. Thus, $\Theta r \in \Theta(B_1) \cap \Theta(B_2)$. As Θ is one-to-one, so $r \in B_1 \cap B_2$. Similarly, for $r \in B_2$, we have $r \in B_1 \cap B_2$. Hence, $r \in \mathrm{C}(\Omega, \Theta)$ gives that $r \in B_1 \cap B_2$.

Let $\kappa, \vartheta \in \mathrm{C}(\Omega, \Theta)$, then $\kappa, \vartheta \in B_1 \cap B_2$ and $\Omega\kappa = \Theta\kappa$, $\Omega\vartheta = \Theta\vartheta$. Let $\Omega\kappa = r = \Theta\kappa$. As (Ω, Θ) is **WC**, so, $\Omega r = \Omega\Theta\kappa = \Theta\Omega\kappa = \Theta r$, i.e., $r \in \mathrm{C}(\Omega, \Theta)$. From the above discussion, $r \in B_1 \cap B_2$. Also, Θ is one-to-one, so, $\Theta\kappa, \Theta\vartheta, \Theta r \in \Theta(B_1 \cap B_2) = \Theta(B_1) \cap \Theta(B_2)$, i.e., $(\Theta\kappa, \Theta r), (\Theta\vartheta, \Theta r) \in (\Theta(B_1) \times \Theta(B_2)) \cup (\Theta(B_2) \times \Theta(B_1))$. Definition of $\alpha \implies \alpha(\Theta\kappa, \Theta r) \geq 1$ and $\alpha(\Theta\vartheta, \Theta r) \geq 1$. Hence, applying Theorem 14.2.6, we conclude that $r \in B_1 \cap B_2$ is a unique point belonging to $\mathrm{Fix}(\Omega, \Theta)$. □

14.4 Application to a Nonlinear Fractional Differential Equation with Singular Kernel

Let $\mathrm{p} \in \mathbb{R}^+$ and Γ be a gamma function. The "Caputo derivative of **FDE** of order p" of a continuous function $\mathcal{A} : [0, \infty) \to \mathbb{R}$ is defined as

$$\left({}_0^c\mathcal{D}^{\mathrm{p}}\mathcal{A}\right)(t) = \frac{1}{\Gamma(i-\mathrm{p})}\int_0^t (t-\xi)^{i-\mathrm{p}-1}\mathcal{A}^i(\xi)d\xi, \quad i = [\mathrm{p}] + 1. \tag{14.19}$$

It is clear that the kernel in equation (14.19) has a singularity.

In this section, we apply Theorem 14.2.7 to obtain the existence of solutions for the following nonlinear **FDE**:

$$\left({}_0^c\mathcal{D}^{\mathrm{p}}\mathrm{u}\right)(t) = \mathrm{M}(t, \mathrm{u}(t)), \quad t \in \mathcal{I} = (0, 1) \quad \text{and } \mathrm{p} \in (0, 2], \tag{14.20}$$

with the integral boundary conditions:

$$\mathrm{u}(0) = 0, \quad \mathrm{u}(1) = \int_0^{\mathrm{q}} \mathrm{u}(\xi)d\xi, \quad \mathrm{q} \in (0, 1), \tag{14.21}$$

where $\mathrm{u} \in C[0, 1]$ (the set of all real-valued continuous functions on $[0, 1]$) and $\mathrm{M} : \mathcal{I} \times \mathbb{R} \to \mathbb{R}$ is a continuous function.

Let $\Xi = C[0, 1]$. Define $\mathrm{m} : \Xi \times \Xi \to \mathbb{R}^+$ by $\mathrm{m}(\mathrm{u}, \nu) = \|\mathrm{u} - \nu\|_\infty, \quad \mathrm{u}, \nu \in \Xi$, where $\|\mathrm{u}\|_\infty = \sup_{t\in[0,1]} |\mathrm{u}(t)|$. Obviously, (Ξ, m) is a **CMS**.

Theorem 14.4.1 *Let $\Omega : \Xi \to \Xi$ be an operator defined by*

$$\Omega\mathrm{u}(t) = \frac{1}{\Gamma(\mathrm{p})}\int_0^t (t-\xi)^{\mathrm{p}-1}\mathrm{M}(\xi, \mathrm{u}(\xi))d\xi - \frac{2t}{(2-\mathrm{q}^2)\Gamma(\mathrm{p})}\int_0^1 (1-\xi)^{\mathrm{p}-1}\mathrm{M}(\xi, \mathrm{u}(\xi))d\xi$$
$$+\frac{2t}{(2-\mathrm{q}^2)\Gamma(\mathrm{p})}\int_0^{\mathrm{q}}\left(\int_0^{\xi}(\xi-\mathfrak{z})^{\mathrm{p}-1}\mathrm{M}(\mathfrak{z}, \mathrm{u}(\mathfrak{z}))d\mathfrak{z}\right)d\xi, \tag{14.22}$$

for $\mathrm{u} \in \Xi$ and $t \in \mathcal{I}$. Also, let $\tau : \mathbb{R}^2 \to \mathbb{R}$ be a given function and $\mathrm{M} : \mathcal{I} \times \mathbb{R} \to \mathbb{R}$ be a continuous function. Suppose that the following assertions hold:

(F_1) *$\exists$ a function $H \in \mathbb{H}$ s.t. $\forall\, t \in \mathcal{I}, \gamma, \delta \in \mathbb{R}$ with $\tau(\gamma, \delta) \geq 0$,*

$$\left|\mathrm{M}(t, \gamma) - \mathrm{M}(t, \delta)\right| \leq \mathfrak{N}_1\sigma\left(\max\{|\gamma-\delta|, |\Omega\gamma-\gamma|, |\Omega\delta-\delta|, |\Omega\gamma-\delta|, |\Omega\delta-\gamma|\}\right), \tag{14.23}$$

where $\mathfrak{N}_1 = \dfrac{\Gamma(\mathrm{p}+2)}{3\beta+5}$;

(F_2) $\exists\, \mathrm{u}_0 \in \Xi$ *s.t.* $\tau(\mathrm{u}_0(t), \Omega\mathrm{u}_0(t)) \geq 0, \forall\, t \in \mathcal{I}$;

(F_3) $\forall\, t \in \mathcal{I}$ *and* $\mathrm{u}, \nu \in \Xi, \tau(\mathrm{u}(t), \nu(t)) \geq 0$ *gives that* $\tau(\Omega\mathrm{u}(t), \Omega\nu(t)) \geq 0$;

(F_4) *let* $\{\mathrm{u}_i\}$ *be a sequence in* Ξ *s.t.* $\lim_{i\to\infty} \mathrm{u}_i = \mathrm{u} \in \Xi$. $\tau(\mathrm{u}_i(t), \mathrm{u}_{i+1}(t)) \geq 0$ *gives that* $\tau(\mathrm{u}_i(t), \mathrm{u}(t)) \geq 0, \forall\, t \in \mathcal{I}, i \in \mathbb{N}$.

Then, the differential equation (14.20) along with the integral boundary conditions (14.21) has at least one solution $\mathrm{u}' \in \Xi$.

Proof Clearly, assertions $(F_2) - (F_4)$ imply conditions (i)- (iii) of Theorem 14.2.7. Let $\mathrm{u}, \nu \in \Xi$ be s.t. $\forall\, t \in \mathcal{I}$, $\tau(\mathrm{u}(t), \nu(t)) \geq 0$. From (F_1), we have

$$
\begin{aligned}
|\Omega\mathrm{u}(t) - \Omega\nu(t)| =& \Big| \frac{1}{\Gamma(\mathrm{p})} \int_0^t (t-\xi)^{\mathrm{p}-1} \big(\mathrm{M}(\xi, \mathrm{u}(\xi)) - \mathrm{M}(\xi, \nu(\xi)\big) d\xi \\
& - \frac{2t}{(2-\mathrm{q}^2)\Gamma(\mathrm{p})} \int_0^1 (1-\xi)^{\mathrm{p}-1} \big(\mathrm{M}(\xi, \mathrm{u}(\xi)) - \mathrm{M}(\xi, \nu(\xi)\big) d\xi \\
& + \frac{2t}{(2-\mathrm{q}^2)\Gamma(\mathrm{p})} \int_0^{\mathrm{q}} \Big(\int_0^{\xi} (\xi - \mathfrak{z})^{\mathrm{p}-1} \big(\mathrm{M}(\mathfrak{z}, \mathrm{u}(\mathfrak{z})) - \mathrm{M}(\mathfrak{z}, \nu(\mathfrak{z})\big) d\mathfrak{z} \Big) d\xi \Big| \\
\leq & \frac{1}{\Gamma(\mathrm{p})} \int_0^t |(t-\xi)|^{\mathrm{p}-1} \big|\mathrm{M}(\xi, \mathrm{u}(\xi)) - \mathrm{M}(\xi, \nu(\xi))\big| d\xi \\
& - \frac{2t}{(2-\mathrm{q}^2)\Gamma(\mathrm{p})} \int_0^1 |1-\xi|^{\mathrm{p}-1} \big|\mathrm{M}(\xi, \mathrm{u}(\xi)) - \mathrm{M}(\xi, \nu(\xi))\big| d\xi \\
& + \frac{2t}{(2-\mathrm{q}^2)\Gamma(\mathrm{p})} \int_0^{\mathrm{q}} \Big(\int_0^{\xi} |\xi - \mathfrak{z}|^{\mathrm{p}-1} \big|\mathrm{M}(\mathfrak{z}, \mathrm{u}(\mathfrak{z})) - \mathrm{M}(\mathfrak{z}, \nu(\mathfrak{z}))\big| d\mathfrak{z} \Big) d\xi \\
\leq & \frac{1}{\Gamma(\mathrm{p})} \int_0^t |(t-\xi)|^{\mathrm{p}-1} \mathfrak{N}_1 \sigma\big(\max\{|\mathrm{u}(\xi) - \nu(\xi)|, |\Omega\mathrm{u}(\xi) - \mathrm{u}(\xi)|, \\
& |\Omega\nu(\xi) - \nu(\xi)|, |\Omega\mathrm{u}(\xi) - \nu(\xi)|, |\Omega\nu(\xi) - \mathrm{u}(\xi)|\}\big) d\xi \\
& - \frac{2t}{(2-\mathrm{q}^2)\Gamma(\mathrm{p})} \int_0^1 |1-\xi|^{\mathrm{p}-1} \mathfrak{N}_1 \sigma\big(\max\{|\mathrm{u}(\xi) - \nu(\xi)|, |\Omega\mathrm{u}(\xi) - \mathrm{u}(\xi)|, \\
& |\Omega\nu(\xi) - \nu(\xi)|, |\Omega\mathrm{u}(\xi) - \nu(\xi)|, |\Omega\nu(\xi) - \mathrm{u}(\xi)|\}\big) d\xi \\
& + \frac{2t}{(2-\mathrm{q}^2)\Gamma(\mathrm{p})} \int_0^{\mathrm{q}} \Big(\int_0^{\xi} |\xi - \mathfrak{z}|^{\mathrm{p}-1} \mathfrak{N}_1 \sigma\big(\max\{|\mathrm{u}(\mathfrak{z}) - \nu(\mathfrak{z})|, |\Omega\mathrm{u}(\mathfrak{z}) - \mathrm{u}(\mathfrak{z})|, \\
& |\Omega\nu(\mathfrak{z}) - \nu(\mathfrak{z})|, |\Omega\mathrm{u}(\mathfrak{z}) - \nu(\mathfrak{z})|, |\Omega\nu(\mathfrak{z}) - \mathrm{u}(\mathfrak{z})|\}\big) d\mathfrak{z} \Big) d\xi
\end{aligned}
$$

$$\leq \mathfrak{N}_1\sigma\big(\max\{\|u(\xi)-\nu(\xi)\|, \|\Omega u(\xi)-u(\xi)\|, \|\Omega\nu(\xi)-\nu(\xi)\|, \|\Omega u(\xi)-\nu(\xi)\|,$$

$$\|\Omega\nu(\xi)-u(\xi)\|\}\big) \times \sup_{t\in\mathcal{I}}\Big(\frac{1}{\Gamma(\mathrm{p})}\int_0^t |(t-\xi)|^{\mathrm{p}-1}d\xi + \frac{2t}{(2-\mathrm{q}^2)\Gamma(\mathrm{p})}\int_0^1 |1-\xi|^{\mathrm{p}-1}d\xi$$

$$+\frac{2t}{(2-\mathrm{q}^2)\Gamma(\mathrm{p})}\int_0^{\mathrm{q}}\int_0^{\xi} |\xi-\mathfrak{z}|^{\mathrm{p}-1}d\mathfrak{z}\,d\xi\Big)$$

$$\leq \sigma\big(\max\{\mathrm{m}(u,\nu), \mathrm{m}(\Omega u, u), \mathrm{m}(\Omega\nu,\nu), \mathrm{m}(\Omega u,\nu), \mathrm{m}(\Omega\nu, u)\}\big).$$

If we define a function $\alpha : \Xi \times \Xi \to [0,\infty)$ by

$$\alpha(u,\nu) = \begin{cases} 1, & \tau(u(t),\nu(t)) \geq 0,\ \forall\, t\in\mathcal{I}; \\ 0, & \text{otherwise}. \end{cases}$$

and consider $H(\mu_1,\mu_2,\mu_3,\mu_4,\mu_5,\mu_6) = \mu_1 - \sigma(\max\{\mu_2,\mu_3,\mu_4,\mu_5,\mu_6\})$, then $H \in \mathbb{H}$ and condition (14.15) is satisfied. Hence, applying Theorem 14.2.7, we conclude that $\exists$ an **FP** $u' \in \Xi$. Hence, u' is also a solution of **FDE** (14.20) with boundary conditions (14.21). □

14.5 Conclusion

This chapter sheds light on the innovation of a new implicit type contractive condition and its utilization to yield a unique **CFP** for a pair of α-**Adm** mappings satisfying this condition. The results obtained here extend and unify several findings in the literature on **FP** theory. Furthermore, as an application, a sufficient condition is studied for the existence of solutions to **FDE** in the frame of "Caputo type fractional operators."

Acknowledgements The authors are very thankful to the referees for their valuable comments and suggestions.

References

Agarwal, R. P., Alghamdi, M. A., & Shahzad, N. (2012). Fixed point theory for cyclic generalized contractions in partial metric spaces. *Fixed Point Theory Applications, 2012*, Art id 40, 11 pages.

Agarwal, R., Baleanu, D., Hedayati, V., & Rezapour, S. (2012). Two fractional derivative inclusion problems via integral boundary condition. *Applied Mathematics and Computation,257*, 205–212.

Aliouche, A., & Djoudi, A. (2007). Common fixed point theorems for mappings satisfying an implicit relation without decreasing assumption. *Hacettepe Journal of Mathematics and Statistics,36*(1), 11–18.

Alsaedi, A., Baleanu, D., Etemad, S. & Rezapour, S. (2016). On coupled systems of time-fractional differential problems by using a new fractional derivative. *Journal of Function Spaces, 2016*, Art id 4626940, 8 pages.

Aydi, H., Karapinar, E. , Erhan, I. M., & Salimi , P. (2014). Best proximity points of generalized almost ψ-Geraghty contractive non-self-mappings. *Fixed Point Theory Applications, 2014*, Art id 32, 13 pages.

Baleanu, D., Mohammadi, H. & Rezapour, S. (2013). The existence of solutions for a nonlinear mixed problem of singular fractional differential equations. *Advances in Differential Equations, 2013*, Art id 359, 12 pages.

Banach, S. (1922). Surles opérations dans les ensembles abstraites et leur applications aux équations intégrales. *Fundamenta Mathematicae,3*, 133–181.

Bhaskar, T. G., & Lakshmikantham, V. (2006). Fixed point theory in partially ordered metric spaces and applications. *Nonlinear Analysis,65*, 1379–1393.

Ćirić, L., & B. (1971). Generalized contractions and fixed-point theorems. *Publications de l'Institut Mathématique (Belgrade),12*(26), 19–26.

Ćirić, L., Cakić, N., Rajović, M., & Ume, J. S. (2008). Monotone generalized nonlinear contractions in partially ordered metric *d* metric spaces. *Fixed Point Theory Applications, 2008*, Art id 131294, 11 pages.

Chugh, R., & Kumar, S. (2001). Common fixed points for weakly compatible maps. *Proceedings of the Indian Academy of Sciences (Mathematical Sciences), 111*(2), 241–247.

Hardy, G. E., & Rogers, T. D. (1973). A generalization of a fixed point theorem of Reich. *Canadian Mathematical Bulletin,16*, 201–206.

Jain, M., Tas, K., & Gupta N. (2013). Coupled common fixed point results involving (ψ, ϕ)-contractions on ordered generalized metric spaces with application to integral equations. *Journal of Inequalities and Applications, 2013*, Art id 372, 22 pages.

Jain, R., Nashine, H. K., & Kumar, S. (2021). An implicit relation approach in metric spaces under $w-$distance and application to the fractional differential equation. *Journal of Function Spaces, 2021*, Art id 9928881, 12 pages.

Jungck, G. (1996). Common fixed points for non continuous non-self maps on non-metric spaces. *Far East Journal of Mathematical Sciences,4*(2), 199–215.

Karapinar, E. (2011). Fixed point theory for cyclic weak $\phi-$contraction. *Applied Mathematics Letters,24*, 822–825.

Karapinar, E., & Samet, B. (2012). Generalized $\alpha - \psi$ contractive type mappings and related fixed point theorems with applications. *Abstract and Applied Analysis, 2012*, Art id 793486, 17 pages.

Karapinar, E., Kumam, P. & Salimi, P. (2013). On $\alpha - \psi-$Meir-Keeler contractive mappings. *Fixed Point Theory Applications, 2013*, Art id 94, 12 pages.

Karapinar, E., Abdeljawad, T., & Jarad, F. (2019). Applying new fixed point theorems on fractional and ordinary differential equations. *Advances in Difference Equations, 2019*, Art id 421, 25 pages.

Kirk, W. A., Srinivasan, P. S., & Veeramani, P. (2003). Fixed points for mappings satisfying cyclical contractive conditions. *Fixed Point Theory,4*, 79–89.

Nieto, J. J., & Lopez, R. R. (2005). Contractive mapping theorems in partially ordered sets and applications to ordinary differential equation. *Order,22*, 223–239.

Pacurar, M., & Rus, I. A. (2010). Fixed point theory for cyclic ψ-contractions. *Nonlinear Analysis,72*, 1181–1187.

Petric, M. A. (2010). Some results concerning cyclic contractive mappings. *General Mathematics,18*, 213–226.

Popa, V. (1999). Some fixed point theorems for compatible mappings satisfying an implicit relation. *Demonstratio Mathematica,32*(1), 157–163.

Popa, V., & Mocanu, M. (2009). Altering distance and common fixed points under implicit relations. *Hacettepe Journal of Mathematics and Statistics,38*(3), 329–337.

Ran, A. C. M., & Reurings, M. C. B. (2004). A fixed point theorem in partially ordered sets and some applications to matrix equations. *Proceedings of the American Mathematical Society,132*, 1435–1443.

Sahi, P., Kaur, J., & Bhatia, S. S. (2015). Coincidence and common fixed point results for generalized $\alpha - \psi$ contractive type mappings with applications. *Bulletin of the Belgian Mathematical Society Simon Stevin,22*, 299–318.

Samet, B. (2010). Coupled fixed point theorems for a generalized Meir-Keeler contraction in partially ordered metric spaces. *Nonlinear Analysis,72*(12), 4508–4517.

Samet, B., Vetro, C., & Vetro, P. (2012). Fixed point theorem for $\alpha - \psi$ contractive type mappings. *Nonlinear Analysis,75*, 2154–2165.

Singh, S. L., & Tomar, A. (2003). Weaker forms of commuting maps and existence of fixed points. *Journal of the Korean Society of Mathematical Education, Series B: Pure and Applied Mathematics, 10*(03), 145–161.

Turinici, M. (1986). Abstract comparison principles and multivariable Gronwall-Bellman inequalities. *Journal of Mathematical Analysis and Applications,117*, 100–127.

Chapter 15
New Fixed-Point Results with Hybrid-Interpolative Reich–Istratescu-Type Contractions on S-Metric Spaces

Hande Poşul, Elif Kaplan, and Nihal Taş

Abstract Metric fixed-point theory is one of the most extensively researched mathematics subjects. We express and prove fixed-point theorems using various approaches, both on the structure of metric spaces and on generalized metric space structures. Along with the help of the hybrid-interpolative Reich–Istratescu-type $S - a\mu)$-contraction structure, new fixed-point results are found on S-metric spaces in this chapter. Every new result obtained sheds light on another research topic. These results significantly advance fixed-point theory by generalizing existing results in the literature.

Keywords Fixed point · S-metric space · Hybrid-interpolative Reich–Istratescu-type $(S\text{-}\alpha\text{-}\mu)$-contraction

AMS Subject Classification 47H10 · 54H25

15.1 Introduction and History

The concept of fixed points dates back to the nineteenth century, with early exploration by mathematicians in the context of topology and analysis. A fixed point is a point that remains unchanged when a function is applied to it, such as if

$$Tx = x,$$

H. Poşul
Department of Mathematics, Kilis 7 Aralık University, Kilis, Türkiye
e-mail: handeposul@kilis.edu.tr

E. Kaplan
Department of Mathematics, Faculty of Science, Ondokuz Mayıs University, Samsun, Türkiye
e-mail: elifaydin@omu.edu.tr

N. Taş (✉)
Department of Mathematics, Balıkesir University, Balıkesir, Türkiye
e-mail: nihaltas@balikesir.edu.tr

© The Author(s), under exclusive license to Springer Nature Singapore Pte Ltd. 2026
A. Tomar and M. Jain (eds.), *Applications of Fixed-Point Theorem*, Industrial and Applied Mathematics, https://doi.org/10.1007/978-981-95-5160-6_15

then x is a fixed point of the self-mapping $T : X \to X$. One of the most important milestones in the development of fixed-point theory was the formulation of Brouwer's Fixed-Point Theorem by the Dutch mathematician L.E.J. Brouwer in 1911 Brouwer (1911). This result laid the foundation for much of the modern understanding of topology and dynamical systems.

In the 1920s, Stefan Banach extended the fixed-point theory by formulating the Banach Fixed-Point Theorem (also known as the Contraction Mapping Theorem) Banach (1922). This theorem provides conditions under which a unique fixed point exists for a contraction mapping in a complete metric space. This theorem is crucial in the analysis of iterative methods for solving equations, such as in numerical methods and differential equations.

Throughout the twentieth century, fixed-point theorems found applications across various fields, including nonlinear analysis, dynamical systems, game theory, and economics. For example, the Nash Equilibrium in game theory, formulated by John Nash in 1950, relies on a fixed-point argument Nash (1950). Nash's theorem states that every game with a finite number of players and strategies has at least one equilibrium point, a concept that heavily depends on the fixed-point theory.

The fixed-point theory has been further developed, with generalizations such as Krasnoselskii's Fixed-Point Theorem Krasnoselskii (1955) and applications in computer science, dynamical systems, and optimization theory. Researchers continue to explore fixed points in various contexts, including in spaces that are not necessarily compact or convex, and they also study the behavior of iterated functions and their convergence properties.

Metric Fixed-Point Theory is a branch of mathematics that deals with the existence and properties of fixed points for functions defined on metric spaces. It is a fundamental area of study in topology, functional analysis, and many applied fields. Research in this area has had a significant impact on a variety of disciplines.

The foundation of these works is based on Banach's fixed-point theory and its generalizations. The "Banach Fixed-Point Theorem", also known as the "Contraction Mapping Theorem," is one of the most important results in metric space theory and mathematical analysis. Its significance lies in its wide range of applications and its fundamental role in understanding the existence and uniqueness of fixed points in mathematical systems.

Recently, a majority of fixed-point theory research continues to be conducted with three different approaches. The oldest approach is the generalization of the contraction condition, which can be considered the foundation of the generalization logic. Examples of generalized contraction conditions include the Kannan-type contraction Kannan (1968), Chatterjea-type contraction Chatterjea (1972), Reich-type contraction Reich (1971), Edelstein-type contraction Edelstein (1960), and Rhoades-type contraction Rhoades (1977).

Another important approach used in generalization is the generalization of the metric space being studied. By updating the conditions of the metric space being studied, new generalized metric spaces are obtained, which is another crucial area of study in fixed-point theory. As a result, fixed point studies conducted on the structure of metric spaces are also continued in newly defined generalized metric space structures. Examples of generalized metric space structures include the G-metric space and S-metric space Sedghi (2012), which are just two of the generalized metric spaces defined for this purpose.

The geometric-based fixed-circle problem Özgür and Taş (2019) is the most recent approach, aiming to determine whether a given function has multiple fixed points and leaves all points of this geometric figure fixed. After then, in the literature, new studies are continuing to be conducted using different approaches (for example, Joshi and Tomar (2024), Khairul et al. (2025), Khairul et al. (2025) and the references therein).

In this book chapter, we plan to obtain new fixed-point results on S-metric spaces in light of all the approaches mentioned above. To do this, inspired by the results obtained in Karapınar et al. (2022) and Jiddah et al. (2022), fixed-point results will be derived in complete S-metric spaces with the help of a new contraction condition. First, the definition of a hybrid-interpolative Reich–Istratescu-type $(S\text{-}\alpha\text{-}\mu)$-contraction will be given. The fact that this study is conducted on S-metric spaces is of particular importance. This is because the concept of S-metric spaces generalizes the concept of metric spaces, thereby extending the work of Karapınar et al. (2022). Moreover, since the concept of S-metric spaces is independent of the concept of G-metric spaces, this work differentiates itself from Jiddah et al. (2022).

15.2 Preliminaries

In this section, we recall some fundamental concepts and results that assist in the work to be done throughout the paper.

Definition 15.2.1 (Sedghi 2012) Let X be a nonempty set. An S-metric on X is a function $S : X \times X \times X \to [0, +\infty)$ that satisfies the following conditions for all $x, y, z, a \in X$:

$(S1)$ $S(x, y, z) = 0$ if and only if $x = y = z$,

$(S2)$ $S(x, y, z) \leq S(x, x, a) + S(y, y, a) + S(z, z, a)$.

The pair (X, S) is called an S-metric space.

Remark 15.2.1 (Sedghi 2012) For an S-metric space, the symmetry condition can be considered as follows:

$$S(x, x, y) = S(y, y, x)$$

for all $x, y \in X$.

Example 15.2.1 (Sedghi 2012) Let $\mathbb{R}$ be the set of real numbers. Then

$$S(x, y, z) = |x - z| + |y - z|$$

for all $x, y, z \in \mathbb{R}$ is an S-metric on $\mathbb{R}$. This S-metric on $\mathbb{R}$ is called the usual S-metric.

Definition 15.2.2 (Sedghi 2012) Let (X, S) be an S-metric space and $\{x_n\}$ be a sequence in (X, S).
(1) A sequence $\{x_n\} \subset X$ converges to $x \in X$ if $S(x_n, x_n, x) \to 0$ as $n \to \infty$, that is, for each $\varepsilon > 0$, there exists $n_0 \in \mathbb{N}$ such that for all $n \geq n_0$ we have $S(x_n, x_n, x) < \varepsilon$.
(2) A sequence $\{x_n\} \subset X$ is an S-Cauchy sequence if $S(x_n, x_n, x_m) \to 0$ as $n, m \to \infty$, that is, for each $\varepsilon > 0$, there exists $n_0 \in \mathbb{N}$ such that for all $n, m \geq n_0$ we have $S(x_n, x_n, x_m) < \varepsilon$.
(3) The S-metric space (X, S) is complete if every an S-Cauchy sequence is a convergent sequence.

Definition 15.2.3 (Sedghi and Dung 2014) Let $T : X \to Y$ be a map from an S-metric space X to an S-metric space Y. Then T is continuous at $x \in X$ if and only if $Tx_n \to Tx$ whenever $x_n \to x$.

The relationships between a metric and an S-metric were studied in different papers (see Gupta (2013), Hieu et al. (2015) and Özgür and Taş (2017) for more details). In Hieu et al. (2015), a formula of an S-metric space which is generated by a metric d was given as follows:

Let (X, d) be a metric space. Then the function $S_d : X \times X \times X \to [0, \infty)$ defined by

$$S_d(x, y, z) = d(x, z) + d(y, z),$$

for all $x, y, z \in X$ is an S-metric on X.

The S-metric S_d is called the S-metric generated by d Özgür and Taş (2017). There exists an S-metric which is not generated by any metric d as seen in the following examples:

Example 15.2.2 (Özgür and Taş 2017) Let $X = \mathbb{R}$. If we consider the function $S : X \times X \times X \to [0, \infty)$ defined by

$$S(x, y, z) = |x - z| + |x + z - 2y|,$$

for all $x, y, z \in X$, then S is an S-metric on X which is not generated by any metric d.

Example 15.2.3 Let $X = \mathbb{R}$ and

$$S(x, y, z) = |y + z - 2x| + |y - z|$$

for all $x, y, z \in X$. Then, (X, S) is an S-metric space Sedghi (2012). But this S-metric cannot be generated by any metric Hieu et al. (2015).

Also in Gupta (2013), it was shown that every S-metric defines a metric as follows:

$$d_S(x, y) = S(x, x, y) + S(y, y, x),$$

for all $x, y \in X$. But the function d_S does not always define a metric since the triangle inequality does not satisfied for all elements of X everywhen (see Özgür and Taş (2017) for more details) as seen in the following example:

Example 15.2.4 (Özgür and Taş 2017) Let $X = \{1, 2, 3\}$. If we consider the function $S : X \times X \times X \to [0, \infty)$ defined by

$$S(1, 1, 2) = S(2, 2, 1) = 5,$$

$$S(2, 2, 3) = S(3, 3, 2) = S(1, 1, 3) = S(3, 3, 1) = 2,$$

$$S(x, y, z) = 0 \text{ if } x = y = z,$$

$$S(x, y, z) = 1 \text{ otherwise},$$

for all $x, y, z \in X$, then S is an S-metric on X which is not generated by any metric d and does not generate a metric d_S.

The relationship between a b-metric defined in Bakhtin (1989) and an S-metric was proved in the following theorem:

Theorem 15.2.1 (Sedghi and Dung 2014) *Let (X, S) be an S-metric space and*

$$d^S(x, y) = S(x, x, y),$$

for all $x, y \in X$. Then we have
(1) d^S is a b-metric on X,
(2) $x_n \to x$ in (X, S) if and only if $x_n \to x$ in (X, d^S),
(3) $\{x_n\}$ is a S-Cauchy sequence in (X, S) if and only if $\{x_n\}$ is a Cauchy sequence in (X, d^S).
The metric d^S is called the b-metric generated by S.

Definition 15.2.4 (Mustafa and Sims 2006) Let X be a nonempty set and $G : X \times X \times X \to [0, \infty)$ be a function such that for all $x, y, z, a \in X$:
$(G1)$ $G(x, y, z) = 0$ if $x = y = z$,
$(G2)$ $0 < G(x, x, y)$ if $x \neq y$,
$(G3)$ $G(x, x, y) \leq G(x, y, z)$ if $y \neq z$,
$(G4)$ The symmetry on three variables

$$G(x, y, z) = G(x, z, y) = G(y, x, z) = G(y, z, x) = G(z, x, y) = G(z, y, x),$$

$(G5)$ The rectangle inequality

$$G(x, y, z) \leq G(x, a, a) + G(a, y, z).$$

Then G is called a G-metric on X and the pair (X, G) is called a G-metric space.

Example 15.2.5 (Mustafa and Sims 2006) Let $X = \{x, y\}$ and the function $G : X \times X \times X \to [0, \infty)$ be defined as

$$G(x, x, x) = G(y, y, y) = 0,$$

$$G(x, x, y) = 1,$$

$$G(x, y, y) = 2$$

and extend G to all of $X \times X \times X$ by symmetry in the variables. Then G is a G-metric on X.

In the following examples, we see that the class of all S-metrics and the class of all G-metrics are distinct:

Example 15.2.6 (Dung et al. 2014) Let us consider the G-metric defined as in 15.2.5. Then we have

$$2 = G(x, y, y) > 1 = G(x, y, y) + G(y, y, y) + G(y, y, y)$$

and so we say that G is not an S-metric on X.

Example 15.2.7 (Dung et al. 2014) Let us consider the S-metric defined as in 15.2.3. Then we have

$$S(1, 0, 2) = 2,$$

$$S(2, 0, 1) = 4$$

and

$$S(1, 0, 2) \neq S(2, 0, 1).$$

Hence, we say that S is not a G-metric.

From the above relationships, it is important to study new fixed-point results on S-metric spaces.

15.3 Main Results

In this section, we obtain new fixed-point results on S-metric spaces.

Popescu (2014) provided the subsequent definition within the context of metric space.

Definition 15.3.1 (Popescu 2014) Let $\alpha : X \times X \to \mathbb{R}_+$ be a function. A self-mapping $T : X \to X$ is referred as α-orbital admissible if for all $x \in X$,

$$\alpha(x, Tx) \geq 1 \Longrightarrow \alpha\left(Tx, T^2x\right) \geq 1.$$

We change the following definitions in the context of S-metric space as follows:

Definition 15.3.2 Let $\alpha : X \times X \times X \to \mathbb{R}_+$ be a function. A self-mapping $T : X \to X$ is referred as (S-α)-orbital admissible if for all $x \in X$,

$$\alpha(x, x, Tx) \geq 1 \Longrightarrow \alpha\left(Tx, Tx, T^2x\right) \geq 1.$$

Definition 15.3.3 Let $\alpha : X \times X \times X \to \mathbb{R}_+$ be a function. The set X is called regular with respect to α if and only if for a sequence $\{x_n\}$ in X such that $\alpha(x_n, x_n, x_{n+1}) \geq 1$ for all n and $x_n \to x \in X$ as $n \to \infty$, we have $\alpha(x_n, x_n, x) \geq 1$ for all n.

Proposition 15.3.1 (Jiddah et al. 2022) *Given* $C \in [0, 1)$, *let* $\{x_n\} \subset \mathbb{R}_+$ *be a sequence such that*

$$x_{n+2} \leq C \max\{x_n, x_{n+1}\}, \ \forall n \in \mathbb{N}. \tag{15.1}$$

Let $K = \max\{x_0, x_1\}$. *Then,*

$$x_{2n} \leq C^n K,$$
$$x_{2n+1} \leq C^n K, \ \forall n \geq 1.$$

Lemma 15.3.1 *Let* $\{x_n\}$ *be a sequence in an S-metric space* (X, S). *Suppose that there exists* $C \in [0, 1)$ *such that*

$$S(x_{n+2}, x_{n+2}, x_{n+3}) \leq C \max\{S(x_n, x_n, x_{n+1}), S(x_{n+1}, x_{n+1}, x_{n+2})\}, \ \forall n \in \mathbb{N}.$$

Then $\{x_n\}$ *is an S-Cauchy sequence in* (X, S).

Proof Consider the sequence $\{x_n\}$ in X defined by

$$x_n = S(x_n, x_n, x_{n+1}), \ \forall n \in \mathbb{N}.$$

Then by the hypothesis, $\{x_n\}$ satisfies (15.1). Hence, by Proposition 15.3.1, we obtain

$$S(x_{2n}, x_{2n}, x_{2n+1}) = x_{2n} \leq C^n K;$$
$$S(x_{2n+1}, x_{2n+1}, x_{2n+2}) = x_{2n+1} \leq C^n K$$

for all $n \in \mathbb{N}$, where $K = \max\{x_0, x_1\}$.

In particular,

$$S(x_{2n}, x_{2n}, x_{2n+1}) + S(x_{2n+1}, x_{2n+1}, x_{2n+2}) \leq 2C^n K, \ \forall n \in \mathbb{N}. \tag{15.2}$$

If $C = 0$ or $K = 0$, then $\{x_n\}_{n\geq 2}$ is constant, hence S-Cauchy. Assume that $C > 0$ and $K > 0$. To see that $\{x_n\}$ is S-Cauchy in (X, S), take arbitrary $\varepsilon > 0$. Since $\frac{\varepsilon}{4K} > 0$ and $0 < C < 1$, then we can find $n_0 \in \mathbb{N}$ such that

$$\sum_{i=n_0}^{\infty} C^i < \frac{\varepsilon}{4K}.$$

In particular,

$$\begin{aligned}
4K \sum_{i=n_0}^{p} C^i &< 4K \sum_{i=n_0}^{\infty} C^i < \varepsilon, \quad \forall p \in \mathbb{N}, \quad p \geq n_0 \\
&\leq 2 \sum_{q=n_0}^{p-1} \left[S\left(x_{2q}, x_{2q}, x_{2q+1}\right) + S\left(x_{2q+1}, x_{2q+1}, x_{2q+2}\right)\right] \\
&\leq 2 \sum_{q=n_0}^{p-1} 2C^q K \\
&= 4K \sum_{q=n_0}^{p} C^q \\
&< 4K \sum_{q=n_0}^{\infty} C^q \\
&< \varepsilon.
\end{aligned}$$

This completes the proof. □

Proposition 15.3.2 (Jiddah et al. 2022) *Given $C \in [0, 1)$ and $\sigma \in (0, 1)$, let $\{x_n\} \subset (0, 1)$ be a sequence such that*

$$x_{n+2} \leq C \max\{x_n, x_{n+1}\}^{\sigma}, \quad \forall n \in \mathbb{N}.$$

Let $K = \max\{x_0, x_1\}$. Then

$$\begin{aligned}
x_{2n} &\leq C^{1+\sigma+\cdots+\sigma^{n-1}} K^{\sigma^n}, \\
x_{2n+1} &\leq C^{1+\sigma+\cdots+\sigma^{n-1}} K^{\sigma^n}, \quad \forall n \geq 1.
\end{aligned}$$

Therefore,

$$\lim_{n\to\infty} \sup x_n \leq C^{\frac{1}{1-\sigma}}.$$

Corollary 15.3.1 *Let $\{x_n\}$ be a sequence in an S-metric space (X, S). Assume that there exist $C \in [0, 1)$ and $\theta, \eta \in [0, 1]$ with $\theta + \eta = 1$ such that,*

$$S\left(x_{n+2}, x_{n+2}, x_{n+3}\right) \leq C S\left(x_n, x_n, x_{n+1}\right)^{\theta} S\left(x_{n+1}, x_{n+1}, x_{n+2}\right)^{\eta}$$

for all $n \in \mathbb{N}$. Then $\{x_n\}$ is an S-Cauchy sequence in (X, S).

Proof Notice that, for all $n \in \mathbb{N}$,

$$\begin{aligned} S(x_{n+2}, x_{n+2}, x_{n+3}) &\le CS(x_n, x_n, x_{n+1})^{\theta} S(x_{n+1}, x_{n+1}, x_{n+2})^{\eta} \\ &\le C \max\{S(x_n, x_n, x_{n+1}), S(x_{n+1}, x_{n+1}, x_{n+2})\}^{\theta} \\ &\max\{S(x_n, x_n, x_{n+1}), S(x_{n+1}, x_{n+1}, x_{n+2})\}^{\eta} \\ &= C \max\{S(x_n, x_n, x_{n+1}), S(x_{n+1}, x_{n+1}, x_{n+2})\}^{\theta+\eta} \\ &= C \max\{S(x_n, x_n, x_{n+1}), S(x_{n+1}, x_{n+1}, x_{n+2})\}. \end{aligned}$$

Then, the result follows from Lemma 15.3.1. □

In metric space, the hybrid-interpolative Reich–Istratescu-type contraction was defined by Karapinar et al. (2022), and in G-metric spaces, it was modified by Jiddah et al. (2022). For S-metric spaces, we have introduced the hybrid-interpolative Reich–Istratescu-type (S-α-μ) contractive condition.

Definition 15.3.4 Let (X, S) be an S-metric space and let $\alpha : X \times X \times X \to \mathbb{R}_+$ be a function. A self-mapping $T : X \to X$ is called hybrid-interpolative Reich–Istratescu-type (S-α-μ) contraction if for some $q \in \mathbb{R}_+$, there exist constant $\mu \in (0, 1)$, $\delta \ge 0$ and $\lambda_i \ge 0$ with $i = 1, 2, 3, 4, 5$ such that for all $x, y \in X \setminus Fix(T)$,

$$\alpha(x, x, y) S\left(T^2x, T^2x, T^2y\right) \le \mu N(x, x, y) \tag{15.3}$$

where

$$N(x, x, y) = \begin{cases} [A]^{\frac{1}{q}}, & for\ all\ q > 0\ with\ \sum_{i=1}^{5} \lambda_i + \delta \le 1 \\ B, & for\ q = 0\ with\ \sum_{i=1}^{5} \lambda_i + \delta = 1 \end{cases}, \tag{15.4}$$

where

$$A = \begin{array}{c} \lambda_1 S(x, x, y)^q + \lambda_2 S(x, x, Tx)^q + \lambda_3 S(y, y, Ty)^q + \lambda_4 S(Tx, Tx, Ty)^q \\ +\lambda_5 S\left(Tx, Tx, T^2x\right)^q + \delta S\left(Ty, Ty, T^2y\right)^q \end{array}$$

and

$$B = \begin{array}{c} [S(x, x, y)]^{\lambda_1} [S(x, x, Tx)]^{\lambda_2} [S(y, y, Ty)]^{\lambda_3} [S(Tx, Tx, Ty)]^{\lambda_4} \\ \left[S\left(Tx, Tx, T^2x\right)\right]^{\lambda_5} \left[S\left(Ty, Ty, T^2y\right)\right]^{\delta}. \end{array}$$

with

$$Fix(T) = \{x \in X : Tx = x\}.$$

Theorem 15.3.1 *Let (X, S) be a complete S-metric space and let $T : X \to X$ be a hybrid-interpolative Reich–Istratescu-type (S-α-μ) contraction satisfying the following conditions:*

(i) T is an (S-α)-orbital admissible,
(ii) T is continuous,
(iii) There exists $x_0 \in X$ such that $\alpha(x_0, x_0, Tx_0) \geq 1$.
Then T has at least a fixed point in X.

Proof Let $x_0 \in X$ be such that $\alpha(x_0, x_0, Tx_0) \geq 1$. Since T is (S-α)-orbital admissible, then $\alpha\left(Tx_0, Tx_0, T^2x_0\right) \geq 1$ and by induction, we have $\alpha\left(T^n x_0, T^n x_0, T^{n+1}x_0\right) \geq 1$ for any $n \in \mathbb{N}$. Let $\{x_n\}$ be a sequence in X defined by $x_n = T^n x_0$ for all $n \in \mathbb{N}$. If there exists some $m \in \mathbb{N}$ such that $Tx_m = x_{m+1} = x_m$, then clearly, x_m is a fixed point of T. Assume now that $x_n \neq x_{n+1}$ for any $n \in \mathbb{N}$. Since T is a hybrid-interpolative Reich–Istratescu-type (S-α-μ) contraction, then we have from (15.3) that;

$$\begin{aligned} S(x_{n+2}, x_{n+2}, x_{n+3}) &\leq \alpha(x_n, x_n, x_{n+1})\, S\left(T^2x_n, T^2x_n, T^2x_{n+1}\right) \\ &\leq \mu N(x_n, x_n, x_{n+1}). \end{aligned} \tag{15.5}$$

We now consider the following cases:
Case 1: For $q > 0$, we have

$$\begin{aligned} &N(x_n, x_n, x_{n+1}) \\ &= \begin{bmatrix} \lambda_1 S(x_n, x_n, x_{n+1})^q + \lambda_2 S(x_n, x_n, Tx_n)^q \\ +\lambda_3 S(x_{n+1}, x_{n+1}, Tx_{n+1})^q + \lambda_4 S(Tx_n, Tx_n, Tx_{n+1})^q \\ +\lambda_5 S\left(Tx_n, Tx_n, T^2x_n\right)^q + \delta S\left(Tx_{n+1}, Tx_{n+1}, T^2x_{n+1}\right)^q \end{bmatrix}^{\frac{1}{q}} \\ &= \begin{bmatrix} \lambda_1 S(x_n, x_n, x_{n+1})^q + \lambda_2 S(x_n, x_n, x_{n+1})^q \\ +\lambda_3 S(x_{n+1}, x_{n+1}, x_{n+2})^q + \lambda_4 S(x_{n+1}, x_{n+1}, x_{n+2})^q \\ +\lambda_5 S(x_{n+1}, x_{n+1}, x_{n+2})^q + \delta S(x_{n+2}, x_{n+2}, x_{n+3})^q \end{bmatrix}^{\frac{1}{q}} \\ &= \begin{bmatrix} (\lambda_1 + \lambda_2) S(x_n, x_n, x_{n+1})^q + (\lambda_3 + \lambda_4 + \lambda_5) S(x_{n+1}, x_{n+1}, x_{n+2})^q \\ +\delta S(x_{n+2}, x_{n+2}, x_{n+3})^q \end{bmatrix}^{\frac{1}{q}} \\ &\leq \begin{bmatrix} (\lambda_1 + \lambda_2 + \lambda_3 + \lambda_4 + \lambda_5) \max\left\{S(x_n, x_n, x_{n+1})^q, S(x_{n+1}, x_{n+1}, x_{n+2})^q\right\} \\ +\delta S(x_{n+2}, x_{n+2}, x_{n+3})^q \end{bmatrix}^{\frac{1}{q}}. \end{aligned}$$

Therefore, (15.5) becomes

$$\begin{aligned} &S\left(x_{n+2}, x_{n+2}, x_{n+3}\right)^q \\ &\leq \mu^q \begin{bmatrix} (\lambda_1 + \lambda_2 + \lambda_3 + \lambda_4 + \lambda_5) \max\left\{S\left(x_n, x_n, x_{n+1}\right)^q, S\left(x_{n+1}, x_{n+1}, x_{n+2}\right)^q\right\} \\ +\delta S\left(x_{n+2}, x_{n+2}, x_{n+3}\right)^q \end{bmatrix}. \end{aligned}$$

So that,

$$\begin{aligned}&\left(1-\mu^{q}\delta\right)S\left(x_{n+2},x_{n+2},x_{n+3}\right)^{q}\\&\leq\mu^{q}\left[\left(\lambda_{1}+\lambda_{2}+\lambda_{3}+\lambda_{4}+\lambda_{5}\right)\max\left\{S\left(x_{n},x_{n},x_{n+1}\right)^{q},S\left(x_{n+1},x_{n+1},x_{n+2}\right)^{q}\right\}\right]\\&\leq\mu^{q}\left(1-\delta\right)\max\left\{S\left(x_{n},x_{n},x_{n+1}\right),S\left(x_{n+1},x_{n+1},x_{n+2}\right)\right\}^{q}\end{aligned}$$

implying that for all $n\in\mathbb{N}$,

$$S\left(x_{n+2},x_{n+2},x_{n+3}\right)^{q}\leq\left(\frac{\mu^{q}\left(1-\delta\right)}{1-\mu^{q}\delta}\right)\max\left\{S\left(x_{n},x_{n},x_{n+1}\right),S\left(x_{n+1},x_{n+1},x_{n+2}\right)\right\}^{q}$$

which is equivalent to

$$S\left(x_{n+2},x_{n+2},x_{n+3}\right)\leq C.\max\left\{S\left(x_{n},x_{n},x_{n+1}\right),S\left(x_{n+1},x_{n+1},x_{n+2}\right)\right\}$$

where

$$C=\left(\frac{\mu\left(1-\delta\right)}{1-\mu^{q}\delta}\right)\in\left(0,1\right).$$

Hence, by Lemma 15.3.1 $\{x_n\}_{n\in\mathbb{N}}$ is an S-Cauchy sequence in (X,S).
Case 2: For $q=0$, we have

$$\begin{aligned}&N\left(x_{n},x_{n},x_{n+1}\right)\\&=\left[\begin{array}{c}S\left(x_{n},x_{n},x_{n+1}\right)^{\lambda_{1}}S\left(x_{n},x_{n},Tx_{n}\right)^{\lambda_{2}}S\left(x_{n+1},x_{n+1},Tx_{n+1}\right)^{\lambda_{3}}\\S\left(Tx_{n},Tx_{n},Tx_{n+1}\right)^{\lambda_{4}}S\left(Tx_{n},Tx_{n},T^{2}x_{n}\right)^{\lambda_{5}}S\left(Tx_{n+1},Tx_{n+1},T^{2}x_{n+1}\right)^{\delta}\end{array}\right]\\&=\left[\begin{array}{c}S\left(x_{n},x_{n},x_{n+1}\right)^{\lambda_{1}}S\left(x_{n},x_{n},x_{n+1}\right)^{\lambda_{2}}S\left(x_{n+1},x_{n+1},x_{n+2}\right)^{\lambda_{3}}\\S\left(x_{n+1},x_{n+1},x_{n+2}\right)^{\lambda_{4}}S\left(x_{n+1},x_{n+1},x_{n+2}\right)^{\lambda_{5}}S\left(x_{n+2},x_{n+2},x_{n+3}\right)^{\delta}\end{array}\right]\\&=\left[S\left(x_{n},x_{n},x_{n+1}\right)^{\lambda_{1}+\lambda_{2}}S\left(x_{n+1},x_{n+1},x_{n+2}\right)^{\lambda_{3}+\lambda_{4}+\lambda_{5}}S\left(x_{n+2},x_{n+2},x_{n+3}\right)^{\delta}\right].\end{aligned}$$

Therefore, (15.5) becomes

$$\begin{aligned}&S\left(x_{n+2},x_{n+2},x_{n+3}\right)\\&\leq\mu S\left(x_{n},x_{n},x_{n+1}\right)^{\lambda_{1}+\lambda_{2}}S\left(x_{n+1},x_{n+1},x_{n+2}\right)^{\lambda_{3}+\lambda_{4}+\lambda_{5}}S\left(x_{n+2},x_{n+2},x_{n+3}\right)^{\delta}.\end{aligned}\tag{15.6}$$

If $\delta=1$, then $\lambda_1=\lambda_2=\lambda_3=\lambda_4=\lambda_5=0$, and so we obtain

$$0<S\left(x_{n+2},x_{n+2},x_{n+3}\right)\leq\mu S\left(x_{n+2},x_{n+2},x_{n+3}\right)$$

which is a contraction. Therefore, $\delta<1$, so that $\sum_{i=1}^{5}\lambda_i=1-\delta>0$, implying that

$$\theta=\frac{\lambda_1+\lambda_2}{1-\delta},\quad\eta=\frac{\lambda_3+\lambda_4+\lambda_5}{1-\delta}$$

satisfying $\theta+\eta=1$. Hence 15.6 becomes

$$S(x_{n+2}, x_{n+2}, x_{n+3})^{1-\delta} \le \mu S(x_n, x_n, x_{n+1})^{\lambda_1+\lambda_2} S(x_{n+1}, x_{n+1}, x_{n+2})^{\lambda_3+\lambda_4+\lambda_5}$$

so that

$$\begin{aligned} & S(x_{n+2}, x_{n+2}, x_{n+3}) \\ & \le \mu^{\left(\frac{1}{1-\delta}\right)} S(x_n, x_n, x_{n+1})^{\left(\frac{\lambda_1+\lambda_2}{1-\delta}\right)} S(x_{n+1}, x_{n+1}, x_{n+2})^{\left(\frac{\lambda_3+\lambda_4+\lambda_5}{1-\delta}\right)} \\ & = \mu^{\left(\frac{1}{1-\delta}\right)} S(x_n, x_n, x_{n+1})^{\theta} S(x_{n+1}, x_{n+1}, x_{n+2})^{\eta} \end{aligned}$$

for all $n \in \mathbb{N}$. Noting that $\mu \in (0, 1)$, we have

$$0 < 1 - \delta \le 1 \Longrightarrow 1 \le \frac{1}{1-\delta} \Longrightarrow \mu^{\left(\frac{1}{1-\delta}\right)} \le \mu < 1.$$

Hence, since $0 < \mu^{\left(\frac{1}{1-\delta}\right)} < 1$ and $\theta + \eta = 1$, then by Corollary 15.3.1, we can conclude that $\{x_n\}_{n\in\mathbb{N}}$ is an S-Cauchy in (X, S).

Therefore, for all $q \ge 0$, we have established that $\{x_n\}_{n\in\mathbb{N}}$ is an S-Cauchy sequence in (X, S) and so by the completeness of (X, S), there exists a point c in X such that $lim_{n\to\infty} x_n = c$. Moreover, since T is continuous, then we can conclude that $Tc = c$, that is, c is a fixed point of T. □

Theorem 15.3.2 *Let (X, S) be a complete S-metric space and let $T : X \to X$ be a hybrid-interpolative Reich–Istratescu-type $(S\text{-}\alpha\text{-}\mu)$ contraction satisfying the following conditions:*
(i) T is an $(S\text{-}\alpha)$-orbital admissible,
(ii) There exists $x_0 \in X$ such that $\alpha(x_0, x_0, Tx_0) \ge 1$,
(iii) T^2 is continuous and $\alpha(x, x, T) \ge 1$ for any $x \in Fix(T^2)$.
Then T has at least a fixed point in X.

Proof Let $x_0 \in X$ be arbitrary and define a sequence $\{x_n\}_{n\in\mathbb{N}}$ in X by $x_n = T^n x_0$. We have shown in Theorem 15.3.1 that there exists $c \in X$ such that $x_n \to c$. Since T^2 is continuous, then

$$T^2 c = \lim_{n\to\infty} T^2 x_n = c, \tag{15.7}$$

that is, c is a fixed point of T^2. This implies that T^2 has at least one fixed point in X, that is, $Fix(T^2)$ is nonempty. Moreover,

$$T^3 c = Tc. \tag{15.8}$$

To see that c is a fixed point of T, assume contrary that $Tc \ne c$. In this case, Tc is not a fixed point T^2 either, since $Tc = T^2 c = c$, which is a contradiction. Also, by (15.3), we have

$$\begin{aligned} S(c,c,Tc) &\leq \alpha(c,c,Tc)\, S(c,c,Tc) \\ &= \alpha(c,c,Tc)\, S\left(T^2c,T^2c,T^3c\right) \\ &\leq \mu N(c,c,Tc)\,. \end{aligned} \tag{15.9}$$

Considering Case 1, we obtain

$$\begin{aligned} N(c,c,Tc) &= \begin{bmatrix} \lambda_1 S(c,c,Tc)^q + \lambda_2 S(c,c,Tc)^q \\ +\lambda_3 S\left(Tc,Tc,T^2c\right)^q + \lambda_4 S\left(Tc,Tc,T^2c\right)^q \\ +\lambda_5 S\left(Tc,Tc,T^2c\right)^q + \delta S\left(T^2c,T^2c,T^3c\right)^q \end{bmatrix}^{\frac{1}{q}} \\ &= \begin{bmatrix} \lambda_1 S(c,c,Tc)^q + \lambda_2 S(c,c,Tc)^q \\ +\lambda_3 S\left(T^2c,T^2c,Tc\right)^q + \lambda_4 S\left(T^2c,T^2c,Tc\right)^q \\ +\lambda_5 S\left(T^2c,T^2c,Tc\right)^q + \delta S(c,c,Tc)^q \end{bmatrix}^{\frac{1}{q}} \\ &= \begin{bmatrix} \lambda_1 S(c,c,Tc)^q + \lambda_2 S(c,c,Tc)^q \\ +\lambda_3 S(c,c,Tc)^q + \lambda_4 S(c,c,Tc)^q \\ +\lambda_5 S(c,c,Tc)^q + \delta S(c,c,Tc)^q \end{bmatrix}^{\frac{1}{q}} \\ &= \left[(\lambda_1+\lambda_2+\lambda_3+\lambda_4+\lambda_5)\, S(c,c,Tc)\right] \\ &\leq S(c,c,Tc) \end{aligned}$$

implying that

$$0 < S(c,c,Tc) \leq \mu S(c,c,Tc)$$

which is a contradiction. Hence $Tc = c$.

Similarly, for Case 2, we obtain

$$\begin{aligned} N(c,c,Tc) &= \begin{bmatrix} S(c,c,Tc)^{\lambda_1}\, S(c,c,Tc)^{\lambda_2}\, S\left(Tc,Tc,T^2c\right)^{\lambda_3} \\ S\left(Tc,Tc,T^2c\right)^{\lambda_4}\, S\left(Tc,Tc,T^2c\right)^{\lambda_5}\, S\left(T^2c,T^2c,T^3c\right)^{\delta} \end{bmatrix} \\ &= \begin{bmatrix} S(c,c,Tc)^{\lambda_1}\, S(c,c,Tc)^{\lambda_2}\, S\left(T^2c,T^2c,Tc\right)^{\lambda_3} \\ S\left(T^2c,T^2c,Tc\right)^{\lambda_4}\, S\left(T^2c,T^2c,Tc\right)^{\lambda_5}\, S(c,c,Tc)^{\delta} \end{bmatrix} \\ &= \begin{bmatrix} S(c,c,Tc)^{\lambda_1}\, S(c,c,Tc)^{\lambda_2}\, S(c,c,Tc)^{\lambda_3} \\ S(c,c,Tc)^{\lambda_4}\, S(c,c,Tc)^{\lambda_5}\, S(c,c,Tc)^{\delta} \end{bmatrix} \\ &= \left[S(c,c,Tc)^{(\lambda_1+\lambda_2+\lambda_3+\lambda_4+\lambda_5+\delta)}\right] \\ &= [S(c,c,Tc)] \end{aligned}$$

so that (15.9) becomes

$$0 < S(c,c,Tc) \leq \mu S(c,c,Tc)$$

which is also a contradiction. Hence, $Tc = c$.

Therefore, c is a fixed point of T in X. □

Theorem 15.3.3 *If in addition to the hypothesis of Theorem 15.3.2, we assume that* $\alpha(x, x, y) \geq 1$ *for any* $x, y \in Fix(T)$, *then the fixed point of* T *is unique.*

Proof Let $u, v \in X$ be any two fixed point of T with $u \neq v$. By replacing this in (15.3) and noting the additional hypothesis, we have

$$\begin{aligned} S(u, u, v) &\leq \alpha(u, u, v) S(u, u, v) \\ &= \alpha(u, u, v) S\left(T^2u, T^2u, T^2v\right) \\ &\leq \mu N(u, u, v). \end{aligned} \tag{15.10}$$

By case 1, we obtain

$$\begin{aligned} N(u, u, v) &= \begin{bmatrix} \lambda_1 S(u, u, v)^q + \lambda_2 S(u, u, Tu)^q \\ +\lambda_3 S(v, v, Tv)^q + \lambda_4 S(Tu, Tu, Tv)^q \\ +\lambda_5 S\left(Tu, Tu, T^2u\right)^q + \delta S\left(Tv, Tv, T^2v\right)^q \end{bmatrix}^{\frac{1}{q}} \\ &= \begin{bmatrix} \lambda_1 S(u, u, v)^q + \lambda_2 S(u, u, u)^q \\ +\lambda_3 S(v, v, v)^q + \lambda_4 S(u, u, v)^q \\ +\lambda_5 S(u, u, u)^q + \delta S(v, v, v)^q \end{bmatrix}^{\frac{1}{q}} \\ &= \left[(\lambda_1 + \lambda_4) S(u, u, v)^q\right]^{\frac{1}{q}} \\ &= (\lambda_1 + \lambda_4)^{\frac{1}{q}} S(u, u, v) \\ &\leq S(u, u, v). \end{aligned}$$

Hence, (15.10) becomes

$$0 < S(u, u, v) \leq \mu S(u, u, v)$$

which is a contradiction. Therefore, $u = v$, and so the fixed point of T is unique. □

In the following result, we examine the existence of fixed point of T, when the S-metric space (X, S) is regular.

Theorem 15.3.4 *Let* (X, S) *be a complete* S*-metric space and let* $T : X \to X$ *be a hybrid-interpolative Reich–Istratescu-type* (S-α-μ) *contraction for* $q = 0$ *such that* $\lambda_1 > 0$, $\lambda_2 > 0$ *and* $\lambda_5 > 0$. *Suppose further that*:
(i) *T is an* (S-α)*-orbital admissible,*
(ii) *There exists* $x_0 \in X$ *such that* $\alpha(x_0, x_0, Tx_0) \geq 1$,
(iii) (X, S) *is regular with respect to* α.
Then T *has a fixed point in* X.

Proof In Theorem 15.3.1, we have established that for any $q \geq 0$, the sequence $\{x_n\}_{n\in\mathbb{N}}$ is S-Cauchy and by the completeness of the S-metric space (X, S), there exists a point u in X such that $x_n \to u$. To prove that u is a fixed point of T, suppose contrary that $Tu \neq u$.

Assume $\{x_n\}_{n\in\mathbb{N}}$ is such that $x_n \neq x_m$ whenever $n \neq m$, for all $n, m \in \mathbb{N}$. Then, there exists $n_0 \in \mathbb{N}$ such that x_n and u are distinct and not in $Fix(T)$ for all $n \geq n_0$. We will verify that u is a fixed point of T^3. Indeed, for all $n \geq n_0$,

$$\begin{aligned}
&S\left(x_{n+2}, x_{n+2}, T^2u\right)\\
&\leq \alpha(x_n, x_n, u)\, S\left(T^2x_n, T^2x_n, T^2u\right)\\
&\leq \mu N(x_n, x_n, u)\\
&= \begin{bmatrix} \mu S(x_n, x_n, u)^{\lambda_1}\, S(x_n, x_n, Tx_n)^{\lambda_2}\, S(u, u, Tu)^{\lambda_3} \\ S(Tx_n, Tx_n, Tu)^{\lambda_4}\, S\left(Tx_n, Tx_n, T^2x_n\right)^{\lambda_5}\, S\left(Tu, Tu, T^2u\right)^{\delta} \end{bmatrix}\\
&= \begin{bmatrix} \mu S(x_n, x_n, u)^{\lambda_1}\, S(x_n, x_n, x_{n+1})^{\lambda_2}\, S(u, u, Tu)^{\lambda_3} \\ S(x_{n+1}, x_{n+1}, Tu)^{\lambda_4}\, S(x_{n+1}, x_{n+1}, x_{n+2})^{\lambda_5}\, S\left(Tu, Tu, T^2u\right)^{\delta} \end{bmatrix}
\end{aligned}$$

Since $\lambda_1 > 0$, $\lambda_2 > 0$ and $\lambda_5 > 0$, then letting $n \to \infty$, it is verified that u is a fixed point of T^2. Hence, by (15.7) and (15.8), we obtain a contradiction. Therefore $Tu = u$, implying that u is a fixed point of T.

□

Example 15.3.1 Let $X = [-2, 2]$ and let $T : X \to X$ be a self-mapping on X defined by $Tx = \frac{x}{3}$ for all $x \in X$. Define $S : X^3 \to \mathbb{R}^+$ by

$$S(x, y, z) = |x - z| + |x + z - 2y|, \quad \forall x, y, z \in X.$$

Then S is a complete S-metric on X which is not generated by any metric d Özgür and Taş (2017). Define

$$\alpha : X \times X \times X \to \mathbb{R}^+, \text{ by}$$

$$\alpha(x, y, z) = \begin{cases} 1, & if\ x, y, z \in \{-2\} \cup [0, 2] \\ 0, & otherwise \end{cases}$$

Then obviously, T is an (S-α)-orbital admissible and T is continuous for all $x \in X$. Also, there exists $x_0 = \frac{1}{4} \in X$ such that $\alpha\left(\frac{1}{4}, \frac{1}{4}, T\left(\frac{1}{4}\right)\right) = \alpha\left(\frac{1}{4}, \frac{1}{4}, \frac{1}{12}\right) \geq 1$. Hence, conditions of Theeorem 15.3.1 and 15.3.2 are satisfied. To see T is a hybrid-interpolative Reich–Istratescu-type (S-α-μ)-contraction, let $\mu = \frac{1}{2}$. Notice that $\alpha(x, x, y) = 0$ for all $x, y \in (-2, 0]$. Hence, inequality (15.3) holds for all $x, y \in (-2, 0)$. Now, for all $x, y \in \{-2, 2\}$, if $x = y$, then it is obvious. Firstly, let us take $x = 2$ and $y = -2$. For $\lambda_5 = \delta = \frac{1}{5}$, $\lambda_1 = \lambda_2 = \lambda_3 = \lambda_4 = 0$, and $q = 1$, we obtain

$$\begin{aligned}
\alpha(x, x, y).S\left(T^2x, T^2x, T^2y\right) &= \alpha(2, 2, -2).S\left(\frac{2}{9}, \frac{2}{9}, -\frac{2}{9}\right)\\
&\leq \frac{1}{2}N(2, 2, -2).
\end{aligned}$$

Similarly, for $x = -2$ and $y = 2$, the inequality (15.3) holds. Also, if $q = 0$, we have same inequality for two cases. Finally, for all $x, y \in (0, 2)$, we take $\lambda_1 = 1$, $\lambda_2 = \lambda_3 = \lambda_4 = \lambda_5 = \delta = 0$. Then,

$$\alpha(x, x, y) . S\left(T^2x, T^2x, T^2y\right) = S\left(\frac{x}{9}, \frac{x}{9}, \frac{y}{9}\right) = \frac{2}{9}|x - y| \leq \frac{1}{2}N(x, x, y)$$

for all $q \geq 0$. Hence inequality (15.3) is satisfied for all $x, y \in X \setminus Fix(T)$. Therefore, T is a hybrid-interpolative-Reich–Istratescu-type (S-α-μ)-contraction which satisfies all the assumptions of Theorem 15.3.1 and 15.3.2. The point $x = 0$ is the fixed point of T in X.

Corollary 15.3.2 *Let (X, S) be a complete S-metric space and $T : X \to X$ be a continuous self-mapping such that there exist $\eta, \lambda \in (0, 1)$ with $\eta + \lambda < 1$, satisfying*

$$S\left(T^2x, T^2x, T^2y\right) \leq \eta S(x, x, y) + \lambda S(Tx, Tx, Ty)$$

for all $x, y \in X$. Then T has a unique fixed point.

Proof Let $\mu = \eta + \lambda \in (0, 1)$. Consider Definition 15.3.4 and let $\alpha(x, x, y) = 1$, $q = 1$, $\lambda_1 = \frac{\eta}{\mu}$, $\lambda_4 = \frac{\lambda}{\mu}$, $\lambda_2 = \lambda_3 = \lambda_5 = \delta = 0$. Then for all $x, y \in X$, we have

$$\begin{aligned}
\alpha(x, x, y) S\left(T^2x, T^2x, T^2y\right) &\leq \mu N(x, x, y) \\
&\leq \mu\left[\lambda_1 S(x, x, y)^q + 0 + 0 + \lambda_4 S(Tx, Tx, Ty)^q + 0 + 0\right]^{\frac{1}{q}} \\
&= \mu\left[\frac{\eta}{\mu} S(x, x, y) + \frac{\lambda}{\mu} S(Tx, Tx, Ty)\right] \\
&= \eta S(x, x, y) + \lambda S(Tx, Tx, Ty)
\end{aligned} \tag{15.11}$$

implying that inequality (15.3) holds for all $x, y \in X \setminus Fix(T)$. Let $x_0 \in X$ be arbitrary and define a sequence $\{x_n\}_{n \in \mathbb{N}}$ in X by $x_n = T^n x_0$. Then by 15.3.4, we have

$$\begin{aligned}
S(x_{n+2}, x_{n+2}, x_{n+3}) &\leq \eta S(x_n, x_n, x_{n+1}) + \lambda S(x_{n+1}, x_{n+1}, x_{n+2}) \\
&\leq \begin{array}{l} \eta \max\{S(x_n, x_n, x_{n+1}), S(x_{n+1}, x_{n+1}, x_{n+2})\} \\ +\lambda \max\{S(x_n, x_n, x_{n+1}), S(x_{n+1}, x_{n+1}, x_{n+2})\} \end{array} \\
&= (\lambda + \eta) \max\{S(x_n, x_n, x_{n+1}), S(x_{n+1}, x_{n+1}, x_{n+2})\} \\
&= \mu \max\{S(x_n, x_n, x_{n+1}), S(x_{n+1}, x_{n+1}, x_{n+2})\}
\end{aligned} \tag{15.12}$$

Hence, by Lemma 15.3.1, $\{x_n\}$ is a S-Cauchy sequence in (X, S) and since this space is complete, then there exists $c \in X$ such that $x_n \to c$. Since T is continuous, then c is a fixed point of T.

□

15.4 Conclusions

This chapter presents notable contributions to fixed-point theory, addresses critical research gaps, and enhances the existing literature in several ways:
(i) Introduction of hybrid-interpolative Reich–Istratescu-type $(S\text{-}\alpha\text{-}\mu)$-contraction:
This study adds a new idea to the traditional framework of contraction mappings: the hybrid-interpolative Reich–Istratescu-type $(S\text{-}\alpha\text{-}\mu)$-contraction. This generalization addresses a gap in the literature by providing a more flexible and inclusive method for studying fixed-point problems.
(ii) Existence and Uniqueness of Fixed Points:
This part shows that fixed points exist for hybrid-interpolative Reich–Istratescu-type $(S\text{-}\alpha\text{-}\mu)$-contractions in S-metric spaces and are unique. This result addresses a significant gap in the literature by offering a cohesive framework for the analysis of fixed points across a wider range of maps and spaces.

15.5 Future Works

(i) Extension to Other Spaces:
Investigate the applicability of hybrid-interpolative Reich–Istratescu-type $(S\text{-}\alpha\text{-}\mu)$ contractions in diverse generalized metric spaces, such as S_b-metric spaces, partial metric spaces, modular metric spaces, modular S-metric spaces, and controlled S-metric type spaces. This investigation may yield new insights regarding the wider applicability of the proposed concepts.
(ii) Generalization of Contraction Conditions:
To make fixed-point theory more useful, look at broader versions of hybrid-interpolated contraction conditions by adding more parameters or loosening up on assumptions.
(iii) Multi-valued Mappings:
Extend the findings to multi-valued mappings, which hold considerable relevance in optimization, game theory, and economics. This task involves the examination of common fixed points for hybrid-interpolative Reich–Istratescu-type $(S - \alpha - \mu)$-multi-valued contractions.

By following these directions, researchers can expand upon the contributions of this work, thereby enhancing the field of fixed-point theory and its applications across various disciplines.

References

Bakhtin, I. A. (1989). The contraction principle in quasimetric spaces. *Func. An., Ulianowsk, Gos. Ped. Ins., 30*, 26–37.

Banach, S. (1922). Sur les opérations dans les espaces vectoriels normés. *Fundamenta Mathematicae, 3*, 133–181.

Brouwer, L. E. J. (1911). Über die topologischen Eigenschaften der linearen Mengen. *Mathematische Annalen, 71*(1), 97–115.

Chatterjea, S. K. (1972). Fixed-point theorems. *C. R. Academie Bulgare Des Sciences, 25*, 727–730.

Dung, N. V., Hieu, N. T., & Radojevic, S. (2014). Fixed point theorems for g-Monotone Maps on partially ordered S-metric spaces. *Filomat, 28*(9), 1885–1898.

Edelstein, M. (1960). An extension of Banach's contraction principle. *Proceedings of the American Mathematical Society, 12*, 7–10.

Gupta, A. (2013). Cyclic contraction on S-metric space. *International Journal of Analysis and Applications, 3*(2), 119–130.

Hieu, N. T., Ly, N. T. & , N. V. A. (2015). generalization of Ćirić quasi-contractions for maps on S-metric spaces. *Thai Journal of Mathematics, 13*(2), 369–380. Aralık, Mayıs

Jiddah, J. A., Noorwali, M., Shagari, M. S., Rashid, S., & Jarad, F. (2022). point results of a new family of hybrid contractions in generalised metric space with applications. *AIMS Mathematics, 7*(10), 17894–17912.

Joshi, M., & Tomar, A. (2024). On set-valued fixed points, its geometry and solution of boundary value problem for a differential inclusion. *Indian Journal of Pure and Applied Mathematics*. https://doi.org/10.1007/s13226-024-00688-z.

Kannan, R. (1968). Some results on fixed points. *Bulletin of Calcutta Mathematical Society, 60*, 71–76.

Karapınar, E., Fulga, A., Shahzad, N., & Roldán López de Hierro, A. F. (2022). Solving integral equations by means of fixed point theory. *Journal of Function Spaces, 2022*, Article ID 7667499, 16 pages.

Khairul, H. A., Yumnam, R., & Tomar, A. (2025). Fixed point and its geometry and application for multivalued integral type contractions in m_{bv}-metric spaces. *Statistics: Hacettepe Journal of Mathematics*.

Khairul, H. A., Yumnam, R., Tomar, A., & Mohammad, S. (2025). On geometry of fixed figures via ϕ-interpolative contractions and application of activation functions in neural networks and machine learning models. *Ain Shams Engineering Journal, 16*(1).

Krasnoselskii, M. A. (1955). *Topological Methods in the Theory of Nonlinear Integral Equations*, P. Noordhoff Ltd.

Mustafa, Z., & Sims, B. (2006). A new approach to generalized metric spaces. *Journal of Nonlinear and Convex Analysis, 7*(2), 289–297.

Nash, J. F. (1950). Equilibrium Points in N-person Games. *Proceedings of the National Academy of Sciences, 36*(1), 48–49.

Özgür, N. Y., & Taş, N. (2017). Some new contractive mappings on S-metric spaces and their relationships with the mapping ($S25$). *Mathematical Sciences (Springer), 11*(1), 7–16.

Özgür, N. Y., & Taş, N. (2019). Some fixed-circle theorems on metric spaces. *Bulletin of the Malaysian Mathematical Sciences Society, 42*, 1433–1449.

Popescu, O. (2014). Some new fixed point theorems for -geraghty contraction type maps in metric spaces. *Fixed Point Theory Application, 2014*, 190.

Reich, S. (1971). Some remarks concerning contraction mappings. *Canadian Mathematical Bulletin, 14*, 121–124.

Rhoades, B. E. (1977). A comparison of various definitions of contractive mappings. *Transactions of the American Mathematical Society, 226*, 257–290.

Sedghi, S., Shobe, N., & Aliouche, A. (2012). A generalization of fixed point theorems in S-metric spaces. *Matematički Vesnik, 64*(3), 258–266.

Sedghi, S., & Dung, N. V. (2014). Fixed point theorems on S-metric spaces. *Matematički Vesnik, 66*(1), 113–124.

Chapter 16
On Fixed Points of $(\alpha_{qs^p}, \beta_{qs^p})$–$\phi$–Hybrid Contraction in b–Metric Spaces

Loitongbam Melei Singh and Yumnam Rohen

Abstract In this paper, we extend the well-known definitions of α-admissible and (α, β)-admissible and introduce new definitions of $(\alpha_{qs^p}, \beta_{qs^p})$-admissible, $(\alpha_{qs^p}, \beta_{qs^p})$-orbital admissible, triangular $(\alpha_{qs^p}, \beta_{qs^p})$-orbital admissible. With these new definitions we introduce a new hybrid contraction which we named as $(\alpha_{qs^p}, \beta_{qs^p})-\phi$-hybrid contraction. With the help of this new contraction we prove some fixed point theorems in b-metric spaces. Further, we study the general Ulam-type stability and well-posedness for the newly introduced hybrid contraction.

Keywords Fixed point · b-metric space · $(\alpha_{qs^p}, \beta_{qs^p})$-admissible · $(\alpha_{qs^p}, \beta_{qs^p})$-orbital admissible · Hybrid contraction · Ulam stability

MSC 47H10 · 54H25

16.1 Introduction and Preliminaries

In the advancement of Mathematics, fixed point theory plays an important role by serving as an essential tool. In 1922 Banach (1922) introduced Banach contraction principle taking the field a significant leap forward. The introduction of b-metric space (Bakhtin 1989; Czerwik 1993) as generalization of metric space made to generalize Banach contraction principle in a large scale. b-metric space is also generalized in different directions. Some of them can be found in Tomar et al. (2021, 2022), Alam et al. (2024a, b) and Joshi et al. (2025) and references therein. Samet et al. (2012) introduced α-admissible with $\alpha-\psi$-contraction to generalize Banach contraction. Chandok (2015) extended the work of Samet et al. and introduced (α, β)-admissible mappings. Zoto et al. (2017) also extended Samet et al. and introduced α_{qs^p}-admissible mappings. Some other important extensions of Samet et al. can be found in Devi et al. (2021a, b). In 1973, Geraghty (1973) introduced Geraghty

L. M. Singh · Y. Rohen (✉)
Department of Mathematics, Manipur University, Canchipur, India
e-mail: ymnehor2008@yahoo.com; yumnam.rohen@manipuruniv.ac.in

L. M. Singh
e-mail: meleiloitongbam@gmail.com
© The Author(s), under exclusive license to Springer Nature Singapore Pte Ltd. 2026

A. Tomar and M. Jain (eds.), *Applications of Fixed-Point Theorem*, Industrial and Applied Mathematics, https://doi.org/10.1007/978-981-95-5160-6_16

contraction and generalized Banach contractions. There are various works in the literature concerning Geraghty contraction (example see Khan et al. (2019), Priyobarta et al. (2021), and Alam et al. (2023, 2024) and the references therein).

In another important achievement Karapinar and Fulga (2019) proposed a novel hybrid contraction. There are various generalization of this concept like p-hybrid Wardowski contractions by Karapinar et al. (2020), admissible hybrid Z-contraction by Chifu and Karapinar (2020), p-hybrid L-fuzzy contraction by Mohammed et al. (2021), admissible hybrid $(G-\alpha, \phi)$-contraction by Jiddah et al. (2023)

Motivated by the works mentioned above as well as the work of Popescu (2014), who introduced α-orbital and triangular α-orbital admissible, in this paper, we introduce a new version of α-admissible hybrid contraction. In order to achieve the new contraction, we introduce various versions of α-admissible like $(\alpha_{qs^p}, \beta_{qs^p})$-admissible, $(\alpha_{qs^p}, \beta_{qs^p})$-orbital admissible, triangular $(\alpha_{qs^p}, \beta_{qs^p})$-orbital admissible etc. We use our newly obtained hybrid contraction in b-metric space to study fixed points. Further, we study the general Ulam-type stability and well-posedness for the newly introduced hybrid contraction.

Definition 16.1.1 (*Bakhtin* (1989)) Let $X \neq \phi$. $d\colon X \times X \to [0, +\infty)$ is called a b-metric if $\forall\, x, y, z \in X$ and for some $s \geq 1$ hold

1. $d(x, y) = 0$ iff $x = y$;
2. $d(x, y) = d(y, x)$;
3. $d(x, y) \leq s[d(x, z) + d(z, y)]$.

(X, d) is called a b-metric space with parameter s.

Czerwik (1993) made b-metric space popular among researchers in fixed point theory. We present below various types of admissible mappings.

Definition 16.1.2 (*Samet et al.* (2012), *Chandok* (2015), *Zoto et al.* (2017) *and Popescu* (2014)) Let $X \neq \phi$, $\alpha, \beta\colon X \times X \to [0, +\infty)$. $T\colon X \to X$ is

1. α-admissible if $\alpha(x, y) \geq 1 \Rightarrow \alpha(Tx, Ty) \geq 1\ \forall\, x, y \in X$.
2. (α, β)-admissible if $\alpha(x, y) \geq 1, \beta(x, y) \geq 1 \Rightarrow \alpha(Tx, Ty) \geq 1, \beta(Tx, Ty) \geq 1\ \forall\, x, y \in X$.
3. α_{qs^p}-admissible if $\alpha(x, y) \geq qs^p \Rightarrow \alpha(Tx, Ty) \geq qs^p\ \forall\, x, y \in X$, $q \geq 1$, and $p \geq 2$.
4. α-orbital admissible if $\alpha(x, Tx) \geq 1 \Rightarrow \alpha(Tx, T^2x) \geq 1\ \forall\, x \in X$.
5. triangular α-orbital admissible if T is α-orbital admissible and $\alpha(x, y) \geq 1$, $\alpha(y, Ty) \geq 1 \Rightarrow \alpha(x, Ty) \geq 1\ \forall x \in X$.

16.2 Main Results

In this paper, we use the following terms, which are defined below.

1. $F(T)$: family of fixed points of T.
2. Φ: family of functions and non-decreasing function $\phi\colon [0, +\infty) \to [0, +\infty)$,

   $\lim_{r\to+\infty} \phi^r = 0$ and $\phi(t) < t\ \forall\, t \in (0, +\infty)$, $\sum_{i=1}^{\infty} \phi^i(t)$ is convergent for $t \geq 0$.
3. $\mathcal{G}$: family of functions $g\colon [0, +\infty) \to [0, 1)$, $g(t_r) \to 1 \Rightarrow t_r \to 0$, $\{t_r\}$ is bounded sequence of positive reals.

Here we define $(\alpha_{qs^p}, \beta_{qs^p})-$hybrid contraction in b-metric space.

Definition 16.2.1 In b-metric space (X, d), $s \geq 1$, let $\alpha, \beta\colon X \times X \to [0, +\infty)$; $T\colon X \to X$ is $(\alpha_{qs^p}, \beta_{qs^p})$-admissible mapping if $\alpha(x, y) \geq qs^p$ and $\beta(x, y) \geq qs^p$ $\Rightarrow \alpha(Tx, Ty) \geq qs^p$ and $\beta(Tx, Ty) \geq qs^p\ \forall\, x, y \in X, q \geq 1,\ p \geq 2$ are constants.

Example 16.2.1 Let $X = [0, +\infty)$, $\alpha, \beta\colon X \times X \to [0, +\infty)$, $T\colon X \to X$ by $Tx = 4x\ \forall\, x \in X$,

$$\alpha(x, y) = \begin{cases} 9 & if\ x \neq y \\ 0 & if\ x = y \end{cases}$$

and

$$\beta(x, y) \begin{cases} 50 & if\ x \neq y \\ 0 & if\ x = y \end{cases}$$

$\forall\, x, y \in X$.
$(\alpha_{qs^p}, \beta_{qs^p})$-admissible holds by T.

Example 16.2.2 Let $X = [0, +\infty)$, $\alpha, \beta\colon X \times X \to [0, +\infty)$, $T\colon X \to X$ by $Tx = e^x\ \forall\, x \in X$ and

$$\alpha(x, y) = \begin{cases} 3\,s^2 & if\ x \neq y \\ 0 & if\ x = y \end{cases}$$

and

$$\beta(x, y) = \begin{cases} 3\,s^2 & if\ x \neq y \\ 0 & if\ x = y \end{cases}$$

$\forall\, x, y \in X, s \geq 1$.
$(\alpha_{qs^p}, \beta_{qs^p})$-admissible holds by T.

Samet et al. (2012) and Chandok (2015) introduced α-regular and (α, β)-regular respectively. From the idea of α-regular and (α, β)-regular, we introduce $(\alpha_{qs^p}, \beta_{qs^p})$-regular.

Definition 16.2.2 Let (X, d) be a b-metric space with $s \geq 1$, $T: X \to X$ and $\alpha, \beta: X \times X \to [0, +\infty)$. X is $(\alpha_{qs^p}, \beta_{qs^p})$-regular if $\{x_r\} \in X$, $x_r \to x \in X$, $\alpha(x_r, x_{r+1}) \geq qs^p, \beta(x_r, x_{r+1}) \geq qs^p \ \forall\ r$, $\exists\ \{x_{r_k}\}$ of $\{x_r\}$ s.t. $\alpha(x_{r_k}, x_{r_k+1}) \geq qs^p, \beta(x_{r_k}, x_{r_k+1}) \geq qs^p \ \forall\ k \in \mathbb{N}$; $\alpha(x, Tx) \geq qs^p, \beta(x, Tx) \geq qs^p$ where $q \geq 1$ and $p \geq 2$.

Definition 16.2.3 Considering T, α, β, p, q and (X, d) as in Definition 16.2.1, T is called

1. $(\alpha_{qs^p}, \beta_{qs^p})$-orbital admissible if $\forall\ x \in X, \alpha(x, Tx) \geq qs^p \Rightarrow \alpha(Tx, T^2x) \geq qs^p, \beta(x, Tx) \geq qs^p \Rightarrow \beta(Tx, T^2x) \geq qs^p$.
2. triangular $(\alpha_{qs^p}, \beta_{qs^p})$-orbital admissible if $\forall\ x \in X$, T is $(\alpha_{qs^p}, \beta_{qs^p})$-orbital admissible, $\alpha(x, y) \geq qs^p, \alpha(y, Ty) \geq qs^p \Rightarrow \alpha(x, Ty) \geq qs^p$ and $\beta(x, y) \geq qs^p, \beta(y, Ty) \geq qs^p \Rightarrow \beta(x, Ty) \geq qs^p$.

Definition 16.2.4 Considering T, α, β, p, q and (X, d) as in Definition 16.2.1, T is called $(\alpha_{qs^p}, \beta_{qs^p})$-hybrid contraction if

$$\alpha(x, Tx)\beta(y, Ty)d(Tx, Ty) \leq \phi(M(x, y)), \tag{16.1}$$

where

$$M(x, y) = \begin{cases} \left[k_1(d(x, y))^\lambda + k_2(d(x, Tx))^\lambda + k_3(d(y, Ty))^\lambda + k_4\left(\frac{d(y,Ty)(1+d(x,Tx))}{1+d(x,y)}\right)^\lambda + \right. \\ \left. k_5\left(\frac{d(y,Tx)(1+d(x,Ty))}{1+d(x,y)}\right)^\lambda\right]^{\frac{1}{\lambda}} \qquad for\ x > 0, x \neq y \\ (d(x, y))^{k_1}.(d(x, Tx))^{k_2}.(d(y, Ty))^{k_3}.\left[\frac{d(y,Ty)(1+d(x,Tx))}{1+d(x,y)}\right]^{k_4}. \\ \left[\frac{d(x,Ty)+(d(y,Tx))}{2s}\right]^{k_5} \qquad for\ \lambda = 0, x, y \in X \setminus Fix(T), \end{cases}$$

$\forall\ x, y \in X, \lambda \geq 0$ and $k_i \geq 0, i = 1, 2, 3, 4, 5$ s.t. $k_1 + k_2 + k_3 + k_4 + k_5 = 1$.

Theorem 16.2.1 *In complete b-metric space (X, d), $T: X \to X$ be continuous or T^2 is continuous and $\alpha(Tx, x) \geq qs^p$, $\beta(Tx, x) \geq qs^p\ \forall\ x \in Fix_{T^2}(X)$, triangular $(\alpha_{qs^p}, \beta_{qs^p})$-orbital admissible, and $(\alpha_{qs^p}, \beta_{qs^p})$-hybrid contraction with $\exists\ x_0 \in X$ s.t. $\alpha(x_0, Tx_0) \geq qs^p$, $\beta(x_0, Tx_0) \geq qs^p$. Then $Tx = x,\ x \in X$.*

Proof For $x_0 \in X$, construct sequence $\{x_r\}_{r \in \mathbb{N}} \in X$ by $x_r = T^r x_0\ \forall\ r \in \mathbb{N}$. Suppose $\exists m \in \mathbb{N}$ s.t. $Tx_m = x_{m+1} = x_m$, x_m is a fixed point of T. Suppose $x_r \neq x_{r-1}\ \forall\ r \in \mathbb{N}$. As T is $(\alpha_{qs^p,\beta_{qs^p}})$-hybrid contraction, by (16.1)

$$\begin{aligned} d(x_r, x_{r+1}) &= d(Tx_{r-1}, Tx_r) \\ &\leq (qs^p)^2 d(Tx_{r-1}, Tx_r) \end{aligned}$$

$$\leq \alpha(x_{r-1}, Tx_{r-1})\beta(x_r, Tx_r)d(Tx_{r-1}, Tx_r)$$
$$\leq \phi(M(x_{r-1}, x_r)). \tag{16.2}$$

Case 1: If $\lambda > 0$,

$$\begin{aligned} M(x_{r-1}, x_r) &= \Big[k_1 d(x_{r-1}, x_r)^\lambda + k_2 d(x_{r-1}, Tx_{r-1})^\lambda + k_3 d(x_r, Tx_r)^\lambda \\ &+k_4\left(\frac{d(x_r, Tx_r)(1+d(x_{r-1}, Tx_{r-1}))}{1+d(x_{r-1}, x_r)}\right)^\lambda + k_5\left(\frac{d(x_r, Tx_{r-1})(1+d(x_{r-1}, Tx_r))}{1+d(x_{r-1}, x_r)}\right)^\lambda\Big]^{\frac{1}{\lambda}} \\ &= \Big[k_1(d(x_{r-1}, x_r))^\lambda + k_2(d(x_{r-1,x_r}))^\lambda + k_3(d(x_r, x_{r+1}))^\lambda + k_4(d(x_r, x_{r+1}))^\lambda\Big]^{\frac{1}{\lambda}} \\ &= \Big[(k_1+k_2)(d(x_{r-1}, x_r))^\lambda + (k_3+k_4)(d(x_r, x_{r+1}))^\lambda\Big]^{\frac{1}{\lambda}}. \end{aligned} \tag{16.3}$$

Since ϕ is non-decreasing, suppose

$$d(x_{r-1}, x_r) \leq d(x_r, x_{r+1}),$$

then (16.2) becomes

$$\begin{aligned} d(x_r, x_{r+1}) &\leq \phi\left(\left[(k_1+k_2)(d(x_{r-1}, x_r))^\lambda + (k_3+k_4)(d(x_r, x_{r+1}))^\lambda\right]^{\frac{1}{\lambda}}\right) \\ &= \phi\left(\left[(k_1+k_2+k_3+k_4)(d(x_r, x_{r+1}))^\lambda\right]^{\frac{1}{\lambda}}\right) \\ &< \left[(k_1+k_2+k_3+k_4)(d(x_r, x_{r+1}))^\lambda\right]^{\frac{1}{\lambda}} \\ &= (1-k_5)^{\frac{1}{\lambda}} d(x_r, x_{r+1}) \\ &\leq d(x_r, x_{r+1}), \end{aligned}$$

a contraction. Therefore, $d(x_{r-1}, x_r) > d(x_r, x_{r+1})\ \forall\, r \in \mathbb{N}$.

Hence, (16.2) becomes

$$\begin{aligned} d(x_r, x_{r+1}) &\leq \phi\left(\left[(k_1+k_2)(d(x_{r-1}, x_r))^\lambda + (k_3+k_4)(d(x_r, x_{r+1}))^\lambda\right]^{\frac{1}{\lambda}}\right) \\ &< \phi\left(\left[(k_1+k_2)(d(x_{r-1}, x_r))^\lambda + (k_3+k_4)(d(x_{r-1}, x_r))^\lambda\right]^{\frac{1}{\lambda}}\right) \\ &\leq \phi(d(x_{r-1}, x_r)) \\ &< d(x_{r-1}, x_r). \end{aligned}$$

Continuing inductively,

$$d(x_r, x_{r+1}) \leq \phi^r d(x_0, x_1) \tag{16.4}$$

$\forall\, r \in \mathbb{N}$ with $x_{r+1} \neq x_{r+2}$, then $\forall\, r, m \in \mathbb{N}, r < m$.

$$\begin{aligned} d(x_r, x_m) &\le sd(x_r, x_{r+1}) + s^2 d(x_{r+1}, x_{r+2}) + s^3 d(x_{r+2}, x_{r+3}) + \ldots + s^{m-r} d(x_{m-1}, x_m) \\ &= s\phi^r(d(x_0, x_1)) + s^2\phi^{r+1}(d(x_0, x_1)) + s^3\phi^{r+2}(d(x_0, x_1)) + \ldots + s^{m-r}\phi^{m-1}(d(x_0, x_1)) \\ &= \frac{1}{s^{r-1}}(s^r\phi^r + s^{r+1}\phi^{r+1} + \ldots + s^{m-1}\phi^{m-1})d(x_0, x_1) \\ &= \frac{1}{s^{r-1}}\sum_{i=r}^{m-1} s^i\phi^i(d(x_0, x_1)). \end{aligned} \tag{16.5}$$

The series $\sum_{i=0}^{\infty} s^i\phi^i(d(x_0, x_1))$ is convergent where $S_r = \sum_{i=0}^{r} s^i\phi^i(d(x_0, x_1))$, from (16.5)

$$d(x_r, x_m) \le \frac{1}{s_{r-1}}(S_{m-1} - S_{r-1}).$$

By $r, m \to \infty$,

$$d(x_r, x_m) \to 0. \tag{16.6}$$

Thus $\{x_r\}_{r\in\mathbb{N}}$ is a Cauchy, $\exists\, u \in X$ s.t. $\{x_r\}_{r\in\mathbb{N}}$ is convergent to u, i.e.,

$$\lim_{r\to\infty} x_r = u \quad ((X, d)\ complete).$$

And

$$u = \lim_{r\to\infty} x_{r+1} = \lim_{r\to\infty} Tx_r = T\lim_{r\to\infty} x_r = Tu \quad (T\ continuous).$$

Also, $T^2u = \lim_{r\to\infty} T^2x_r = u\,(T^2$ continuous). Suppose $Tu \ne u$ on the contrary. Then by (16.1)

$$\begin{aligned} d(u, Tu) = d(T^2u, Tu) &= (qs^p)^2 d(T^2u, Tu) \\ &\le \alpha(Tu, T^2u)\beta(u, Tu)d(T^2u, Tu) \\ &= \alpha(Tu, u)\beta(u, Tu)d(u, Tu) \\ &\le \phi(M(u, Tu)) \\ &< M(u, Tu) \\ &= \Big[k_1(d(u, Tu))^\lambda + k_2(d(u, Tu))^\lambda + k_3(d(Tu, T^2u))^\lambda \\ &\quad + k_4\left(\frac{d(T^2u, Tu)(1 + d(u, Tu))}{1 + d(u, Tu)}\right)^\lambda + k_5\left(\frac{d(Tu, Tu)(1 + d(u, T^2u))}{1 + d(u, Tu)}\right)^\lambda\Big]^{\frac{1}{\lambda}} \\ &= \Big[k_1(d(u, Tu))^\lambda + k_2(d(u, Tu))^\lambda + k_3(d(Tu, u))^\lambda + k_4(d(Tu, u))^\lambda\Big]^{\frac{1}{\lambda}} \\ &= \Big[(k_1 + k_2 + k_3 + k_4)d(u, Tu)^\lambda\Big]^{\frac{1}{\lambda}} \\ &\le d(u, Tu), \end{aligned}$$

a contradiction, thus $Tu = u$.

Case 2: For $\lambda = 0$, we have

$$M(x_{r-1}, x_r) = [d(x_{r-1}, x_r)]^{k_1}.[d(x_{r-1}, Tx_{r-1})]^{k_2}.[d(x_r, Tx_r)]^{k_3}.\left[\frac{d(x_r, Tx_r)(1 + d(x_{r-1}, Tx_{r-1}))}{1 + d(x_{r-1}, x_r)}\right]^{k_4}.$$
$$\left[\frac{d(x_{r-1}, Tx_r)(1 + d(x_r, Tx_{r-1}))}{2s}\right]^{k_5}$$
$$\leq \left[d(x_{r-1}, x_r)\right]^{k_1+k_2} \left[d(x_r, x_{r+1})\right]^{k_3+k_4} \left[\frac{d(x_{r-1}, x_r) + d(x_r, x_{r+1})}{2}\right]^{k_5}.$$

Now, if $d(x_{r-1}, x_r) \leq d(x_r, x_{r+1})$, then

$$\begin{aligned} M(x_{r-1}, x_r) &\leq \left[d(x_{r-1}, x_r)\right]^{k_1+k_2} . \left[d(x_r, x_{r+1})\right]^{k_3+k_4} . \left[\frac{d(x_{r-1}, x_r) + d(x_r, x_{r+1})}{2}\right]^{k_5} \\ &= \left[d(x_r, x_{r+1})\right]^{k_1+k_2+k_3+k_4+k_5} \\ &= d(x_r, x_{r+1}). \end{aligned}$$

From (16.2),

$$\begin{aligned} d(x_r, x_{r+1}) &\leq \phi(M(x_{r-1}, x_r)) \\ &< d(x_r, x_{r+1}), \end{aligned}$$

a contradiction. Therefore, $d(_{r-1}, x_r) > d(x_r, x_{r+1})$.

Hence, by (16.2)

$$\begin{aligned} d(x_r, x_{r+1}) &< \phi(d(x_{n-1}, x_n)) \\ &< \phi^2(d(x_{r-2}, x_{r-1})) \\ &\vdots \\ &< \phi^r(d(x_0, x_1)). \end{aligned}$$

By using the case $q > 0$, $\{x_r\}$ is Cauchy so, $\lim_{r \to \infty} x_r = u$ (b-metric space is complete).

By continuity of T, then

$$u = \lim_{r \to \infty} x_{r+1} = \lim_{r \to \infty} Tx_r = T \lim_{r \to \infty} x_r = Tu.$$

If T^2 is continuous as in case 1, we have $T^2u = u$.

$$\begin{aligned} d(u, Tu) = d(T^2u, Tu) &\leq (qs^p)^2 d(T^2u, Tu) \\ &\leq \alpha(Tu, T^2u)\beta(u, Tu)d(T^2u, Tu) \\ &\leq \phi(M(T^2u, Tu)) \end{aligned}$$

$$\begin{aligned} &= \phi(M(u, Tu)) \\ &\leq \phi\left([d(u, Tu)]^{k_1+k_2+k_3+k_4+k_5}\right) \\ &< d(u, Tu), \end{aligned}$$

a contradiction. Therefore, $Tu = u$.

Example 16.2.3 Let $X = [-1, 1], T: X \times X \to [0, +\infty)$ and $d(x, y) = |x - y|\ \forall\ x, y \in X$. Then (X, d) is a complete b-metric space with parameter $s = 1$.

Define ϕ: $[0, +\infty) \to [0, +\infty)$ by $\phi(t) = \frac{t}{3}\ \forall\, t \geq 0$, $T: X \to X$ by

$$Tx = \begin{cases} \frac{1}{3}, & if\ x \in (-1, 1) \\ \frac{x}{7}, & if\ x \in \{-1, 1\} \end{cases}$$

$\forall\, x \in X$; α, β: $X \times X \to [0, +\infty)$ by

$$\alpha(x, y) = \begin{cases} 5 & if\ x, y \in \{-1\} \cup [0, 1] \\ 0 & otherwise \end{cases}$$

and

$$\beta(x, y) = \begin{cases} 7 & if\ x, y \in \{-1\} \cup [0, 1] \\ 0 & otherwise. \end{cases}$$

Clearly, T is $(\alpha_{qs^p}, \beta_{qs^p})$-orbital admissible, T is continuous $\forall\ x \in X$ and T^2 is continuous $\forall\ x \in Fix(T^2)$. $\exists\ x_0 = \frac{1}{4} \in X$ s.t. $\alpha(\frac{1}{4}, T\frac{1}{4}) = \alpha(\frac{1}{4}, \frac{1}{3}) \geq 5$ and $\beta(\frac{1}{4}, T\frac{1}{4}) = \beta(\frac{1}{4}, \frac{1}{3}) \geq 7$. Hence, all conditions of Theorem 16.2.1 holds.

We know that $\alpha(x, Tx)\beta(y, Ty)d(Tx, Ty) = 0\ \forall\ x, y \in X$. Hence, inequality (16.1) is satisfied $\forall\, x, y \in X$. So, T is $(\alpha_{qs^p}, \beta_{qs^p})$-hybrid contraction. Hence, $x = \frac{1}{3}$ is the fixed point of T.

Theorem 16.2.2 *In complete b-metric space (X, d), $T: X \to X$ be triangular $(\alpha_{qs^p}, \beta_{qs^p})$-orbital admissible and $(\alpha_{qs^p}, \beta_{qs^p})$-hybrid contraction with*

1. *$\exists\, x_0 \in X$ s.t. $\alpha(x_0, Tx_0) \geq qs^p$ and $\beta(x_0, Tx_0) \geq qs^p$;*
2. *(X, d) is $(\alpha_{qs^p}, \beta_{qs^p})$-regular.*

Then $Tx = x,\ x \in X$.

Proof From Theorem 16.2.1, we already prove that for any $\lambda \geq 0$, $\{x_r\}$ is Cauchy, and by property of completeness, $\exists\, u$ s.t. $\lim_{r\to\infty} d(x_r, u) = 0$. Since X is $(\alpha_{qs^p}, \beta_{qs^p})$-regular. Therefore, $\exists$ a sub-sequence $\{x_{r_k}\}$ of $\{x_r\}$ s.t. $\alpha(x_{r_{k-1}}, x_{r_k}) \geq qs^p$ and $\beta(x_{r_{k-1}}, x_{r_k}) \geq qs^p\ \forall\, k \in \mathbb{N}, \alpha(u, Tu) \geq qs^p$ and $\beta(u, Tu) \geq qs^p$. Now, from (16.1) with $(x, y) = (x_{r_k}, u)$,

$$\begin{aligned} d(Tx_{r_k}, Tu) &\leq (qs^p)^2 d(Tx_{r_k}, Tu) \\ &\leq \alpha(x_{r_k}, Tx_{r_k})\beta(u, Tu)d(Tx_{r_k}, Tu) \\ &\leq \phi(M(x_{r_k}, u)). \end{aligned} \tag{16.7}$$

Again, there are two cases arise. If $\lambda > 0$,

$$\begin{aligned} M(x_{r_k}, u) = &\Big[k_1 \left(d(x_{r_k}, u)\right)^\lambda + k_2 \left(d(x_{r_k}, Tx_{r_k})\right)^\lambda + k_3 \left(d(u, Tu)\right)^\lambda \\ &+k_4 \left(\frac{d(u, Tu)(1 + d(x_{r_k}, Tx_{r_k}))}{1 + d(x_{r_k}, u)}\right)^\lambda + k_5 \left(\frac{d(u, Tx_{r_k})(1 + d(x_{r_k}, Tu))}{1 + d(x_{r_k}, u)}\right)^\lambda\Big]^{\frac{1}{\lambda}} \\ &= \Big[k_1 \left(d(x_{r_k}, u)\right)^\lambda + k_2 \left(d(x_{r_k}, x_{r_k+1})\right)^\lambda + k_3 \left(d(u, Tu)\right)^\lambda \\ &+k_4 \left(\frac{d(u, Tu)(1 + d(x_{r_k}, x_{r_k+1}))}{1 + d(x_{r_k}, u)}\right)^\lambda + k_5 \left(\frac{d(u, x_{r_k+1})(1 + d(x_{r_k}, Tu))}{1 + d(x_{r_k}, u)}\right)^\lambda\Big]^{\frac{1}{\lambda}}. \end{aligned}$$

Since

$$\begin{aligned} \lim_{r\to\infty} M(x_{r_k}, u) &= \left[(k_3 + k_4)(d(u, Tu))^\lambda\right]^{\frac{1}{\lambda}} \\ &= (k_3 + k_4)^{\frac{1}{\lambda}} \left(d(u, Tu)\right) \\ &\leq d(u, Tu). \end{aligned}$$

By $r \to \infty$ in (16.7),

$$\begin{aligned} d(u, Tu) &\leq \phi(d(u, Tu)) \\ &< d(u, Tu), \end{aligned}$$

which implies that $Tu = u$.

Similarly, if $q = 0$, we obtain $\lim_{r\to\infty} M(x_{r_k}, u) = 0$ and $d(u, Tu) = 0$.

Theorem 16.2.3 *In Theorems 16.2.1 and 16.2.2, if we add* $\alpha(x, Tx) \geq qs^p$, $\alpha(y, Ty) \geq qs^p$ *and* $\beta(x, Tx) \geq qs^p$, $\beta(y, Ty) \geq qs^p$ $\forall x, y \in Fix_T(X)$ *then fixed point will be unique.*

Proof Let u, v be fixed points of T s.t. $u \neq v$ and $\alpha(u, Tu) \geq qs^p$, $\alpha(v, Tv) \geq qs^p$ and $\beta(u, Tu) \geq qs^p$, $\beta(v, Tv) \geq qs^p$. By (16.1)

$$\begin{aligned} d(u, v) = d(Tu, Tv) &\leq (qs^p)^2 d(Tu, Tv) \\ &\leq \alpha(u, Tu)\beta(v, Tv)d(Tu, Tv) \\ &\leq \phi(M(u, v)) \\ &< M(u, v) \\ &= \Big[k_1 \left(d(u, v)\right)^\lambda + k_2(d(u, Tu))^\lambda + k_3(d(v, Tv))^\lambda \end{aligned}$$

$$+k_4\left(\frac{d(v,Tv)(1+d(u,Tu))}{(1+d(u,v))}\right)^{\lambda}+k_5\left(\frac{d(v,Tu)(1+d(u,Tv))}{1+d(u,v)}\right)^{\lambda}\Bigg]^{\frac{1}{\lambda}}$$
$$=\Big[k_1\,(d(u,v))^{\lambda}+k_2(d(u,u))^{\lambda}+k_3(d(v,v))^{\lambda}$$
$$+k_4\left(\frac{d(v,v)(1+d(u,u))}{(1+d(u,v))}\right)^{\lambda}+k_5\left(\frac{d(v,u)(1+d(u,v))}{1+d(u,v)}\right)^{\lambda}\Bigg]^{\frac{1}{\lambda}}$$
$$=\Big[k_1\,(d(u,v))^{\lambda}+k_5(d(v,u))^{\lambda}\Big]^{\frac{1}{\lambda}}$$
$$=(k_1+k_5)^{\frac{1}{\lambda}}\,d(u,v)$$
$$\le d(u,v),$$

a contradiction. So, $u=v$.

Remark 16.2.1 Taking $\beta(x,y)=1$ and $s=1$, Theorem 16.2.1 becomes Theorems 16.2.2 and 16.2.3 of Karapinar and Fulga (2019). So, our result extends results of Karapinar and Fulga (2019).

16.3 Application to Ulam-Type Stability

Ulam stability was established considering as a type of data dependence (Ulam 1940, 1968) and generalized by Hyers (1941), Rassias (1978), etc. Here, we discuss the general Ulam-type stability and well-posedness to establish a fixed point problem in b-metric space.

Suppose $T\colon X\to X$, mapping on b-metric space (X,d) with parameter $s\ge 1$. Then

$$Tx=x \tag{16.8}$$

has the general Ulam-type stability iff $\exists$ an increasing function $f\colon[0,+\infty)\to[0,+\infty)$, continuous at 0, $f(0)=0$ s.t. $\forall\,\varepsilon>0$, $\forall\,y^*\in X$

$$d(y^*,Ty^*)\le\varepsilon \tag{16.9}$$

$\exists\,u\in X$ of (16.8) s.t.

$$d(u,y^*)\le f(\varepsilon). \tag{16.10}$$

For $C>0$, consider $f(t)=Ct\ \forall\,t\le 0$, Eq. (16.8) is called Ulam-type stable.

Equation (16.8) is called well-posed on a b-metric space (X,d), if

1. T has a unique fixed point u in X;
2. $d(x_r,u)=0\ \forall\,\{x_r\}\in X$ s.t. $\lim_{r\to\infty} d(x_r,Tx_r)=0$.

Theorem 16.3.1 *Let (X, d) be a complete b-metric space with $s \geq 1$. Adding $k_1 + k_5 < \frac{1}{C^2(\lambda)}$, where $C(\lambda) = max\{1, 2^\lambda\}$, to Theorem 16.2.3, then Eq. (16.8) is*

1. *Ulam–Hyers stable if $\alpha(v, w) \geq qs^p$ and $\beta(v, w) \geq qs^p$ $\forall$ v, w satisfying (16.10);*
2. *well-posed if $\alpha(x_r, Tx_r) = 0$ and $Fix_T(X) = u$.*

Proof From Theorem 16.2.3, there is a unique $u \in X$ with $Tu = u$, let $y^* \in X$ s.t.

$$d(y^*, Ty^*) \leq \varepsilon, \quad \varepsilon > 0.$$

Then obviously, u satisfies (16.9). So, $\alpha(y^*, u) \geq qs^p$ and $\beta(y^*, u) \geq qs^p$. Hence, by triangle inequality

$$\begin{aligned}
d(u, y^*) &\leq sd(Tu, Ty^*) + sd(Ty^*, y^*) \\
&\leq (qs^p)^2 d(Ty^*, Tu) + sd(Ty^*, y^*) \\
&\leq \alpha(y^*, Ty^*)\beta(u, Tu)d(Ty^*, Tu) + sd(Ty^*, y^*) \\
&\leq \phi(M(y^*, u)) + sd(Ty^*, y^*) \\
&< M(y^*, u) + sd(Ty^*, y^*) \\
&\leq \Bigg[k_1(d(y^*, u))^\lambda + k_2(d(y^*, Ty^*))^\lambda + k_3(d(u, Tu))^\lambda + k_4\left(\frac{d(u, Tu)(1 + d(y^*, Ty^*))}{1 + d(y^*, u)}\right)^\lambda \\
&\quad + k_5\left(\frac{d(u, Ty^*)(1 + d(y^*, Tu))}{1 + d(y^*, u)}\right)^\lambda\Bigg]^{\frac{1}{\lambda}} + s\varepsilon \\
&= \left[k_1(d(y^*, u))^\lambda + k_2(d(y^*, Ty^*))^\lambda + k_5(d(u, Ty^*))^\lambda\right]^{\frac{1}{\lambda}} + s\varepsilon \\
&\leq \left[k_1(d(y^*, u))^\lambda + k_2\varepsilon^\lambda + k_5(d(u, Ty^*))^\lambda\right]^{\frac{1}{\lambda}} + s\varepsilon \\
&\leq \left[k_1(d(y^*, u))^\lambda + k_2\varepsilon^\lambda + k_5(sd(u, y^*) + sd(y^*, Ty^*))^\lambda\right]^{\frac{1}{\lambda}} + s\varepsilon \\
&\leq \left[k_1(d(y^*, u))^\lambda + k_2\varepsilon^\lambda + k_5(sd(u, y^*) + s\varepsilon)^\lambda\right]^{\frac{1}{\lambda}} + s\varepsilon.
\end{aligned}$$

Therefore,

$$\begin{aligned}
d^\lambda(u, y^*) &\leq C(\lambda)\left[k_1(d(y^*, u))^\lambda + k_2\varepsilon^\lambda + k_5(sd(u, y^*) + s\varepsilon)^\lambda + s^\lambda\varepsilon^\lambda\right] \\
&\leq C(\lambda)\left[k_1(d(y^*, u))^\lambda + k_2\varepsilon^\lambda + k_5 s^\lambda C(\lambda)(d^\lambda(u, y^*) + \varepsilon^\lambda) + s^\lambda\varepsilon^\lambda\right],
\end{aligned} \tag{16.11}$$

where $C(\lambda) = \max\{1, 2^{\lambda-1}\}$. Hence, (16.11) reduces to $d^\lambda(u, y^*) \leq \frac{(k_2+k_5 s^\lambda C(\lambda)+s^\lambda)C(\lambda)}{1-k_1 C(\lambda)-k_5 s^\lambda C^2(\lambda)}\varepsilon^\lambda$, and similar to

$$d(u, y^*) \leq C\varepsilon,$$

where $C = \left[\frac{(k_2+k_5s^\lambda C(\lambda)+s^\lambda)C(\lambda)}{1-k_1C(\lambda)-k_5s^\lambda C^2(\lambda)}\right]^{\frac{1}{\lambda}}$, for any $\lambda > 0$ and $s \geq 1$, $k_1, k_5 \in [0, 1]$ s.t. $k_1 + k_5 < \frac{1}{C^2(\lambda)}$.

2. Taking the supplementary condition and since $Fix_T(X) = u$,

$$
\begin{aligned}
d(x_r, u) &\leq sd(x_r, Tx_r) + sd(Tx_r, Tu)\\
&\leq sd(x_r, Tx_r) + (qs^p)d(Tx_r, Tu)\\
&\leq s(x_r, Tx_r) + \alpha(x_r, Tx_r)\beta(u, Tu)d(Tx_r, Tu)\\
&\leq sd(x_r, Tx_r) + \phi(M(x_r, u))\\
&\leq \left[k_1d^\lambda(x_r, u) + k_2d^\lambda(x_r, Tx_r) + k_3d^\lambda(u, Tu) + k_4\left(\frac{d(u, Tu)(1 + d(x_r, Tx_r))}{1 + d(x_r, u)}\right)^\lambda + \right.\\
&\left. k_5\left(\frac{d(u, Tx_r)(1 + d(x_r, Tu))}{1 + d(x_r, u)}\right)^\lambda\right]^{\frac{1}{\lambda}} + sd(x_r, Tx_r)\\
&= \left[k_1d^\lambda(x_r, u) + k_2d^\lambda(x_r, Tx_r) + k_5(d(u, Tx_r))^\lambda\right]^{\frac{1}{\lambda}} + sd(x_r, Tx_r)\\
&\leq \left[k_1d^\lambda(x_r, u) + k_2d^\lambda(x_r, Tx_r) + k_5(sd(u, x_r) + sd(x_r, Tx_r))^\lambda\right]^{\frac{1}{\lambda}} + sd(x_r, Tx_r)\\
&\leq \left[k_1d^\lambda(x_r, u) + k_2d^\lambda(x_r, Tx_r) + k_5s^\lambda C(\lambda)(d(u, x_r) + d(x_r, Tx_r))\right]^{\frac{1}{\lambda}} + sd(x_r, Tx_r).
\end{aligned}
$$

Therefore,

$$
\begin{aligned}
d^\lambda(x_r, u) \leq C(\lambda)&\left[k_1d^\lambda(x_r, u) + k_2d^\lambda(x_r, Tx_r) + k_5s^\lambda C(\lambda)d(u, x_r) + k_5s^\lambda C(\lambda)d(x_r, Tx_r))\right.\\
&\left. + s^\lambda d(x_r, Tx_r)\right],
\end{aligned}
$$

where $C(\lambda) = \max\{1, 2^{\lambda-1}\}$.

Therefore,

$$
d^\lambda(u, x_r) \leq \frac{\left(k_2 + k_5s^\lambda C(\lambda) + s^\lambda\right)C(\lambda)}{1 - k_1C(\lambda) - k_5s^\lambda C^2(\lambda)}d(x_r, Tx_r). \tag{16.12}
$$

By $r \to +\infty$ in (16.12),

$$
\lim_{r\to+\infty} d(x_r, u) = 0.
$$

Hence, Eq. (16.8) is well-posed.

16.4 Conclusions

Utilizing the newly proposed notion of $(\alpha_{qs^p}, \beta_{qs^p})$-hybrid contraction, we investigate the existence of unique fixed points in a b-metric space. In addition, we examined Ulam-type stability and well-posedness for the new contraction proposed herein. Further, we studied application for finding solution in general Ulam-type stability. We claim that our findings represent an advancement over the current results in the literature. In future work, one of our results may be used to establish new existence conditions for solutions of a class of integral and differential equations. Additionally, our newly obtained hybrid contraction may be used to examine the existence of fixed points in various spaces.

References

Alam, K. H., Rohen, Y., & Tomar, A. (2024). (α, F)- Geraghty-type generalized F-contractions on non-Archimedean fuzzy metric-unlike spaces. *Demonstratio Mathematica,57*, 20240046.

Alam, K. H., Rohen, Y., & Saleem, N. (2023). Fixed points of (α, β, F^*) and (α, β, F^{**}) Weak Geraghty contractions with an application. *Symmetry, 15*, 243.

Alam, K. H., Rohen, Y., & Tomar, A. (2024a). On fixed point and its application to the spread of infectious disease model in M_v^b-metric space. *Mathematical Methods in the Applied Sciences, 47*, 6489–6503.

Alam, K. H., Rohen, Y., Tomar, A., & Sajid, Md. (2024b). On fixed point and solution to the nonlinear matrix equations related to beam theory in M_v^b-metric space. *Journal of Nonlinear and Convex Analysis, 25*(9), 2149–2171.

Bakhtin, I. A. (1989). The contraction mapping principle in quasimetric spaces. *Functional Analysis,30*, 26–37. Ulyanovsk Gos. Ped. Inst.

Banach, S. (1922). Sur les opérations dans les ensembles abstraits et leur application aux équations integrals. *Fundamenta Mathematicae, 3*, 133–181.

Chandok, S. (2015). Some fixed point theorems for (α, β)-admissible Geraghty type contractive mappings and related results. *Mathematical Sciences, 9*, 127–135.

Chifu, I. C., & Karapinar, E. (2020). Admissible hybrid Z-contractions in b-metric spaces. *Axioms,4*(1).

Czerwik, S. (1993). Contraction mappings in b-metric spaces. *Acta Mathematica et Informatica Universitatis Ostraviensis, 1*, 5–11.

Devi, M. B., Khomdram, B., & Rohen, Y. (2021). Fixed point of theorems of generalised α-rational contractive mappings on rectangular b-metric spaces. *Journal of Mathematics and Computer Science,11*, 991–1010.

Devi, M. B., Priyobarta, N., & Rohen, Y. (2021). Fixed point of theorems for $(\alpha, \beta) - (\phi, \psi)$-rational contractive type mappings. *Journal of Mathematics and Computer Science,11*, 955–969.

Geraghty, M. (1973). On contractive mappings. *Proceedings of the American Mathematical Society, 40*, 604–608.

Hyers, D. H. (1941). On the stability of linear functional equations. *Proceedings of the National Academy of Sciences USA,27*, 222–224.

Jiddah, J. A., Shagari, M. S., & Imam, A. T. (2023). On fixed points of a general class of hybrid contractions with Ulam-type stability. *Sahand Communications in Mathematical Analysis, 20*(2), 39–64.

Joshi, M., Tomar, A., & Padaliya, S. K. (2021). On geometric properties of non-unique fixed points in b-metric spaces. In *Fixed point theory and its applications to real word problem*. New York, USA. ISBN: 978-1-53619-336-7.
Karapinar, E., Aydi, H., & Fulga, A. (2020). On p-hybrid Wardowski contraction. *Journal of Mathematics,1155*, 1632526.
Karapinar, E., & Fulga, A. (2019). An admissible hybrid contraction with an Ulam type contraction. *Demonstratio Mathematica, 52*, 428–436.
Khan, M. S., Priyobarta, N., & Rohen, Y. (2019). Fixed point of generalised rational $\alpha^* - \psi$-Geraghty contraction for multivalued mappings. *Journal of Advanced Mathematical Studies, 12*, 156–169.
Mohammed, S. S., Fulatan, I., & Sirajo, Y. (2021). Fixed points of p-hybrid Z-fuzzy contractions. *Sahand Communications in Mathematical Analysis, 18*(3), 1–25.
Popescu, O. (2014). Some new fixed point theorems for α-Geraghty contraction type maps in metric spaces. *Fixed Point Theory and Applications, 1*(190), 1–12.
Priyobarta, N., Khomdram, B., Rohen, Y., & Saleem, N. (2021). On generalized rational α-Geraghty contraction mapping in G-metric spaces. *Journal of Mathematics,2021*, Article ID 6661045.
Rassias, T. M. (1978). On the stability of linear mapping in Banach spaces. *Proceedings of the American Mathematical Society, 72*, 297–300.
Samet, B., Vetro, P., & Vetro, C. (2012). Fixed point theorems for $\alpha - \psi$-contractive type mappings. *Nonlinear Analysis,75*, 2154–2165.
Tomar, A., Joshi, M., & Bhatt, V. (2021). Fixed point and a cantilever beam problem in a partial b-metric space. *Acta Universitatis Sapientiae, Mathematica,13*(2), 506–518.
Tomar, A., Joshi, M., & Padaliya, S. K. (2022). Relation theoretic contractions and its applications in b-metric like spaces. *Journal of Nonlinear and Convex Analysis, 28*, 295–309.
Ulam, S. M. (1940). *Problems in modern mathematics*. New York, NY, USA: John Wiley and Sons.
Ulam, S. M. (1968). *A collection of mathematical problems*. New York, NY, USA: Interscience.
Zoto, K., Rhoades, B. E., & Radenovic, S. (2017). Some generalizations for $(\alpha - \psi, \phi)$-contractions in b-metric-like spaces and an application. *Fixed Point Theory and Applications, 2017*, 26.

Chapter 17
Best Proximity Points Theorems of Generalized α_{ψ}^{ϕ}−Proximal Contractive Mappings with Applications

Solomon Zerfu Degefa, Mujahid Abbas, Vizender Singh, and Yohannes Gebru Aemro

Abstract The aim of this paper is to introduce the concepts of generalized α_{ψ}^{ϕ}−proximal contractive mappings and generalized α_{η}^{λ}−proximal contractive mappings in fuzzy metric spaces. The existence and uniqueness of the best proximity point of such mappings are proved. The main results extend the comparable results in the existing literature in two ways: by generalizing the underlying spaces and by broadening the class of contractive mappings. Moreover, an illustrative example is provided to support the results presented herein. Finally, as an application, the approach of the best proximity point for solving Urysohn integral equations and boundary value problems for second-order differential equations is demonstrated.

Keywords Best proximity point · Proximal contractive mapping · α-proximal admissible · Urysohn integral equation · Boundary value problem · Differential equation

AMS Subject Classification 47H10 · 45D05 · 54H25

S. Z. Degefa
Research Scholar, Department of Mathematics, Guru Jambheshwar University of Science and Technology, Hisar, Haryana, India

M. Abbas
Department of Mechanical Engineering Science, Faculty of Engineering and the Built Environment, University of Johannesburg, Johannesburg 2092, South Africa

V. Singh (✉)
Department of Mathematics, Guru Jambheshwar University of Science and Technology, Hisar, Haryana, India
e-mail: vsihag3@gmail.com

S. Z. Degefa · Y. G. Aemro
Department of Mathematics, Wolkite University, Wolkite, Ethiopia
e-mail: yohannes.gebru@wku.edu.et

© The Author(s), under exclusive license to Springer Nature Singapore Pte Ltd. 2026
A. Tomar and M. Jain (eds.), *Applications of Fixed-Point Theorem*, Industrial and Applied Mathematics, https://doi.org/10.1007/978-981-95-5160-6_17

17.1 Introduction

Fixed point theory plays a vital role in nonlinear analysis and has been generalized in various directions, one of which is the concept of the best approximation. An important aspect of this generalization is that the mappings under consideration are non-self mappings. In 1969, Fan (1969) proved a classical best approximation theorem which deals with the existence of approximate solutions. Subsequently, the best approximation theorems were generalized to guarantee both existence and uniqueness, leading to the development of the best proximity point theorems. The theory of best proximity points represents a natural extension of fixed point theory, as the best proximity point becomes a fixed point when the underlying mapping is a self-mapping. In the late 1990s and early 2000s, many researchers started to explore the concept of the best proximity point, which combines the results of fixed point theory and best approximation. For more details, we refer to [Reich (1978); Prolla (1983); Sehgal & Singh (1988); Basha & Veeramani (1997); Tomar et al. (2023); Petwal (2023); Beloul et al. (2021); Prasad et al. (2020)] and references cited therein.

In 2005, Eldred et al. (2005) proved the existence of the best proximity point for relatively non-expansive mappings. In 2006, Eldred & Veeramani (2006) not only established the existence of the best proximity point but also provided a constructive way to find the best proximity point for a cyclic contraction mapping within the framework of a uniformly convex Banach space. In 2011, Basha (2011) extended Banach's contraction principle to the case of non-self mappings and established best proximity point theorems for proximal contraction mappings. In 2013, Jleli & Samet (2013) introduced the new class of mappings called $\alpha - \psi$ proximal contractive mapping for non-self mappings.

In 2018, Paknazar (2018) developed the best proximity point theorems for new classes of proximal contraction mappings within the framework of a non-Archimedean fuzzy metric space. Pragadeeswarar & Gopi (2019) extended the notion of proximal coincidence and proximal CLRg property to non-self mappings. In 2022, Zhou et al. (2022) introduced new types of $(\psi - \phi)$ weak proximal contractions and investigated the existence of the unique best proximity point for such mappings.

Motivated by the work of Paknazar (2018), in the present work, the concept of modified α_{ϕ}^{ψ} −proximal contractive mappings is generalized and applied to derive the best proximity point in the setting of fuzzy metric spaces. Moreover, an example is provided to validate and support the best proximity point theorems presented in the main results. Finally, as an application, we demonstrate the approach of the best proximity point for solving Urysohn integral equations and the existence and uniqueness of solutions of the boundary value problem for the second-order differential equations.

17.2 Preliminary

The following definitions, terminology, and related concepts are employed for further use.

Definition 17.2.1 (*Shweizer and Skala* (1960)) A binary operation $* : [0, 1]^2 \longrightarrow [0, 1]$ is called a continuous triangular norm if it is commutative, associative, continuous, and for each w, x, y, z in $[0, 1]$, we have $w * 1 = w$ and $w * x \leq z * y$ when $w \leq z$ and $x \leq y$

George & Veeramani (1994) improved the idea of fuzzy metric space presented in Kramosil & Michalek (1975) using the ideas of triangular norm.

Definition 17.2.2 (*George & Veeramani* (1994)) A triplet $(X, M, *)$ of any arbitrary set $X \neq \emptyset$, a continuous triangular norm $*$ with a fuzzy set M on $X \times X \times (0, \infty)$ is known as "fuzzy metric space," if for each w, x, y in X and every $s, t > 0$, we have:

GV-1: $M(w, x, s) > 0$,
GV-2: $M(w, x, s) = 1$ if and only if $w = x$,
GV-3: $M(w, x, s) = M(x, w, s)$,
GV-4: $M(w, y, s + t) \geq M(w, x, s) * M(x, y, t)$,
GV-5: $M(w, x, \cdot) : (0, \infty) \longrightarrow [0, 1]$ is continuous.

Let B and D be nonempty subsets of a fuzzy metric space $(X, M, *)$.

Definition 17.2.3 (*Shayanpour & Nematizadeh* (2017)) The distance of B and D is defined as

$$M(B, D, s) = \sup_{w \in B, x \in D} M(w, x, s), s \geq 0. \tag{17.1}$$

Definition 17.2.4 (*Shayanpour & Nematizadeh* (2017)) Define the following sets:

$$B_0 = \{w \in B : \exists x \in D \text{ such that } \forall s \geq 0, M(w, x, s) = M(B, D, s)\},$$
$$D_0 = \{x \in D : \exists w \in B \text{ such that } \forall s \geq 0, M(w, x, s) = M(B, D, s)\}.$$

Definition 17.2.5 (*Shayanpour & Nematizadeh* (2017)) Let $T : B \longrightarrow D$. An element $w^* \in B$ is called "the best proximity point of the mapping T" if

$$M(w^*, Tw^*, s) = M(B, D, s), \tag{17.2}$$

for all $s > 0$.

Definition 17.2.6 (*Shayanpour & Nematizadeh* (2017)) Let $F, T : B \longrightarrow D$. A point $w \in B$ is known as a common best proximity point of the mappings F and T if

$$M(w, Fw, s) = M(B, D, s) = M(w, Tw, s), \tag{17.3}$$

for all $s \geq 0$.

Definition 17.2.7 (*Pragadeeswarar & Gopi* (2019)) Mappings $F, T : B \longrightarrow D$ are said to possess the proximal CLRg property if a point $y \in B$ and a sequence $\{w_n\} \subseteq B$, we have

$$\begin{aligned} M(u_n, Fw_n, s) = M(B, D, s) = M(v_n, Tw_n, s), \\ M(a, Fy, s) = M(B, D, s) = M(b, Ty, s), \end{aligned} \tag{17.4}$$

such that $u_n \to a$ and $v_n \to b$, where $u_n, v_n, a, b \in B$, for all $s \geq 0$.

Definition 17.2.8 (*Pragadeeswarar & Gopi* (2019)) Let $F, T : B \longrightarrow D$. The mapping F and T are commute proximally if

$$M(u, Fw, s) = M(B, D, s) = M(v, Tx, s),$$

for any $s > 0$, then $Tu = Fv$, where $w, x, u, v \in B$.

Definition 17.2.9 (*Shayanpour & Nematizadeh* (2017)) Let $F, T : B \longrightarrow D$. The mapping F dominates T proximally if

$$M(u_1, Fw_1, s) = M(u_2, Fw_2, s) = M(B, D, s) = M(v_1, Tx_1, s) = M(v_2, Tx_2, s), \tag{17.5}$$

for any $s > 0$, then there exists $\alpha \in (0, 1)$ so that

$$M(u_1, u_2, \alpha s) \geq M(v_1, v_2, s),$$

where $u_1, u_2, v_1, v_2, w_1, w_2, x_1, x_2 \in B$.

Definition 17.2.10 (*Shayanpour & Nematizadeh* (2017)) Let $F, T : B \longrightarrow D$. The mapping F weakly dominates T proximally if

$$M(u_1, Fw_1, s) = M(u_2, Fw_2, s) = M(B, D, s) = M(v_1, Tw_1, s) = M(v_2, Tw_2, s),$$

for any $s > 0$, then there exists $\alpha \in (0, 1)$ so that

$$M(u_1, u_2, \alpha s) \geq \min\{M(v_1, v_2, s), M(v_1, u_1, s), M(v_1, u_2, s), M(v_2, u_1, s)\},$$

where $u_1, u_2, v_1, v_2, w_1, w_2, x_1, x_2 \in B$.

Definition 17.2.11 (*Paknazar* (2018)) Let $T : B \longrightarrow D$ and $\alpha : B^2 \times (0, \infty) \longrightarrow (0, \infty)$. A mapping T is known as a triangular α-proximal admissible if the following conditions are met:

(T1) For every $w, x, u, v \in B$ and $s > 0$,

$$\begin{cases} \alpha(w, x, s) \geq 1 \\ M(u, Tw, s) = M(B, D, s) \Rightarrow \alpha(u, v, s) \geq 1. \\ M(v, Tx, s) = M(B, D, s) \end{cases} \tag{17.6}$$

(T2) For every $w, x, y \in B$ and $s > 0$,

$$\alpha(w, x, s) \geq 1, \alpha(x, y, s) \geq 1 \Rightarrow \alpha(w, y, s) \geq 1.$$

Suppose that Φ is the collection of mappings φ from [0, 1] to [0, 1] satisfying the following properties:

(a) φ is continuous and decreasing,
(b) $\varphi(s) = 0$ iff $s = 1$.

Definition 17.2.12 (*Paknazar* (2018)) Let $T : B \longrightarrow D$ and $\alpha : B^2 \times (0, \infty) \longrightarrow [0, \infty)$. For any $w, x, u, v \in B$ and $s > 0$, T is known as a "modified $\varphi - \omega-$proximal contractive mapping,"

$$\begin{cases} \alpha(w, x, s) \geq 1 \\ M(u, Tw, s) = M(B, D, s) \Rightarrow \varphi(M(u, v, s)) \leq \omega(s)\varphi(M(w, x, s)), \\ M(v, Tx, s) = M(B, D, s) \end{cases} \tag{17.7}$$

where $\varphi \in \Phi$ and a function $\omega : (0, +\infty) \longrightarrow (0, 1)$ are provided.

Definition 17.2.13 (*Paknazar* (2018)) Let $T : B \longrightarrow D$ and $\alpha : B^2 \times (0, \infty) \longrightarrow [0, \infty)$. A mapping T is called "a modified $\alpha - \beta-$proximal contractive mapping if there is a function $\beta : [0, 1] \longrightarrow [0, +\infty)$ such that, for any sequence $\{s_n\} \subseteq [0, 1]$ of positive reals, $\beta(s_n) \to 1$ implies $s_n \to 1$ and for all $w, x, u, v \in B$ and $s > 0$, we have

$$\begin{cases} \alpha(w, x, s) \geq 1 \\ M(u, Tw, s) = M(B, D, s) \Rightarrow M(u, v, s) \geq \beta(M(w, x, s))M(w, x, s). \\ M(v, Tx, s) = M(B, D, s) \end{cases} \tag{17.8}$$

Paknazar (2018) proved the best proximity point theorems based on the positive parameter s in the setting of non-Archimedean fuzzy metric spaces.

Theorem 17.2.1 (Paknazar (2018), Theorem 2.6) *Let B and D be nonempty closed subsets of a complete non-Archimedean fuzzy metric space $(X, M, *)$ such that $B_0(s)$ is nonempty for all $s > 0$. If $T : B \longrightarrow D$ satisfies the following conditions:*

(i) T is a triangular $\alpha-$proximal admissible mapping and $T(B_0(s)) \subseteq D_0(s)$ for all $s > 0$,
(ii) T is a modified $\alpha - \varphi - \omega-$proximal contractive mapping,
(iii) if $\{x_n\}$ is a sequence in $D_0(s)$ and $w \in B$ is such that $S(w, x_n, s) \to M(B, D, s)$, as $n \to \infty$, then $w \in B_0(s)$ for all $s > 0$,
(iv) there exists $w_0, w_1 \in B_0(s)$ such that $M(w_1, Tw_0, s) = M(B, D, s), \alpha(w_0, w_1, s) \geq 1$ for all $s > 0$,

(v) *if* $\{w_n\}$ *is a sequence in* X *such that* $\alpha(w_n, w_{n+1}, s) \geq 1$ *for all* $n \geq 1$ *and* $w_n \to w$ *as* $n \to \infty$, *then* $\alpha(w_n, w, s) \geq 1$ *for all* $n \geq 1$ *and* $s > 0$.

Then, there exists $w^* \in B$ *such that* $M(w^*, Tw^*, s) = M(B, D, s)$ *for all* $s > 0$.

Theorem 17.2.2 (Paknazar (2018), Theorem 2.10) *Let* B *and* D *be nonempty closed subsets of a complete non-Archimedean fuzzy metric space* $(X, M, *)$ *such that* $B_0(s)$ *is nonempty for all* $s > 0$. *If* $T : B \longrightarrow D$ *satisfies the following assertions:*

(i) T *is a triangular* α*−proximal admissible mapping and* $T(B_0(s)) \subseteq D_0(s)$ *for all* $s > 0$,
(ii) T *is a modified* $\alpha - \beta$*−proximal contractive mapping,*
(iii) *if* $\{x_n\}$ *is a sequence in* $D_0(s)$ *and* $w \in B$ *is such that* $M(w, x_n, s) \to M(B, D, s)$, *as* $n \to \infty$, *then* $w \in B_0(s)$ *for all* $s > 0$,
(iv) *there exist* $w_0, w_1 \in B_0(s)$ *such that* $M(w_1, Tw_0, s) = M(B, D, s), \alpha(w_0, w_1, s) \geq 1$ *for all* $s > 0$,
(v) *if* $\{w_n\}$ *is a sequence in* X *such that* $\alpha(w_n, w_{n+1}, s) \geq 1$ *for all* $n \geq 1$ *and* $w_n \to w$ *as* $n \to \infty$, *then* $\alpha(w_n, w, s) \geq 1$ *for all* $n \geq 1$ *and* $s > 0$.

Then, there exists $w^* \in B$ *such that* $M(w^*, Tw^*, s) = M(B, D, s)$ *for all* $s > 0$, *that is,* T *has a best proximity point* $w^* \in B$.

Moreover, if $M(w, Tw, s) = M(B, D, s)$ *and* $M(x, Tx, s) = M(B, D, s)$ *imply* $\alpha(w, x, s) \geq 1$, *then* T *has a unique best proximity point.*

17.3 Main Results

Let Ψ be the collection of functions $\psi : [0, 1] \to [0, 1]$ such that ψ is monotone increasing and continuous, with $\psi(w) = 0 \Leftrightarrow w = 1$ and $\psi(w) \geq w$ for all $w \in [0, 1]$.

Let Λ be the collection of non-decreasing functions $\lambda : [0, 1] \to [0, 1]$ such that $\lambda(w_n) \to 1 \Rightarrow w_n \to 1$ for every sequence of positive reals $\{w_n\} \subseteq [0, 1]$.

Definition 17.3.1 Let $\alpha : B^2 \times (0, \infty) \longrightarrow [0, \infty)$. A mapping $T : B \longrightarrow D$ is called a generalized α_{ψ}^{ϕ}−proximal contractive mapping if for every $w, x \in B$, and $s > 0$, we have

$$\begin{cases} \alpha(w, x, s) \geq 1 \\ M(w, Tw, s) = M(B, D, s) \Rightarrow \phi(w, x, s)M(Tw, Tx, s) \geq \psi(Z(w, x, s)), \\ M(x, Tx, s) = M(B, D, s) \end{cases} \tag{17.9}$$

where $\phi : B^2 \times (0, \infty) \longrightarrow [1, \infty)$ is a continuous function and

$$Z(w, x, s) = \min\{M(w, x, s), \max\{M(w, Tw, s), M(w, Tx, s), M(x, Tx, s), M(x, Tw, s)\}\}.$$

Definition 17.3.2 Let $T : B \longrightarrow D$ and $\alpha, \eta : B^2 \times (0, \infty) \longrightarrow [0, \infty)$. A mapping T is called a generalized α_{λ}^{η}−proximal contractive mapping if for each $w, x \in B$, and $s > 0$, we have

$$\begin{cases} \alpha(w, x, s) \geq \eta(w, x, s) \\ M(w, Tw, s) = M(B, D, s) \Rightarrow \eta(w, x, s)M(Tw, Tx, s) \geq \lambda(M(w, x, s))Z(w, x, s)Y(w, x, s), \\ M(x, Tx, s) = M(B, D, s) \end{cases} \tag{17.10}$$

where $Z(w, x, s) = \min\{M(w, x, s), \max\{M(w, Tw, s), M(x, Tx, s)\}\}$ and

$$Y(w, x, s) = \max\{M(w, x, s), M(w, Tw, s), M(x, Tx, s), M(w, Tx, s), M(x, Tw, s)\}.$$

Remark 17.3.1 If $\eta(w, x, s) = 1$ and $Z(w, x, s)Y(w, x, s) = M(w, x, s)$, the above definition reduces to the definition 2.12 of Paknazar (2018).

Theorem 17.3.1 *Suppose that B and D are nonempty closed subsets of a complete fuzzy metric space $(X, M, *)$ such that B_0 is nonempty for every $s > 0$. If $T : B \longrightarrow D$ satisfies the following conditions:*

(a) T represents a triangular α−proximal admissible mapping, $T(B_0) \subseteq D_0$,
(b) T represents a generalized α_{ϕ}^{ψ}−proximal contractive mapping,
(c) there exist elements w_0 and w_1 in B_0 such that $M(w_1, Tw_0, s) = M(B, D, s)$, $\alpha(w_0, w_1, s) \geq 1$ for all $s > 0$.
(d) if $\{x_n\}$ is a sequence in D_0 and $w \in B$ is so that $M(w, x_n, s) \to M(B, D, s)$, as $n \to \infty$, then $w \in B_0$ for any $s > 0$,
(e) if $\{w_n\}$ is a sequence in X such that $\alpha(w_n, w_{n+1}, s) \geq 1$ for all $n \geq 1$ and $w_n \to w$ as $n \to \infty$, then $\alpha(w_n, w, s) \geq 1$ for all $n \geq 1, s > 0$.

Then, there exists $w^* \in B$ such that $M(w^*, Tw^*, s) = M(B, D, s)$ for every $s > 0$, that is, $w^* \in B$ is the best proximity point of T. Moreover, if $M(w, Tw, s) = M(B, D, s)$ and $M(x, Tx, s) = M(B, D, s)$ imply $\alpha(w, x, s) \geq 1$, then the best proximity point for T is unique.

Proof By our assumption, there exist w_0 and w_1 in B_0 such that

$$M(w_1, Tw_0, s) = M(B, D, s), \alpha(w_0, w_1, s) \geq 1 \text{ for all } s > 0. \tag{17.11}$$

As $T(B_0) \subseteq D_0$, there exists $w_2 \in B_0$ such that

$$M(w_2, Tw_1, s) = M(B, D, s).$$

Given that T is a triangular α−proximal admissible mapping, hence $\alpha(w_1, w_2, s) \geq 1$. Thus, we have

$$M(w_2, Tw_1, s) = M(B, D, s) \text{ and } \alpha(w_1, w_2, s) \geq 1. \tag{17.12}$$

Again, $T(B_0) \subseteq D_0$, there exists $w_3 \in B_0$ such that

$$M(w_3, Tw_2, s) = M(B, D, s).$$

Combining (17.12) with the above, we have

$$\begin{aligned} M(w_2, Tw_1, s) &= M(B, D, s),\\ M(w_3, Tw_2, s) &= M(B, D, s),\\ \alpha(w_1, w_2, s) &\geq 1. \end{aligned} \tag{17.13}$$

By our assumption on T, we obtain that $\alpha(w_2, w_3, s) \geq 1$. Thus, we conclude that

$$M(w_3, Tw_2, s) = M(B, D, s) \text{ and } \alpha(w_2, w_3, s) \geq 1. \tag{17.14}$$

Continuing in the same way, we arrive at

$$M(w_{n+1}, Tw_n, s) = M(B, D, s) \text{ and } \alpha(w_n, w_{n+1}, s) \geq 1, \text{ for each } n \geq 0 \text{ and } s > 0. \tag{17.15}$$

Using $\alpha(w_n, w_{n+1}, s) \geq 1$, $\alpha(w_{n+1}, w_{n+2}, s) \geq 1$ and the condition (T2), we obtain that

$$\alpha(w_n, w_{n+2}, s) \geq 1.$$

Again, by $\alpha(w_n, w_{n+2}, s) \geq 1$, $\alpha(w_{n+2}, w_{n+3}, s) \geq 1$ and the condition (T2), we get that

$$\alpha(w_n, w_{n+3}, s) \geq 1.$$

Finally, we obtain that

$$M(w_{n+1}, Tw_n, s) = M(B, D, s), \alpha(w_n, w_m, s) \geq 1, \text{ for each } m, n \geq 1 \text{ with } n < m. \tag{17.16}$$

By condition (b) and Eq. (17.9), we have

$$\begin{aligned} &\alpha(w_n, w_{n+1}, s) \geq 1\\ &M(w_n, Tw_n, s) = M(B, D, s) \Rightarrow \phi(w_n, w_{n+1}, s)M(Tw_n, Tw_{n+1}, s) \geq \psi(Z(w_n, w_{n+1}, s)).\\ &M(w_{n+1}, Tw_{n+1}, s) = M(B, D, s) \end{aligned} \tag{17.17}$$

Defining the sequence $\{w_n\}$ recursively by $w_n = Tw_{n-1}$ with the property of ϕ and assumption (b), i.e., assuming $\alpha(w_{n-1}, w_n, s) \geq 1$, we observe

$$\begin{aligned} M(w_n, w_{n+1}, s) &= M(Tw_{n-1}, Tw_n, s)\\ &\leq \phi(w_{n-1}, w_n, s)M(Tw_{n-1}, Tw_n, s)\\ &\geq \psi(Z(w_{n-1}, w_n, s)) \end{aligned} \tag{17.18}$$

where

$$
\begin{aligned}
&Z(w_{n-1}, w_n, s)\\
&\quad = \min\{M(w_{n-1}, w_n, s), \max\{M(w_{n-1}, Tw_{n-1}, s), M(w_{n-1}, Tw_n, s), M(w_n, Tw_n, s), M(w_n, Tw_{n-1}, s)\}\}\\
&\quad = \min\{M(w_{n-1}, w_n, s), \max\{M(w_{n-1}, w_n, s), M(w_{n-1}, w_{n-1}, s), M(w_n, w_{n-1}, s), M(w_n, w_n, s)\}\}\\
&\quad = \min\{M(w_{n-1}, w_n, s), \max\{M(w_{n-1}, w_n, s), M(w_n, w_{n-1}, s), 1, 1\}\}\\
&\quad = \min\{M(w_{n-1}, w_n, s), \max\{M(w_{n-1}, w_n, s), 1\}\}\\
&\quad = M(w_{n-1}, w_n, s).
\end{aligned}
\tag{17.19}
$$

Equation (17.18) then becomes

$$M(w_n, w_{n+1}, s) \geq \psi(M(w_{n-1}, w_n, s)),$$

which implies that the sequence $\{M(w_n, w_{n+1}, s)\}$ is increasing in [0, 1] for every $n \geq 1$ and $s > 0$. Thus, for every $s > 0$, there exists $t(s)$ such that

$$\lim_{n\to\infty} M(w_{n-1}, w_n, s) = t(s) \leq 1. \tag{17.20}$$

Now, we show that $s(t) = 1$ for every $s > 0$. Suppose that $s > 0$; from

$$M(w_n, w_{n+1}, s) \geq \psi(M(w_{n-1}, w_n, s)) \geq 1,$$

we get

$$\lim_{n\to\infty} \psi(M(w_{n-1}, w_n, s)) = 1. \tag{17.21}$$

Hence, by the continuity of ψ,

$$\lim_{n\to\infty} M(w_{n-1}, w_n, s) = 1, \tag{17.22}$$

which implies $s(t) = 1$.

Next, we prove that $\{w_n\}$ is a Cauchy sequence. If it is not, then there exists $\delta \in (0, 1)$ and $s_0 > 0$ such that, for every $p \geq 1$, there exists $n(p), m(p) \in \mathbb{N}$ with $m(p) > n(p) \geq p$, and

$$M(w_{m(p)}, w_{n(p)}, s_0) \leq 1 - \delta. \tag{17.23}$$

Assuming $m(p)$ is the smallest integer greater than $n(p)$, then we have

$$M(w_{m(p)-1}, w_{n(p)}, s_0) > 1 - \delta. \tag{17.24}$$

Then for all p and any $k > 0$, we obtain

$$\begin{aligned} 1-\delta &\geq M(w_{m(p)}, w_{n(p)}, s_0) \\ &\geq M(w_{m(p)-1}, w_{m(p)}, \frac{s_0}{k}) * M(w_{m(p)-1}, w_{n(p)}, \frac{s_0}{k}) \\ &> (1-\delta)^2 > (1-\delta), \end{aligned} \tag{17.25}$$

which is contradiction. Proceeding with the limit as $p \to +\infty$, we deduce

$$M(w_{m(p)}, w_{n(p)}, s_0) = 1-\delta. \tag{17.26}$$

From

$$\begin{aligned} M(w_{m(p)+1}, w_{n(p)+1}, s_0) \geq{}& M(w_{m(p)+1}, w_{m(p)}, \frac{s_0}{k}) * M(w_{m(p)}, w_{n(p)}, \frac{s_0}{k}) \\ &* M(w_{n(p)}, w_{n(p)+1}, \frac{s_0}{k}), \end{aligned} \tag{17.27}$$

and

$$\begin{aligned} M(w_{m(p)}, w_{n(p)+1}, s_0) \geq{}& M(w_{m(p)}, w_{m(p)+1}, \frac{s_0}{k}) * M(w_{m(p)+1}, w_{n(p)+1}, \frac{s_0}{k}) \\ &* M(w_{n(p)+1}, w_{n(p)}, \frac{s_0}{k}), \end{aligned} \tag{17.28}$$

it follows that

$$\lim_{p\to+\infty} M(w_{m(p)+1}, w_{n(p)+1}, s_0) = 1-\delta. \tag{17.29}$$

From (17.16), we have

$$\begin{cases} \alpha(w_{n(p)}, w_{m(p)}, s) \geq 1, \\ M(w_{m(p)+1}, Tw_{m(p)}, s_0) = M(B, D, s_0), \\ M(w_{n(p)+1}, Tw_{n(p)}, s_0) = M(B, D, s_0). \end{cases} \tag{17.30}$$

Thus, by (17.9) and (17.30), we obtain

$$\phi(w_{m(p)+1}, w_{n(p)+1}, s_0) M(Tw_{m(p)}, Tw_{n(p)}, s_0) \geq \psi(Z(w_{m(p)+1}, w_{n(p)+1}, s_0)).$$

From similar inequalities and limits, we eventually get

$$\phi(1-\delta)(1-\delta) \geq \psi(1-\delta),$$

which leads to a contradiction since either $\delta = 0$ or $\delta \geq 1$. Hence, $\{w_n\}$ is a Cauchy sequence. Due to the completeness of $(X, M, *)$, for every $s > 0$, the sequence $\{w_n\}$ converges to some $w^* \in X$, i.e.,

$$\lim_{n\to+\infty} M(w_n, w^*, s) = 1.$$

We also have

$$\begin{aligned} M(B, D, s) &= M(w_{n+1}, Tw_n, s) \\ &\geq M\left(w_{n+1}, w^*, \frac{s}{k}\right) * M\left(w^*, Tw_n, \frac{s}{k}\right) \\ &\geq M\left(w_{n+1}, w^*, \frac{s}{k}\right) * M\left(w^*, w_{n+1}, \frac{s}{k}\right) * M\left(w_{n+1}, Tw_n, \frac{s}{k}\right) \\ &= M\left(w_{n+1}, w^*, \frac{s}{k}\right) * M\left(w^*, w_{n+1}, \frac{s}{k}\right) * M(B, D, s), \end{aligned} \tag{17.31}$$

which implies (taking the limit)

$$\lim_{n\to+\infty} M(w^*, Tw_n, s) = M(B, D, s).$$

So, $w^* \in B_0$. Since $T(B_0) \subseteq D_0$, there exists $w \in B_0$ such that

$$M(w, Tw^*, s) = M(B, D, s).$$

Combining this with earlier results yields

$$\phi(M(w, w_{n+1}, s))M(Tw, Tw_{n+1}, s) \geq \psi(M(w^*, w_n, s)). \tag{17.32}$$

Taking the limit as $n \to +\infty$, we find

$$\lim_{n\to+\infty} \psi(M(w, w_{n+1}, s)) = 0,$$

so that

$$\lim_{n\to+\infty} M(w, w_{n+1}, s) = 1.$$

Since limits are unique, $w = w^*$, and thus

$$M(w^*, Tw^*, s) = M(B, D, s). \tag{17.33}$$

This proves that w^* is a best proximity point of T. Lastly, to show uniqueness, suppose there is another point $x \neq w^*$ such that

$$M(x, Tx, s) = M(B, D, s).$$

Then by Assumption (e),

$$\phi(M(w^*, x, s)) > \phi(M(w^*, x, s))M(Tw^*, Tx, s) \geq \psi(M(w^*, x, s)),$$

which leads to a contradiction unless $M(w^*, x, s) = 1$, i.e., $x = w^*$. Hence the best proximity point is unique. □

Remark 17.3.2 Theorem 17.3.1 is the generalization of Theorem 2.6. of Paknazar (2018).

Theorem 17.3.2 *Suppose that a triplet $(X, M, *)$ is a complete fuzzy metric space with nonempty closed subsets B and D such that B_0 is nonempty for every $s > 0$. Consider a non-self-mapping $T : B \longrightarrow D$ that satisfies the following conditions:*

(a) there exist elements w_0 and w_1 in B_0 such that $M(w_1, Tw_0, s) = M(B, D, s)$, $\alpha(w_0, w_1, s) \geq 1$ for all $s > 0$.
(b) T represents a triangular $\alpha-$proximal admissible mapping, $T(B_0) \subseteq D_0$,
(c) T represents a generalized $\alpha_\lambda^\eta-$proximal contractive mapping,
(d) if $\{x_n\}$ is a sequence in D_0 and $w \in B$ is so that $M(w, x_n, s) \to M(B, D, s)$ as $n \to \infty$, then $w \in B_0$ for each $s > 0$,
(e) if $\{w_n\}$ is a sequence in X so that $\alpha(w_n, w_{n+1}, s) \geq 1$ for any $n \geq 1$ and $w_n \to w$ as n goes to infinity, then $\alpha(w_n, w, s) \geq 1$ in place of every $n \geq 1, s > 0$.

Then, the map T has a unique best proximity point $w^* \in B$.

Proof By our assumption, there exist elements w_0 and w_1 in B_0 such that

$$M(w_1, Tw_0, s) = M(B, D, s), \quad \alpha(w_0, w_1, s) \geq 1 \text{ for all } s > 0. \tag{17.34}$$

As $T(B_0) \subseteq D_0$, there exists $w_2 \in B_0$ such that

$$M(w_2, Tw_1, s) = M(B, D, s).$$

Given that T is a triangular α-proximal admissible mapping, hence $\alpha(w_1, w_2, s) \geq 1$. Thus, we have

$$M(w_2, Tw_1, s) = M(B, D, s), \quad \alpha(w_1, w_2, s) \geq 1. \tag{17.35}$$

Again, $T(B_0) \subseteq D_0$, there exists $w_3 \in B_0$ such that

$$M(w_3, Tw_2, s) = M(B, D, s).$$

Combining (17.35) with above, we have

$$\begin{aligned} M(w_2, Tw_1, s) &= M(B, D, s), \\ M(w_3, Tw_2, s) &= M(B, D, s), \\ \alpha(w_1, w_2, s) &\geq 1. \end{aligned} \tag{17.36}$$

By our assumption on T, we obtain that $\alpha(w_2, w_3, s) \geq 1$. Thus, we conclude that

$$M(w_3, Tw_2, s) = M(B, D, s) \text{ and } \alpha(w_2, w_3, s) \geq 1. \tag{17.37}$$

Continuing this procedure, we obtain a sequence $\{w_n\} \subseteq B_0$ satisfying

$$M(w_{n+1}, Tw_n, s) = M(B, D, s) \text{ and } \alpha(w_n, w_{n+1}, s) \geq 1, \text{ for each } n \geq 0 \text{ and } s > 0. \tag{17.38}$$

Using $\alpha(w_n, w_{n+1}, s) \geq 1, \alpha(w_{n+1}, w_{n+2}, s) \geq 1$ and the condition (T2), we obtain that

$$\alpha(w_n, w_{n+2}, s) \geq 1.$$

Again, by $\alpha(w_n, w_{n+2}, s) \geq 1, \alpha(w_{n+2}, w_{n+3}, s) \geq 1$ and condition (T2), we get that

$$\alpha(w_n, w_{n+3}, s) \geq 1.$$

Finally, we obtain

$$M(w_{n+1}, Tw_n, s) = M(B, D, s), \alpha(w_n, w_m, s) \geq 1, \text{ for each } m, n \geq 1 \text{ with } n < m. \tag{17.39}$$

By condition (c) and the above, we obtain

$$\begin{aligned}
\alpha(w_n, w_{n+1}, s) &\geq \eta(w_n, w_{n+1}, s) \\
M(w_n, Tw_n, s) &= M(B, D, s) \\
M(w_{n+1}, Tw_{n+1}, s) &= M(B, D, s) \\
\Rightarrow \eta(w_n, w_{n+1}, s)M(Tw_n, Tw_{n+1}, s) &\geq \lambda(M(w_n, w_{n+1}, s))Z(w_n, w_{n+1}, s)Y(w_n, w_{n+1}, s).
\end{aligned} \tag{17.40}$$

Defining the sequence $\{w_n\}$ recursively by $w_n = Tw_{n-1}$ with the property of η and assumption (c), we observe

$$\begin{aligned}
M(w_n, w_{n+1}, s) &= M(Tw_{n-1}, Tw_n, s) \\
&\geq \eta(w_{n-1}, w_n, s)M(Tw_{n-1}, Tw_n, s) \\
&\geq \lambda(M(w_{n-1}, w_n, s))Z(w_{n-1}, w_n, s)Y(w_{n-1}, w_n, s).
\end{aligned} \tag{17.41}$$

where

$$\begin{aligned}
&Z(w_{n-1}, w_n, s) \\
&\quad = \min\{M(w_{n-1}, w_n, s), \max\{M(w_{n-1}, Tw_{n-1}, s), M(w_n, Tw_n, s)\}\} \\
&\quad = M(w_{n-1}, w_n, s).
\end{aligned} \tag{17.42}$$

and

$$\begin{aligned}
&Y(w_{n-1}, w_n, s) \\
&\quad = \max\{M(w_{n-1}, Tw_{n-1}, s), M(w_{n-1}, Tw_n, s), M(w_n, Tw_n, s), M(w_n, Tw_{n-1}, s)\} \\
&\quad = \max\{M(w_{n-1}, w_n, s), M(w_{n-1}, w_{n-1}, s), M(w_n, w_{n-1}, s), M(w_n, w_n, s)\} \\
&\quad = \max\{M(w_{n-1}, w_n, s), M(w_n, w_{n-1}, s), 1, 1\} \\
&\quad = 1.
\end{aligned} \tag{17.43}$$

Therefore,

$$\begin{aligned} M(w_n, w_{n+1}, s) &\geq \lambda(M(w_{n-1}, w_n, s))M(w_{n-1}, w_n, s) \\ &\geq M(w_{n-1}, w_n, s). \end{aligned} \tag{17.44}$$

This inequality implies that $\{M(w_n, w_{n+1}, s)\}$ is an increasing sequence in [0, 1] for every $n \geq 1$ and $s > 0$. Hence, there exists $t(s)$ such that

$$\lim_{n\to\infty} M(w_n, w_{n+1}, s) = t(s) \leq 1. \tag{17.45}$$

Now we show that $s(t) = 1$. From the above,

$$\frac{M(w_n, w_{n+1}, s)}{M(w_{n-1}, w_n, s)} \geq \lambda(M(w_{n-1}, w_n, s)) \geq 1,$$

and taking the limit yields

$$\lim_{n\to\infty} \lambda(M(w_{n-1}, w_n, s)) = 1, \tag{17.46}$$

which implies by continuity of λ that

$$\lim_{n\to\infty} M(w_{n-1}, w_n, s) = 1. \tag{17.47}$$

That is, $t(s) = 1$. Next, we prove that $\{w_n\}$ is a Cauchy sequence. Suppose not. Then $\delta \in (0, 1)$ and $s_0 > 0$ exist such that for every $p \geq 1$, there are $n(p), m(p) \in \mathbb{N}$ with $m(p) > n(p) \geq p$, satisfying

$$M(w_{m(p)}, w_{n(p)}, s_0) \leq 1 - \delta. \tag{17.48}$$

Assume $m(p)$ is the smallest integer greater than $n(p)$, then

$$M(w_{m(p)-1}, w_{n(p)}, s_0) > 1 - \delta. \tag{17.49}$$

So for all $k > 0$, we obtain

$$\begin{aligned} 1 - \delta &\geq M(w_{m(p)}, w_{n(p)}, s_0) \\ &\geq M(w_{m(p)-1}, w_{m(p)}, \frac{s_0}{k}) * M(w_{m(p)-1}, w_{n(p)}, \frac{s_0}{k}) \\ &> (1-\delta)^2 > (1-\delta), \end{aligned} \tag{17.50}$$

which is a contradiction. Proceeding with the limit as $p \to +\infty$, we deduce

$$M(w_{m(p)}, w_{n(p)}, s_0) = 1 - \delta. \tag{17.51}$$

From

$$M(w_{m(p)+1}, w_{n(p)+1}, s_0) \geq M(w_{m(p)+1}, w_{m(p)}, \frac{s_0}{k}) * M(w_{m(p)}, w_{n(p)}, \frac{s_0}{k}) \\ * M(w_{n(p)}, w_{n(p)+1}, \frac{s_0}{k}), \tag{17.52}$$

and

$$M(w_{m(p)}, w_{n(p)+1}, s_0) \geq M(w_{m(p)}, w_{m(p)+1}, \frac{s_0}{k}) * M(w_{m(p)+1}, w_{n(p)+1}, \frac{s_0}{k}) \\ * M(w_{n(p)+1}, w_{n(p)}, \frac{s_0}{k}), \tag{17.53}$$

it follows that

$$\lim_{p\to+\infty} M(w_{m(p)+1}, w_{n(p)+1}, s_0) = 1 - \delta. \tag{17.54}$$

From (17.39), we know that

$$\begin{cases} \alpha(w_{n(p)}, w_{m(p)}, s_0) \geq \eta(w_{n(p)}, w_{m(p)}, s_0), \\ M(w_{m(p)+1}, Tw_{m(p)}, s_0) = M(B, D, s_0), \\ M(w_{n(p)+1}, Tw_{n(p)}, s_0) = M(B, D, s_0). \end{cases} \tag{17.55}$$

$$\Rightarrow \eta(w_{m(p)+1}, w_{n(p)+1}, s_0) M(Tw_{m(p)}, Tw_{n(p)}, s_0) \\ \geq \lambda(M(w_{m(p)+1}, w_{n(p)+1}, s_0)) Z(w_{m(p)+1}, w_{n(p)+1}, s_0) Y(w_{m(p)+1}, w_{n(p)+1}, s_0).$$

Again applying the same fashion as in (17.40) to (17.44), we get

$$M(w_{m(p)+1}, w_{n(p)+1}, s_0) \geq \lambda(M(w_{m(p)}, w_{n(p)}, s_0)) M(w_{m(p)}, w_{n(p)}, s_0),$$

which implies

$$\frac{M(w_{m(p)+1}, w_{n(p)+1}, s_0)}{M(w_{m(p)}, w_{n(p)}, s_0)} \geq \lambda(M(w_{m(p)}, w_{n(p)}, s_0)) \geq 1.$$

Proceeding with the limit as $p \to +\infty$, we get

$$\lim_{p\to+\infty} \lambda(M(w_{m(p)}, w_{n(p)}, s_0)) = 1.$$

It follows that

$$(1 - \delta) = \lim_{p\to+\infty} \lambda(M(w_{m(p)}, w_{n(p)}, s_0)) = 1,$$

yields $\delta = 0$ which is a contradiction. Therefore, a sequence $\{w_n\}$ is Cauchy. Since $(X, M, *)$ is complete, there exists $w^* \in X$ such that

$$\lim_{n\to+\infty} M(w_n, w^*, s) = 1.$$

Moreover,

$$\begin{aligned} M(B, D, s) &= M(w_{n+1}, Tw_n, s) \\ &\geq M(w_{n+1}, w^*, \frac{s}{k}) * M(w^*, Tw_n, \frac{s}{k}) \\ &\geq M(w_{n+1}, w^*, \frac{s}{k}) * M(w^*, w_{n+1}, \frac{s}{k}) * M(w_{n+1}, Tw_n, \frac{s}{k}) \\ &= M(w_{n+1}, w^*, \frac{s}{k}) * M(w^*, w_{n+1}, \frac{s}{k}) * M(B, D, s). \end{aligned} \tag{17.56}$$

This implies

$$\begin{aligned} M(B, D, s) &\geq M(w_{n+1}, w^*, \frac{s}{k}) * M(w^*, Tw_n, \frac{s}{k}) \\ &\geq M(w_{n+1}, w^*, \frac{s}{k}) * M(w^*, w_{n+1}, \frac{s}{k}) * M(B, D, s). \end{aligned} \tag{17.57}$$

Taking the limit yields

$$\lim_{n\to+\infty} M(w^*, Tw_n, s) = M(B, D, s).$$

Thus, by Assumption (d), $w^* \in B_0$. Since $T(B_0) \subseteq D_0$, there exists $w' \in B_0$ such that

$$M(w', Tw^*, s) = M(B, D, s).$$

Combining this with earlier results, we get

$$\eta(w', w_{n+1}, s)M(Tw', Tw_{n+1}, s) \geq \lambda(M(w^*, w_n, s))M(w^*, w_n, s). \tag{17.58}$$

Taking the limit as $n \to +\infty$, we find

$$\lim_{n\to+\infty} \lambda(M(w', w_{n+1}, s)) = 0, \Rightarrow \lim_{n\to+\infty} M(w', w_{n+1}, s) = 1,$$

so, $w' = w^*$, and

$$M(w^*, Tw^*, s) = M(B, D, s) \text{ for every } s > 0. \tag{17.59}$$

Therefore, w^* is a best proximity point of T. Lastly, we show that w^* is the unique best proximity point of T. Suppose that for some $s > 0$, we have $0 < M(w^*, x, s) < 1$, and $x \neq w^*$ is another best proximity point of T, that is,

$$M(w^*, Tw^*, s) = M(B, D, s)$$

and

$$M(x, Tx, s) = M(B, D, s).$$

Then, by Assumption (e) together with Eq. (17.10), it follows that

$$\eta(w^*, x, s) > \eta(w^*, x, s)M(Tw^*, Tx, s) \geq \lambda(M(w^*, x, s))M(w^*, x, s) > M(w^*, x, s),$$

which leads to a contradiction. Therefore, $M(w^*, x, s) = 1$ for all $s > 0$, implying that $x = w^*$. Hence, the proof is complete. □

Remark 17.3.3 Theorem 17.3.2 is the generalization of Theorem 2.10 of Paknazar (2018).

Example 17.3.1 Suppose that $(X, M, *)$ is a complete fuzzy metric space with two nonempty subsets B and D.

Let $B = [0, 4/5] \cup \{1, \infty\}$, $D = [0, +\infty)$, and $S(w, x, s) = \dfrac{s}{s + m(w, x)}$, where $m(w, x) = |w - x|$ for all w, x in X and $*$ is the product continuous t-norm.

Let $T : B \to D$ be a map defined by

$$Tw = \begin{cases} \dfrac{w}{4}, & \text{if } w \in [0, 4/5], \\ 0, & \text{otherwise.} \end{cases}$$

Suppose that (a) $\alpha, \eta : B^2 \times (0, \infty) \to [0, \infty)$ defined by

$$\alpha(w, x, s) = \begin{cases} 3, & \text{if } w, x \in [0, 4/5] \cup \{1\}, \\ 1, & \text{otherwise.} \end{cases}$$

and

$$\eta(w, x, s) = \begin{cases} 2, & \text{if } w, x \in [0, 4/5] \cup \{1\}, \\ 1, & \text{otherwise.} \end{cases}$$

(b) $\phi : B^2 \times (0, \infty) \longrightarrow [1, \infty)$ defined by $\phi(w, x, s) = 2$ for all $w, x \in [0, 4/5] \cup \{1\}$.

Now, for all $w, x, \varsigma \in [0, 4/5] \cup \{1\}$, we have that $\alpha(w, x, s) \geq 1$ and $\alpha(x, \varsigma, s) \geq 1$, which implies $\alpha(w, \varsigma, s) \geq 1$. Hence, the map T represents a "triangular $\alpha-$proximal admissible."

Suppose that $\{w_n\}$ is in X such that $\alpha(w_n, w_{n+1}, s) \geq 1$ and $\alpha(w_n, w_{n+1}, s) \geq \eta(w_n, w_{n+1}, s)$ for all $n \geq 0$, $w_n \to w$ as $n \to \infty$, then $\{w_n\} \subseteq [0, 1]$ and hence $w \in [0, 4/5] \cup \{1\}$ shows that $\alpha(w, Tw, s) \geq \eta(w, Tw, s)$ for all $n \in \mathbb{N}$ and $s > 0$.

Again for all $w, x \in [0, 4/5] \cup \{1\}$, we have

$$M(Tw, Tx, s) = \frac{s}{s + m(Tw, Tx)} = \frac{s}{s + |Tw - Tx|} = \frac{s}{s + |\frac{w}{4} - \frac{x}{4}|} = \frac{s}{s + \frac{1}{4}|w - x|},$$

$$M(w, Tw, s) = \frac{s}{s + m(w, Tw)} = \frac{s}{s + |w - Tw|} = \frac{s}{s + |w - \frac{w}{4}|} = \frac{s}{s + \frac{3}{4}|w|},$$

$$M(w, Tx, s) = \frac{s}{s + m(w, Tx)} = \frac{s}{s + |w - Tx|} = \frac{s}{s + |w - \frac{x}{4}|} = \frac{s}{s + \frac{1}{4}|4w - x|},$$

$$M(x, Tx, s) = \frac{s}{s + m(x, Tx)} = \frac{s}{s + |x - Tx|} = \frac{s}{s + |x - \frac{x}{4}|} = \frac{s}{s + \frac{3}{4}|x|},$$

$$M(x, Tw, s) = \frac{s}{s + m(x, Tw)} = \frac{s}{s + |x - Tw|} = \frac{s}{s + |x - \frac{w}{4}|} = \frac{s}{s + \frac{1}{4}|4x - w|},$$

$$M(w, x, s) = \frac{s}{s + m(w, x)} = \frac{s}{s + |w - x|}.$$

$$\begin{aligned} Z(w, x, s) &= \min\{M(w, x, s), \max\{M(Tw, Tx, s), M(w, Tw, s), M(w, Tx, s), M(x, Tx, s), M(x, Tw, s)\}\} \\ &\min\{M(w, x, s), M(w, Tx, s)\} \\ &= M(w, x, s). \end{aligned}$$

And also we have $Tw \leq 1$, $Tx \leq 1$ implies $T(B) \subseteq D$. Hence, all the conditions of Theorems 17.3.1 and 17.3.2 are satisfied.

Now, we can verify that the contractive condition of Theorems 2.6 and 2.10 of Paknazar (2018) fails to hold.

For all $w, x \in B$ and $s > 0$,

$$\psi(w, x, s) \leq \omega(s)\psi(w, x, s)$$

implies

$$\frac{s}{s + |w - x|} \leq \omega(s)\frac{s}{s + |w - x|},$$

which is a contradiction as $\omega(s)$ lies in (0, 1).

Therefore, Theorems 17.3.1 and 17.3.2 are generalizations of Theorems 2.6 and 2.10 of Paknazar (2018), respectively.

17.4 Application

In this section, we present the approach of the best proximity point for solving Urysohn integral equations and second-order differential equations. In the study of ordinary and partial differential equations, best proximity points serve to identify solutions that are close to a desired behavior or state, especially in cases where a solution may not exist or may be difficult to find.

Best proximity point theory represents a significant area in nonlinear analysis that focuses on the approximation of fixed points of nonlinear mappings. This theory proves particularly useful when dealing with mappings that do not necessarily have fixed points but can be closely approximated. It applies to solving nonlinear functional equations where traditional fixed point theorems may not hold.

17.4.1 Application to Urysohn Integral Equations

A Urysohn integral equation named after the Russian mathematician Pavel Urysohn in 1920 Arkhangelskii & Tikhomirov (1998) is a type of integral equation that arises in various fields of mathematics, particularly in functional analysis and the theory of integral equations.

Definition 17.4.1 A Urysohn integral equation typically has the following form:

$$w(u) = f(u) + \lambda \int_{IR} K(u, v, w(v))dv \tag{17.60}$$

where: $w(u)$ is the unknown function we want to solve for, $f(u)$ is a given function, λ is a parameter (scalar), $K(u, v, w(v))$ is a kernel function that depends on u, v and the unknown function $w(u)$.

Suppose an integration region $(IR) = [a, b]$ is a set of finite measures, and define the Hilbert space:

$$l_{IR}^2 = \left\{ w \int_{IR} |w(v)|^2 dv < \infty \right\}.$$

Define the norm $\|\cdot\|$ on l_{IR}^2 as

$$\|w\| = \sqrt{\int_{IR} |w(u)|^2 du}, \text{ for all } w \in l_{IR}^2. \tag{17.61}$$

Equivalently it is given as:

$$\|w\| = \sqrt{\sup \left\{ e^{-t \int_{IR} \alpha(v)dv} \int_{IR} |w(v)|^2 dv \right\}} \text{ for all } w \in l_{IR}^2, t > 1 \tag{17.62}$$

Then, $C\left(l_{IR}^2, \|\cdot\|_{2,v}\right)$ is a Banach space. Let $X = \left\{w \in l_{IR}^2 : w(u) > 0 \text{ for almost every } u\right\}$. The fuzzy metric M_v associated with the norm $\|.\|_{2,v}$ is given by $M_v(w, x, s) = e^{-\frac{\|w-x\|}{s}}$ for all $w, x \in X$. Then $(X, M_v, *)$ is a complete fuzzy metric spaces. Now, consider the following conditions satisfied:

(i) the Kernal $K : [a, b] \times [a, b] \times \mathbb{R} \to \mathbb{R}$ satisfies Carthodory conditions, and

$$|K(u, v, w(v))| \leq w(u, v) + x(u, v),\ w, x \in ([a, b] \times [a, b])^2,\ x(u, s) > 0$$

(ii) the function $f : [a, b] \to [1, \infty)$ is continuous and bounded on $[a, b]$.

(iii) there exists a positive constant C such that

$$\sup_{u \in [a,b]} \int_a^b |K(u, v)| dv \leq C.$$

(iv) there exists a nonnegative and measurable function $q : [a, b] \times [a, b] \to \mathbb{R}$ such that

$$\alpha(u) = \int_a^b q^2(u, v) dv \leq \frac{1}{B^2}$$

and integrable over $[a, b]$ with

$$e^{-\frac{|K(u,v,w(v))-K(u,v,x(v))|}{s}} \geq e^{-\frac{q(u,v)}{s}} e^{-\frac{|w(v)-x(v)|}{s}}$$

for all $u, v \in [a, b]$ and $w, x \in B$.

Theorem 17.4.1 *Suppose the mappings f and K satisfy conditions (i)–(iv). Then, the Urysohn integral equation (17.60) has a unique solution.*

Proof Let define the operator $T : B \longrightarrow D$, in accordance with the above mentioned notations, by

$$Tw(u) = f(u) + \int_a^b K(u, v, w(v)) dv$$

Assume that $w \in B$, since $f(u) \geq 1$, for almost every $u \in IR$, we have

$$Tw(u) = f(u) + \int_a^b K(u, v, w(v)) dv \geq 1$$

By Conditions (i) and (ii), the mapping T is continuous and compact from B to D. Using condition (iv) and the Holder inequality, we have

$$\begin{aligned}
e^{-\frac{|Tw(u)-Tx(u)|^2}{s}} &= e^{-\frac{|\int_{IR} K(u,v,w(v))dv - \int_{IR} K(u,v,x(v))dv|^2}{s}} \\
&\geq e^{\frac{-\left(\int_{IR} |K(u,v,w(v))-K(u,v,x(v))|dv\right)^2}{s}} \\
&\geq e^{-\frac{(\int_{IR} q(u,v)|w(v)-x(v)|dv)^2}{s}} \\
&\geq e^{-\frac{\int_{IR} q^2(u,v)dv}{s}} \cdot e^{\frac{-\int_{IR} |w(v)-x(v)|^2 dv}{s}} \\
&\geq e^{-\frac{\alpha(u)}{s}} \cdot e^{-\frac{\int_{IR} |w(v)-x(v)|^2 dv}{s}}.
\end{aligned}$$

Integrating both sides with respect to $u \in IR$, we obtain

$$\begin{aligned} e^{-\frac{|Tw(u)-Tx(u)|^2 du}{s}} &\geq e^{-\frac{\int_{IR}\left(\alpha(u)\cdot\int_{IR}|w(v)-x(v)|^2 dv\right)du}{s}} \\ &= e^{-\frac{\int_{IR}\left(\alpha(u)e^{v\int_{IR}\alpha(v)dv}.e^{-v\int_{IR}\alpha(v)dv}\int_{IR}|w(v)-x(v)|^2 dv\right)du}{s}} \\ &\geq e^{-\frac{\|w-x\|_{2,v}\int_{IR}\alpha(u)e^{v\int_{IR}\alpha(v)dv}du}{s}} \\ &\geq e^{-\frac{\|w-x\|_{2,v}^2 e^{v}\int_{IR}\alpha(v)dv}{vB^2 s}}. \end{aligned}$$

Thus, we have

$$e^{-\frac{B^2 e^{-v\int_{IR}\alpha(v)dv}|Tw(u)-Tx(u)|^2 du}{vs}} \geq e^{-\frac{\|w-x\|_{2,v}^2}{s}}.$$

This implies that

$$e^{-\frac{B^2\|Tw-Tx\|_{2,v}^2}{vs}} \geq e^{-\frac{\|w-x\|_{2,v}^2}{s}}.$$

Now, defining $\phi(w, x, s) = \frac{1}{\ln(w,x,s)}$ and $\psi(w, x, s) = (w, x, s)$ with $\alpha(w, x, s) = e^{\|w-x\|_{2,v}^2}$ (or define it as $\alpha(w, x, s) \geq 1$) then, we have

$$\phi(w, x, s)M_v(Tw, Tx, s) \geq \psi(M_v(w, x, s)), \text{ for some } \psi \in \Psi,$$

whenever $M_v(w, Tw, s) = M_v(B, D, s)$ and $M_v(x, Tx, s) = M_v(B, D, s)$.

Thus, T satisfies the condition of generalized α_{ψ}^{ϕ}-proximal contractive mapping in the fuzzy metric space $(X, M_v, *)$. Hence, from Theorem 17.3.1, the operator T admits a unique best proximity point. Since $(X, M_v, *)$ is a complete fuzzy metric space, and under assumptions (i)–(iv), the function T maps bounded subsets of l_{IR}^2 into relatively compact ones, the Urysohn integral equation (17.60) has a unique solution in X. □

17.4.2 Application to Second-Order Differential Equations

Now we discuss the existence and uniqueness of solutions of the boundary value problem for the second-order differential equation

$$-\frac{d^2w}{ds^2} = f(s, w(s)), \quad s \in [0, 1] \tag{17.63}$$

with boundary conditions $w(0) = w(1) = 0$, where $I = [0, 1]$ and $f : I \times \mathbb{R} \longrightarrow \mathbb{R}$ is a continuous function.

Consider $X = C(I, \mathbb{R})$ be the space of all continuous function $g : I \longrightarrow \mathbb{R}$. It is well known that X, endowed with

$$\alpha_\infty(u, v) = \sup_{s\in[0,1]} |u(s) - v(s)|$$

for all $u, v \in X$, is a complete metric space. Define fuzzy set M on $X^2 \times (0, +\infty)$ by

$$M(u, v, s) = e^{\frac{\alpha_\infty(u,v)}{s}},$$

for all $u, v \in X$ and $s > 0$. Then $(X, M, *)$ is a complete fuzzy metric space. Moreover,

$$\lim_{s\to+\infty} M(u, v, s) = 1,$$

for all $u, v \in X$, that is, $(X, M, *)$ is a fuzzy metric space.

Theorem 17.4.2 *Suppose that the following condition holds for all $w, x \in \mathbb{R}$ and for all $s \in I$:*

$$|f(s, w) - f(s, x)| < |w - x|. \tag{17.64}$$

Then problem (17.63) has a unique solution in $u^* \in C(I, \mathbb{R})$.

Proof It is commonly understood that problem (17.63) is identical to the following integral equation:

$$u(s) = \int_0^1 G(s, t) f(t, u(t))dt$$

for all $s \in [0, 1]$, where $G(s, t)$ is Green function associated to problem (17.63) and given by

$$G(s, t) = \begin{cases} s(1 - t) & \text{for } 0 \leqslant s \leqslant t \leqslant 1 \\ t(1 - s) & \text{for } 0 \leqslant t \leqslant s \leqslant 1 \end{cases}$$

Then u in $C(I, \mathbb{R})$ is a solution of Eq. (17.63) if and only if $u \in C(I, \mathbb{R})$ is a solution of the integral equation (17.63).

Now we can define a mapping $T : B \longrightarrow D$ as follows:

$$Tu(s) = \int_0^1 G(s, t) f(t, u(t))dt,$$

for all $s \in [0, 1]$ and $u \in B$. Let $u, v \in B$ such that $Tu \neq Tv$,

$$\begin{aligned} |Tu(s) - Tv(s)| &= \Big|\int_0^s G(s, t) f(t, u(t))dt - \int_0^s G(s, t) f(t, v(t))dt\Big| \\ &= \int_0^s G(s, t)|f(t, u(t)) - f(t, v(t))|dt \\ &\leqslant |u(t) - v(t)| \int_0^1 G(s, t)dt \end{aligned}$$

Since $\sup_{s\in[0,1]} \int_0^1 G(s,t)dt = 1/8$, we have

$$\begin{aligned}
&\sup_{s\in[0,1]} |Tu(s) - Tv(s)| \leqslant \frac{1}{8} \sup_{s\in[0,1]} |u(t) - v(t)| \\
&\Longrightarrow \alpha_\infty(Tu, Tv) \leqslant \frac{1}{8}\alpha_\infty(u, v) \\
&\Longrightarrow -\frac{\alpha_\infty(u, v)}{t} \leqslant -\frac{\alpha_\infty(Tu, Tv)}{\frac{1}{8}t} \\
&\Longrightarrow e^{-\left(\frac{\alpha_\infty(u,v)}{t}\right)} \leqslant e^{-\left(\frac{\alpha_\infty(Tu,Tv)}{\frac{1}{8}t}\right)}
\end{aligned}$$

Now, let us define $\alpha(u, v, s) = \eta(u, v, s) = 1$ implies $\alpha \geq \eta$. Choose $\lambda(t) = t^\mu$, with $\mu = \frac{1}{8} \in (0, 1)$ for $\lambda \in \Lambda$, then

$$e^{-\left(\frac{\alpha_\infty(u,v)}{t}\right)} \leqslant e^{-\left(\frac{\alpha_\infty(Tu,Tv)}{\mu t}\right)}.$$

It follows that $M(u, v, t) \leqslant M^{1/\mu}(Tu, Tv, t)$, which implies

$$\eta(u, v, t)M(Tu, Tv, t) \geq \lambda(M(u, v, t))Z(u, v, t)Y(u, v, t),$$

whenever $M(u, Tu, t) = M(B, D, t)$ and $M(v, Tv, t) = M(B, D, t)$, where $Z(u, v, t) = \min\{M(u, v, t), \max\{M(u, Tu, t), M(v, Tv, t)\}\}$ and

$$Y(u, v, t) = \max\{M(u, v, t), M(u, Tu, t), M(v, Tv, t), M(u, Tv, t), M(v, Tu, t)\}.$$

Thus, all the conditions of the generalized α_λ^η-proximal contractive mapping are satisfied by T in $(X, M, *)$. Hence, from Theorem (17.3.2), the operator T possesses a one and only one best proximity point. This implies the problem (17.63) has a unique solution $u^* \in C(I, \mathbb{R})$. □

17.5 Conclusion

In this work, we have presented best proximity point results for generalized α_ψ^ϕ-proximal contractive mappings and generalized α_η^λ-proximal contractive mappings in the setting of complete fuzzy metric spaces. As a natural extension of fixed point theory, best proximity point theory offers a valuable framework for analyzing non-self-mappings by identifying points that minimize distances between disjoint subsets of a metric space. This strategy is very important for ensuring convergence of iterative algorithms when no fixed point exists.

Unlike the approach by Paknazar (2018), our results are established without relying on the parameter s, offering a more generalized and flexible formulation. Additionally, an illustrative example is provided to demonstrate the applicability and validity of the established theorems.

To highlight the practical significance of our results, we have applied the developed results to solve Urysohn integral equations and to prove the existence and uniqueness of solutions for boundary value problems involving second-order differential equations. These applications underscored the broader impact of the best proximity point theory in addressing nonlinear problems in analysis.

References

Arkhangelskii, A. V., & Tikhomirov, V. M. (1998). Pavel Samuilovich Urysohn (1898–1924). *Russian Mathematical Surveys,53*(5), 875.

Basha, S. S. (2011). Best proximity point theorems. *Journal of Approximation Theory,163*(11), 1772–1781.

Basha, S. S., & Veeramani, P. (1997). Best proximity pairs and best approximations. *Acta Scientiarum Mathematicarum,63*(1), 289–300.

Beloul, S., Tomar, A., & Ritu, S. (2021). Weak subsequential continuity in fuzzy metric spaces and application. *International Journal of Nonlinear Analysis and Applications,12*(2), 1485–1496.

Eldred, A. A., Kirk, W. A., & Veeramani, P. (2005). Proximal normal structure and relatively nonexpansive mappings. *Studia Mathematica,171*, 283–293.

Eldred, A. A., & Veeramani, P. (2006). Existence and convergence of best proximity points. *Journal of Mathematical Analysis and Applications,323*(2), 1001–1006.

Fan, K. (1969). Extensions of two fixed point theorems of FE Browder. *Mathematische zeitschrift,112*(3), 234–240.

George, A., & Veeramani, P. (1994). On some results in fuzzy metric spaces. *Fuzzy Sets and Systems,64*(3), 395–399.

Jleli, M., & Samet, B. (2013). Best proximity points for $\alpha - \psi -$proximal contractive type mappings and applications. *Bulletin des Sciences Mathématiques,137*(8), 977–995.

Kramosil, I., & Michalek, J. (1975). Fuzzy metric and statistical metric space. *Kybernetica,11*(5), 326–334.

Paknazar, M. (2018). Non-Archimedean fuzzy metric spaces and best proximity point theorems. *Sahand Communications in Mathematical Analysis,9*(1), 85–112.

Petwal, S., Tomar, A., & Joshi, M. (2023). Best proximant for set-valued maps via proximal relations. *International Journal of Nonlinear Analysis and Applications,14*(1), 1237–1247.

Pragadeeswarar, V., & Gopi, R. (2019). Existence and uniqueness of a common best proximity point on fuzzy metric space. *Fuzzy Information and Engineering,11*(1), 54–63.

Prasad, G., Tomar, A., Dimri, R. C., & Bartwal, A. (2020). Coincidence theorems via contractive mappings in ordered non-Archimedean fuzzy metric spaces. *The Pure and Applied Mathematics,27*(4), 187–205.

Prolla, J. B. (1983). Fixed-point theorems for set-valued mappings and existence of best approximants. *Numerical Functional Analysis and Optimization,5*(4), 449–455.

Reich, S. (1978). Approximate selections, best approximations, fixed points, and invariant sets. *Journal of Mathematical Analysis and Applications,62*(1), 104–113.

Sehgal, V. M., & Singh, S. P. (1988). A generalization to multifunctions of Fan's best approximation theorem. *Proceedings of the American Mathematical Society,102*(3), 534–537.

Shayanpour, H., & Nematizadeh, A. (2017). Some results on common best proximity point in fuzzy metric spaces. *Boletim da Sociedade Paranaense de Matematica,35*, 177–194.

Shweizer, B., & Skala, A. (1960). Statistical metric spaces. *Pacific Journal of Mathematics,10*(1), 314–334.

Tomar, A., Joshi, M., & Padaliya, S. K. (2023). On relation-theoretic F-contractions and applications in F-metric space. *Acta Universitatis Sapientiae Mathematica,15*(2), 314–336.

Zhou, M., Saleem, N., López, R., de Hierro, A. F., & Liu, X. (2022). Best proximity point theorems without fuzzy p-property for several $(\psi - \phi)$−weak contractions in non-archimedean fuzzy metric spaces. *Mathematics,10*(21), 4031.

Chapter 18
Applications of Fixed Points for Delay Fractional Differential Equations and Inclusions Involving Fractional Operators

Ahsan Abbas, Akbar Azam, and Nayyar Mehmood

Abstract In this chapter the study with the applications of fixed point results for delay fractional differential equations and inclusions involving Caputo-Fabrizio operator of order $1 < \varkappa \leq 2$ is presented. Schauder-type fixed point theorem is applied to obtain the existence results. Banach Contraction principle is used to find the uniqueness of the solution. The Hyers-Ulam stability is also investigated for delay fractional differential equations. For the solutions of the delay fractional differential inclusions, we have used Bohnenblust-Karlin's theorem. In the last, examples are furnished to illustrate our results.

Keywords Caputo-Fabrizio · Existence results · Uniqueness · Stability · Inclusion

18.1 Introduction

During the last 30 years, fractional calculus (the study of integrals and derivatives of arbitrary real or complex orders) has gained attention in both relevance and popularity. It has become popular because it can be used in many different fields such as science, engineering, finance, and physics (Baleanu, 2009; Efe, 2012; Herrmann, 2008; Naber, 2004; Stanislavsky, 2004). Recently, fractional calculus has known as a useful tool for explaining processes that have long-term memory. The key models in this area are based on differential equations that include derivatives of fractional order (Yong et al., 2016).

A. Abbas (✉)
National University of Technology (School of Applied Sciences and Humanities), Islamabad, Pakistan
e-mail: ahsanabbas@nutech.edu.pk

A. Azam
Department of Mathematics, Grand Asian University, Sialkot, Pakistan
e-mail: akbarazam@gaus.edu.pk

N. Mehmood
International Islamic University, Islamabad, Pakistan

© The Author(s), under exclusive license to Springer Nature Singapore Pte Ltd. 2026
A. Tomar and M. Jain (eds.), *Applications of Fixed-Point Theorem*, Industrial and Applied Mathematics, https://doi.org/10.1007/978-981-95-5160-6_18

Fractional calculus is a rapidly growing field in mathematical analysis that includes special types of fractional operators, such as Caputo and Riemann-Liouville (R-L). The main difference among them is the order of operations. In the R-L method, integration is done before the differentiation , but on the other hand, the Caputo fractional operator performs differentiation first, then the integration process (Kilbas, 2006; Losta & Opial, 1965; Oldham & Spanier, 1974; Samko et al., 1993).

Recently Caputo-Fabrizio (CF) fractional derivative operator was defined in 2015 by Caputo and Fabrizio (2015), which based upon an exponential kernel. This derivative is also known as the fractional derivative with non-singular approximation of the space-time.

Delay differential equations involve unknown functions with the past time and its derivatives, e.g.,

$$\acute{z}(\ell) = \zeta(\ell,\ z(\ell),\ z(\ell - \hat{\tau})),\ \hat{\tau} > 0.$$

Delay differential equations also tell us about past things like life history, gestation, duration of events, etc. Here we study delay fractional differential equations as given by Skóra (2012) in the sense of CF. In fact first we study the existence of the following delay fractional differential equation

$$({}^{CFC}D^{\varkappa}z)(\ell) = \zeta(\ell, z_\ell),\ \text{and } \ell \in J = [0, \Upsilon],\ \Upsilon > 0 \tag{18.1.1}$$

with boundary conditions,

$$z_0 = \phi,\ z'(\Upsilon) = \eta\, z'(0) \text{ where } \eta > 1, \tag{18.1.2}$$

where $\zeta : J \times C([-\hat{\tau}, 0], \mathbb{R}) \longrightarrow \mathbb{R},\ z \in C([-\hat{\tau}, \Upsilon], \mathbb{R})$ and $\phi\ \in C([-\hat{\tau}, 0], \mathbb{R})$, for $\hat{\tau} > 0$. For any $\ell \in J$, z_ℓ denotes the element of $C([-\hat{\tau}, 0], \mathbb{R})$ defined by

$$z_\ell(\omega) = z(\ell + \omega),\ \omega \in [-\hat{\tau}, 0].$$

Here $z_\ell(.)$ represents history of the state from time "$\ell - \hat{\tau}$" up to time "ℓ", and "ℓ" denotes the present time. Condition $z_0 = \phi$ implies that $z(\omega) = \phi(\omega),\ \omega \in [-\hat{\tau}, 0]$. Define for $\phi \in C([-\hat{\tau}, 0], \mathbb{R})$,

$$\|\phi\|_0 = \sup_{-\hat{\tau} \le \omega \le 0} |\phi(\omega)|\,.$$

Properties of stability for different types of equations has attained attention of many mathematicians (Wang et al., 2011). Stability of fractional differential equations was inverstigated by Ulam (1964) in 1940 and later on in 1941 Hyers (1941) conformed it. Rassias than in Rassias (1978) introduced Hyer's–Ulam stability for linear and non-linear fractional differentials equations (see also for details) (Haq et al., 2017; Rus, 2009, 2010).

Differential inclusions (Smirnov, 2002) serve as models for many dynamical systems. They are utilized to establish sufficient optimality conditions, play a crucial role in control theory while dealing with conditions of uncertainty, in economics, and differential games theory. Secondly we discuss the delay fractional differential inclusions of the following type;

$$({}^{CFC}D^{\varkappa}z)(\ell) \in \zeta(\ell, z_{\ell}), \text{ and } \ell \in J = [0, \Upsilon]. \tag{18.1.3}$$

Differential inclusions are a generalization of differential equations (DEs) where derivatives of functions are not defined by a single expression but can be taken any value from a specified set. This approach is helpful for various types of modeling systems having non-deterministic or uncertain behavior. In this study, we explore the existence of solutions for such inclusions using some famous fixed point theorems. It also provides an important tool for analyzing complex dynamical systems. Differential inclusion in different ways are studied by many authors for details, which can be seen in Abbasbandy et al. (2005), Ahmad and Otero-Espinar (2009), Chang et al. (2007, 2009), Frigon (2007), Henderson and Ouahab (2009), Li et al. (2009), Nieto and Rodríguez-López (2007), Ouahab (2008), Smirnov (2002).

This chapter is divided into five sections. Section 18.1, contains some basic concept of fractional calculus. Section 18.2, we use Schauder-type fixed point theorem for the existence of a solution of CFC fractional delay boundary value problem (BVP) (18.1.1)–(18.1.2). Section 18.3, uniqueness is proved using Banach contraction theorem. Section 18.4, we discuss the sufficient conditions for Hyers-Ulam stability of given BVP (18.1.1)–(18.1.2). Section 18.5, contains the existence result for CFC delay fractional differential inclusion (18.1.3) by using Bohnenblust-Karlin fixed point theorem. Examples are furnished to illustrate the uniqueness and Hyers-Ulam stability of the solution.

Definition 1 (Abdeljawad, 2017; Frigon, 2007) Let $z \in H^1(a, b)$, and $\varkappa \in [0, 1]$, then the definition of left and right CFC fractional derivative are

$$\left({}_{a}^{CFC}D^{\varkappa}z\right)(\ell) = \frac{\Im(\varkappa)}{1-\varkappa}\int_{a}^{\ell}\left(\acute{z}(\omega)\exp\left[-\varkappa\frac{(\ell-\omega)}{1-\varkappa}\right]\right)d\omega,$$

$$\left({}^{CFC}D_{b}^{\varkappa}z\right)(\ell) = \frac{-\Im(\varkappa)}{1-\varkappa}\int_{\ell}^{b}\left(\acute{z}(\omega)\exp\left[-\varkappa\frac{(\omega-\ell)}{1-\varkappa}\right]\right)d\omega.$$

where $\acute{z}$ is an ordinary derivative of z. The corresponding integrals are

$$\left({}_{a}^{CF}I^{\varkappa}z\right)(\ell) = \frac{1-\varkappa}{\Im(\varkappa)}z(\ell) + \frac{\varkappa}{\Im(\varkappa)}\int_{a}^{\ell}z(\omega)\,d\omega$$

and

$$\left({}^{CF}I_b^{\varkappa}z\right)(\ell) = \frac{1-\varkappa}{\Im(\varkappa)} z(\ell) + \frac{\varkappa}{\Im(\varkappa)} \int_{\ell}^{b} z(\omega)\, d\omega$$

respectively. Where $\Im(\varkappa)$ (normalization function) satisfy with

$$\Im(0) = \Im(1) = 1.$$

Definition 2 (Abdeljawad, 2017) Let $\hat{n} < \varkappa \leq \hat{n}+1$, when $\hat{n} \in \mathbb{N}$ and $z^{\hat{n}} \in H^1(a,b)$. Set $\gamma = \varkappa - \hat{n}$, then $\gamma \in (0,1]$ and we define higher order CFC fractional derivatives as

$$({}_{a}^{CFC}D^{\varkappa}z)(\ell) = ({}_{a}^{CFC}D^{\gamma}z^{\hat{n}})(\ell)$$

and

$$({}^{CFC}D_b^{\varkappa}z)(\ell) = ({}^{CFC}D_b^{\gamma}(-1)^{\hat{n}}z^{\hat{n}})(\ell).$$

now define the associated fractional integral as

$$({}_{a}^{CF}I^{\varkappa}z)(\ell) = ({}_{a}I^{\hat{n}}\,{}_{a}^{CF}I^{\gamma}z)(\ell).$$

The following Schauder's fixed point theorem is cruicial result for our main result.

Theorem 1 (Yong et al., 2016, Schauder fixed point theorem) *Let Z be a Banach space and $\Omega \subset Z$ is a convex, closed, and bounded set. If $P : \Omega \to \Omega$ is a continuous operator such that $P\Omega \subset Z$ and $P\Omega$ is relatively compact. Then P has at least one fixed point in Ω.*

Lemma 1 (Yong et al., 2016, Arzela-Ascoli theorem) *A subset F in $C(J,\mathbb{R})$ is relatively compact if and only if it is uniformly bounded and equi-continuous on J.*

Lemma 2 (Abdeljawad, 2017) *For $0 < \varkappa < 1$, we obtain*

$$({}_{a}^{CF}I^{\varkappa}\,{}_{a}^{CFC}D^{\varkappa}z)(\ell) = z(\ell) - z(a)$$

and

$$({}^{CF}I_b^{\varkappa}\,{}^{CFC}D_b^{\varkappa}z)(\ell) = z(\ell) - z(b).$$

Proposition 1 (Atangana & Alqahtani, 2016) *For $z(\ell)$ defined on $[a,b]$ and $\varkappa \in (m, m+1)$ for some $m \in \mathbb{N}$, we have*

(1) $({}_{a}^{CFR}D^{\varkappa}\,{}_{a}^{CF}I^{\varkappa}z)(\ell) = z(\ell)$.

(2) $({}_{a}^{CF}I^{\varkappa}\,{}_{a}^{CFR}D^{\varkappa}z)(\ell) = z(\ell) - \sum_{\omega=0}^{m-1} \frac{z^{\omega}(a)}{\omega!}(\ell-a)^{\omega}$.

(3) $({}_{a}^{CF}I^{\varkappa}\,{}_{a}^{CFC}D^{\varkappa}z)(\ell) = z(\ell) - \sum_{\omega=0}^{m} \frac{z^{\omega}(a)}{\omega!}(\ell-a)^{\omega}$.

18.2 Existence Result

We establish the existence result for the CFC delay fractional BVP (18.1.1)–(18.1.2) by applying Theorem 1.

First consider the linear CFC delay fractional BVP

$$({}^{CFC}D^{\varkappa}z)(\ell) = \xi(\ell),\ 1 < \varkappa \le 2 \text{ and } \ell \in J = [0\,,\Upsilon],\ \Upsilon > 0, \tag{18.2.1}$$

with boundary conditions (18.1.2).

Onwards from here we consider $H = C([-\hat{\tau}, \Upsilon], \mathbb{R}) \cap C^2([0, \Upsilon], \mathbb{R})$. The next lemma play cruicial role in supporting our results.

Lemma 3 *Suppose a function $z \in H$ is considered as solution of CFC delay fractional BVP (18.2.1) with (18.1.2) as*

$$z(\ell) = \begin{cases} \phi(\ell)\,,\ \ell \in [-\hat{\tau}, 0] \\ \phi(0) + \ell\frac{(2-\varkappa)}{\Im(\varkappa-1)(\eta-1)}\xi(\Upsilon) + \ell\frac{(\varkappa-1)}{\Im(\varkappa-1)(\eta-1)}\int_0^{\Upsilon}\xi(\omega)d\omega \\ +\frac{(2-\varkappa)}{\Im(\varkappa-1)}\int_0^{\ell}\xi(\omega)d\omega + \frac{(\varkappa-1)}{\Im(\varkappa-1)}\int_0^{\ell}(\ell-\omega)\xi(\omega)d\omega,\ \ell \in J, \end{cases}$$

where $\xi : J \to \mathbb{R}$ is a given function.

Proof Given

$$({}^{CFC}D^{\varkappa}z)(\ell) = \xi(\ell),\ 1 < \varkappa \le 2 \text{ and } \ell \in J = [0\,,\Upsilon],\ \Upsilon > 0,$$

from Proposition 1, we obtain

$$z(\ell) = C_1 + C_2\ell + \frac{(2-\varkappa)}{\Im(\varkappa-1)}\int_0^{\ell}\xi(\omega)d\omega + \frac{(\varkappa-1)}{\Im(\varkappa-1)}\int_0^{\ell}(\ell-\omega)\xi(\omega)d\omega \tag{18.2.2}$$

differentiating both sides w.r.t "ℓ", we have

$$z'(\ell) = C_2 + \frac{(2-\varkappa)}{\Im(\varkappa-1)}\xi(\ell) + \frac{(\varkappa-1)}{\Im(\varkappa-1)}\int_0^{\ell}\xi(\omega)d\omega \tag{18.2.3}$$

using $z_0 = \phi$ in (18.2.2), we get

$$z(0) = C_1$$

implies

$$C_1 = \phi(0).$$

Now using $z'(\Upsilon) = \eta z'(0)$ in (18.2.3), we have

$$z'(\Upsilon) = C_2 + \frac{(2-\varkappa)}{\Im(\varkappa-1)}\xi(\Upsilon) + \frac{(\varkappa-1)}{\Im(\varkappa-1)}\int_0^{\Upsilon}\xi(\omega)d\omega \tag{18.2.4}$$

and

$$\eta z'(0) = \eta C_2 \tag{18.2.5}$$

comparing (18.2.4) and (18.2.5), we get

$$C_2 = \frac{(2-\varkappa)}{\Im(\varkappa-1)(\eta-1)}\xi(\Upsilon) + \frac{(\varkappa-1)}{\Im(\varkappa-1)(\eta-1)}\int_0^{\Upsilon}\xi(\omega)d\omega$$

putting value of C_1 and C_2 in (18.2.2), finally we get

$$\begin{aligned} z(\ell) = \phi(0) &+ \ell\frac{(2-\varkappa)}{\Im(\varkappa-1)(\eta-1)}\xi(\Upsilon) + \ell\frac{(\varkappa-1)}{\Im(\varkappa-1)(\eta-1)}\int_0^{\Upsilon}\xi(\omega)d\omega \\ &+ \frac{(2-\varkappa)}{\Im(\varkappa-1)}\int_0^{\ell}\xi(\omega)d\omega + \frac{(\varkappa-1)}{\Im(\varkappa-1)}\int_0^{\ell}(\ell-\omega)\xi(\omega)d\omega. \end{aligned} \tag{18.2.6}$$

This proves the lemma. □

To prove the existence of (18.1.1)–(18.1.2), we transform the CFC delay fractional BVP (18.1.1)–(18.1.2) in the view of Lemma 3, as a solution of following fixed point problem

$$z = Pz,$$

where $P : H \to H$ is given as

$$\begin{aligned} (Pz)(\ell) = \phi(0) &+ \ell\frac{(2-\varkappa)}{\Im(\varkappa-1)(\eta-1)}\zeta(\Upsilon, z_{\Upsilon}) \\ &+ \ell\frac{(\varkappa-1)}{\Im(\varkappa-1)(\eta-1)}\int_0^{\Upsilon}\zeta(\omega, z_{\omega})d\omega \\ &+ \frac{(2-\varkappa)}{\Im(\varkappa-1)}\int_0^{\ell}\zeta(\omega, z_{\omega})d\omega \\ &+ \frac{(\varkappa-1)}{\Im(\varkappa-1)}\int_0^{\ell}(\ell-\omega)\zeta(\omega, z_{\omega})d\omega. \end{aligned} \tag{18.2.7}$$

Now we prove our existence result of CFC delay fractional BVP (18.1.1)–(18.1.2) using the Schauder-type fixed point theorem (Theorem 1).

Theorem 2 *Assume* $\zeta \in C\left(J \times C([-\hat{\tau}, 0], \mathbb{R}), \mathbb{R}\right)$ *is continuous and satisfying: (i)* $|\zeta(\ell, z_{\ell})| \le m(\ell)\ \forall\ (\ell, z) \in [0, \Upsilon] \times \mathbb{R}$*, for some* $m \in L^1([0, \Upsilon], \mathbb{R}_+)$*. Therefore (18.1.1)–(18.1.2) has at least one possible solution.*

Proof Take β_{ε} (closed ball), which is defined as,

$$\beta_{\varepsilon} = \left\{ z \in H : \|z\| \le \varepsilon,\ \text{where}\ \varepsilon \ge |\phi(0)| + \left(\frac{\Upsilon^2(1+\eta) + 2\Upsilon\eta}{2\Im(\varkappa-1)(\eta-1)}\right)\|m\| \right\},$$

where $\Psi = \sup_{\ell\in[0,\Upsilon]} \{|f(\ell,0)|$. Clearly β_ε is the bounded and closed subset of H. First we show that P defined in (18.2.7) is uniformly bounded, for this $z \in \beta_\varepsilon$,

$$|(Pz)(\ell)| = \left| \begin{array}{c} \phi(0) + \ell\frac{(2-\varkappa)}{\Im(\varkappa-1)(\eta-1)}\zeta(\Upsilon, z_\Upsilon) \\ +\ell\frac{(\varkappa-1)}{\Im(\varkappa-1)(\eta-1)}\int_0^\Upsilon \zeta(\omega, z_\omega)d\omega + \frac{(2-\varkappa)}{\Im(\varkappa-1)}\int_0^\ell \zeta(\omega, z_\omega)d\omega \\ +\frac{(\varkappa-1)}{\Im(\varkappa-1)}\int_0^\ell (\ell-\omega)\zeta(\omega, z_\omega)d\omega \end{array} \right|$$

$$\begin{aligned} |(Pz)(\ell)| \le |\phi(0)| &+ \ell\frac{(2-\varkappa)}{\Im(\varkappa-1)(\eta-1)}|\zeta(\Upsilon, z_\Upsilon)| \\ &+ \ell\frac{(\varkappa-1)}{\Im(\varkappa-1)(\eta-1)}\int_0^\Upsilon |\zeta(\omega, z_\omega)|\, d\omega \\ &+ \frac{(2-\varkappa)}{\Im(\varkappa-1)}\int_0^\ell |\zeta(\omega, z_\omega)|\, d\omega \\ &+ \frac{(\varkappa-1)}{\Im(\varkappa-1)}\int_0^\ell (\ell-\omega)\, |\zeta(\omega, z_\omega)|\, d\omega \\ &\le |\phi(0)| + \ell\frac{(2-\varkappa)}{\Im(\varkappa-1)(\eta-1)}m(\Upsilon) \\ &+ \ell\frac{(\varkappa-1)}{\Im(\varkappa-1)(\eta-1)}\int_0^\Upsilon m(\omega)d\omega \\ &+ \frac{(2-\varkappa)}{\Im(\varkappa-1)}\int_0^\ell (m(\omega)d\omega \\ &+ \frac{(\varkappa-1)}{\Im(\varkappa-1)}\int_0^\ell (\ell-\omega)m(\omega)d\omega, \end{aligned}$$

which implies

$$\begin{aligned} \|(Pz)(\ell)\| \le |\phi(0)| &+ \Upsilon\frac{1}{\Im(\varkappa-1)(\eta-1)}\|m\| \\ &+ \frac{1}{\Im(\varkappa-1)(\eta-1)}\Upsilon^2\|m\| \\ &+ \frac{1}{\Im(\varkappa-1)}\Upsilon\|m\| \\ &+ \frac{1}{\Im(\varkappa-1)}\|m\|\frac{\Upsilon^2}{2} \end{aligned}$$

$$\|(Pz)(\ell)\| \le |\phi(0)| + \left(\begin{array}{c} \frac{\Upsilon}{\Im(\varkappa-1)(\eta-1)} + \frac{\Upsilon^2}{\Im(\varkappa-1)(\eta-1)} \\ +\frac{\Upsilon}{\Im(\varkappa-1)} + \frac{\Upsilon^2}{2\Im(\varkappa-1)} \end{array} \right)\|m\|$$

$$\|(Pz)(\ell)\| \leq |\phi(0)| + \left(\frac{\Upsilon^2(1+\eta)+2\Upsilon\eta}{2\Im(\varkappa-1)(\eta-1)}\right)\|m\| \leq \varepsilon$$

Hence

$$P(\beta_\varepsilon) \subseteq \beta_\varepsilon.$$

The continuity of ζ implies the continuity of P, and $\|Pz\| \leq \varepsilon$, so P (define in 18.2.7) is uniformly bounded on β_ε.

Now we prove that P is equi-continuous. For this take $\ell_1 < \ell_2 \in J$ and each $z \in \beta_\varepsilon$,

$$|(Pz)(\ell_1)-(Pz)(\ell_2)| \leq \left|\begin{pmatrix} \ell_1\frac{(2-\varkappa)}{\Im(\varkappa-1)(\eta-1)}\zeta(\Upsilon, z_\Upsilon)+\ell_1\frac{(\varkappa-1)}{\Im(\varkappa-1)(\eta-1)}\int_0^\Upsilon\zeta(\omega, z_\omega)d\omega \\ +\frac{(2-\varkappa)}{\Im(\varkappa-1)}\int_0^{\ell_1}\zeta(\omega, z_\omega)d\omega+\frac{(\varkappa-1)}{\Im(\varkappa-1)}\int_0^{\ell_1}(\ell_1-\omega)\zeta(\omega, z_\omega)d\omega \\ -\ell_2\frac{(2-\varkappa)}{\Im(\varkappa-1)(\eta-1)}\zeta(\Upsilon, z_\Upsilon)-\ell_2\frac{(\varkappa-1)}{\Im(\varkappa-1)(\eta-1)}\int_0^\Upsilon\zeta(\omega, z_\omega)d\omega \\ -\frac{(2-\varkappa)}{\Im(\varkappa-1)}\int_0^{\ell_2}\zeta(\omega, z_\omega)d\omega-\frac{(\varkappa-1)}{\Im(\varkappa-1)}\int_0^{\ell_2}(\ell_2-\omega)\zeta(\omega, z_\omega)d\omega \end{pmatrix}\right|$$

$$\begin{aligned} &\leq \frac{(2-\varkappa)}{\Im(\varkappa-1)(\eta-1)}(\ell_1-\ell_2)\,|\zeta(\Upsilon, z_\Upsilon)| + \frac{(\varkappa-1)}{\Im(\varkappa-1)(\eta-1)}(\ell_1-\ell_2)\int_0^\Upsilon |\zeta(\omega, z_\omega)|\,d\omega \\ &+ \frac{(2-\varkappa)}{\Im(\varkappa-1)}\left(\int_0^{\ell_1}|\zeta(\omega, z_\omega)|\,d\omega - \int_0^{\ell_2}|\zeta(\omega, z_\omega)|\,d\omega\right) \\ &+ \frac{(\varkappa-1)}{\Im(\varkappa-1)}\left(\int_0^{\ell_1}(\ell_1-\omega)\,|\zeta(\omega, z_\omega)|\,d\omega - \int_0^{\ell_2}(\ell_2-\omega)\,|\zeta(\omega, z_\omega)|\,d\omega\right) \end{aligned}$$

in the view of hypothesis (i), set $\max\limits_{(\ell,z)\in\Im_{\hat{r}}} \zeta(\ell, z_\ell) = \zeta$, we have

$$\begin{aligned} &\leq \frac{(2-\varkappa)}{\Im(\varkappa-1)(\eta-1)}(\ell_1-\ell_2)\,\zeta + \frac{(\varkappa-1)}{\Im(\varkappa-1)(\eta-1)}(\ell_1-\ell_2)\zeta\int_0^\Upsilon d\omega \\ &+ \frac{(2-\varkappa)}{\Im(\varkappa-1)}\left(\zeta\int_0^{\ell_1} d\omega - \zeta\int_0^{\ell_2} d\omega\right) \\ &+ \frac{(\varkappa-1)}{\Im(\varkappa-1)}\left(\zeta\int_0^{\ell_1}(\ell_1-\omega)d\omega - \zeta\int_0^{\ell_2}(\ell_2-\omega)d\omega\right) \\ &= \frac{(2-\varkappa)\zeta}{\Im(\varkappa-1)(\eta-1)}(\ell_1-\ell_2) + \frac{(\varkappa-1)\Upsilon\zeta}{\Im(\varkappa-1)(\eta-1)}(\ell_1-\ell_2) \\ &+ \frac{(2-\varkappa)\zeta}{\Im(\varkappa-1)}(\ell_1-\ell_2) + \frac{(\varkappa-1)\zeta}{2\Im(\varkappa-1)}\left(\ell_1^2-\ell_2^2\right) \end{aligned}$$

we have

$$\begin{aligned} \|(Pz)(\ell_1)-(Pz)(\ell_2)\| \leq\ & \frac{\zeta}{\Im(\varkappa-1)(\eta-1)}(\ell_1-\ell_2) + \frac{\Upsilon\zeta}{\Im(\varkappa-1)(\eta-1)}(\ell_1-\ell_2) \\ &+ \frac{\zeta}{\Im(\varkappa-1)}(\ell_1-\ell_2) + \frac{\zeta(\ell_1+\ell_2)}{2\Im(\varkappa-1)}(\ell_1-\ell_2), \end{aligned}$$

as $\ell_1 \to \ell_2$ on the right hand side, then above inequality also approaches to zero. Now using Lemma 1, we obtain that P is relatively compact. Consequently from Theorem 1, there exists at least one solution of (18.1.1)–(18.1.2). □

18.3 Unique Solution

In previous Sect. 18.1, we prove the existence result of the proposed CFC delay fractional BVP (18.1.1)–(18.1.2) via Schauder-type fixed point theorem, which does not guarantee the uniqueness. Therefore, we use the Banach contraction principle to prove the uniqueness of CFC delay fractional BVP (18.1.1)–(18.1.2) in this section.

Theorem 3 *Assume* $\zeta \in C\left(J \times C(\left[-\hat{\tau}, 0\right], \mathbb{R}), \mathbb{R}\right)$ *and* $\exists\ m \in L^1([0, \Upsilon], \mathbb{R}_+)$ *such that*

$$|\zeta(\ell, u_\ell) - \zeta(\ell, v_\ell)| \leq m(\ell)\, \|u - v\|_0 \qquad (A)$$

$\forall\, \ell \in [0, \Upsilon]$, $u, v \in C\left[-\hat{\tau}, 0\right]$, *where* $\mu \in (0, 1]$ *and* $M(\ell) = \int_0^\ell m(s)ds$. *If*

$$\frac{\Upsilon + 1}{\mu \Im(\varkappa - 1)(\eta - 1)}[\{\mu m(\Upsilon) + \eta\} e^{\mu M} - \eta] < 1, \tag{18.3.1}$$

holds, then (18.1.1)–(18.1.2) has a unique solution.

Proof For $z \in C([0, \Upsilon], \mathbb{R})$

$$\|z\| = \max_{\omega \in [0, \Upsilon]} \{e^{-\mu M(\omega)} \max_{\hat{r} \in [0, \omega]} \left|z(\hat{r})\right|\},$$

where $\mu \in (0, 1]$, $z_{\hat{r}}(\omega) = z(\omega + \hat{r}) = \phi(\omega + \hat{r})$ for $\omega + \hat{r} \leq 0$ and $\max_{\ell \in J} M(\ell) = M$. We need to find a unique fixed point of the operator P defined in (18.2.7).

For this, take any $u, v \in C([0, \Upsilon], \mathbb{R})$ and $\ell \in J$,

$$|(Pu)(\ell) - (Pv)(\ell)| \leq \left| \begin{array}{l} \left(\begin{array}{c} \phi(0) + \ell \frac{(2-\varkappa)}{\Im(\varkappa-1)(\eta-1)} \zeta(\Upsilon, u_\Upsilon) + \ell \frac{(\varkappa-1)}{\Im(\varkappa-1)(\eta-1)} \int_0^\Upsilon \zeta(\omega, u_\omega) d\omega \\ + \frac{(2-\varkappa)}{\Im(\varkappa-1)} \int_0^\ell \zeta(\omega, u_\omega) d\omega + \frac{(\varkappa-1)}{\Im(\varkappa-1)} \int_0^\ell (\ell - \omega) \zeta(\omega, u_\omega) d\omega \end{array} \right) \\ - \left(\begin{array}{c} \phi(0) + \ell \frac{(2-\varkappa)}{\Im(\varkappa-1)(\eta-1)} \zeta(\Upsilon, v_\Upsilon) + \ell \frac{(\varkappa-1)}{\Im(\varkappa-1)(\eta-1)} \int_0^\Upsilon \zeta(\omega, v_\omega) d\omega \\ + \frac{(2-\varkappa)}{\Im(\varkappa-1)} \int_0^\ell \zeta(\omega, v_\omega) d\omega + \frac{(\varkappa-1)}{\Im(\varkappa-1)} \int_0^\ell (\ell - \omega) \zeta(\omega, v_\omega) d\omega \end{array} \right) \end{array} \right|$$

$$\begin{aligned} &\leq \left(\ell \frac{(2 - \varkappa)}{\Im(\varkappa - 1)(\eta - 1)}\right) |\zeta(\Upsilon, u_\Upsilon) - \zeta(\Upsilon, v_\Upsilon)| \\ &+ \left(\ell \frac{(\varkappa - 1)}{\Im(\varkappa - 1)(\eta - 1)}\right) \int_0^\Upsilon |\zeta(\omega, u_\omega) - \zeta(\omega, v_\omega)|\, d\omega \end{aligned}$$

$$+\left(\frac{(2-\varkappa)}{\Im(\varkappa-1)}\right)\int_0^\ell |\zeta(\omega,u_\omega)-\zeta(\omega,v_\omega)|\,d\omega$$
$$+\left(\frac{(\varkappa-1)}{\Im(\varkappa-1)}\right)\int_0^\ell (\ell-\omega)\,|\zeta(\omega,u_\omega)-\zeta(\omega,v_\omega)|\,d\omega$$

$$\leq \left(\ell\frac{(2-\varkappa)}{\Im(\varkappa-1)(\eta-1)}\right)m(\Upsilon)\,\|u_\Upsilon-v_\Upsilon\|_0$$
$$+\left(\ell\frac{(\varkappa-1)}{\Im(\varkappa-1)(\eta-1)}\right)\int_0^\Upsilon m(\omega)\,\|u_\omega-v_\omega\|_0\,d\omega$$
$$+\left(\frac{(2-\varkappa)}{\Im(\varkappa-1)}\right)\int_0^\ell m(\omega)\,\|u_\omega-v_\omega\|_0\,d\omega$$
$$+\left(\frac{(\varkappa-1)}{\Im(\varkappa-1)}\right)\int_0^\ell \ell m(\omega)\,\|u_\omega-v_\omega\|_0\,d\omega$$

notice that if $\omega\in[0,\hat{\tau}]$ then

$$\|u_\omega-v_\omega\|_0=\sup_{\hat{r}\in[-\hat{\tau},0]}\left\|u(\omega+\hat{r})-v(\omega+\hat{r})\right\|$$
$$=\max_{\hat{r}\in[-\omega,0]}\left\|u(\omega+\hat{r})-v(\omega+\hat{r})\right\|\leq\max_{\hat{r}\in[0,\omega]}\left\|u(\hat{r})-v(\hat{r})\right\|,$$

if $\omega\in[\hat{\tau},\Upsilon]$, then

$$\|u_\omega-v_\omega\|_0=\sup_{\hat{r}\in[-\hat{\tau},0]}\left\|u(\omega+\hat{r})-v(\omega+\hat{r})\right\|$$
$$=\max_{\hat{r}\in[\omega-\hat{\tau},\omega]}\left\|u(\omega+\hat{r})-v(\omega+\hat{r})\right\|\leq\max_{\hat{r}\in[0,\omega]}\left\|u(\hat{r})-v(\hat{r})\right\|.$$

which implies

$$\|(Pu)(\ell)-(Pv)(\ell)\|\leq\frac{\Upsilon}{\Im(\varkappa-1)(\eta-1)}m(\Upsilon)\left\{\max_{\hat{r}\in[0,\omega]}\left\|u(\hat{r})-v(\hat{r})\right\|\right\}$$
$$+\left(\frac{\Upsilon+1}{\Im(\varkappa-1)(\eta-1)}\right)\int_0^\Upsilon m(\omega)\left\{\max_{\hat{r}\in[0,\omega]}\left\|u(\hat{r})-v(\hat{r})\right\|\right\}d\omega$$
$$+\left(\frac{\Upsilon+1}{\Im\,(\varkappa-1)}\right)\int_0^\ell m(\omega)\left\{\max_{\hat{r}\in[0,\omega]}\left\|u(\hat{r})-v(\hat{r})\right\|\right\}d\omega$$

$$\leq\frac{\Upsilon+1}{\Im(\varkappa-1)(\eta-1)}m(\Upsilon)\left\{\max_{\hat{r}\in[0,\omega]}\left\|u(\hat{r})-v(\hat{r})\right\|\right\}$$
$$+\left(\frac{\Upsilon+1}{\Im(\varkappa-1)(\eta-1)}\right)\int_0^\Upsilon m(\omega)\left\{\max_{\hat{r}\in[0,\omega]}\left\|u(\hat{r})-v(\hat{r})\right\|\right\}d\omega$$

$$+\left(\frac{\Upsilon+1}{\Im\,(\varkappa-1)}\right)\int_0^{\ell} m(\omega)\left\{\max_{\hat{r}\in[0,\omega]}\|u(\hat{r})-v(\hat{r})\|\right\}d\omega$$

$$\begin{aligned}\|(Pu)(\ell)-(Pv)(\ell)\| \le{}& \frac{\Upsilon+1}{\Im(\varkappa-1)(\eta-1)}m(\Upsilon)e^{\mu M(\Upsilon)}e^{-\mu M(\Upsilon)}\left\{\max_{\hat{r}\in[0,\omega]}\|u(\hat{r})-v(\hat{r})\|\right\}\\ &+\left(\frac{\Upsilon+1}{\Im(\varkappa-1)(\eta-1)}\right)\int_0^{\Upsilon} m(\omega)e^{\mu M(\omega)}e^{-\mu M(\omega)}\left\{\max_{\hat{r}\in[0,\omega]}\|u(\hat{r})-v(\hat{r})\|\right\}d\omega\\ &+\left(\frac{(\Upsilon+1)(\eta-1)}{\Im(\varkappa-1)(\eta-1)}\right)\int_0^{\ell} m(\omega)e^{\mu M(\omega)}e^{-\mu M(\omega)}\left\{\max_{\hat{r}\in[0,\omega]}\|u(\hat{r})-v(\hat{r})\|\right\}d\omega\end{aligned}$$

$$=\frac{\Upsilon+1}{\mu\Im(\varkappa-1)(\eta-1)}\|u-v\|\,[\mu m(\Upsilon)e^{\mu M(\Upsilon)}+e^{\mu M(\Upsilon)}-1+\{(\eta-1)(e^{\mu M(\ell)}-1)\}],$$

set

$$\max_{\ell\in J} M(\ell)=M$$

finally we have

$$\|(Pu)(\ell)-(Pv)(\ell)\|\le\frac{\Upsilon+1}{\mu\Im(\varkappa-1)(\eta-1)}\|u-v\|\,[\{\mu m(\Upsilon)+\eta\}\,e^{\mu M}-\eta].$$

As

$$\frac{\Upsilon+1}{\mu\Im(\varkappa-1)(\eta-1)}[\{\mu m(\Upsilon)+\eta\}\,e^{\mu M}-\eta]<1,$$

hence P is a contraction, therefore via Banach contraction principle, P has unique fixed point. □

In the following example, we illustrate the applicability of Theorem 3 to determine a unique solution of a given CFC delay fractional BVP.

Example 1 Consider,

$$({}^{CFC}D_b^{1.5}z)(\ell)=\begin{cases}\ell^3+\frac{1}{10e^{\ell}}z\left(\ell-\frac{1}{2}\right), & \ell\in[0,1]\\ z(\omega)=0, \ \omega\in\left[-\frac{1}{2},0\right], & \acute{z}(1)=8\acute{z}(0).\end{cases}$$

Here we have $\Upsilon=1,\ \eta=8,\ \tau=\frac{1}{2}$ and $\varkappa=1.5$, $\Im\,(\varkappa-1)=1$ for all $1<\varkappa\le 2$ and $\zeta(\ell,z_\ell)=\ell^3+\frac{1}{10e^{\ell}}z\left(\ell-\frac{1}{2}\right)$.

For $u,v\in C([-\frac{1}{2},0],\mathbb{R})$ consider;

$$\begin{aligned}|\zeta(\ell,u_\ell)-\zeta(\ell,v_\ell)| &= \left|\ell^3+\frac{1}{10e^{\ell}}u\left(\ell-\frac{1}{2}\right)-\ell^3-\frac{1}{10e^{\ell}}v\left(\ell-\frac{1}{2}\right)\right|\\ &=\left|\frac{1}{10e^{\ell}}u\left(\ell-\frac{1}{2}\right)-\frac{1}{10e^{\ell}}v\left(\ell-\frac{1}{2}\right)\right|\end{aligned}$$

$$= \frac{1}{10e^{\ell}} \left| u\left(\ell - \frac{1}{2}\right) - v\left(\ell - \frac{1}{2}\right) \right|$$

$$\|\zeta(\ell, u_\ell) - \zeta(\ell, v_\ell)\| \leq \frac{1}{10e^{\ell}} \|u - v\|$$

for all $\ell \in [0, 1]$. By usual calculation, we find that

$$\frac{\Upsilon + 1}{\mu \Im(\varkappa - 1)(\eta - 1)} [\{\mu m(\Upsilon) + \eta\} e^{\mu M} - \eta] < 1 = 0.264 < 1$$

by taking $\mu = \frac{1}{2} \in (0, 1]$.

All conditions of Theorem 3 are satisfied, therefore we obtain a unique solution of the given CFC delay fractional BVP.

18.4 Hyers-Ulam Stability

The conditions for Hyers-Ulam stability of the CFC delay fractional BVP (18.1.1)–(18.1.2) are discussed in this section. Before going to prove our main result, we discuss some important definitions and remark which are very helpful in our upcoming results.

Remark 1 From Lemma 3, we can write $z \in H$ is solution of CFC delay fractional BVP (18.1.1)–(18.1.2) as;

$$z(\ell) = g(\ell) + \int_0^{\Upsilon} G(\ell, \omega)\xi(v, u_\omega)dr,$$

where

$$g(\ell) = \phi(0) + \ell \frac{(2 - \varkappa)}{\Im(\varkappa - 1)(\eta - 1)} \xi(\Upsilon)$$

and

$$G(\ell, \omega) = \begin{cases} \ell \frac{(\varkappa-1)}{\Im(\varkappa-1)(\eta-1)} + \frac{(2-\varkappa)}{\Im(\varkappa-1)} + \frac{(\varkappa-1)(\ell-\omega)}{\Im(\varkappa-1)}, & 0 \leq \omega \leq \ell \\ \ell \frac{(\varkappa-1)}{\Im(\varkappa-1)(\eta-1)}, & \ell \leq \omega \leq \Upsilon. \end{cases}$$

$$|G(\ell, \omega)| \leq \left| \frac{2\Upsilon}{\Im(\varkappa - 1)(\eta - 1)} \right| + \left| \frac{1}{\Im(\varkappa - 1)} \right| + \left| \frac{\Upsilon}{\Im(\varkappa - 1)} \right| = \kappa.$$

Definition 3 The CFC delay fractional BVP (18.1.1)–(18.1.2) is Hyers-Ulam stable if $\exists$ a constant $\hbar > 0$, such that for every $\epsilon > 0$ and for each solution $z(\ell) \in H$ of

$$\left| ({}^{CFC}D_b^{\varkappa} z)(\ell) - \zeta(\ell, z_\ell) \right| \leq \epsilon, \ \ell \in [0, \Upsilon] \qquad (18.4.1)$$

then $\exists$ a solution $y(\ell) \in H$ of (18.1.1)–(18.1.2) such that

$$|z(\ell) - y(\ell)| \leq \hbar\epsilon,\ \ell \in [0, \Upsilon]. \tag{18.4.2}$$

Remark 2 $z(\ell) \in H$ is solution of (4.1) if and only if $\exists$ a function $\delta \in H$ (depending upon y only) such that
(i) $|\delta(\ell)| \leq \epsilon,\ \ell \in [0, \Upsilon]$,
(ii) $({}^{CFC}D_b^{\varkappa} z)(\ell) = \zeta(\ell, z_\ell) + \delta(\ell),\ \ell \in [0, \Upsilon]$.

Next result identifies the conditions that tell us about the stability of CFC delay fractional BVP (18.1.1)–(18.1.2).

Theorem 4 *Suppose* $\zeta : [0, \Upsilon] \times \mathbb{R} \to \mathbb{R}$ *is continuous and satisfying*

$$|\zeta(\ell, u_\ell) - \zeta(\ell, v_\ell)| \leq m(\ell)\, \|u - v\|_0 \qquad (A)$$

If $(1 \neq \kappa\Upsilon m(\Upsilon))$ *then CFC delay fractional BVP (18.1.1)–(18.1.2) is Hyers-Ulam stable.*

Proof Let $\zeta : [0, \Upsilon] \times \mathbb{R} \to \mathbb{R}$ be continuous and satisfying (A). Let $z(\ell) \in H$ be solution of (18.4.1) and from using Remark 2,

$$({}^{CFC}D_b^{\varkappa} z)(\ell) = \zeta(\ell, z_\ell) + \delta(\ell),\ \text{ for all } \ell \in [0, \Upsilon].$$

Now use Remark 1 then we can write

$$z(\ell) = g(\ell) + \int_0^{\Upsilon} G(\ell, \omega)\zeta(\omega, u_\omega)\, d\omega + \int_0^{\Upsilon} G(\ell, \omega)\delta(\omega)\, d\omega,$$

which gives

$$\left| z(\ell) - g(\ell) - \int_0^{\Upsilon} G(\ell, \omega)\zeta(\omega, u_\omega)\, d\omega \right| \leq \kappa\Upsilon\epsilon, \tag{18.4.3}$$

where κ and $G(\ell, \omega)$ are presented in Remark 1. Now suppose $y(\ell) \in H$ is unique solution of (18.1.1)–(18.1.2), and consider

$$\begin{aligned}
|z(\ell) - y(\ell)| &= \left| z(\ell) - g(\ell) - \int_0^{\Upsilon} G(\ell, \omega)\zeta(\omega, y_\omega)\, d\omega \right| \\
&\leq \left| z(\ell) - g(\ell) - \int_0^{\Upsilon} G(\ell, \omega)\zeta(\omega, z_\omega)\, d\omega \right| \\
&\quad + \left| \int_0^{\Upsilon} G(\ell, \omega)\zeta(\omega, z_\omega)\, d\omega - \int_0^{\Upsilon} G(\ell, \omega)\zeta(\omega, y_\omega)\, d\omega \right|
\end{aligned}$$

from (18.4.3) we have

$$|z(\ell) - y(\ell)| \leq \kappa\Upsilon\epsilon + \left| \int_0^{\Upsilon} G(\ell, \omega)\zeta(\omega, z_\omega)\, d\omega - \int_0^{\Upsilon} G(\ell, \omega)\zeta(\omega, y_\omega)\, d\omega \right|$$

using (A) we get

$$|z(\ell) - y(\ell)| \leq \kappa\Upsilon\epsilon + \left[\left|\frac{2\ell(\varkappa-1)}{\Im(\varkappa-1)(\eta-1)}\right|\ell + \left|\frac{(2-\varkappa)}{\Im(\varkappa-1)}\right|\ell + \left|\frac{\ell(\varkappa-1)}{2\Im(\varkappa-1)}\right|\ell\right] m(\ell)\,\|z-y\|_0$$
$$\|z-y\| \leq \kappa\Upsilon\epsilon + \left[\left|\frac{2\Upsilon}{\Im(\varkappa-1)(\eta-1)}\right|\Upsilon + \left|\frac{1}{\Im(\varkappa-1)}\right|\Upsilon + \left|\frac{\Upsilon}{2\Im(\varkappa-1)}\right|\Upsilon\right] m(\Upsilon)\,\|z-y\|$$

$$\|z-y\| \leq \kappa\Upsilon\epsilon + \kappa\Upsilon m(\Upsilon)\,\|z-y\|$$

which implies

$$\|z-y\| \leq \frac{\kappa\Upsilon\epsilon}{1-\kappa\Upsilon m(\Upsilon)} = \hbar\epsilon$$

where $\hbar = \frac{\kappa\Upsilon}{1-\kappa\Upsilon m(\Upsilon)}$. Since $1 \neq \kappa\Upsilon m(\Upsilon)$, hence CFC delay fractional BVP (18.1.1)–(18.1.2) is stabe. □

18.5 Differential Inclusion

As we examine the multivalued form of the previously discussed CFC delay fractional BVP (18.1.3)–(18.1.2). Now in this section, the existence of a solution of (18.1.3)–(18.1.2), with the help of the Bohnenblust-Karlin theorem are derived. Before going forward, we use the following assumptions, which are helpful in our next discussion.

(A_1). Let $\zeta \in C\left(J \times C(\left[-\hat{\tau}, 0\right], \mathbb{R}\right) \longrightarrow K(\mathbb{R})$; $(\ell, z) \longrightarrow \zeta(\ell, z)$ be measurable w.r.t ℓ for every $z \in C([0, \Upsilon], \mathbb{R})$ and upper semi (us) continuous w.r.t z for a.e $\ell \in [0, \Upsilon]$. For each fixed $z \in C([0, \Upsilon], \mathbb{R})$ and we set

$$S_{\zeta,z} = \{\xi \in L^1([0,\Upsilon],\mathbb{R}) : \xi(\ell) \in \zeta(\ell, z_\ell)\},$$

$\forall\, \ell \in [0, \Upsilon]$. Where $K(\mathbb{R})$ represents set of nonempty, convex and compact subset of $\mathbb{R}$.

(A_2). For every $r > 0$, $\exists$ function $m_r \in L^1([0,\Upsilon],\mathbb{R}^+)$ then $\|\zeta(\ell, z_\ell)\| = \sup_{\ell\in J}\{|v| : v(\ell) \in \zeta(\ell, z_\ell) \leq m_r(\ell)$ for every $(\ell, z) \in ([0,\Upsilon] \times C([0,\Upsilon],\mathbb{R}))\}$ with $|z| \leq r$, and

$$\lim_{r\longrightarrow\infty} \inf \left(\frac{\int_0^\Upsilon m_r(\ell)d\ell}{r}\right) = \gamma.$$

Definition 4 A function $z\in(C[0,\Upsilon],\mathbb{R})$ is solution of (18.1.3) *if* $\exists\xi \in L^1([0,\Upsilon],\mathbb{R})$ such that $({}^{CFC}D_b^{\varkappa}\xi)(\ell) \in \zeta(\ell, z(\ell))$ a.e on $[0, \Upsilon]$ and

$$z(\ell) = \phi(0) + \frac{(2-\varkappa)\ell}{\Im(\varkappa-1)(\eta-1)}\xi(\Upsilon) + \int_0^\Upsilon G(\ell,\omega)\xi(\omega)d\omega \tag{18.5.1}$$

where

$$G(\ell, s) = \begin{cases} \ell\frac{(\varkappa-1)}{\Im(\varkappa-1)(\eta-1)} + \frac{(2-\varkappa)}{\Im(\varkappa-1)} + \frac{(\varkappa-1)(\ell-\omega)}{\Im(\varkappa-1)}, & 0 \le \omega \le \ell \\ \ell\frac{(\varkappa-1)}{\Im(\varkappa-1)(\eta-1)}, & \ell \le \omega \le \Upsilon. \end{cases} \tag{18.5.2}$$

The following two lemmas will be crucial for our main result of this section.

Lemma 4 (Bohnenblust & Karlin, 1950). *Assume* D *is nonempty subset of* Z *which is Banach space, is convex, bounded and closed. Suppose* $G : D \longrightarrow 2^Z\backslash\{0\}$ *is upper semi (us) continuous with convex and closed values then* $G(D) \subset D$ *and* $\overline{G(D)}$ *is a compact. Therefore* G *has a fixed point.*

Lemma 5 (Losta & Opial, 1965) *Assume* J *is real interval, which is also compact and* ζ *is multivalued map and hold* (A_1). *Suppose* $\Theta : L^1(J, \mathbb{R}) \longrightarrow C(J)$, *which is linear continuous then the operator* $\Theta \circ S_\zeta : C(J) \longrightarrow K(C(J)), z \longmapsto (\Theta \circ S_\zeta)(z)$ *is closed graph operator in* $C(J) \times C(J)$.

Now we prove the main existence result for CF fractional differential inclusion of (18.1.3)–(18.1.2).

Theorem 5 *Suppose that assumptions* (A_1) *and* (A_2) *are satisfied, and*

$$\gamma < \frac{\eta(1+\Upsilon)-1}{\Im(\varkappa-1)(\eta-1)} \tag{18.5.3}$$

then (18.1.3) has at least one possible solution on $[0, \Upsilon]$.

Proof First we transform our problem (18.1.3) into fixed point problem and define multivalued map as

$$\Omega(z) = \left\{ \begin{array}{l} h \in C([0,\Upsilon]) : h(\ell) = \phi(0) + \ell\frac{(2-\varkappa)}{\Im(\varkappa-1)(\eta-1)}\xi(\Upsilon) + \frac{(2-\varkappa)}{\Im(\varkappa-1)}\int_0^\ell \xi(\omega)d\omega \\ \quad + \frac{(\varkappa-1)}{\Im(\varkappa-1)}\int_0^\ell (\ell-\omega)\xi(\omega)d\omega + \ell\frac{(\varkappa-1)}{\Im(\varkappa-1)(\eta-1)}\int_0^\Upsilon \xi(\omega)d\omega \end{array} \right\}.$$

Now, we show that Ω satisfies all assumptions which are given in Lemma 4, and therefore Ω is solution of our problem (18.1.3) which is fixed point.

First, we prove $\Omega(z)$ is convex for every $z \in C([0, \Upsilon], \mathbb{R})$. For this assume that $h_1, h_2 \in \Omega(z)$, then $\exists \xi_1, \xi_2 \in S_{\zeta,z}$ then we have

$$\begin{aligned} h_i(\ell) &= \phi(0) + \ell\frac{(2-\varkappa)}{\Im(\varkappa-1)(\eta-1)}\xi_i(\Upsilon) + \frac{(2-\varkappa)}{\Im(\varkappa-1)}\int_0^\ell \xi_i(\omega)d\omega \\ &\quad + \frac{(\varkappa-1)}{\Im(\varkappa-1)}\int_0^\ell (\ell-\omega)\xi_i(\omega)d\omega + \ell\frac{(\varkappa-1)}{\Im(\varkappa-1)(\eta-1)}\int_0^\Upsilon \xi_i(\omega)d\omega \end{aligned}$$

$\forall\, \ell \in [0, \Upsilon]$. Suppose $0 \leq \lambda \leq 1$, then

$$
\begin{aligned}
[\lambda h_1 + (1-\lambda)h_2](\ell) &= \phi(0) + \ell \frac{(2-\varkappa)}{\Im(\varkappa-1)(\eta-1)} [\lambda\xi_1(\Upsilon) + (1-\lambda)\xi_2(\Upsilon)] \\
&+ \frac{(2-\varkappa)}{\Im(\varkappa-1)} \int_0^\ell [\lambda\xi_1(\omega) + (1-\lambda)\xi_2(\omega)]\, d\omega \\
&+ \frac{(\varkappa-1)}{\Im(\varkappa-1)} \int_0^\ell (\ell-\omega)\, [\lambda\xi_1(\omega) + (1-\lambda)\xi_2(\omega)]\, d\omega \\
&+ \ell \frac{(\varkappa-1)}{\Im(\varkappa-1)(\eta-1)} \int_0^\Upsilon [\lambda\xi_1(\omega) + (1-\lambda)\xi_2(\omega)]\, d\omega
\end{aligned}
$$

$\forall \ell \in [0, \Upsilon]$. Since $S_{\zeta,z}$ is convex and ζ contain convex values, therefore it implies $\lambda h_1 + (1-\lambda)h_2 \in \Omega(z)$. To establish that $\Omega(z)$ is closed for every $z \in C([0, \Upsilon], \mathbb{R})$. Take $\{u_n\}_{n\geq 0} \in \Omega(z)$ such that $u_n \longrightarrow u$ when $n \longrightarrow \infty$ in $C([0, \Upsilon], \mathbb{R})$. So $u \in C([0, \Upsilon], \mathbb{R})$ and $\exists$ a $v_n \in S_{\zeta,z}$ such that

$$
\begin{aligned}
z_n(\ell) &= \phi(0) + \ell \frac{(2-\varkappa)}{\Im(\varkappa-1)(\eta-1)} v_n + \frac{(2-\varkappa)}{\Im(\varkappa-1)} \int_0^\ell v_n d\omega \\
&+ \frac{(\varkappa-1)}{\Im(\varkappa-1)} \int_0^\ell (\ell-\omega) v_n d\omega + \ell \frac{(\varkappa-1)}{\Im(\varkappa-1)(\eta-1)} \int_0^\Upsilon v_n d\omega,
\end{aligned}
$$

as ζ contains compact values, now passing to a suitable subsequence to get v_n converges v.

$$
\begin{aligned}
u_n(\ell) \longrightarrow u(\ell) &= \phi(0) + \ell \frac{(2-\varkappa)}{\Im(\varkappa-1)(\eta-1)} v + \frac{(2-\varkappa)}{\Im(\varkappa-1)} \int_0^\ell v d\omega \\
&+ \frac{(\varkappa-1)}{\Im(\varkappa-1)} \int_0^\ell (\ell-\omega) v d\omega + \ell \frac{(\varkappa-1)}{\Im(\varkappa-1)(\eta-1)} \int_0^\Upsilon v d\omega
\end{aligned}
$$

hence $u \in \Omega(z)$.

Next, we establish $\Omega(\Im_r) \subseteq \Im_r$, where $r > 0$ and

$$\Im_r = \{z \in C([0, \Upsilon], \mathbb{R}) : \|z\| \leq r\}.$$

Clearly $\Im_r$ in $C([0, \Upsilon], \mathbb{R})$. If it does not hold then $\exists$ a function $z_r \in \Im_r$, $h_r \in \Omega(z_r)$ with $\|\Omega(z_r)\| > r$, such that

$$
\begin{aligned}
h_r(\ell) &= \phi(0) + \ell \frac{(2-\varkappa)}{\Im(\varkappa-1)(\eta-1)} \xi_r(\Upsilon) + \frac{(2-\varkappa)}{\Im(\varkappa-1)} \int_0^\ell \xi_r(\omega) d\omega \\
&+ \frac{(\varkappa-1)}{\Im(\varkappa-1)} \int_0^\ell (\ell-\omega)\xi_r(\omega) d\omega + \ell \frac{(\varkappa-1)}{\Im(\varkappa-1)(\eta-1)} \int_0^\Upsilon \xi_r(\omega) d\omega
\end{aligned}
$$

on other hand, consider

$$
\begin{aligned}
r &< \|\Omega(z_r)\| \\
&\leq \phi(0) + \ell\frac{(2-\varkappa)}{\Im(\varkappa-1)(\eta-1)}\xi_r(\Upsilon) + \frac{(2-\varkappa)}{\Im(\varkappa-1)}\int_0^{\ell}|\xi_r(\omega)|\,d\omega \\
&+ \frac{(\varkappa-1)}{\Im(\varkappa-1)}\int_0^{\ell}|(\ell-\omega)|\,|\xi_r(\omega)|\,d\omega + \ell\frac{(\varkappa-1)}{\Im(\varkappa-1)(\eta-1)}\int_0^{\Upsilon}|\xi_r(\omega)|\,d\omega \\
&\leq \phi(0) + \frac{\Upsilon}{\Im(\varkappa-1)(\eta-1)}m_r(\Upsilon) + \frac{1}{\Im(\varkappa-1)}\int_0^{\Upsilon}m_r(\omega)d\omega \\
&+ \frac{\Upsilon}{\Im(\varkappa-1)}\int_0^{\Upsilon}m_r(\omega)d\omega + \frac{\Upsilon}{\Im(\varkappa-1)(\eta-1)}\int_0^{\Upsilon}m_r(\omega)d\omega \\
&+ \frac{\Upsilon}{\Im(\varkappa-1)(\eta-1)}\int_0^{\Upsilon}m_r(\omega)d\omega \\
&\leq \phi(0) + \frac{\Upsilon}{\Im(\varkappa-1)(\eta-1)}m_r(\Upsilon) + \left[\frac{\eta(1+\Upsilon)-1}{\Im(\varkappa-1)(\eta-1)}\right]\int_0^{\Upsilon}m_r(\omega)d\omega
\end{aligned}
$$

now divide r on both side and taking limit $r \longrightarrow \infty$, we obtain

$$
\gamma > \frac{\eta(1+\Upsilon)-1}{\Im(\varkappa-1)(\eta-1)},
$$

which is not possible because it contradicts (18.5.3). Hence $\Omega(\Im_r) \subseteq \Im_r$.

Now we verify continuity of $\Omega(\Im_r)$. For this we take $\ell_1,\ \ell_2 \in [0,\Upsilon]$ with $\ell_1 < \ell_2$. Suppose $z \in \Im_r$ and $h \in \Omega(z)$ then $\exists \xi \in S_{\varsigma,z}$ such that for every $\ell \in [0,\Upsilon]$, we obtain

$$
\begin{aligned}
h(\ell) &= \phi(0) + \ell\frac{(2-\varkappa)}{\Im(\varkappa-1)(\eta-1)}\xi(\Upsilon) + \frac{(2-\varkappa)}{\Im(\varkappa-1)}\int_0^{\ell}\xi(\omega)d\omega \\
&+ \frac{(\varkappa-1)}{\Im(\varkappa-1)}\int_0^{\ell}(\ell-\omega)\xi(\omega)d\omega + \ell\frac{(\varkappa-1)}{\Im(\varkappa-1)(\eta-1)}\int_0^{\Upsilon}\xi(\omega)d\omega
\end{aligned}
\tag{18.5.4}
$$

using (18.5.4)

$$
|h(\ell_2)-h(\ell_1)| = \left|\begin{array}{c}
\ell_2\frac{(2-\varkappa)}{\Im(\varkappa-1)(\eta-1)}\xi(\Upsilon) + \frac{(2-\varkappa)}{\Im(\varkappa-1)}\int_0^{\ell_2}\xi(\omega)d\omega + \frac{(\varkappa-1)}{\Im(\varkappa-1)}\int_0^{\ell_2}(\ell_2-\omega)\xi(\omega)d\omega \\
+\ell_2\frac{(\varkappa-1)}{\Im(\varkappa-1)(\eta-1)}\int_0^{\Upsilon}\xi(\omega)d\omega - \ell_1\frac{(2-\varkappa)}{\Im(\varkappa-1)(\eta-1)}\xi(\Upsilon) - \frac{(2-\varkappa)}{\Im(\varkappa-1)}\int_0^{\ell_1}\xi(\omega)d\omega \\
-\frac{(\varkappa-1)}{\Im(\varkappa-1)}\int_0^{\ell_1}(\ell_1-\omega)\xi(\omega)d\omega - \ell_1\frac{(\varkappa-1)}{\Im(\varkappa-1)(\eta-1)}\int_0^{\Upsilon}\xi(\omega)d\omega
\end{array}\right|
$$

$$
\leq \left|\begin{array}{c}
(\ell_2-\ell_1)\frac{(2-\varkappa)}{\Im(\varkappa-1)(\eta-1)}\xi(\Upsilon) + \frac{(2-\varkappa)}{\Im(\varkappa-1)}\int_0^{\ell_1}\xi(\omega)d\omega + \frac{(2-\varkappa)}{\Im(\varkappa-1)}\int_{\ell_1}^{\ell_2}\xi(\omega)d\omega \\
+\frac{(\varkappa-1)}{\Im(\varkappa-1)}\int_0^{\ell_1}(\ell_2-\omega)\xi(\omega)d\omega + \frac{(\varkappa-1)}{\Im(\varkappa-1)}\int_{\ell_1}^{\ell_2}(\ell_2-\omega)\xi(\omega)d\omega + \ell_2\frac{(\varkappa-1)}{\Im(\varkappa-1)(\eta-1)}\int_0^{\Upsilon}\xi(\omega)d\omega \\
-\frac{(2-\varkappa)}{\Im(\varkappa-1)}\int_0^{\ell_1}\xi(\omega)d\omega - \frac{(\varkappa-1)}{\Im(\varkappa-1)}\int_0^{\ell_1}(\ell_1-\omega)\xi(\omega)d\omega - \ell_1\frac{(\varkappa-1)}{\Im(\varkappa-1)(\eta-1)}\int_0^{\Upsilon}\xi(\omega)d\omega
\end{array}\right|
$$

$$
\begin{aligned}
&|(\ell_2-\ell_1)|\,\tfrac{1}{\Im(\varkappa-1)(\eta-1)}m_\epsilon(\Upsilon)+\tfrac{1}{\Im(\varkappa-1)}\textstyle\int_0^{\ell_2} m_\epsilon(\omega)d\omega\\
\leq\ &+\tfrac{1}{\Im(\varkappa-1)}\textstyle\int_0^{\ell_1}|(\ell_2-\ell_1)|\,m_\epsilon(\omega)d\omega+\tfrac{1}{\Im(\varkappa-1)}\textstyle\int_{\ell_1}^{\ell_2}|(\ell_2-\omega)|\,m_\epsilon(\omega)d\omega\\
&+\tfrac{1}{\Im(\varkappa-1)(\eta-1)}|(\ell_2-\ell_1)|\textstyle\int_0^{\Upsilon} m_\epsilon(\omega)d\omega,
\end{aligned}
$$

as $\ell_2 \longrightarrow \ell_1$, so right side tends to zero which is independent of $z \in \Im_\epsilon$. Therefore Ω is equicontinuous. Hence (A_1) and (A_2) satisfies, so it follows that Ω is compact multivalued map.

In the last, we verify Ω has a closed graph. For this take $z_n \longrightarrow z_*$, $h_n \in \Omega(z_n)$ and $h_n \longrightarrow h_*$. For this we will show that $h_* \in \Omega(z)$. This relation implies $\exists\ \xi_n \in S_{\zeta,z_n}$ such that

$$
\begin{aligned}
h_n(\ell) &= \phi(0)+\ell\frac{(2-\varkappa)}{\Im(\varkappa-1)(\eta-1)}\xi_n(\Upsilon)\\
&\quad+\frac{(\varkappa-1)}{\Im(\varkappa-1)}\int_0^{\ell}(\ell-\omega)\xi_n(\omega)d\omega+\ell\frac{(\varkappa-1)}{\Im(\varkappa-1)(\eta-1)}\int_0^{\Upsilon}\xi_n(\omega)d\omega
\end{aligned}
$$

for every $\ell \in [0,\Upsilon]$. Now if $\exists \xi_* \in S_{\zeta,z_*}$ such that

$$
\begin{aligned}
h_*(\ell) &= \phi(0)+\ell\frac{(2-\varkappa)}{\Im(\varkappa-1)(\eta-1)}\xi_*(\Upsilon)+\frac{(2-\varkappa)}{\Im(\varkappa-1)}\int_0^{\ell}\xi_*(\omega)d\omega\\
&\quad+\frac{(\varkappa-1)}{\Im(\varkappa-1)}\int_0^{\ell}(\ell-\omega)\xi_*(\omega)d\omega+\ell\frac{(\varkappa-1)}{\Im(\varkappa-1)(\eta-1)}\int_0^{\Upsilon}\xi_*(\omega)d\omega
\end{aligned}
$$

for every $\ell \in [0,\Upsilon]$. Consider $\Theta : L^1([0,\Upsilon],\mathbb{R}) \longrightarrow C([0,\Upsilon],\mathbb{R})$ is linear continuous operator, so that

$$
\begin{aligned}
\xi \longrightarrow \Theta(\xi)(\ell) &= \phi(0)+\ell\frac{(2-\varkappa)}{\Im(\varkappa-1)(\eta-1)}\xi(\Upsilon)+\frac{(2-\varkappa)}{\Im(\varkappa-1)}\int_0^{\ell}\xi(\omega)d\omega\\
&+\frac{(\varkappa-1)}{\Im(\varkappa-1)}\int_0^{\ell}(\ell-\omega)\xi(\omega)d\omega+\ell\frac{(\varkappa-1)}{\Im(\varkappa-1)(\eta-1)}\int_0^{\Upsilon}\xi(\omega)d\omega
\end{aligned}
$$

observe that

$$
\|h_n(\ell)-h_*(\ell)\| = \left\| \begin{array}{c}
\ell\frac{(2-\varkappa)}{\Im(\varkappa-1)(\eta-1)}\xi_n(\Upsilon)+\frac{(2-\varkappa)}{\Im(\varkappa-1)}\int_0^{\ell}\xi_n(\omega)d\omega+\frac{(\varkappa-1)}{\Im(\varkappa-1)}\int_0^{\ell}(\ell-\omega)\xi_n(\omega)d\omega\\
+\ell\frac{(\varkappa-1)}{\Im(\varkappa-1)(\eta-1)}\int_0^{\Upsilon}\xi_n(\omega)d\omega-\ell\frac{(2-\varkappa)}{\Im(\varkappa-1)(\eta-1)}\xi_*(\Upsilon)-\frac{(2-\varkappa)}{\Im(\varkappa-1)}\int_0^{\ell}\xi_*(\omega)d\omega\\
-\frac{(\varkappa-1)}{\Im(\varkappa-1)}\int_0^{\ell}(\ell-\omega)\xi_*(\omega)d\omega-\ell\frac{(\varkappa-1)}{\Im(\varkappa-1)(\eta-1)}\int_0^{\Upsilon}\xi_*(\omega)d\omega
\end{array} \right\| \longrightarrow 0
$$

as $n \longrightarrow \infty$. Thus, it follows by Lemma 5 that $\Theta\ o\ S_\zeta$ is a closed graph operator. Further, we have $h_n(\ell) \in \Theta(S_{\zeta,z_n})$. Since $z_n \longrightarrow z_*$, therefore, Lemma 5 implies that

$$h_*(\ell) = \phi(0) + \ell \frac{(2-\varkappa)}{\Im(\varkappa-1)(\eta-1)} \xi_*(\Upsilon) + \frac{(2-\varkappa)}{\Im(\varkappa-1)} \int_0^\ell \xi_*(\omega) d\omega$$
$$+ \frac{(\varkappa-1)}{\Im(\varkappa-1)} \int_0^\ell (\ell-\omega)\xi_*(\omega) d\omega + \ell \frac{(\varkappa-1)}{\Im(\varkappa-1)(\eta-1)} \int_0^\Upsilon \xi_*(\omega) d\omega$$

for some $\xi_* \in S_{\zeta,z}$. It follows that Ω is a compact multivalued map, which is upper semi (us) continuous along with the closed convex values. Hence all assumptions hold, so Ω has a fixed point z. □

Conclusion

In this chapter, we investigated the solution of CFC delay fractional BVP with order $1 < \varkappa \leq 2$. To establish the existence result, we employed Schauder-type fixed point theorem, and for uniqueness, the Banach contraction principle is applied. Furthermore, we derived some sufficient conditions under which the CFC delay fractional BVP satisfies the Hyers-Ulam stability criteria. In addition, we extended our analysis to determine the existence of solutions for the CFC delay fractional differential inclusion. For this purpose, we apply the Bohnenblust-Karlin fixed point theorem, which is helpful for multivalued mappings in Banach spaces. Overall, this study helps to improve the basic concept of fractional equations that include delays and non-local boundary conditions.

References

Abbasbandy, S., Nieto, J. J., & Alavi, M. (2005). Tuning of reachable set in one dimensional fuzzy differential inclusions. *Chaos, Solitons and Fractals, 26*(5), 1337–1341.

Abdeljawad, T. (2017). Fractional operators with exponential kernels and a Lyapunov type inequality. *Advances in Difference Equations, 2017*(1), 313.

Ahmad, B., & Otero-Espinar, V. (2009). Existence of solutions for fractional differential inclusions with antiperiodic boundary conditions. *Boundary Value Problems, 2009*(1), Article 625347.

Ali, F., Saqib, M., Khan, I., & Sheikh, N. A. (2016). Application of Caputo-Fabrizio derivatives to MHD free convection flow of generalized Walters'-B fluid model. *The European Physical Journal Plus, 131*(10), 377.

Atangana, A., & Alqahtani, R. T. (2016). Numerical approximation of the space-time Caputo-Fabrizio fractional derivative and application to groundwater pollution equation. *Advances in Difference Equations, 2016*(1), 156.

Baleanu, D. (2009). Fractional variational principles in action. *Physica Scripta, 2009*(T136), Article 014006.

Bohnenblust, H. F., & Karlin, S. (1950). On a theorem of Ville. *Contributions to the Theory of Games, 1*, 155–160.

Caputo, M., & Fabrizio, M. (2015). A new definition of fractional derivative without singular kernel. *Progress in Fractional Differentiation and Applications, 1*(2), 1–13.

Chang, Y. K., Li, W. T., & Nieto, J. J. (2007). Controllability of evolution differential inclusions in Banach spaces. *Nonlinear Analysis: Theory, Methods and Applications, 67*(2), 623–632.

Chang, Y. K., & Nieto, J. J. (2009). Existence of solutions for impulsive neutral integro-differential inclusions with nonlocal initial conditions via fractional operators. *Numerical Functional Analysis and Optimization, 30*(3–4), 227–244.

Chang, Y. K., Nieto, J. J., & Li, W. S. (2009). On impulsive hyperbolic differential inclusions with nonlocal initial conditions. *Journal of Optimization Theory and Applications, 140*(3), 431.

Efe, M. Ö. (2012). Battery power loss compensated fractional order sliding mode control of a quadrotor UAV. *Asian Journal of Control, 14*(2), 413–425.

Frigon, M. (2007). Systems of first order differential inclusions with maximal monotone terms. *Nonlinear Analysis: Theory, Methods and Applications, 66*(9), 2064–2077.

Haq, F., Shah, K., ur Rahman, G., & Shahzad, M. (2017). Hyers-Ulam stability to a class of fractional differential equations with boundary conditions. *International Journal of Applied and Computational Mathematics, 3*(1), 1135–1147.

Henderson, J., & Ouahab, A. (2009). Fractional functional differential inclusions with finite delay. *Nonlinear Analysis: Theory, Methods and Applications, 70*(5), 2091–2105.

Herrmann, R. (2008). Gauge invariance in fractional field theories. *Physics Letters A, 372*(34), 5515–5522.

Hyers, D. H. (1941). On the stability of the linear functional equation. *Proceedings of the National Academy of Sciences of the United States of America, 27*(4), 222.

Kilbas, A. A., Srivastava, H. M., & Trujillo, J. J. (2006). *North-Holland mathematics studies. Theory and applications of fractional differential equations* (Vol. 204).

Li, W. S., Chang, Y. K., & Nieto, J. J. (2009). Solvability of impulsive neutral evolution differential inclusions with state-dependent delay. *Mathematical and Computer Modelling, 49*(9–10), 1920–1927.

Losta, A., & Opial, Z. (1965). Application of the Kakutani-Ky-Fan theorem in the theory of ordinary differential equations or noncompact acyclic-valued map. *Bulletin of the Polish Academy of Sciences, Series: Mathematics, Astronomy, Physics, 13*, 781–786.

Miller, K. S., & Ross, B. (1993). *An introduction to the fractional calculus and fractional differential equations.*

Naber, M. (2004). Time fractional Schrödinger equation. *Journal of Mathematical Physics, 45*(8), 3339–3352.

Nieto, J. J., & Rodríguez-López, R. (2007). Euler polygonal method for metric dynamical systems. *Information Sciences, 177*(20), 4256–4270.

Oldham, K., & Spanier, J. (1974). *The fractional calculus theory and applications of differentiation and integration to arbitrary order* (Vol. 111). Elsevier.

Ouahab, A. (2008). Some results for fractional boundary value problem of differential inclusions. *Nonlinear Analysis: Theory, Methods and Applications, 69*(11), 3877–3896.

Rassias, T. M. (1978). On the stability of the linear mapping in Banach spaces. *Proceedings of the American Mathematical Society, 72*(2), 297–300.

Rus, I. A. (2009). *Ulam stability of ordinary differential equation* (p. 4). Mathematica: Studia Universitatis Babes-Bolyai.

Rus, I. A. (2010). Ulam stabilities of ordinary differential equations in a Banach space. *Carpathian Journal of Mathematics,* 103–107.

Samko, S. G., Kilbas, A. A., & Marichev, O. I. (1993). *Fractional integrals and derivatives, translated from the 1987 Russian original.* Gordon and Breach.

Skóra, L. (2012). Boundary value problems for second order delay differential equations. *Opuscula Mathematica, 32*, 551–558.

Smirnov, G. V. (2002). *Introduction to the theory of differential inclusions* (Vol. 41). American Mathematical Soc.

Stanislavsky, A. A. (2004). Fractional oscillator. *Physical Review E, 70*(5), Article 051103.

Ulam, S. M. (1964). *Problems in modern mathematics, science editions.*

Wang, J., Lv, L., & Zhou, Y. (2011). Ulam stability and data dependence for fractional differential equations with Caputo derivative. *Electronic Journal of Qualitative Theory of Differential Equations, 2011*(63), 1–10.

Yong, Z., Jinrong, W., & Lu, Z. (2016). *Basic theory of fractional differential equations.* World Scientific.

Chapter 19
Stability of Logarithmic and Generalized Logarithmic Set-Valued Functional Equations

Sudheer Petwal, Anita Tomar, and Meena Joshi

Abstract We introduce logarithmic and generalized logarithmic set−valued functional equations and inequalities to establish the Hyers−Ulam−Rassias stability theorems via the fixed point technique in a real Banach space. We offer substantial illustrations to demonstrate the significance of our investigations concerning the stability of these functional equations.

Keywords Generalized logarithmic functional inequality · Fixed point · Functional equation

AMS Subject Classification 47H10 · 54H25

19.1 Introduction

In 1941, Hyers (1941) solved partially the question of the stability of functional equations, concerning the stability of group homomorphisms, initiated by Ulam (1940) in 1940. Subsequently, this is called Hyers-Ulam stability. It is noteworthy to mention that (Hyers, 1941) considered only bounded control functions for the Cauchy additive functional equation. Later on, Rassias (1978) tried to weaken the assertion of boundedness of the norm to set up the stability of linear mappings in a generalized context. Today, his result is recognized as the Hyers−Ulam−Rassias stability theorem. The elegancy of this result and its proof fascinated several researchers

Sudheer Petwal, Anita Tomar and Meena Joshi—These authors contributed equally to this work.

S. Petwal
Department of Mathematics, APB Government Post Graduate College, Agastyamuni, India

A. Tomar (✉)
Deparment of Mathematics, Pt. L. M. S. Campus, Sridev Suman Uttarakhand University, Rishikesh, India
e-mail: anitatmr@yahoo.com

M. Joshi
Department of Mathematics, L.M.S. Campus, Soban Singh Jeena University, Pithoragarh, India

© The Author(s), under exclusive license to Springer Nature Singapore Pte Ltd. 2026
A. Tomar and M. Jain (eds.), *Applications of Fixed-Point Theorem*, Industrial and Applied Mathematics, https://doi.org/10.1007/978-981-95-5160-6_19

and consequently, these are further solved and improved under different conditions for additive mappings, linear mappings, exploiting different control functions (see, for instance, Aumann, 1965, Cadariu & Radu, 2008, Hyers & Isac, 1998, Rassias, 2000, Ulam, 1940 and references therein). In 1991, Baker (1991) initiated the utilization of fixed points in proving the stability of functional equations through the exploitation of the Banach contraction principle (Banach, 1922). It is significant to highlight that the functional equations we use these days emerged almost with the modern theory of functions. In 1747, the three papers on functional equations of D'Alembert (1747, 1750a, b) may be considered to initiate the notion of the functional equation. Motivated by the problems emerging in probability theory, control theory, vector optimization, and mathematical economics, we focus on logarithmic and generalized logarithmic set−valued functional equations of the form

$$\mathfrak{f}(\mathfrak{v}^{\mathfrak{a}}\mathfrak{w}^{\mathfrak{b}}) = \mathfrak{f}(\mathfrak{v}) \oplus \mathfrak{f}(\mathfrak{w}) \quad \text{and} \quad \mathfrak{n} \cdot \mathfrak{f}(\mathfrak{v}^{\mathfrak{a}}\mathfrak{w}^{\mathfrak{b}})^{\frac{1}{\mathfrak{n}}} = \mathfrak{f}(\mathfrak{v}) \oplus \mathfrak{f}(\mathfrak{w}),$$

$\mathfrak{n} \in \mathbb{N}$; $\mathfrak{a}$ and $\mathfrak{b}$ are non-zero and non-negative elements of a given field. Solutions of the logarithmic and the generalized logarithmic set−valued functional equations are known as logarithmic and generalized logarithmic set−valued functional mappings. We explore the Hyers−Ulam−Rassias stability of these functional equations utilizing the Hausdorff distance. For this, we introduce and exploit logarithmic and generalized logarithmic functional inequalities in a real Banach space and vector space. Noticeably, the stability of a functional equation is equivalent to a fixed point of an appropriate mapping.

Definition 1 (Luxemburg, 1958) A function $\mathfrak{d} : \mathscr{X} \times \mathscr{X} \to [0, \infty]$ defined on an abstract set $\mathscr{X}$ is characterized as a generalized metric on $\mathscr{X}$ when

(i) $\mathfrak{d}(\mathfrak{v}, \mathfrak{w}) = 0$ iff $\mathfrak{v} = \mathfrak{w}$;
(ii) $\mathfrak{d}(\mathfrak{v}, \mathfrak{w}) = \mathfrak{d}(\mathfrak{w}, \mathfrak{v})$;
(iii) $\mathfrak{d}(\mathfrak{v}, \mathfrak{w}) \leq \mathfrak{d}(\mathfrak{v}, \mathfrak{u}) + \mathfrak{d}(\mathfrak{u}, \mathfrak{w})$, for all $\mathfrak{v}, \mathfrak{w}, \mathfrak{u} \in \mathscr{X}$.

If each $\mathfrak{d}$−Cauchy sequence is considered $\mathfrak{d}$−convergent in $\mathscr{X}$, that is, for a sequence $\{\mathfrak{v}_{\mathfrak{n}}\}$ in $\mathscr{X}$, $\lim_{\mathfrak{n},\mathfrak{m}\to\infty} \mathfrak{d}(\mathfrak{v}_{\mathfrak{n}}, \mathfrak{v}_{\mathfrak{m}}) = 0$ implies that we have an element $\mathfrak{v}$ in $\mathscr{X}$ so that $\lim_{\mathfrak{n}\to\infty} \mathfrak{d}(\mathfrak{v}, \mathfrak{v}_{\mathfrak{n}}) = 0, \mathfrak{n} = 1, 2, ...$, then $(\mathscr{X}, \mathfrak{d})$ is a complete generalized metric space. Noticeably, $\mathfrak{v}$ is unique in view of (i) and (iii).

Noticeably, the generalized metric is more fascinating than the usual metric in view of the fact that the distance between any two of its points is permitted to be infinity.

Theorem 1 (Diaz and Margolis, 1968) *Consider* $\mathscr{M} : \mathscr{X} \to \mathscr{X}$ *as a mapping in a complete generalized metric space* $(\mathscr{X}, \mathfrak{d})$*, where for* $\mathfrak{v}, \mathfrak{w} \in \mathscr{X}$

(i) $\mathfrak{d}(\mathfrak{v}, \mathfrak{w}) < \infty$ *implies that* $\mathfrak{d}(\mathscr{M}\mathfrak{v}, \mathscr{M}\mathfrak{w}) \leq \mathscr{L}\mathfrak{d}(\mathfrak{v}, \mathfrak{w}), 0 < \mathscr{L} < 1$,
(ii) If $\mathfrak{v}_0 \in \mathscr{X}$ *is the initial element and the sequence of successive approximations is* $\mathfrak{v}_0, \mathscr{M}\mathfrak{v}_0, \mathscr{M}^2\mathfrak{v}_0, \cdots, \mathscr{M}^N\mathfrak{v}_0, \cdots$. *Then either* $\mathfrak{d}(\mathscr{M}^{\mathscr{N}}\mathfrak{v}, \mathscr{M}^{\mathscr{N}+1}\mathfrak{v}) = \infty, \forall$ *integer* $\mathscr{N} = 0, 1, 2, 3 \cdots$, *or the successive approximations is* $\mathfrak{d}$−*convergent to a fixed point of* $\mathcal{M}$.

Remark 1 The above theorem demonstrates that if

(i) Given that $\mathscr{M}\mathfrak{v} = \mathfrak{v}$ and $\mathscr{M}\mathfrak{w} = \mathfrak{w}$, then it follows that $\mathfrak{d}(\mathfrak{v}, \mathfrak{w}) < \infty$
(ii) The sequence $\{\mathscr{M}^{n}\mathfrak{v}\}$ converges to a unique fixed point $\mathfrak{w}$ of $\mathscr{M}$ in the set $\mathscr{Y} = \{\mathfrak{w} \in \mathscr{X} \mid \mathfrak{d}(\mathscr{M}^{n\circ}\mathfrak{v}, \mathfrak{w}) < \infty\}$
(iii) If $\mathcal{N} \to \infty$, $\mathfrak{d}(\mathfrak{v}, \mathfrak{v}_n) < \frac{\mathscr{L}^{n-\mathcal{N}}}{1-\mathscr{L}}\mathfrak{d}(\mathfrak{v}_{\mathcal{N}}, \mathscr{M}\mathfrak{v}_{\mathcal{N}})$. In fact, it provides an estimate of the rate of convergence of the sequence $\{\mathfrak{v}_n\}$ to its limit $\mathfrak{v}$.

In a Banach space $\mathscr{Y}$,

1. on $2^{\mathscr{Y}}$, the collection of subsets of $\mathscr{Y}$,

$$\mathscr{C} + \mathscr{C}^{'} = \{\mathfrak{v} + \mathfrak{v}^{'} : \mathfrak{v} \in \mathscr{C}, \mathfrak{v}^{'} \in \mathscr{C}^{'}\}, \quad \lambda\mathscr{C} = \{\lambda\mathfrak{v} : \mathfrak{v} \in \mathscr{C}\},$$

$\mathscr{C}, \mathscr{C}^{'} \in 2^{\mathscr{Y}}$ and $\lambda \in \mathbb{R}$. The distance function

$$\mathfrak{d}(\mathfrak{v}, \mathscr{C}) = inf\{\|\mathfrak{v} - \mathfrak{w}\| : \mathfrak{w} \in \mathscr{C}\}, \quad \mathfrak{v} \in \mathscr{Y}$$

and the support function

$$(\mathfrak{v}^*, \mathscr{C}) = \sup\{\langle \mathfrak{v}^*, \mathfrak{v}\rangle : \mathfrak{v} \in \mathscr{C}\}, \quad \mathfrak{v}^* \in \mathscr{Y}^* \ \ \mathscr{Y} \in 2^{\mathscr{Y}}.$$

2. Further, in the context of the collection of closed convex subsets of $\mathscr{Y}$, $\mathscr{C}\mathfrak{c}(\mathscr{Y})$,

$$\mathscr{C} \oplus \mathscr{C}^{'} = \overline{\mathscr{C} + \mathscr{C}^{'}},$$

$$\lambda\mathscr{C} + \lambda\mathscr{C}^{'} = \lambda(\mathscr{C} + \mathscr{C}^{'}) \quad \text{and} \quad (\lambda + \mu)\mathscr{C} \subseteq \lambda\mathscr{C} + \mu\mathscr{C}, \ \ \mathscr{C}, \mathscr{C}^{'} \in \mathscr{C}\mathfrak{c}(\mathscr{Y})$$

3. When $\mathscr{C}$ is convex, we get $(\lambda + \mu)\mathscr{C} = \lambda\mathscr{C} + \mu\mathscr{C}$, $\lambda, \mu \in \mathbb{R}^+$.
4. Regarding the definition of the Hausdorff distance on the set of closed bounded subsets of $\mathscr{Y}$, $\mathscr{C}_{\mathfrak{b}}(\mathscr{Y})$, it is defined as
$$\mathfrak{h}(\mathscr{C}, \mathscr{C}^{'}) = \inf\{\lambda > 0 : \mathscr{C} \subseteq \mathscr{C}^{'} + \lambda\mathscr{B}_{\mathscr{Y}}, \mathscr{C}^{'} \subseteq \mathscr{C} + \lambda\mathscr{B}_{\mathscr{Y}}\}, \ \ \mathscr{C}, \mathscr{C}^{'} \in \mathscr{C}^{'}_{\mathfrak{b}}(\mathscr{Y}),$$

$\mathscr{B}_{\mathscr{Y}}$ represents a closed unit ball in $\mathscr{Y}$, that is, $\mathscr{B}_{\mathscr{Y}} = \{\mathfrak{w} \in \mathscr{Y} : \mathscr{Y} \leq 1\}$.

Proposition 2 *In the set of all closed convex bounded subsets of $\mathscr{Y}$, denoted as $\mathscr{C}_{cb}(\mathscr{Y})$ the following properties hold:*

(a) $\mathfrak{h}(\mathscr{C} \oplus \mathscr{C}^{'}, \mathscr{K} \oplus \mathscr{K}^{'}) \leq \mathfrak{h}(\mathscr{C}, \mathscr{K}) + \mathfrak{h}(\mathscr{C}^{'}, \mathscr{K}^{'})$;
(b) $\mathfrak{h}(\lambda\mathscr{C}, \lambda\mathscr{K}) = \lambda\mathfrak{h}(\mathscr{C}, \mathscr{K})$, $\mathscr{C}, \mathscr{C}^{'}, \mathscr{K}, \mathscr{K}^{'} \in \mathscr{C}_{\mathfrak{cb}}(\mathscr{Y})$ *and* $\lambda > 0$.

Debreu (1966) proved that a complete metric semigroup $(\mathscr{C}_{\mathfrak{cb}}(\mathscr{Y}), \oplus, \mathfrak{h})$ equipped with the Hausdorff distance $\mathfrak{h}$ is isometrically embedded in a Banach space $\mathscr{Y}$.

Lemma 1 (Debreu, 1966) *If $\mathscr{C}(\mathscr{B}_{\mathscr{Y}^*})$ is the Banach space of continuous real-valued mappings defined on $\mathscr{B}_{\mathscr{Y}^*}$ equipped with a uniform norm $\|\cdot\|_{\mathfrak{u}}$. Then the mapping $\mathfrak{k} : (\mathscr{C}_{\mathfrak{cb}}(\mathscr{Y}), \oplus, \mathfrak{h}) \to \mathscr{C}(\mathscr{B}_{\mathscr{Y}^*})$, defined as $\mathfrak{k}(\mathscr{A}) = \mathfrak{s}(\cdot, \mathscr{A})$, verifies*

(a) $\mathfrak{k}(\mathscr{A} \oplus \mathscr{B}) = \mathfrak{k}(\mathscr{A}) + \mathfrak{k}(\mathscr{B})$;
(b) $\mathfrak{k}(\lambda\mathscr{A}) = \lambda\mathfrak{k}(\mathscr{A})$;
(c) $\mathfrak{h}(\mathscr{A}, \mathscr{B}) = \|\mathfrak{k}(\mathscr{A}) - \mathfrak{k}(\mathscr{B})\|_{\mathfrak{u}}$;
(d) $\mathfrak{k}(\mathscr{C}_{\mathfrak{cb}}(\mathscr{Y}))$ *is closed in* $\mathscr{C}(\mathscr{B}_{\mathscr{Y}^*})$, $\mathscr{A}, \mathscr{B} \in \mathscr{C}_{\mathfrak{cb}}(\mathscr{Y})$, $\lambda \geq 0$.

19.2 Main Results

In the following context, $\mathscr{X} = (0, \infty)$ represents a real vector space, $\mathscr{Y}$ is a real Banach space, and ρ a positive real number, which is less than 1, and $(\mathscr{C}_{\mathfrak{cb}}(\mathscr{Y}), \oplus, \mathfrak{h})$ is a complete metric semigroup.

Definition 2 A mapping $\mathfrak{f} : \mathscr{X} \to \mathscr{C}_{\mathfrak{cb}}(\mathscr{Y})$ is considered a logarithmic set−valued functional equation, if

$$\mathfrak{f}(\mathfrak{v}^{\mathfrak{a}}\mathfrak{w}^{\mathfrak{b}}) = \mathfrak{a}\mathfrak{f}(\mathfrak{v}) \oplus \mathfrak{b}\mathfrak{f}(\mathfrak{w}) \tag{19.1}$$

$\mathfrak{v}, \mathfrak{w} \in \mathscr{X}$, where $\mathfrak{a}$ and $\mathfrak{b}$ are non-zero elements of a field.

Definition 3 A mapping $\mathfrak{f} : \mathscr{X} \to \mathscr{C}_{\mathfrak{cb}}(\mathscr{Y})$ is considered a generalized logarithmic set−valued functional equation, if

$$\mathfrak{n} \cdot \mathfrak{f}(\mathfrak{v}^{\mathfrak{a}}\mathfrak{w}^{\mathfrak{b}})^{\frac{1}{\mathfrak{n}}} = \mathfrak{a}\mathfrak{f}(\mathfrak{v}) \oplus \mathfrak{b}\mathfrak{f}(\mathfrak{w}) \tag{19.2}$$

$\mathfrak{v}, \mathfrak{w} \in \mathscr{X}$, $\mathfrak{n} \in \mathbb{N}$, $\mathfrak{a}$, and $\mathfrak{b}$ are non-zero and non-negative elements of a field.

Remark 2 For a single-valued mapping $\mathfrak{f} : \mathbb{R}^+ \to \mathbb{R}$, taking $\mathfrak{a} = \mathfrak{b} = 1$ in Definition 1 and $\mathfrak{a} = \mathfrak{b} = \mathfrak{n} = 1$ in Definition 2 we get the classical logarithmic functional equation which is also known as the Cauchy logarithmic functional equation. One of its solutions is the natural logarithmic function $\log \mathfrak{v}$. However, it has many solutions.

Now, we define logarithmic and generalized logarithmic set−valued functional inequalities.

Definition 4 For a mapping $\mathfrak{f} : \mathscr{X} \to \mathscr{C}_{\mathfrak{cb}}(\mathscr{Y})$ a logarithmic and a generalized logarithmic set−valued ρ−functional inequalities are defined as

$$\mathfrak{h}(\mathfrak{f}(\mathfrak{v}^{\mathfrak{a}}\mathfrak{w}^{\mathfrak{b}}), \mathfrak{a}\mathfrak{f}(\mathfrak{v}) \oplus \mathfrak{b}\mathfrak{f}(\mathfrak{w})) \leq \rho \cdot \mathfrak{h}(2\mathfrak{f}(\sqrt{\mathfrak{v}^{\mathfrak{a}}\mathfrak{w}^{\mathfrak{b}}}), \mathfrak{a}\mathfrak{f}(\mathfrak{v}) \oplus \mathfrak{b}\mathfrak{f}(\mathfrak{w})),$$

and

$$\mathfrak{h}(\mathfrak{n} \cdot \mathfrak{f}(\mathfrak{v}^{\mathfrak{a}}\mathfrak{w}^{\mathfrak{b}})^{\frac{1}{\mathfrak{n}}}, a\mathfrak{f}(\mathfrak{v}) \oplus \mathfrak{b}\mathfrak{f}(\mathfrak{w})) \leq \rho \cdot \mathfrak{h}(\mathfrak{f}(\mathfrak{v}^{\mathfrak{a}}\mathfrak{w}^{\mathfrak{b}}), \mathfrak{a}\mathfrak{f}(\mathfrak{v}) \oplus \mathfrak{b}\mathfrak{f}(\mathfrak{w})),$$

respectively, $\mathfrak{n} \in \mathbb{N}$, $\rho \in [0, 1)$ $\mathfrak{a}$, and $\mathfrak{b}$ are non-zero and non-negative elements of a field.

Lemma 2 *Let a mapping* $\mathfrak{f}: \mathscr{X} \to \mathscr{C}_{\mathfrak{cb}}(\mathscr{Y}), \mathfrak{h})$ *verifies a logarithmic set−valued functional inequality*

$$\mathfrak{h}(\mathfrak{f}(\mathfrak{v}^{\mathfrak{a}}\mathfrak{w}^{\mathfrak{b}}), \mathfrak{a}\mathfrak{f}(\mathfrak{v}) \oplus \mathfrak{b}\mathfrak{f}(\mathfrak{w})) \leq \rho \cdot \mathfrak{h}(2\mathfrak{f}(\sqrt{\mathfrak{v}^{\mathfrak{a}}\mathfrak{w}^{\mathfrak{b}}}), \mathfrak{a}\mathfrak{f}(\mathfrak{v}) \oplus \mathfrak{b}\mathfrak{f}(\mathfrak{w})), \tag{19.3}$$

$\mathfrak{v}, \mathfrak{w} \in \mathscr{X}$, *where* $\mathfrak{a} > 0$, $\mathfrak{b} > 0$ *such that* $\mathfrak{a} + \mathfrak{b} = 2$. *Then* $\mathfrak{f}$ *is a logarithmic set−valued function.*

Proof Taking $\mathfrak{w} = \mathfrak{v}$ in inequality (19.3), we attain

$$\mathfrak{h}(\mathfrak{f}(\mathfrak{v}^{\mathfrak{a}+\mathfrak{b}}), 2\mathfrak{f}(\mathfrak{v})) \leq \rho \cdot \mathfrak{h}(2\mathfrak{f}(\sqrt{\mathfrak{v}^{\mathfrak{a}+\mathfrak{b}}}), 2\mathfrak{f}(\mathfrak{v})),$$

$\mathfrak{v}, \mathfrak{w} \in \mathscr{X}$. Since $\mathfrak{a} + \mathfrak{b} = 2$, we obtain

$$\mathfrak{h}(\mathfrak{f}(\mathfrak{v}^2), 2\mathfrak{f}(\mathfrak{v})) \leq \rho \cdot \mathfrak{h}(2\mathfrak{f}(\sqrt{\mathfrak{v}^2}), 2\mathfrak{f}(\mathfrak{v})) = 0,$$

that is, $\mathfrak{h}(\mathfrak{f}(\mathfrak{v}^2), 2\mathfrak{f}(\mathfrak{v})) = 0$, that is, $\mathfrak{f}(\mathfrak{v}^2) = 2\mathfrak{f}(\mathfrak{v})$ and so $\frac{1}{2}\mathfrak{f}(\mathfrak{v}) = \mathfrak{f}(\sqrt{\mathfrak{v}})$, $\mathfrak{v} \in \mathscr{X}$. Hence, using logarithmic set−valued functional inequality (19.3)

$$\begin{aligned}\mathfrak{h}(\mathfrak{f}(\mathfrak{v}^{\mathfrak{a}}\mathfrak{w}^{\mathfrak{b}}), \mathfrak{a}\mathfrak{f}(\mathfrak{v}) \oplus \mathfrak{b}\mathfrak{f}(\mathfrak{w})) &\leq \rho \cdot \mathfrak{h}(2\mathfrak{f}(\sqrt{\mathfrak{v}^{\mathfrak{a}}\mathfrak{w}^{\mathfrak{b}}}), \mathfrak{a}\mathfrak{f}(\mathfrak{v}) \oplus \mathfrak{b}\mathfrak{f}(\mathfrak{w})) \\ &= \rho \cdot \mathfrak{h}(\mathfrak{f}(\mathfrak{v}^{\mathfrak{a}}\mathfrak{w}^{\mathfrak{b}}), \mathfrak{a}\mathfrak{f}(\mathfrak{v}) \oplus \mathfrak{b}\mathfrak{f}(\mathfrak{w})).\end{aligned}$$

Since $\rho < 1$. Therefore

$$\mathfrak{h}(\mathfrak{f}(\mathfrak{v}^{\mathfrak{a}}\mathfrak{w}^{\mathfrak{b}}), \mathfrak{a}\mathfrak{f}(\mathfrak{v}) \oplus \mathfrak{b}\mathfrak{f}(\mathfrak{w})) = 0,$$

that is,

$$\mathfrak{f}(\mathfrak{v}^{\mathfrak{a}}\mathfrak{w}^{\mathfrak{b}}) = \mathfrak{a}\mathfrak{f}(\mathfrak{v}) \oplus \mathfrak{b}\mathfrak{f}(\mathfrak{w}).$$

□

Lemma 3 *Let a mapping* $\mathfrak{f}: \mathscr{X} \to \mathscr{C}_{\mathfrak{cb}}(\mathscr{Y}), \mathfrak{h})$ *verifies* $\mathfrak{f}(1) = \{0\}$ *and a generalized logarithmic set−valued functional inequality*

$$\mathfrak{h}(\mathfrak{n} \cdot \mathfrak{f}(\mathfrak{v}^{\mathfrak{a}}\mathfrak{w}^{\mathfrak{b}})^{\frac{1}{\mathfrak{n}}}, \mathfrak{a}\mathfrak{f}(\mathfrak{v}) \oplus \mathfrak{b}\mathfrak{f}(\mathfrak{w})) \leq \rho \cdot \mathfrak{h}(\mathfrak{f}(\mathfrak{v}^{\mathfrak{a}}\mathfrak{w}^{\mathfrak{b}}), \mathfrak{a}\mathfrak{f}(\mathfrak{v}) \oplus \mathfrak{b}\mathfrak{f}(\mathfrak{w})), \tag{19.4}$$

$\mathfrak{v}, \mathfrak{w} \in \mathscr{X}$, $\mathfrak{a} > 0$, $\mathfrak{b} > 0$, $\mathfrak{n} \in \mathbb{N}$. *Then* $\mathfrak{f}$ *is a generalized logarithmic set−valued function.*

Proof Taking $\mathfrak{w} = 1$ and $\mathfrak{a} = 1$ in inequality (19.4), we have

$$\mathfrak{h}(\mathfrak{n} \cdot \mathfrak{f}(\mathfrak{v})^{\frac{1}{\mathfrak{n}}}, \mathfrak{f}(\mathfrak{v})) \leq \rho \cdot \mathfrak{h}(\mathfrak{f}(\mathfrak{v}), \mathfrak{f}(\mathfrak{v})) \leq \mathfrak{h}(\mathfrak{f}(\mathfrak{v}), \mathfrak{f}(\mathfrak{v})), \quad \text{a contradiction},$$

that is,

$$\mathfrak{h}(\mathfrak{n}\mathfrak{f}(\mathfrak{v})^{\frac{1}{\mathfrak{n}}}, \mathfrak{f}(\mathfrak{v})) = 0, \quad \text{that is,} \quad \mathfrak{n}\mathfrak{f}(\mathfrak{v})^{\frac{1}{n}} = \mathfrak{f}(\mathfrak{v}), \quad \mathfrak{v} \in \mathscr{X}.$$

Utilizing a generalized logarithmic set−valued functional inequality (19.4)

$$\begin{aligned}\mathfrak{h}(\mathfrak{f}(\mathfrak{v}^{\mathfrak{a}}\mathfrak{w}^{\mathfrak{b}}), \mathfrak{a}\mathfrak{f}(\mathfrak{v}) \oplus \mathfrak{b}\mathfrak{f}(\mathfrak{w})) &= \mathfrak{h}(\mathfrak{n} \cdot \mathfrak{f}(\mathfrak{v}^{\mathfrak{a}}\mathfrak{w}^{\mathfrak{b}})^{\frac{1}{\mathfrak{n}}}, \mathfrak{f}(\mathfrak{v}) \oplus \mathfrak{f}(\mathfrak{w})) \\ &\leq \rho \cdot \mathfrak{h}(\mathfrak{f}(\mathfrak{v}^{\mathfrak{a}}\mathfrak{w}^{\mathfrak{b}}), \mathfrak{f}(\mathfrak{v}) \oplus \mathfrak{f}(\mathfrak{w})).\end{aligned}$$

Since $\rho < 1$, we get

$$\mathfrak{h}(\mathfrak{f}(\mathfrak{v}^{\mathfrak{a}}\mathfrak{w}^{\mathfrak{b}}), \mathfrak{a}\mathfrak{f}(\mathfrak{v}) \oplus \mathfrak{b}\mathfrak{f}(\mathfrak{w})) = 0,$$

that is,

$$\mathfrak{f}(\mathfrak{v}^{\mathfrak{a}}\mathfrak{w}^{\mathfrak{b}}) = \mathfrak{a}\mathfrak{f}(\mathfrak{v}) \oplus \mathfrak{b}\mathfrak{f}(\mathfrak{w}).$$

□

The term stability is used in the context of a functional equation whenever the functional equation is replaced by a corresponding inequality performing a perturbation of the equation. In other words, stability means that by what amount the solution of the inequality varies from that of the functional equation under consideration.

Next, we establish the Hyers−Ulam−Rassias stability of the logarithmic set−valued functional equation exploiting logarithmic set−valued functional inequality and exploiting the fixed point approach.

Theorem 2 *Suppose a function $\varphi : \mathscr{X}^2 \to [0, \infty)$ satisfies*

$$\varphi(\mathfrak{v}, \mathfrak{w}) \leq \frac{\mathscr{L}}{2}\varphi(\mathfrak{v}^2, \mathfrak{w}^2), \ \ \mathscr{L} < 1, \ \ \mathfrak{v}, \mathfrak{w} \in \mathscr{X}. \tag{19.5}$$

Let a mapping $\mathfrak{f} : \mathscr{X} \to \mathscr{C}_{cb}(\mathscr{Y}), \mathfrak{h})$ verifies

$$\mathfrak{h}(\mathfrak{f}(\mathfrak{v}^{\mathfrak{a}}\mathfrak{w}^{\mathfrak{b}}), \mathfrak{a}\mathfrak{f}(\mathfrak{v}) \oplus \mathfrak{b}\mathfrak{f}(\mathfrak{w})) \leq \rho \cdot \mathfrak{h}(2\mathfrak{f}(\sqrt{\mathfrak{v}^{\mathfrak{a}}\mathfrak{w}^{\mathfrak{b}}}), \mathfrak{a}\mathfrak{f}(\mathfrak{v}) \oplus \mathfrak{b}\mathfrak{f}(\mathfrak{w})) + \varphi(\mathfrak{v}, \mathfrak{w}), \tag{19.6}$$

$\mathfrak{v}, \mathfrak{w} \in \mathscr{X}$, where $\mathfrak{a} > 0$, $\mathfrak{b} > 0$ such that $\mathfrak{a} + \mathfrak{b} = 2$. Then a distinct logarithmic set-valued mapping $\mathscr{A} : \mathscr{X} \to (\mathscr{C}_{cb}(\mathscr{Y}), \mathfrak{h})$ satisfying

$$\mathfrak{h}(\mathfrak{f}(\mathfrak{v}), \mathscr{A}(\mathfrak{v})) \leq \frac{\mathscr{L}}{2 - 2\mathscr{L}}\varphi(\mathfrak{v}, \mathfrak{v}), \ \ \mathfrak{v} \in \mathscr{X}, \tag{19.7}$$

exists which confirms the stability of the logarithmic set−valued functional equation (19.1).

Proof We have $\mathfrak{f}(\mathfrak{v})$ is convex, taking $\mathfrak{w} = \mathfrak{v}$ in inequality (19.6), we attain

$$\mathfrak{h}(\mathfrak{f}(\mathfrak{v}^{\mathfrak{a}+\mathfrak{b}}), \mathfrak{a}\mathfrak{f}(\mathfrak{v}) \oplus \mathfrak{b}\mathfrak{f}(\mathfrak{v})) \leq \rho \cdot \mathfrak{h}(2\mathfrak{f}(\sqrt{\mathfrak{v}^{\mathfrak{a}+\mathfrak{b}}}), \mathfrak{a}\mathfrak{f}(\mathfrak{v}) \oplus \mathfrak{b}\mathfrak{f}(\mathfrak{v})) + \varphi(\mathfrak{v}, \mathfrak{v}),$$

where $\mathfrak{a} > 0, \mathfrak{b} > 0$ and $\mathfrak{a} + \mathfrak{b} = 2$. Therefore

$$\mathfrak{h}(\mathfrak{f}(\mathfrak{v}^2), 2\mathfrak{f}(\mathfrak{v})) \leq \rho \cdot \mathfrak{h}(2\mathfrak{f}(\sqrt{\mathfrak{v}^2}), 2\mathfrak{f}(\mathfrak{v})) + \varphi(\mathfrak{v}, \mathfrak{v}) = \varphi(\mathfrak{v}, \mathfrak{v})$$

that is,

$$\mathfrak{h}(\mathfrak{f}(\mathfrak{v}), 2\mathfrak{f}(\sqrt{\mathfrak{v}})) \leq \varphi(\sqrt{\mathfrak{v}}, \sqrt{\mathfrak{v}}) \leq \frac{\mathscr{L}}{2} \cdot \varphi(\mathfrak{v}, \mathfrak{v}), \ \mathfrak{v} \in \mathscr{X}. \tag{19.8}$$

Let $\mathscr{P} := \{\mathfrak{g} : \mathscr{X} \to (\mathscr{C}_{\mathfrak{cb}}(\mathscr{Y}), \mathfrak{h}) \ \ g(0) = \{0\}\}$, $\mathfrak{d}(\mathfrak{g}(\mathfrak{v}), \mathfrak{f}(\mathfrak{v})) = \inf\{\mu \in [0, \infty] : \mathfrak{h}(\mathfrak{g}(\mathfrak{v}), \mathfrak{f}(\mathfrak{v})) \leq \mu\varphi(\mathfrak{v}, \mathfrak{v}), \ \mathfrak{v} \in \mathscr{X}\}$ be the generalized metric $\mathfrak{d}$ on $\mathscr{P}$ and $\inf \phi = +\infty$. Noticeably, $(\mathscr{P}, \mathfrak{d})$ is a complete generalized metric space. Consider $\mathscr{M} : \mathscr{P} \to \mathscr{P}$ as

$$\mathscr{M}\mathfrak{g}(\mathfrak{v}) = 2\mathfrak{g}(\sqrt{\mathfrak{v}}), \ \mathfrak{v} \in \mathscr{X},$$

Let $\mathfrak{g}, \mathfrak{f} \in \mathscr{P}$. Therefore using the definition of $\mathfrak{g}$ and inequality (19.8)

$$\begin{aligned}\mathfrak{h}(\mathscr{M}\mathfrak{g}(\mathfrak{v}), \mathscr{M}\mathfrak{f}(\mathfrak{v})) &= \mathfrak{h}(2\mathfrak{g}(\sqrt{\mathfrak{v}}), 2\mathfrak{f}(\sqrt{\mathfrak{v}})) \\ &= 2\mathfrak{h}(\mathfrak{g}(\sqrt{\mathfrak{v}}), \mathfrak{f}(\sqrt{\mathfrak{v}})) \leq 2\mu\varphi(\sqrt{\mathfrak{v}}, \sqrt{\mathfrak{v}}) \leq \mathscr{L}\varphi(\mathfrak{v}, \mathfrak{v}) \leq \mathscr{L}\mathfrak{d}(\mathfrak{g}, \mathfrak{f}),\end{aligned}$$

$\mathfrak{v} \in \mathscr{X}$. So

$$\mathfrak{h}(\mathscr{M}\mathfrak{g}(\mathfrak{v}), \mathscr{M}\mathfrak{f}(\mathfrak{v})) \leq \mathscr{L}\mathfrak{d}(\mathfrak{g}(\mathfrak{v}), \mathfrak{f}(\mathfrak{v})), \ \mathfrak{g}, \mathfrak{f} \in \mathscr{P}.$$

Utilizing inequality (19.8)

$$\mathfrak{h}(\mathfrak{f}(\mathfrak{v}), \mathscr{M}\mathfrak{f}(\mathfrak{v})) \leq \frac{\mathscr{L}}{2}\varphi(\mathfrak{v}, \mathfrak{v}).$$

Using Theorem 1, a mapping $\mathscr{A} : \mathscr{X} \to (\mathscr{C}_{\mathfrak{cb}}(\mathscr{Y}), \mathfrak{h})$ of $\mathscr{M}$ in $\mathscr{P}$ exists so that the following holds:

1. $\mathscr{A}(\mathfrak{v})$ is a unique fixed point of $\mathscr{M}$ in the collection $\mathscr{S} = \{\mathfrak{g} \in \mathscr{P} : \mathfrak{d}(\mathfrak{f}, \mathfrak{g}) < \infty\}$, that is,

$$\mathscr{A}(\mathfrak{v}) = 2\mathscr{A}(\mathfrak{v}^{\frac{1}{2}}), \ \ \mathfrak{v} \in \mathscr{X}, \tag{19.9}$$

so that

$$\mathfrak{h}(\mathfrak{f}(\mathfrak{v}), \mathscr{A}(\mathfrak{v})) \leq \mu\,\varphi(\mathfrak{v}, \mathfrak{v}), \ \mathfrak{v} \in \mathscr{X}, \ \mu \in [0, \infty].$$

2. $\mathfrak{d}(\mathscr{M}^{\mathfrak{n}}\mathfrak{f}(\mathfrak{v}), \mathscr{A}(\mathfrak{v})) \to 0$ whenever $\mathfrak{n} \to \infty$ and we attain

$$\mathscr{A}(\mathfrak{v}) = \lim_{\mathfrak{n}\to\infty} 2^{\mathfrak{n}}\mathfrak{f}(\mathfrak{v}^{\frac{1}{2^{\mathfrak{n}}}}), \ \mathfrak{v} \in \mathscr{X}.$$

3.

$$\mathfrak{d}(\mathfrak{f}(\mathfrak{v}), \mathscr{A}(\mathfrak{v})) \leq \frac{1}{1 - \mathscr{L}}\mathfrak{d}(\mathfrak{f}(\mathfrak{v}), \mathscr{M}\mathfrak{f}(\mathfrak{v})) \leq \frac{\mathscr{L}}{2 - 2\mathscr{L}}\varphi(\mathfrak{v}, \mathfrak{v}).$$

Hence, in view of the above three points inequality (19.7) holds.

From inequality (19.6), we obtain

$$\begin{aligned}
&\mathfrak{h}(2^{\mathfrak{n}}\mathfrak{f}(\mathfrak{v}^{\mathfrak{a}}\mathfrak{w}^{\mathfrak{b}})^{\frac{1}{2^{\mathfrak{n}}}}, 2^{\mathfrak{n}}\mathfrak{a}\mathfrak{f}(\mathfrak{v})^{\frac{1}{2^{\mathfrak{n}}}} \oplus 2^{\mathfrak{n}}\mathfrak{b}\mathfrak{f}(\mathfrak{w})^{\frac{1}{2^{\mathfrak{n}}}}) \\
&\quad \leq \rho \cdot \mathfrak{h}(2^{\mathfrak{n}+1}\mathfrak{f}(\mathfrak{v}^{\mathfrak{a}}\mathfrak{w}^{\mathfrak{b}})^{\frac{1}{2^{\mathfrak{n}+1}}}, 2^{\mathfrak{n}}\mathfrak{a}\mathfrak{f}(\mathfrak{v})^{\frac{1}{2^{\mathfrak{n}}}} \oplus 2^{\mathfrak{n}}\mathfrak{b}\mathfrak{f}(\mathfrak{w})^{\frac{1}{2^{\mathfrak{n}}}}) + 2^{\mathfrak{n}}\varphi(\mathfrak{v}^{\frac{1}{2^{\mathfrak{n}}}}, \mathfrak{w}^{\frac{1}{2^{\mathfrak{n}}}}) \\
&\quad \leq \rho \cdot \mathfrak{h}(2^{\mathfrak{n}+1}\mathfrak{f}(\mathfrak{v}^{\mathfrak{a}}\mathfrak{w}^{\mathfrak{b}})^{\frac{1}{2^{\mathfrak{n}+1}}}, 2^{\mathfrak{n}}\mathfrak{a}\mathfrak{f}(\mathfrak{v})^{\frac{1}{2^{\mathfrak{n}}}} \oplus 2^{\mathfrak{n}}\mathfrak{b}\mathfrak{f}(\mathfrak{w})^{\frac{1}{2^{\mathfrak{n}}}}) + \mathcal{L}^{\mathfrak{n}}\varphi(\mathfrak{v}, \mathfrak{w}),
\end{aligned}$$

that is, as $n \to \infty$

$$\begin{aligned}
\mathfrak{h}(\mathscr{A}(\mathfrak{v}^{\mathfrak{a}}\mathfrak{w}^{\mathfrak{b}}), \mathfrak{a}\mathscr{A}(\mathfrak{v}) \oplus \mathfrak{b}\mathscr{A}(\mathfrak{w})) &\leq \rho \cdot \mathfrak{h}(2\mathfrak{f}(\sqrt{\mathfrak{v}^{\mathfrak{a}}\mathfrak{w}^{\mathfrak{b}}}), \mathfrak{a}\mathscr{A}(\mathfrak{v}) \oplus \mathfrak{b}\mathscr{A}(\mathfrak{w})) \\
&= \rho \cdot \mathfrak{h}(\mathfrak{f}(\mathfrak{v}^{\mathfrak{a}}\mathfrak{w}^{\mathfrak{b}}), \mathscr{A}(\mathfrak{v}) \oplus \mathscr{A}(\mathfrak{w})).
\end{aligned}$$

Since $\rho < 1$, Lemma 2 and we attain

$$A(\mathfrak{v}^{\mathfrak{a}}\mathfrak{w}^{\mathfrak{b}}) = \mathfrak{a}\mathscr{A}(\mathfrak{v}) \oplus \mathfrak{b}\mathscr{A}(\mathfrak{w}), \ \mathfrak{v}, \ \mathfrak{w} \in \mathscr{X}, \text{ as required.}$$

□

Theorem 2 asserts that the logarithmic set−valued functional equation (19.1) is stable, and the next example validates it.

Example 1 Suppose $\mathscr{X} = [8, 10]$, $(\mathscr{C}_{c\mathfrak{b}}(\mathscr{Y}), \oplus, \mathfrak{h})$ be a complete metric semigroup equipped with a Hausdorff distance $\mathfrak{h}$ defined as

$$\mathfrak{d}_{\mathfrak{h}}(\mathfrak{v}, \mathfrak{w}) = \|v - w\|. \tag{19.10}$$

Consider $\varphi : \mathscr{X}^2 \to [0, \infty)$ to be given by $\varphi(\mathfrak{v}, \mathfrak{w}) = \frac{\mathfrak{v}}{2}$, $\mathfrak{v}$, $\mathfrak{w} \in \mathscr{X}$ and $\mathfrak{f}$, $\mathscr{A}$: $\mathscr{X} \to \mathscr{C}_{c\mathfrak{b}}(\mathscr{Y}), \mathfrak{h})$ be mappings defined as

$$\mathfrak{f}(\mathfrak{v}) = \begin{cases} [0, 1], & \text{if } \ \mathfrak{v} \in [8, 9) \\ [1.5, 2], & \text{if } \ \mathfrak{v} \in [9, 10]; \end{cases}$$

and

$$A(\mathfrak{v}) = \begin{cases} [0, 0.5], & \text{if } \ \mathfrak{v} \in [8, 9) \\ [1, 2], & \text{if } \ \mathfrak{v} \in [9, 10]. \end{cases}$$

Therefore, for $\mathscr{L} = \frac{1}{2}$, we attain

$$\varphi(\mathfrak{v}, \mathfrak{w}) \leq \frac{1}{4} \cdot \varphi(\mathfrak{v}^2, \mathfrak{w}^2).$$

Taking $\mathfrak{a} = 1$, $\mathfrak{b} = 1$, $\mathfrak{v} \in \mathcal{X}$ and $\rho = \frac{3}{4}$, the inequality (19.6) holds.

Using inequality (19.7) and we get

$$\mathfrak{h}(\mathfrak{f}(\mathfrak{v}), \mathscr{A}(\mathfrak{v})) \leq \frac{\mathscr{L}}{2-2\mathscr{L}}\varphi(\mathfrak{v}, \mathfrak{v}) \leq \frac{\mathfrak{v}}{2},$$

$\mathfrak{v}, \mathfrak{w} \in \mathscr{X}$, where $\mathscr{L} < 1$. Theorem 2 is verified.

The next two examples verify the single-valued version of the above theorem.

Example 2 Suppose $\mathscr{X} = (0, \infty)$, with a distance $\mathfrak{d}$ defined as $\mathfrak{d}(\mathfrak{v}, \mathfrak{w}) = \mathfrak{v}\mathfrak{w}$. Consider $\varphi : \mathscr{X}^2 \to [0, \infty)$ to be given as $\varphi(\mathfrak{v}, \mathfrak{w}) = 0$, $\mathfrak{v}, \mathfrak{w} \in \mathscr{X}$, $\mathscr{L} = \frac{1}{2}$ and $\mathscr{A}, \mathfrak{f} : \mathbb{R} \to (0, \infty)$ be mappings defined as $\mathfrak{f}(\mathfrak{v}) = \mathfrak{e}^{\mathfrak{m}\mathfrak{v}}$, where $\mathfrak{m} \in \mathbb{R}$, $A(\mathfrak{v}) = 0$, $\mathfrak{v} \in \mathscr{X}$. So, we get

$$\varphi(\mathfrak{v}, \mathfrak{w}) = 0 \leq \frac{1}{2} \cdot \varphi(\mathfrak{v}^2, \mathfrak{w}^2),$$

$\mathfrak{v}, \mathfrak{w} \in \mathscr{X}$. A single-valued variant of inequality (19.5) is

$$\mathfrak{d}(\mathfrak{e}^{\mathfrak{m}(\mathfrak{v}^{\mathfrak{a}}\mathfrak{w}^{\mathfrak{b}})}, \mathfrak{a}\mathfrak{v} + \mathfrak{b}\mathfrak{w}) \leq \rho \cdot \mathfrak{d}(2\sqrt{\mathfrak{e}^{\mathfrak{m}(\mathfrak{v}^{\mathfrak{a}}\mathfrak{w}^{\mathfrak{b}})}}, \mathfrak{a}\mathfrak{v} + \mathfrak{b}\mathfrak{w}). \tag{19.11}$$

Taking $\mathfrak{a} = 1$, $\mathfrak{b} = 1$, $\mathfrak{v} = \frac{1}{2}$, $\mathfrak{w} = \frac{1}{4}$, $\mathfrak{m} = 1$ and $\rho = \frac{3}{4}$ and using inequality (19.11), we obtain

$$\mathfrak{d}(e^{\frac{1}{8}}, \frac{3}{4}) \leq \rho \cdot \mathfrak{d}(2e^{\frac{1}{\sqrt{8}}}, \frac{3}{4}),$$

that is,

$$\frac{3}{4} \cdot e^{\frac{1}{8}} \leq \frac{3}{4} \cdot 2e^{\frac{1}{\sqrt{8}}} \cdot \frac{3}{4}.$$

Using inequality (19.7), we get

$$\mathfrak{d}(\mathfrak{e}^{\mathfrak{m}\mathfrak{v}}, 0) \leq \frac{\mathscr{L}}{2-2\mathscr{L}}\varphi(\mathfrak{v}, \mathfrak{v}) = 0,$$

$\mathfrak{v}, \mathfrak{w} \in \mathscr{X}$, where $\mathscr{L} < 1$. Theorem 2 is verified for single-valued case.

Example 3 Let $\mathscr{X} = [2, \infty)$, with a distance $\mathfrak{d}(\mathfrak{v}, \mathfrak{w}) = \|\mathfrak{v}^2 - \mathfrak{w}^2\|$. Let $\varphi : \mathscr{X}^2 \to [0, \infty)$ be given by $\varphi(\mathfrak{v}, \mathfrak{w}) = \mathfrak{v}^2$, $\mathfrak{v}, \mathfrak{w} \in \mathscr{X}$ and $\mathscr{A}, \mathfrak{f} : \mathscr{X} \to \mathscr{X}$ be mappings defined as $\mathfrak{f}(\mathfrak{v}) = \mathfrak{v}$, $A(\mathfrak{v}) = \mathfrak{v}$, $\mathfrak{v} \in \mathscr{X}$. Let $\mathscr{L} = \frac{1}{2}$. Then, we get

$$\varphi(\mathfrak{v}, \mathfrak{w}) \leq \frac{1}{4} \cdot \varphi(\mathfrak{v}^2, \mathfrak{w}^2),$$

$\mathfrak{v}, \mathfrak{w} \in \mathscr{X}$. Single valued variant of inequality (19.5) is

$$\mathfrak{d}(\mathfrak{v}^{\mathfrak{a}}\mathfrak{w}^{\mathfrak{b}}, \mathfrak{a}\mathfrak{v} + \mathfrak{b}\mathfrak{w}) \leq \rho \cdot \mathfrak{d}(2\sqrt{\mathfrak{v}^{\mathfrak{a}}\mathfrak{w}^{\mathfrak{b}}}, \mathfrak{a}\mathfrak{v} + \mathfrak{b}\mathfrak{w}), \tag{19.12}$$

Taking $\mathfrak{a} = 1$, $\mathfrak{b} = 1$, $\mathfrak{v} = 2$, $\mathfrak{w} = 2$ and $\rho = \frac{3}{4}$ and using inequality (19.12), we get

$$\eth(4,4) \leq \rho \cdot \eth(4,4) + \varphi(2,2) = 4.$$

Using inequality (19.7), we get

$$\eth(2,2) \leq \frac{1}{2} \cdot \varphi(2,2),$$

Theorem 2 is verified.

The following corollary is based on Theorem 2.

Corollary 1 *Let $\mathfrak{k} \geq 0$ be a real number and $\mathscr{X}$ be a real normed space. Let a mapping $\mathfrak{f} : \mathscr{X} \to \mathscr{C}_{\mathfrak{cb}}(\mathscr{Y}), \mathfrak{h})$ satisfies*

$$\mathfrak{h}(\mathfrak{f}(\mathfrak{v}^{\mathfrak{a}}\mathfrak{w}^{\mathfrak{b}}), \mathfrak{a}\mathfrak{f}(\mathfrak{v}) \oplus \mathfrak{b}\mathfrak{f}(\mathfrak{w})) \leq \rho \cdot \mathfrak{h}(2\mathfrak{f}(\sqrt{\mathfrak{v}^{\mathfrak{a}}\mathfrak{w}^{\mathfrak{b}}}), \mathfrak{a}\mathfrak{f}(\mathfrak{v}) \oplus \mathfrak{b}\mathfrak{f}(\mathfrak{w})) + \mathfrak{k}(\|\mathfrak{v}\|^{\mathfrak{n}} + \|\mathfrak{w}\|^{\mathfrak{n}}), \quad (19.13)$$

$\mathfrak{v}, \mathfrak{w} \in \mathscr{X}$, where $\mathfrak{a} > 0$, $\mathfrak{b} > 0$, $\mathfrak{n} \in \mathbb{N}$. Then a distinct logarithmic set−valued mapping $\mathscr{A} : \mathscr{X} \to (\mathscr{C}_{\mathfrak{cb}}(\mathscr{Y}), \mathfrak{h})$ so that

$$\mathfrak{h}(\mathfrak{f}(\mathfrak{v}), A(\mathfrak{v})) \leq \frac{2\mathfrak{k}}{2^{\mathfrak{n}} - 2}\|\mathfrak{v}\|^{\mathfrak{n}}, \quad \mathfrak{v} \in \mathscr{X}, \quad (19.14)$$

exists which confirms the stability of the logarithmic functional equation (19.1).

Proof On the pattern of Theorem 2, taking $\phi(\mathfrak{v}, \mathfrak{w}) := \mathfrak{k}(\|\mathfrak{v}\|^{\mathfrak{n}} + \|\mathfrak{w}\|^{\mathfrak{n}})$ and $\frac{2\mathfrak{k}}{2^{\mathfrak{n}}-2} < 1$, above corollary may be proved. □

Here, we establish the next result for the generalized logarithmic set−valued functional equations utilizing a similar technique as employed in Theorem 2.

Theorem 4 *Suppose $\varphi : \mathscr{X}^2 \to [0, \infty)$ satisfies*

$$\varphi(\mathfrak{v}, \mathfrak{w}) \leq \frac{\mathscr{L}}{\mathfrak{n}}\varphi(\mathfrak{v}^{\mathfrak{n}}, \mathfrak{w}^{\mathfrak{n}}), \quad \mathscr{L} < 1, \quad \mathfrak{v}, \mathfrak{w} \in \mathscr{X}. \quad (19.15)$$

Let for a mapping $\mathfrak{f} : \mathscr{X} \to \mathscr{C}_{\mathfrak{cb}}(\mathscr{Y}), \mathfrak{h})$, $\mathfrak{f}(1) = \{0\}$ and

$$\mathfrak{h}(\mathfrak{n} \cdot \mathfrak{f}(\mathfrak{v}^{\mathfrak{a}}\mathfrak{w}^{\mathfrak{b}})^{\frac{1}{\mathfrak{n}}}, \mathfrak{a}\mathfrak{f}(\mathfrak{v}) \oplus \mathfrak{b}\mathfrak{f}(\mathfrak{w})) \leq \rho \cdot \mathfrak{h}(\mathfrak{f}(\mathfrak{v}^{\mathfrak{a}}\mathfrak{w}^{\mathfrak{b}}), \mathfrak{a}\mathfrak{f}(\mathfrak{v}) \oplus \mathfrak{b}\mathfrak{f}(\mathfrak{w})) + \varphi(\mathfrak{v}, \mathfrak{w}), \quad (19.16)$$

$\mathfrak{v}, \mathfrak{w} \in \mathscr{X}$, $\mathfrak{a} > 0$, $\mathfrak{b} > 0$, $\mathfrak{n} \in \mathbb{N}$. Then a distinct generalized logarithmic set-valued mapping $\mathscr{A} : \mathscr{X} \to (\mathscr{C}_{\mathfrak{cb}}(\mathscr{Y}), \mathfrak{h})$, so that

$$\mathfrak{h}(\mathfrak{f}(\mathfrak{v}), \mathscr{A}(\mathfrak{v})) \leq \frac{\mathscr{L}}{\mathfrak{n} - \mathfrak{n}\mathscr{L}}\varphi(\mathfrak{v}^{\mathfrak{n}}, 1), \quad \mathfrak{v} \in \mathscr{X}, \quad (19.17)$$

exists which confirms the stability of a generalized logarithmic set−valued functional equation (19.2).

Proof We have, $\mathfrak{f}(\mathfrak{v})$ is convex, taking $\mathfrak{w} = 1$ and $\mathfrak{a} = 1$ in inequality (19.16), we attain

$$\mathfrak{h}(\mathfrak{n} \cdot \mathfrak{f}(\mathfrak{v})^{\frac{1}{\mathfrak{n}}}, \mathfrak{f}(\mathfrak{v})) \leq \rho \cdot \mathfrak{h}(\mathfrak{f}(\mathfrak{v}), \mathfrak{f}(\mathfrak{v})) + \varphi(\mathfrak{v}, 1) \leq \varphi(\mathfrak{v}, 1),$$

that is,

$$\mathfrak{h}(\mathfrak{n} \cdot \mathfrak{f}(\mathfrak{v})^{\frac{1}{\mathfrak{n}}}, \mathfrak{f}(\mathfrak{v})) \leq \varphi(\mathfrak{v}, 1) \leq \frac{\mathscr{L}}{\mathfrak{n}}\varphi(\mathfrak{v}^{\mathfrak{n}}, 1). \tag{19.18}$$

Let $\mathscr{P} := \{\mathfrak{g} : \mathscr{X} \to \mathscr{C}_{cb}(\mathscr{Y}), \mathfrak{h}), \quad \mathfrak{g}(1) = \{0\}\}, \ \eth(\mathfrak{g}(\mathfrak{v}), \mathfrak{f}(\mathfrak{v})) = \inf\{\mu \in [0, \infty] : \mathfrak{h}(\mathfrak{g}(\mathfrak{v}), \mathfrak{f}(\mathfrak{v})) \leq \mu\, \varphi(\mathfrak{v}, 1), \quad \mathfrak{v} \in \mathscr{X}\}$ be the generalized metric $\eth$ on $\mathscr{P}$ and $\inf \phi = +\infty$.

Noticeably, $(\mathscr{P}, \eth)$ is a complete generalized metric space. Consider $\mathscr{M} : \mathscr{P} \to \mathscr{P}$ to be a mapping described as $\mathscr{M}\mathfrak{g}(\mathfrak{v}) = \mathfrak{n} \cdot \mathfrak{g}(\mathfrak{v})^{\frac{1}{\mathfrak{n}}}$, $\mathfrak{v} \in \mathscr{X}$. Then using the definition of generalized metric $\eth$ and inequality (19.18)

$$\mathfrak{h}(\mathscr{M}\mathfrak{g}(\mathfrak{v}), \mathscr{M}\mathfrak{f}(\mathfrak{v})) = \mathfrak{h}(\mathfrak{n} \cdot \mathfrak{g}(\mathfrak{v})^{\frac{1}{\mathfrak{n}}}, \mathfrak{n}\mathfrak{f}(\mathfrak{v})^{\frac{1}{\mathfrak{n}}}) \leq \mu \cdot \mathfrak{n}\ \varphi(\mathfrak{v}^{\frac{1}{\mathfrak{n}}}, 1) \leq \mathcal{L}\mu\varphi(\mathfrak{v}, 1) \leq \mathcal{L}\eth(\mathfrak{g}(\mathfrak{v}), \mathfrak{f}(\mathfrak{v})),$$

$\mathfrak{v} \in \mathscr{X}$. Hence,

$$\mathfrak{h}(\mathscr{M}\mathfrak{g}(\mathfrak{v}), \mathscr{M}\mathfrak{f}(\mathfrak{v})) \leq \mathscr{L}\ \eth(\mathfrak{g}(\mathfrak{v}), \mathfrak{f}(\mathfrak{v})), \quad \mathfrak{g}, \mathfrak{f} \in \mathscr{P}.$$

Using inequality (19.18), we attain

$$\mathfrak{h}(\mathfrak{f}(\mathfrak{v}), \mathscr{M}\mathfrak{f}(\mathfrak{v})) \leq \frac{\mathscr{L}}{\mathfrak{n}}\varphi(\mathfrak{v}^{\mathfrak{n}}, 1).$$

Using Theorem 1, $\mathscr{A} : \mathscr{X} \to (\mathscr{C}_{cb}(\mathscr{Y}), \mathfrak{h})$ of $\mathscr{M}$ exists in $\mathscr{P}$ so that the following holds:

1. $\mathscr{A}$ is a unique fixed point of $\mathscr{M}$ in the collection $\mathscr{S} = \{\mathfrak{g} \in \mathscr{P} : \eth(\mathfrak{f}(\mathfrak{v}), \mathfrak{g}(\mathfrak{v})) < \infty\}$, that is,

$$\mathscr{A}(\mathfrak{v}) = 2\mathscr{A}(\mathfrak{v}^{\frac{1}{2}}), \quad \mathfrak{v} \in \mathscr{X}, \tag{19.19}$$

so that

$$\mathfrak{h}(\mathfrak{f}(\mathfrak{v}), \mathscr{A}(\mathfrak{v})) \leq \mu\varphi(\mathfrak{v}, 1), \quad \mathfrak{v} \in \mathscr{X} \text{ and } \mu \in [0, \infty].$$

2. $\eth(\mathscr{M}^{\mathfrak{n}}\mathfrak{f}(\mathfrak{v}), \mathscr{A}(\mathfrak{v})) \to 0$, whenever $\mathfrak{n} \to \infty$ and we attain

$$\mathscr{A}(\mathfrak{v}) = \lim_{\mathfrak{n}\to\infty} 2^{\mathfrak{n}}\mathfrak{f}(\mathfrak{v})^{\frac{1}{2^{\mathfrak{n}}}}, \quad \mathfrak{v} \in \mathscr{X}.$$

3.

$$\eth(\mathfrak{f}(\mathfrak{v}), \mathscr{A}(\mathfrak{v})) \leq \frac{1}{1 - \mathscr{L}}\eth(\mathfrak{f}(\mathfrak{v}), \mathscr{M}\mathfrak{f}(\mathfrak{v})) \leq \frac{\mathscr{L}}{\mathfrak{n} - \mathfrak{n}\mathscr{L}}\varphi(\mathfrak{v}^{\mathfrak{n}}, 1).$$

Hence, the inequality (19.17) holds, in view of the above three points.

Using inequality (19.16), we get

$$\begin{aligned}
&\mathfrak{h}(\mathfrak{n}^{\mathfrak{m}+1}\mathfrak{f}(\mathfrak{v}^{\mathfrak{a}}\mathfrak{w}^{\mathfrak{b}})^{\frac{1}{\mathfrak{n}^{\mathfrak{m}+1}}}, \mathfrak{n}^{\mathfrak{m}}\mathfrak{a}\mathfrak{f}(\mathfrak{v})^{\frac{1}{\mathfrak{n}^{\mathfrak{m}}}} \oplus \mathfrak{n}^{\mathfrak{m}}\mathfrak{b}\mathfrak{f}(\mathfrak{w})^{\frac{1}{\mathfrak{n}^{\mathfrak{m}}}}) \\
&\leq \rho \cdot \mathfrak{h}(\mathfrak{n}^{\mathfrak{m}}\mathfrak{f}(\mathfrak{v}^{\mathfrak{a}}\mathfrak{w}^{\mathfrak{b}})^{\frac{1}{\mathfrak{n}^{\mathfrak{m}}}}, \mathfrak{n}^{\mathfrak{m}}\mathfrak{a}\mathfrak{f}(\mathfrak{v})^{\frac{1}{\mathfrak{n}^{\mathfrak{m}}}} \oplus \mathfrak{n}^{\mathfrak{m}}\mathfrak{b}\mathfrak{f}(\mathfrak{w})^{\frac{1}{\mathfrak{n}^{\mathfrak{m}}}}) + \mathfrak{n}^{\mathfrak{m}}\varphi(\mathfrak{v}^{\frac{1}{\mathfrak{n}^{\mathfrak{m}}}}, \mathfrak{w}^{\frac{1}{\mathfrak{n}^{\mathfrak{m}}}}) \\
&\leq \rho \cdot \mathfrak{h}(\mathfrak{n}^{\mathfrak{m}}\mathfrak{f}(\mathfrak{v}^{\mathfrak{a}}\mathfrak{w}^{\mathfrak{b}})^{\frac{1}{\mathfrak{n}^{\mathfrak{m}}}}, \mathfrak{n}^{\mathfrak{m}}\mathfrak{a}\mathfrak{f}(\mathfrak{v})^{\frac{1}{\mathfrak{n}^{\mathfrak{m}}}} \oplus \mathfrak{n}^{\mathfrak{m}}\mathfrak{b}\mathfrak{f}(\mathfrak{w})^{\frac{1}{\mathfrak{n}^{\mathfrak{m}}}}) + \mathscr{L}^{\mathfrak{m}}\varphi(\mathfrak{v}, \mathfrak{w}).
\end{aligned}$$

Therefore

$$\begin{aligned}
\mathfrak{h}(\mathscr{A}(\mathfrak{v}^{\mathfrak{a}}\mathfrak{w}^{\mathfrak{b}}), \mathfrak{a}\mathscr{A}(\mathfrak{v}) \oplus \mathfrak{b}\mathscr{A}(\mathfrak{w})) &\leq \rho \cdot \mathfrak{h}(\mathfrak{n} \cdot \mathscr{A}(\mathfrak{v}^{\mathfrak{a}}\mathfrak{w}^{\mathfrak{b}})^{\frac{1}{\mathfrak{n}}}, \mathfrak{a}\mathscr{A}(\mathfrak{v}) \oplus \mathfrak{b}\mathscr{A}(\mathfrak{w})) \\
&= \rho \cdot \mathfrak{h}(\mathscr{A}(\mathfrak{v}^{\mathfrak{a}}\mathfrak{w}^{\mathfrak{b}}), \mathfrak{a}\mathscr{A}(\mathfrak{v}) \oplus \mathfrak{b}\mathscr{A}(\mathfrak{w})).
\end{aligned}$$

Since $\rho < 1$ and Lemma 3, we attain

$$\mathscr{A}(\mathfrak{v}^{\mathfrak{a}}\mathfrak{w}^{\mathfrak{b}}) = \mathfrak{a}\mathscr{A}(\mathfrak{v}) \oplus \mathfrak{b}\mathscr{A}(\mathfrak{w}),\ \ \mathfrak{v}, \mathfrak{w} \in \mathscr{X},$$

which is the required result. □

Theorem 4 asserts that the generalized logarithmic set–valued functional equation (19.2) is stable, and the next example verifies it.

Example 4 Suppose $\mathscr{X} = \{1\} \cup [2, \infty)$, $(\mathscr{C}_{\mathfrak{cb}}(\mathscr{Y}), \oplus, \mathfrak{h})$ is a complete metric semigroup equipped with a Hausdorff distance $\mathfrak{h}$ defined as

$$\mathfrak{d}_{\mathfrak{h}}(\mathfrak{v}, \mathfrak{w}) = \min\{\inf_{\mathfrak{v}\in\mathscr{X}} \sup_{\mathfrak{w}\in\mathscr{Y}} \mathfrak{d}(\mathfrak{v}, \mathfrak{w}),\ \sup_{\mathfrak{w}\in\mathscr{X}} \inf_{\mathfrak{v}\in\mathscr{Y}} \mathfrak{d}(\mathfrak{v}, \mathfrak{w})\}. \tag{19.20}$$

Consider $\varphi : \mathscr{X}^2 \to [0, \infty)$ to be given by

$$\varphi(\mathfrak{v}, \mathfrak{w}) = \begin{cases} 0, & \text{if } \mathfrak{v} = 1 \\ \mathfrak{e}^{\mathfrak{v}}, & \text{if } \mathfrak{v} \in [2, \infty), \end{cases}$$

$\mathfrak{v}, \mathfrak{w} \in \mathscr{X}$ and $\mathfrak{f}, \mathscr{A} : \mathscr{X} \to \mathscr{C}_{\mathfrak{cb}}(\mathscr{Y}), \mathfrak{h})$ be mappings defined as

$$\mathfrak{f}(\mathfrak{v}) = \begin{cases} 0, & \text{if } \mathfrak{v} = 1 \\ (0, 1), & \text{if } \mathfrak{v} \in [2, \infty), \end{cases}$$

and

$$\mathscr{A}(\mathfrak{v}) = \begin{cases} 0, & \text{if } \mathfrak{v} = 1 \\ [\frac{1}{2}, 1], & \text{if } \mathfrak{v} \in [2, \infty) \end{cases}.$$

Therefore, for $\mathscr{L} = \frac{1}{2}$ and $\mathfrak{n} = 2$, we attain

$$\varphi(\mathfrak{v}, \mathfrak{w}) \leq \frac{1}{4} \cdot \varphi(\mathfrak{v}^2, \mathfrak{w}^2), \ \ \mathfrak{v}, \mathfrak{w} \in \mathscr{X}.$$

Case I. If $\mathfrak{v} = 1$. Taking $\mathfrak{a} = 1$, $\mathfrak{b} = 1$, $\mathfrak{n} = 2$, $\rho = \frac{3}{4}$ and $\mathscr{L} = \frac{1}{2}$, the inequality (19.16) holds. One may verify that inequality (19.17) holds, on taking $\varphi(\mathfrak{v}, \mathfrak{w}) = 0$, $\mathfrak{f}(\mathfrak{v}) = 0$ and $\mathscr{A}(\mathfrak{v}) = 0$.

Case II. If $\mathfrak{v} \in [2, \infty)$. Taking $\mathfrak{a} = 2$, $\mathfrak{b} = 2$, $\mathfrak{n} = 2$, $\rho = \frac{3}{4}$ and $\mathscr{L} = \frac{1}{2}$, the inequality (19.16) holds. One may verify that inequality (19.17) holds, on taking $\varphi(\mathfrak{v}, \mathfrak{w}) = \mathfrak{e}^{\mathfrak{v}}$, $\mathfrak{f}(\mathfrak{v}) = (0, 1)$ and $\mathscr{A}(\mathfrak{v}) = [\frac{1}{2}, 1]$.

So, Theorem 4 is verified.

Example 5 Let $\mathscr{X} = [8, 10]$, $(\mathscr{C}_{\mathfrak{cb}}(\mathscr{Y}), \oplus, \mathfrak{h})$ be a complete metric semigroup equipped with usual Hausdorff distance $\mathfrak{d}_{\mathfrak{h}}(\mathfrak{v}, \mathfrak{w}) = \|v - w\|$. Suppose $\varphi : \mathscr{X}^2 \to [0, \infty)$ is given by $\varphi(\mathfrak{v}, \mathfrak{w}) = \frac{\mathfrak{v}}{2}$, for every $\mathfrak{v}, \mathfrak{w} \in \mathscr{X}$, and $\mathfrak{f}, \mathscr{A} : \mathscr{X} \to \mathscr{C}_{\mathfrak{cb}}(\mathscr{Y}), \mathfrak{h})$ be defined by $\mathfrak{f}(\mathfrak{v}) = \mathfrak{v}$ and

$$\mathcal{A}\mathfrak{v} = \begin{cases} [8, 8.5], & \text{if } \ \mathfrak{v} \in [8, 9) \\ [9, 9.5], & \text{if } \ \mathfrak{v} \in [9, 10] \end{cases}.$$

Therefore, for $\mathscr{L} = \frac{1}{2}$ and $\mathfrak{n} = 2$, we get

$$\varphi(\mathfrak{v}, \mathfrak{w}) \leq \frac{1}{4}\varphi(\mathfrak{v}^2, \mathfrak{w}^2), \ \ \mathfrak{v}, \ \mathfrak{w} \in \mathscr{X}.$$

From inequality (19.16) and using the defined Hausdorff distance, we have

$$\|\mathfrak{n}(\mathfrak{v}^{\mathfrak{a}}\mathfrak{w}^{\mathfrak{b}})^{\frac{1}{\mathfrak{n}}} - \mathfrak{v} - \mathfrak{w}\| \leq \rho \cdot \mathfrak{h}\|\mathfrak{v}^{\mathfrak{a}}\mathfrak{w}^{\mathfrak{b}} - \mathfrak{v} - \mathfrak{w}\| + \frac{\mathfrak{v}}{2}. \tag{19.21}$$

Case I. If $\mathfrak{v} \in [8, 9)$. Taking $\mathfrak{v} = 8$, $\mathfrak{w} = 8$, $\rho = \frac{3}{4}$, $\mathfrak{n} = 2$, $\mathfrak{a} = 1$, $\mathfrak{b} = 1$ and $\mathscr{L} = \frac{1}{2}$ in inequality (19.21), we have

$$\|2 \cdot (64)^{\frac{1}{2}} - 8 - 8\| \leq \frac{3}{4}\|64 - 8 - 8\|.$$

Now, from inequality (19.17), we obtain

$$\mathfrak{h}\big(8, [8, 8.5])\big) \leq \varphi(8, 1) = 4, \ \ \mathfrak{v} \in \mathcal{X}.$$

Case II. If $\mathfrak{v} \in [9, 10]$. Taking $\mathfrak{v} = 9$, $\mathfrak{w} = 9$, $\rho = \frac{3}{4}$, $\mathfrak{n} = 2$, $\mathfrak{a} = 1$, $\mathfrak{b} = 1$ and $\mathscr{L} = \frac{1}{2}$ in inequality (19.21), we have

$$\|2 \cdot (81)^{\frac{1}{2}} - 9 - 9\| \leq \frac{3}{4}\|81 - 9 - 9\|.$$

Now, from inequality (19.17), we have

$$\mathfrak{h}\big(9, [9, 9.5])\big) \le \varphi(9, 1) = 4.5, \quad \mathfrak{v} \in \mathcal{X}.$$

So, Theorem 4 is verified.

Example 6 Consider $\mathscr{X} = (0, \infty)$, and usual metric $\mathfrak{d}$ in a real Banach space $\mathcal{X}$ defined as $\mathfrak{d}(\mathfrak{v}, \mathfrak{w}) = \|v - w\|$, $\mathfrak{v}, \mathfrak{w} \in \mathscr{X}$. Consider $\varphi : \mathscr{X}^2 \to [0, \infty)$ to be given as $\varphi(\mathfrak{v}, \mathfrak{w}) = 0$, for every $\mathfrak{v}, \mathfrak{w} \in \mathscr{X}$ and $\mathfrak{f}, \mathscr{A} : \mathscr{X} \to \mathscr{C}$ mappings defined as $\mathfrak{f}(\mathfrak{v}) = \mathfrak{v}$, $\mathscr{A}(\mathfrak{v}) = \mathfrak{v}$, $\mathfrak{v} \in \mathscr{X}$. Therefore for $\mathscr{L} = \frac{1}{2}$ and $\mathfrak{n} = 2$, we get

$$\varphi(\mathfrak{v}, \mathfrak{w}) = 0 \le \frac{1}{2} \cdot \varphi(\mathfrak{v}^2, \mathfrak{w}^2).$$

Single valued version of inequality (19.16) is

$$\|\mathfrak{n}(\mathfrak{v}^{\mathfrak{a}}\mathfrak{w}^{\mathfrak{b}})^{\frac{1}{\mathfrak{n}}} - \mathfrak{a}\mathfrak{v} - \mathfrak{b}\mathfrak{w}\| \le \rho \cdot \mathfrak{d}\|\mathfrak{v}^{\mathfrak{a}}\mathfrak{w}^{\mathfrak{b}} - \mathfrak{a}\mathfrak{v} - \mathfrak{b}\mathfrak{w}\|. \tag{19.22}$$

Taking $\mathfrak{v} = \frac{1}{2}$, $\mathfrak{w} = \frac{1}{4}$, $\mathfrak{a} = 1$, $\mathfrak{b} = 1$, $\rho = \frac{3}{4}$, and $\mathfrak{n} = 2$, in inequality (19.22), we have

$$\|2 \cdot (\frac{1}{8})^{\frac{1}{2}} - \frac{1}{2} - \frac{1}{4}\| \le \frac{3}{4}\|\frac{1}{8} - \frac{1}{2} - \frac{1}{4}\|.$$

Now, from inequality (19.17), we have

$$\mathfrak{d}(\frac{1}{2}, \frac{1}{2}) \le \frac{\mathscr{L}}{2 - 2\mathscr{L}}\varphi(\frac{1}{2}, 1), \ \mathfrak{v} \in \mathscr{X}.$$

So, Theorem 4 is verified for a single-valued case.

The next corollary is a specific case of Theorem 4.

Corollary 2 *Suppose* $\mathfrak{k} \ge 0$ *is a real number,* $\mathscr{X}$ *a real normed space, and* $(\mathscr{C}_{\mathfrak{cb}}$ $(\mathscr{Y}), \oplus,$
$\mathfrak{h})$ *is a complete metric semigroup. Let a mapping* $\mathfrak{f} : \mathscr{X} \to \mathscr{C}_{\mathfrak{cb}}(\mathscr{Y}), \mathfrak{h})$ *satisfies* $\mathfrak{f}(1) = \{0\}$ *and*

$$\mathfrak{h}(\mathfrak{n} \cdot \mathfrak{f}(\mathfrak{v}^{\mathfrak{a}}\mathfrak{w}^{\mathfrak{b}})^{\frac{1}{\mathfrak{n}}}, \mathfrak{a}\mathfrak{f}(\mathfrak{v}) \oplus \mathfrak{b}\mathfrak{f}(\mathfrak{w})) \le \rho \cdot \mathfrak{h}(\mathfrak{f}(\mathfrak{v}^{\mathfrak{a}}\mathfrak{w}^{\mathfrak{b}}), \mathfrak{a}\mathfrak{f}(\mathfrak{v}) \oplus \mathfrak{b}\mathfrak{f}(\mathfrak{w})) + \mathfrak{k}(\|\mathfrak{v}\|^{\mathfrak{n}} + \|\mathfrak{w}\|^{\mathfrak{n}}), \tag{19.23}$$

$\mathfrak{v}, \mathfrak{w} \in \mathscr{X}$, $\mathfrak{a} > 0$, $\mathfrak{b} > 0$, $\mathfrak{n} \in \mathbb{N}$. *Then a distinct generalized logarithmic set–valued mapping* $\mathscr{A} : \mathscr{X} \to (\mathscr{C}_{\mathfrak{cb}}(\mathscr{Y}), \mathfrak{h})$ *satisfying*

$$\mathfrak{h}(\mathfrak{f}(\mathfrak{v}), \mathscr{A}(\mathfrak{v})) \le \frac{2^{\mathfrak{n}}\mathfrak{k}}{2^{\mathfrak{n}} - 2}\|\mathfrak{v}\|^{\mathfrak{n}}, \ \mathfrak{v} \in \mathscr{X}, \tag{19.24}$$

exists which confirms the stability of a generalized logarithmic set−valued functional equation.

Proof Continuing the pattern of Theorem 4, taking $\phi(\mathfrak{v}) := \mathfrak{k}(\|\mathfrak{v}\|^{\mathfrak{n}}), \mathfrak{v} \in \mathscr{X}$ and $\mathfrak{k} < \frac{2^{\mathfrak{n}}-2}{2^{\mathfrak{n}}}$, above corollary may be proved. □

Remark 3 Following the pattern of the above results, we may establish the stability of

(i) $\mathfrak{f}(\mathfrak{v}^{a} + \mathfrak{w}^{b}) - \mathfrak{f}(\mathfrak{v}^{a}) - \mathfrak{f}(\mathfrak{w}^{b}) = \mathfrak{f}(\frac{1}{\mathfrak{v}^{a}} + \frac{1}{\mathfrak{w}^{b}})$;
(ii) $\mathfrak{f}(\mathfrak{v}^{a} + \mathfrak{w}^{b}) - \mathfrak{f}(\mathfrak{v}^{a}\mathfrak{w}^{b}) = \mathfrak{f}(\frac{1}{\mathfrak{v}^{a}} + \frac{1}{\mathfrak{w}^{b}})$;
(iii) $\mathfrak{f}(\mathfrak{w}^{b} - \mathfrak{v}^{a}) - \mathfrak{f}(\mathfrak{v}^{a}\mathfrak{w}^{b}) = \mathfrak{f}(\frac{1}{\mathfrak{v}^{a}} - \frac{1}{\mathfrak{w}^{b}})$;
(iv) $\mathfrak{f}(\mathfrak{w}^{b} - \mathfrak{v}^{a}) - \mathfrak{f}(\mathfrak{v}^{a}) - \mathfrak{f}(\mathfrak{w}^{b}) = \mathfrak{f}(\frac{1}{\mathfrak{v}^{a}} - \frac{1}{\mathfrak{w}^{b}})$;

by modifying suitably single-valued versions of inequalities (19.3) and (19.4). The stability of the analogous single-valued functional equations is a special case of our results (see, Examples, 2, 3, and 6). Consequently, the achieved results extend and improve some celebrated and recent pertinent results.

19.3 Conclusion

Motivated by the question, under what conditions a function roughly verifying a functional equation is in the neighborhood of an accurate solution of the equation, we have defined the logarithmic and the generalized logarithmic set−valued functional inequalities to prove the Hyers−Ulam−Rassias stability of the corresponding set−valued functional equations. It is worth mentioning that Logarithms were initially introduced to simplify calculations and soon became indispensable for scientists, navigators, surveyors, engineers, and many others. Our theorems are sharpened versions of the well-known results, as the generalized metric is exploited in proving the results that are fascinating and dissimilar from a standard metric because the distance between any two of its points is permitted to be infinity. It is noteworthy that stability is a significant area that cannot be disregarded whenever we wish to utilize a particular algorithm for a given problem. Fundamentally, these explorations open the door to a different era in Hyers−Ulam−Rassias stability.

Author contributions All authors contributed equally and significantly to writing this article. All authors read and approved the final manuscript.

Funding Not applicable.

Acknowledgements The authors are grateful to the anonymous referee for the constructive comments and suggestions which significantly have improved the content of this paper.
Conflict of interest All the authors declare that there is no conflict of interest.

References

Aumann, R. J. (1965). Integrals of set−valued functions. *Journal of Mathematical Analysis and Applications, 12*, 1–12.

Bahyrycz, A., & Cieplinski, K. (2016). On an equation characterizing multi-Jensen-quadratic mappings and its Hyers-Ulam stability via a fixed point method. *Journal of Fixed Point Theory and Applications, 18*, 737–751. https://doi.org/10.1007/s11784-016-0298-8

Baker, J. A. (1991). The stability of certain functional equations. *Proceedings of the American Mathematical Society,112*(3), 729–732.

Banach, S. (1922). Surles opérations dans les ensembles abstraits etleur applicationaux équations intégrales. *Fundamenta Mathematicae, 3*, 133–181.

Brzdek, J. (2013). Stability of additivity and fixed point methods. *Fixed Point Theory and Application, 2013*(285), 1–9.

Cadariu, L., & Radu, V. (2008). Fixed point methods for the generalized stability of functional equations for a single variable. *Fixed Point Theory Application*. https://doi.org/10.1155/2008/749392

D'Alembert, J. (1747). Recherches sur la courbe que forme une corde tendue mise en vibration. I. *History of the Academy Berlin, 5*(201), 214–219.

D'Alembert, J. (1750a). Addition au Memoire sur la courbe que forme une corde tendue mise en vibration. *History of the Academy Berlin, 5*(117), 201, 355–360.

D'Alembert, J. (1750b). Recherches sur la courbe que forme une corde tendue mise en vibration, 11. *History of the Academy Berlin, 5*(201), 214–249.

Debreu, G. (1966). Integration of correspondences. *Proceedings of fifth Berkeley Symposium on Mathematical Statistical and Probability* (Vol. II, pp. 351–372).

Diaz, J., & Margolis, B. (1968). A fixed point theorem of the alternative for contractions on a generalized complete metric space. *The Bulletin of the American Mathematical Society, 74*, 305–309.

Gajda, Z. (1991). On stability of additive mappings. *The International Journal of Mathematics and Mathematical Sciences, 14*, 431–434.

Hyers, D. H., Isac, G., & Rassias, T. H. (1998). *Stability of Functional Equations in Several Variables*. Birkhauser.

Hyers, D. H. (1941). On the stability of the linear functional equation. *Proceedings of the National Academy of Sciences of the United States of America, 27*, 222–224.

Isac, G., & Rassias, T. H. (1996). Stability of ψ-additive mappings, applications to nonlinear analysis. *The International Journal of Mathematics and Mathematical Sciences, 19*, 219–228.

Luxemburg, W. A. J. (1958). On the convergence of successive approximations in the theory of ordinary differential equations II. In Proceedings of the Koninklijke Nederlandse Akademie Van Wetenschappen, Amsterdam, Series A, 61(5), Indag. Math., 20, (1958), 540–546.

Park, C. (2017). Fixed point method for set−valued functional equations. *Journal of Fixed Point Theory and Applications, 19*, 2297–2308.

Rassias, T. M. (1978). On the stability of the linear mapping in Banach spaces. *Proceedings of the American Mathematical Society, 72*, 297–300. https://doi.org/10.1090/S0002-9939-1978-0507327-1

Rassias, T. H. (2000). On the stability of functional equations and a problem of Ulam. *Acta Applicandae Mathematicae, 62*, 23–130.

Ulam, S. M. (1940). *Problems in modern mathematics, Chap VI, Science edn.*. Wiley.

MIX
Papier aus verantwortungsvollen Quellen
Paper from responsible sources
FSC® C105338

If you have any concerns about our products, you can contact us on
ProductSafety@springernature.com

In case Publisher is established outside the EU, the EU authorized representative is:
Springer Nature Customer Service Center GmbH
Europaplatz 3, 69115 Heidelberg, Germany

Printed by Libri Plureos GmbH
in Hamburg, Germany